Methods in Nucleic Acids Research

Editors

Jim D. Karam, Ph.D.
Lee Chao, Ph.D.
Gregory W. Warr, Ph.D.

Department of Biochemistry and Molecular Biology
Medical University of South Carolina
Charleston, South Carolina

CRC Press
Boca Raton Ann Arbor Boston

Library of Congress Cataloging-in-Publication Data

Methods in nucleic acids research/editors, Jim D. Karam, Lee Chao,
Gregory W. Warr.
p. cm.
Includes bibliographical references.
Includes index.
ISBN 0-8493-5311-4
1. DNA--Research--Technique. 2. Nucleic acids--Research--Technique. 3. Gene expression--Research--Technique. I. Karam, Jim D. II. Chao, Lee. III. Warr, Gregory W.
[DNLM: 1. DNA. 2. Research--methods. 3. RNA. QU 58 M5922]
QP624.M49 1990
574.87′3282--dc20
DNLM/DLC
for Library of Congress 90-2431
CIP

This book represents information obtained from authentic and highly regarded sources. Reprinted material is quoted with permission, and sources are indicated. A wide variety of references are listed. Every reasonable effort has been made to give reliable data and information, but the authors and the publisher cannot assume responsibility for the validity of all materials or for the consequences of their use.

Direct all inquiries to CRC Press, Inc., 2000 Corporate Blvd., N.W., Boca Raton, Florida, 33431.

International Standard Book Number 0-8493-5311-4

Library of Congress Card Number 90-2431
Printed in the United States

PREFACE

We embarked on this project realizing that there are, on the market, several excellent laboratory manuals in molecular biology, and yet feeling that we and our students needed something which paid more attention to principles and also provided background information on the evolution of new technology in this field. The task was made difficult by the incredibly rapid rate of technical progress, which constantly nagged at us and our authors as we tried to complete this manual in a timely manner. We hope that we have been successful in keeping only one or two steps behind progress and that this book will prove to be a valuable addition to libraries of our colleagues and their students. We have tried to select experts in newly emerging techniques as well as in areas (many of them technically demanding) that are not frequently covered in other methods books. As editors, we also tried to fill in some gaps and details that we thought would add balance, i.e., the short chapters on expression plasmids in prokaryotes (Karam) and rapid immunoscreening of expression libraries (Chao et al.). About midway through the process of receiving and editing manuscripts for the book, polymerase chain-reaction (PCR) technology started mushrooming at a pace that we had not anticipated. Considering that there are several good sources of information on this technology and its application to different situations (substantial methods books in their own right), we decided to invite a chapter on an application of PCR that illustrated the basic principle and that most of us could use right away, hence the short chapter on direct amplification of DNA from phage plaques (Miller and Jozwik). Deshler et al. also include a section on the principles and practice of PCR utilizing DNA or RNA as the starting material, and Budowle et al. deal with the use of PCR in forensic applications. Some techniques, although very important, are so well covered elsewhere and such good kits are sold for their execution (e.g., cDNA library construction) that we decided to omit them from this volume. On the other hand, some methods have become so central and important for most investigations in molecular genetics that we wanted to include a good selection of these: genomic DNA library construction (Hinds et al.), electrophoretic separation of large DNA molecules (pulsed-field gel electrophoresis and related methods, Bentley et al.), the use of prokaryotic (Karam) and eukaryotic (Maruyama et al.) expression vectors, and DNA sequencing methods (Slatko).

Many of the chapters include experimental details pertaining to widely used methods of DNA preparation, quantitation, and analysis. In terms of basic approaches to the analysis of gene structure and function, this volume incorporates a survey of approaches that, in addition to those mentioned above, includes the preparation of synthetic DNA for use in a variety of applications (Deshler et al.), *in vitro* site-directed mutagenesis (Menick), and *in vitro* methods for the analysis of eukaryotic transcription (Ruteshouser et al.) and translation (Wahba and Dholakia). A significant portion of the volume is directed to methods for the detection of protein/nucleic acid interactions, including gel-retardation assays (Revzin), DNA footprinting for the analysis of sequence specific proteins (Pauli et al.), and RNA footprinting techniques, a relatively new tool for the analysis of RNA function (McPheeters). We have also included approaches for studies in newly emerging areas of interest in molecular biology, e.g., the chapters on human genome mapping (Lathrop and Nakamura), the use of molecular biology in forensic analysis (Budowle et al.), the detection of mRNA sequences by *in situ* hybridization (Awgulewitsch and Utset), and the production and use of antibodies to synthetic oligopeptides (Schluter).

In all the invited chapters we asked the authors to provide brief surveys of the literature that would allow the reader to explore further not only the principles of the techniques, but also other experimental details. We think that historical or experimental perspectives on how new technology has developed are of importance in the educational process for the next generation of biological scientists and hope that in some measure this volume can be of both practical bench-top utility and educational value.

Jim D. Karam, Ph.D.
Lee Chao, Ph.D.
Gregory W. Warr, Ph.D.

THE EDITORS

Jim D. Karam, Ph.D., is Professor of Biochemistry and Molecular Biology at the Medical University of South Carolina in Charleston. He obtained his undergraduate training at the American University of Beirut in Lebanon, receiving his B.Sc. degree in 1958. He earned the Ph.D. degree in Biochemistry in 1965 from the University of North Carolina at Chapel Hill. As a recipient of a Postdoctoral Fellowship from the National Institute of General Medical Sciences from 1965 to 1967, Dr. Karam carried out studies at the Cold Spring Harbor Laboratory in New York on the role of DNA polymerase in control of mutation rates. He continued these studies as a Research Assistant Professor at the University of Connecticut, Storrs from 1967 to 1968. Before joining the faculty at the Medical University of South Carolina in 1971, Dr. Karam was Research Associate (1968 to 1970) and Associate (1970 to 1971) at the Sloan-Kettering Institute for Cancer Research in New York City.

Dr. Karam is a member of the American Society of Biochemistry and Molecular Biology, the Genetics Society of America, and the American Society for Microbiology. He has served on several grant review panels for the National Institutes of Health and National Science Foundation. His research interests are in the area of control of gene expression and he has published papers dealing with human, *Drosophila,* and microbial genetic systems. From 1974 to 1979 he was recipient of a Research Career Development Award from the National Institute of General Medical Sciences. He has also been the recipient of many research grants during his career. Currently, Dr. Karam's investigations into the genetic control of phage T4 DNA replication and the roles of translational repressors in control of protein biosynthesis are being supported by grants from the National Institutes of Health and the National Science Foundation.

Lee Chao, Ph.D., is Professor of Biochemistry and Molecular Biology at the Medical University of South Carolina, Charleston, South Carolina. He graduated from the National Taiwan University in 1963 and received his Masters degree in 1966 from Utah State University. In 1970 he completed his Ph.D. training at Iowa State University. Beginning in 1970 he moved to the Molecular Biology Laboratory at the University of Connecticut for his postdoctoral training. He moved to the Medical University of South Carolina in 1974 as an Assistant Professor and was appointed Professor in 1985. He is a member of the American Society of Biochemistry and Molecular Biology, the American Society of Hypertension, the American Heart Association, and the American Association for the Advancement of Science. Dr. Chao has been the recipient of research awards from the National Institutes of Health and has served as reviewer for the NIH and as a panelist for the National Research Council.

Dr. Chao is the author of over 50 research articles, reviews, and book chapters. He has presented over 30 invited lectures at national and international meetings and at universities and institutes. His current major research interests include the molecular mechanisms of multigene family regulation and evolution and the genetic basis of hypertensive diseases.

Gregory W. Warr, Ph.D., is Professor of Biochemistry and Molecular Biology at the Medical University of South Carolina, Charleston, South Carolina. He graduated from the University of Cambridge in 1970 and received his Ph.D. degree from the University of London in 1973. After postdoctoral work at the University of Edinburgh and the Walter and Eliza Hall Institute in Melbourne, Australia, Dr. Warr was a staff scientist at the National Cancer Institute Cancer Research Center in Frederick, Maryland. He moved to the Medical University of South Carolina in 1980 as an Associate Professor and was appointed Professor in 1986. Dr. Warr was a recipient of a Senior International Fellowship of the Fogarty International Center (1987 to 1988) for study at the University of Cambridge.

Dr. Warr is a member of the American Association of Immunologists, the American Society of Zoologists, the International Society for Developmental and Comparative Immunology, the Society for Invertebrate Pathology, and the British Society for Immunology. In 1991 and 1992 he will be Chairman of the Division of Comparative Immunology of the American Society of Zoologists. He has served as a member of the editorial board of the *Journal of Invertebrate Pathology* (1983 to 1987) and is currently an Associate Editor of *Developmental and Comparative Immunology*.

Dr. Warr has been the recipient of research awards from the National Science Foundation and the National Institutes of Health and has served as a reviewer for these two agencies. Dr. Warr is the author of over 100 research articles, reviews, and book chapters and has co-edited 3 books. His current research interests center on the evolution of the immunoglobulin molecule and its genes.

CONTRIBUTORS

Dwight E. Adams, Ph.D.
DNA Analysis Unit
Federal Bureau of Investigation
Washington, D.C.

Robert C. Allen, Ph.D.
Department of Pathology and Laboratory Medicine
Medical University of South Carolina
Charleston, South Carolina

Naoko Arai, M.D., Ph.D.
Department of Molecular Biology
DNAX Research Institute
Palo Alto, California
and National Institute of Health
Tokyo, Japan

Alexander Awgulewitsch, Ph.D.
Department of Biochemistry and Molecular Biology
Medical University of South Carolina
Charleston, South Carolina

Kevin L. Bentley, Ph.D.
Department of Biology
Yale University
New Haven, Connecticut

Charles J. Bieberich, Ph.D.
Laboratory of Virology
American Red Cross
Rockville, Maryland

Bruce Budowle, Ph.D.
Forensic Science Research and Training Center
Federal Bureau of Investigation Academy
Quantico, Virginia

Julie Chao, Ph.D.
Department of Biochemistry and Molecular Biology
Medical University of South Carolina
Charleston, South Carolina

Lee Chao, Ph.D.
Department of Biochemistry and Molecular Biology
Medical University of South Carolina
Charleston, South Carolina

Steven Chao
Department of Biochemistry and Molecular Biology
Medical University of South Carolina
Charleston, South Carolina

Benoit de Crombrugghe, M.D.
Department of Molecular Genetics
The University of Texas M. D. Anderson Cancer Center
Houston, Texas

James O. Deshler, Ph.D.
Department of Molecular Genetics
Beckman Research Institute
City of Hope
Duarte, California

Jaydev N. Dholakia, Ph.D.
Department of Biochemistry
The University of Mississippi Medical Center
Jackson, Mississippi

Kristin R. Hinds, Ph.D.
Department of Pediatrics
University of South Florida
All Children's Hospital
St. Petersburg, Florida

Catherine E. Jozwik, B.S.
Department of Microbiology
North Carolina State University
Raleigh, North Carolina

Bruce E. Kaplan, Ph.D.
Department of Molecular Genetics
Beckman Research Institute
City of Hope
Duarte, California

Jim D. Karam, Ph.D.
Department of Biochemistry and
Molecular Biology
Medical University of South Carolina
Charleston, South Carolina

Garrett P. Larson, M.A.
Department of Molecular Genetics
Beckman Research Institute
City of Hope
Duarte, California

G. Mark Lathrop, Ph.D.
Centre d'Etude du Polymorphisme
Humain
Paris, France

Gary W. Litman, Ph.D.
Department of Pediatrics
University of South Florida
All Children's Hospital
St. Petersburg, Florida

Sankar N. Maity, Ph.D.
Department of Molecular Genetics
The University of Texas M. D. Anderson
Cancer Center
Houston, Texas

Kazuo Maruyama, Ph.D.
Department of Molecular Biology
DNAX Research Institute
Palo Alto, California
and The Institute of Medical Science
University of Tokyo
Tokyo, Japan

David S. McPheeters, Ph.D.
Division of Biology
California Institute of Technology
Pasadena, California

Donald R. Menick, Ph.D.
Cardiology Division
Medical University of South Carolina
Charleston, South Carolina

Eric S. Miller, Ph.D.
Department of Microbiology
North Carolina State University
Raleigh, North Carolina

Atsushi Miyajima, Ph.D.
Department of Molecular Biology
DNAX Research Institute
Palo Alto, California
and The Institute of Medical Science
University of Tokyo
Tokyo, Japan

Yasuke Nakamura, M.D., Ph.D.
Division of Biochemistry
Cancer Institute
Tokyo, Japan

Urs Pauli, Ph.D.
Institute of Veterinary Virology
University of Bern
Bern, Switzerland

Arnold Revzin, Ph.D.
Department of Biochemistry
Michigan State University
East Lansing, Michigan

John J. Rossi, Ph.D.
Department of Molecular Genetics
Beckman Research Institute
City of Hope
Duarte, California

Frank H. Ruddle, Ph.D.
Department of Biology
Yale University
New Haven, Connecticut

E. Cristy Ruteshouser, Ph.D.
Department of Molecular Genetics
The University of Texas M. D. Anderson
Cancer Center
Houston, Texas

Samuel F. Schluter, Ph.D.
Department of Microbiology and
Immunology
Health Sciences Center
University of Arizona
Tucson, Arizona

Michael J. Shamblott, B.S.
Center of Marine Biotechnology
University of Maryland
Baltimore, Maryland

Barton E. Slatko, Ph.D.
New England Biolabs, Inc.
Beverly, Massachusetts

Gary S. Stein, Ph.D.
Department of Cell Biology
University of Massachusetts Medical Center
Worcester, Massachusetts

Janet L. Stein, Ph.D.
Department of Cell Biology
University of Massachusetts Medical Center
Worcester, Massachusetts

Yutaka Takebe, M.D., Ph.D.
Department of Molecular Biology
DNAX Research Institute
Palo Alto, California
and National Institute of Health
Tokyo, Japan

Manuel F. Utset, M.D., Ph.D.
Department of Pathology
Johns Hopkins Hospital
Baltimore, Maryland

André van Wijnen, M.S.
Department of Cell Biology
University of Massachusetts Medical Center
Worcester, Massachusetts

Albert J. Wahba, Ph.D.
Department of Biochemistry
The University of Mississippi Medical Center
Jackson, Mississippi

Huey-Mei Wang, Ph.D.
Department of Molecular Biology
DNAX Research Institute
Palo Alto, California
and The Institute of Medical Science
University of Tokyo
Tokyo, Japan

Gregory W. Warr, Ph.D.
Department of Biochemistry and Molecular Biology
Medical University of South Carolina
Charleston, South Carolina

Kenneth L. Wright, Ph.D.
Department of Cell Biology
University of Massachusetts Medical Center
Worcester, Massachusetts

TABLE OF CONTENTS

I. PREPARATIVE METHODS

Chapter 1

PURIFICATION OF CHROMOSOMAL DNA FOR LIBRARY CONSTRUCTION

K. Hinds, M. J. Shamblott, and G. W. Litman

TABLE OF CONTENTS

I. INTRODUCTION

The construction of recombinant genomic DNA libraries is a critical step in the analysis of the genetic makeup of an organism. Early libraries were constructed in plasmid vectors.[1] While this approach was useful for the study of simple prokaryotic organisms, the size of more complex genomes, typically 3×10^9 bp (base pairs) or greater per haploid genome, requires the use of vectors capable of replicating large pieces of insert DNA. The vectors developed to facilitate the cloning of large fragments of DNA are specially engineered λ-phage[2] and constructs known as cosmids that consist of an antibiotic-resistant plasmid into which cos sites (cohesive ends) from λ have been cloned to allow *in vitro* packaging.[3] Many special features have been added to both types of vectors that enhance the analysis of the insert DNA. Certain vectors have been engineered to facilitate deriving hybridization probes for use in further experiments.

The λ vectors available for genomic library construction are known as replacement vectors. In these vectors, the central portion of the phage genome is not needed for lytic replication and can be replaced by genomic DNA from the animal of interest. Vectors are available that incorporate insert DNA of a variety of sizes up to 23 kb, and most currently available vectors have been designed to maximize the size of the insert DNA. The size constraints are provided by the *in vitro* packaging reaction. This reaction packages DNA (vector + insert) into empty phage protein coats when the DNA is between 78 and 105% of the size of wild-type λ and is flanked by two cos sites.[4] Most available vectors also utilize a selective system (based on the spi^+ phenotype) that does not allow growth of nonrecombinant bacteriophage. Recently developed vectors typically are engineered (by double digests of closely linked restriction sites) to prevent religation of the nonessential central fragment, eliminating the need for physical separation. In addition, these vectors contain the promoters for T3 and T7 RNA polymerases that facilitate rapid synthesis of insert-specific probes for use in chromosome walking and restriction mapping. Examples of λ vectors with these properties are Lambda DASH™ II (Stratagene), as illustrated in Figure 1A, and λGEM-11 (Promega).

Similar features have been engineered into cosmid vectors. One such vector system, pWE15 and 16,[5] available from Stratagene, is used as an example (Figure 1B). pWE15 has T3 and T7 promoters and a *Bam*HI cloning site flanked by *Not*I sites that permit excision of the intact genomic insert. The small size, ~8.8 kb, of the vector permits cloning of 30- to 42-kb genomic fragments, and the antibiotic resistance genes allow cloning in both prokaryotic and eukaryotic host systems.

Since vectors capable of replicating large segments of insert DNA are available, the quality, particularly the size, of the genomic DNA to be cloned is of critical importance. Most cloning methods rely on restriction enzymes to generate clonable DNA. Any DNA ends which are caused by breaks, rather than being generated by enzymes, will compromise the cloning process. In order to realize the full potential of the phage and cosmid cloning systems, great care must be exercised in preparing purified high molecular weight genomic DNA for cloning.

Preparing genomic DNA for partial restriction-enzyme digestion requires special consideration. As mentioned above, the DNA must be prepared carefully to insure that it is large enough to yield approximately 20-kb and 30- to 40-kb fragments with two cohesive ends for λ and cosmid cloning, respectively. The cleavage should be random, although this is an ideal situation that cannot be achieved due to the nonuniform distribution of restriction sites in DNA. To enhance the generation of appropriate DNA fragments, the four base-pair cutting enzyme *Sau*3A is used. Partial enzyme cutting is achieved by enzyme dilution or time-course control.

The size of the library (the number of recombinant phage) needed to find a particular

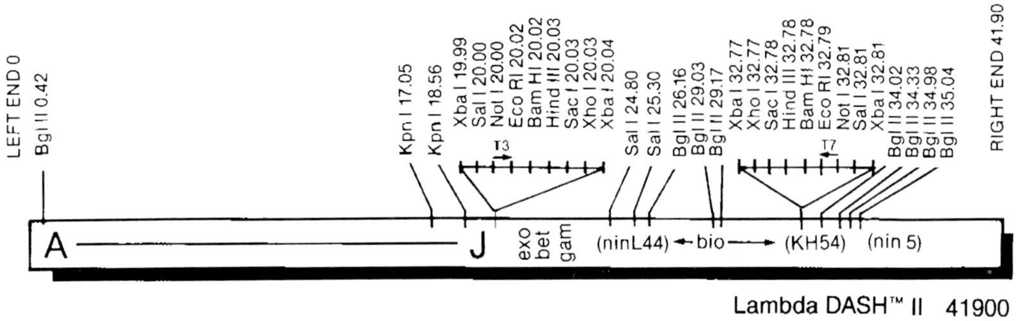

A

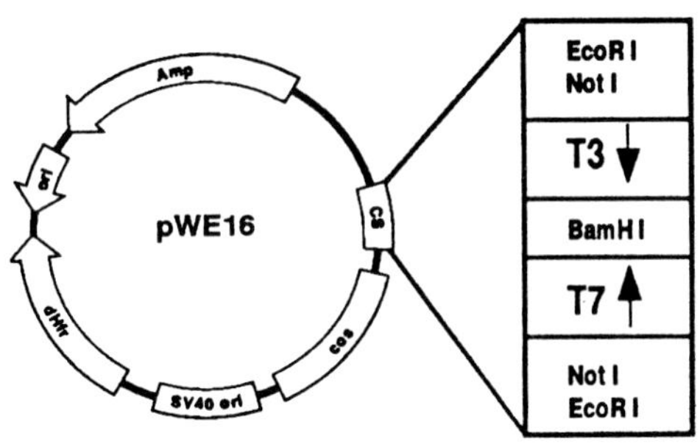

B

FIGURE 1. Cloning vector maps. (A) The Lambda DASH™ II vector showing the restriction enzyme sites used in arm preparation and cloning as well as the location of the promoters for the T3 and T7 RNA polymerases; (B) the pWE16 cosmid vector showing the *Bam*HI cloning site flanked by the T3 and T7 RNA polymerase promoters. (Maps courtesy of John C. Bauer, Stratagene, La Jolla, CA.)

DNA sequence is dependent on the size of the genome of the organism and the size of the insert DNA in each phage. If DNA sequence representation is completely random, the chance of finding a sequence of interest can be calculated using the formula.[1]

$$N = \frac{\ln(1 - P)}{\ln(1 - X/Y)}$$

where X is the insert size and Y is the size of the haploid genome. For a P (probability) of 0.99, it would be necessary to screen 9.2×10^2 λ recombinants (20-kb inserts) or 4.3×10^2 cosmid recombinants (45-kb inserts) if the genome size is 4.2×10^6 bp (*Escherichia coil*). For human DNA (3.3×10^9-bp haploid genome), a P of 0.99 would require 7.6×10^5 λ-recombinants or 3.4×10^5 cosmid recombinants. In these cases, approximately 3.5 genome equivalents would have to be screened.

In doing partial digests of DNA from a variety of species, it has been observed that a variable proportion of the DNA is resistant to digestion with *Sau*3A and therefore is not included in the library. To allow for this excluded DNA, the following formula can be used to determine the included genome equivalents in the library.

- # of genomes represented = (# of recombinant bp recovered)/(# of recombinant bp/genome) × (% digestible DNA)

Where:

- # of recombinant bp recovered = # of recombinant phage observed × insert length in bp
- # of recombinant bp/genome is an experimentally determined value available for many organisms in the *CRC Handbook of Biochemistry*[6]
- % digestible DNA is an experimentally determined value that differs for each enzyme and is estimated by running "complete" digests with a given restriction enzyme (e.g., *Sau*3A) and estimating the amount of DNA that remains undigested

The undigested DNA probably represents highly repetitive sequences and will not be included in the library, thereby lowering the number of recombinant clones that represents one genomic equivalent.

This chapter is intended to provide practical guidelines for the construction of genomic DNA libraries. The emphasis is on techniques that have been shown to work. Other, more extensive descriptions of library construction in λ and cosmid vectors exist.[7-10] The following precautions detail the care which must be taken when working with bacteriophage libraries.

II. GENERAL PRECAUTIONS FOR HANDLING PHAGE LIBRARIES

It is critical that no contamination of library components by exogenous phage or DNA occurs. Both vector and genomic DNA can be contaminated at any stage in their preparation. In addition, contamination can be introduced at any step in library construction. The problem is particularly serious if libraries from several different species are being constructed and screened. These guidelines will help to prevent this type of contamination.

1. Whenever possible disposable plasticware should be used for DNA and bulk phage preparation. If this is not possible, the containers should be scrupulously decontaminated with dilute (5%) bleach and cleaned. Materials used for genomic DNA preparation should be kept separate from those used for purifying phage DNA.
2. Any phage spills should be decontaminated with dilute bleach. If possible, bulk phage work should take place in a room separate from library construction and plating.
3. Do not create aerosols (bubbles) that will facilitate airborne contamination.
4. Be careful when using automatic pipettors. They are capable of causing cross-contamination of phage DNA preparations and λ DNA libraries. Disposable volumetric glass capillary pipettes should be employed at critical stages or the automatic pipettors should be disassembled and exposed to bleach vapor. The exterior of the entire pipette should be cleaned with dilute bleach.
5. Libraries can be constructed using different vectors, allowing easier identification of the source of individual purified clones, either by restriction characteristics or selection properties.
6. Finally, if there is any question about contamination by clones from a different species, the insert fragment of interest can be hybridized to labeled genomic DNA from representative species used in the laboratory. Hybridization should be to the animal from which the library was derived. Alternatively, labeled genomic DNA can be hybridized to the restriction enzyme-digested clone(s) of interest. This check depends on the presence of repetitive DNA in all λ-genomic clones.

There is little that is more discouraging than spending a great deal of time on a DNA sequence from the wrong species.

III. TECHNIQUE I: PREPARATION OF HIGH MOLECULAR WEIGHT GENOMIC DNA

The procedure described is adapted from Little.[9] It is intended as a guide to the isolation of high molecular weight DNA under conditions that allow its use in λ-phage and cosmid library construction. There are a number of somewhat different procedures which also could be employed. In all procedures, precautions should be taken to ensure that the purified DNA will be of high molecular weight. First, it is necessary to prevent shearing of the DNA. This is done by gentle mixing for an extended period of time during the extraction steps and by never using small-bore pipettes (or pipette tips). Second, organic solvents and salt are removed from the DNA by dialysis rather than by ethanol precipitation.

1. For tissue-culture cell lines and small tissue samples (up to 1 to 2 g), resuspend the cells ($\sim 10^8$ to 10^9) in 5 ml of 1X SSC on ice with five to seven passes in a Dounce homogenizer.* Tissue that is difficult to homogenize should be blended (metal blender with open vent cap) in liquid nitrogen prior to the addition of lysis buffer. It is important to work rapidly to get the starting tissue or cells into the lysis buffer. The best quality DNA is obtained from freshly sacrificed animals. If it is not possible to process tissue immediately, it should be frozen in liquid N_2 and stored at -70°C.
2. Add 5 ml of TNE (lysis buffer) and mix gently and carefully for 2 to 3 min to lyse the cells. Add proteinase K to a concentration of 100 μg/ml. (For 10 ml, use 200 μl of a 5 mg/ml stock.) Incubate at 55°C for 2 h. Add 10 ml of TNE (to dilute the viscous mass).
3. Extract with an equal volume of freshly saturated phenol, followed by phenol-chloroform (1:1 v/v) and then chloroform/isoamyl alcohol (24:1), one time each. The extraction is accomplished by shaking the DNA in a capped, gasketed 50-ml centrifuge tube at room temperature for at least 1 h. Good mixing of the phases can be achieved by laying the tube on its side on a rotary platform shaker. If a good emulsion is not achieved, the volume or the extraction time may have to be increased. Centrifuge for 10 min at 3000 rpm (1100 × g) at 4°C in a Sorvall SS34 rotor, or equivalent, to separate the phases after each extraction. When removing the DNA from each extraction, cut the entire tip off a 3-ml disposable transfer pipette and draw the DNA solution up very slowly. This procedure avoids shearing the DNA and pulling the interface into the DNA solution. Patience is critical at this step.
4. Dialyze the aqueous phase overnight at 4°C with gentle stirring vs. TE made 100 m*M* with NaCl. Dialyze against 400 times the sample volume, using commercially available dialysis clips to close the tubing.
5. Repeat the dialysis for 24 h against 4 l of TE (suitable for a 10-ml sample) at 4°C with gentle stirring. This DNA is ready for use in procedures where handling of the DNA has to be minimized, as in the gradient fractionation employed in cosmid (or λ) cloning.
6. A large amount of RNA is present at this stage of preparation and, therefore, the DNA concentration cannot be determined spectrophotometrically, but can be estimated by gel electrophoresis and reference to a known standard. These analyses should be performed on a 0.3% agarose gel run at low voltage with intact λ DNA (~50 kb), bacteriophage T4 DNA (~165 kb), and genomic high molecular weight DNA of known concentration for comparison. High molecular weight DNA should be in a fairly discrete band, migrating more slowly than λ DNA. Alternatively, the concentration

* For very small tissue samples (<1 g) or relatively few cells ($\sim 10^7$), homogenize in 0.5 ml of 1X SSC and make corresponding reductions in step 2, i.e., 0.5 ml of TNE; 20 μl of 5 mg/ml of proteinase K. Do not add the additional TNE used to "dilute viscous mass".

of DNA can be estimated by fluorescence, as described.[7] If 10^8 to 10^9 cells from a mouse were used, the DNA concentration would be 200 to 400 μg/ml in the ~10-ml sample. If it is too dilute, proceed directly to step 10 for ethanol precipitation.

For other purposes, RNAse treatment is desirable.

7. Estimate the volume of the DNA solution in the dialysis bags, open the dialysis clips, and add RNAse to 50 μg/ml. A 20-ml sample would require 1 mg (or 100 μl of a 10 mg/ml stock) of RNAse. The best way to estimate volume is to carefully weigh the dialysis bag in a plastic beaker against an appropriate tare.
8. Reclip the bag and dialyze against 4 l of TE overnight at 4°C with gentle stirring.
9. Gently extract the DNA in a screw-cap polypropylene tube with an equal volume of phenol/chloroform/IAA, then extract with two volumes of ether saturated with water. Centrifuge at 3000 rpm (1000 × g) at 4°C in an SS34 rotor for 3 min after each extraction. Remove and discard the top ether layer.
10. Bubble nitrogen through the solution until the ether smell disappears.
11. Make the solution 0.3 *M* in sodium acetate, pH 5.2 (for a 10-ml sample, use 1 ml of 3 *M* NaOAc), layer two volumes of ice-cold ethanol on top of the DNA solution, gently swirl the tube one or two times, and then wait 60 s to allow the DNA to nucleate; gently rock the tube back and forth for 2 to 3 min. The DNA should form a stringy precipitate, with the strands adhering to each other. If there is no visible precipitate or the precipitate is flocculent, the DNA probably is degraded. It is strongly advised that one does not attempt to construct DNA libraries from this material.
12. Precipitate at −20°C overnight. (This step is optional since good initial nucleation will allow a virtually complete recovery immediately.)
13. Melt the tip of a Pasteur pipette into a hook shape and use it to remove the DNA from the tube. Do not swirl the DNA and under no circumstances allow the DNA to dry. Any ethanol clinging to the DNA mass can be removed by gently touching the drops to the side of the tube. Place the DNA and pipette in 1 ml of TE in an open screw-cap tube and allow the DNA to dissociate from the glass rod. If the DNA clings to the pipette, set the tube (with pipette) on ice for 2 to 3 h to allow the DNA to come off. It is best to resuspend the DNA with gentle rotation of the tube on a 360° "rotator". Depending on the DNA yield, the volume of TE may have to be increased to yield a final DNA concentration of 0.5 mg/ml.
14. Measure the exact concentration of DNA using the OD_{260}. The $OD_{260/280}$ ratio should be >1.8. Sizing is done best on a 0.2 to 0.3% agarose mini-gel, as described in step 6. If these procedures have been followed correctly, there should not be a smear of lower-sized DNA.

Solutions for High Molecular Weight DNA Isolation

1X SSC	0.15 *M* NaCl 0.015 *M* NaCitrate
TNE	100 m*M* Tris, pH 7.5 100 m*M* NaCl 10 m*M* EDTA 1% Sarkosyl (IBI)
TE	10 m*M* Tris, pH 8.0 1 m*M* EDTA

RNAse A stock (10 mg/ml): dissolve 50 mg in 5 ml of H_2O. Boil for 15 min and cool slowly to room temperature. Store at 4°C.

IV. TECHNIQUE II: GENOMIC DNA: PARTIAL ENZYME DIGESTION AND ANALYSIS

Basically, there are two methods for doing partial enzyme digests: serial enzyme dilution[7] and time-course digestion. The former is better for getting a gross digestion "window" for a given DNA, whereas the latter is better for scale up and reproducibility, and is described here.

When working with less than ~150 μg of genomic DNA, a partial digestion series should be carried out. Determine by gel analysis which digests are best, purify these digests via phenol extraction and ethanol precipitation, then proceed to Technique IV (Ligation and Packaging). If there is an ample supply of genomic DNA, a large-scale digest can be carried out, and the partials can be size separated using sucrose-gradient velocity centrifugation. Although gradient separation of partial digests is not essential for library construction, it does assure higher ligation-packaging efficiencies (since more uniform-sized inserts are ligated and packaged) and can facilitate library construction in cases where there may be several DNA fractions that differ in their digestion kinetics. This technique is described at the end of this section. The example illustrated below that uses time-course kinetics has proven to be excellent for deriving libraries where relatively little DNA is available, the digestion kinetics are not known, and the library is to be constructed in as little time as possible.

A. PARTIAL DIGESTS WITH SMALL AMOUNTS OF GENOMIC DNA

1. A rough estimation of the enzyme amount needed to give a partial digest of the genomic DNA should be made by using serial dilutions of enzyme, beginning with 1 U/μg of DNA and allowing the digestion to proceed to completion. Analysis can be carried out on a 0.2 to 0.3% agarose gel to determine the enzyme concentrations where digestion begins. A dilution of 0.1 U/μg is frequently optimal and will be used as an example. An appropriate enzyme concentration is between that giving no digestion and that giving complete digestion.
2. Use ~3.5 μg of genomic DNA at each time point. This is enough DNA for both an analytical gel run and library construction. A six-point digest would require 25 μg of DNA and a useful digest would consist of:

   50 μl DNA at 0.5 μg/μl (25 μg)
   6 μl 10X *Sau*3A buffer
   3 μl H_2O
   1 μl of diluted *Sau*3A (see below)

   60 μl total volume

3. *Sau*3A (*Mbo*I also can be used) is usually supplied (e.g., Bethesda Research Laboratories) at 8 U/μl. In this example, a 1/25 dilution of *Sau*3A into 1X *Sau*3A buffer is made at 4°C immediately before use.
4. On ice, add the enzyme to the reaction mixture and immediately remove 4 μl of the digest and add to it 0.4 μl of 0.2 *M* EDTA. Heat at 65°C for 10 min and freeze (the 0 time-point aliquot).
5. Place the remaining reaction mixture (56 μl) at 37°C. Remove 9 μl aliquots every 10 min for 60 min. Each aliquot is added to a tube containing 1 μl of 0.2 *M* EDTA and inactivated at 65°C for 10 min. All aliquots should be frozen as soon as possible at −20°C. Partials stored any other way will deteriorate rather rapidly. Each aliquot will be the equivalent of 9/60 × 25 μg = 3.75 μg of DNA.

Summary

Enzyme: *Sau*3A (8U/μl): a 1 μl of a 1/25 dilution = 0.32 U was used
DNA: 25 μg in 60 μl
Kinetic sampling intervals: 10 min
Individual 9-μl aliquots = ~3.75 μg
Freeze as soon as possible

6. Take a 0.5-μl aliquot (~187 ng) from each time-point sample (including the 0-time digest) and add 9.5 μl of TE and 1 μl of gel-loading dye. If the samples have been frozen, heat at 65°C for 5 min before loading. Always spin before use to get the condensate to the bottom of the tube. Analyze on a 0.2 to 0.4% agarose mini-gel run at 20 V. After the tracking dye has run off, run the gel 1 to 1.5 h longer. Suitable markers are λ47.1 = (40.6 kb), wild-type λ digested with *Xho*I = (two bands of 33 kb + 17 kb), and the wild-type λ digested with *Hind* III.
7. Photograph the gel without overexposing the film. Using the photograph and the markers, ascertain the digestion time needed to produce the maximum intensity of fluorescence in the 19 to 22-kb* region of the gel. This is accomplished by blocking off all regions of the gel not containing DNA of the desired size. The various gel tracks then are compared to estimate the degree of digestion that produces the maximum amount of DNA of the desired size. The intensity of fluorescence is related to the mass distribution of the DNA. To obtain the maximum number of molecules in this size range, select digestion time points just below that which produces the maximum amount of fluorescence in this area.[11] With DNA exhibiting complex digestion kinetics, also block off the lower fractions to watch "increases" in low molecular weight DNA.

Note: Estimating fragment distributions is useful, especially when working with the same type (e.g., mouse) of DNA. With nonmammalian DNA, it is more useful to package DNA time points that are shorter and longer than the "ideal". This empirical method will identify the best digest based on packaging efficiency.

8. When the appropriate digestion times have been selected, remove the digests from the freezer, add TE to a total volume of 50 μl, and extract the samples with an equal volume of phenol/chloroform. Ethanol precipitate the DNA, collect by centrifugation for 10 min at 12,000 × g at 4°C in a microfuge, and vacuum dry. Wash the precipitate in 70% ethanol, centrifuge for 5 min, and resuspend the samples at 0.5 μg/μl in TE. This wash is very important to remove excess salt that can inhibit the ligation reaction. Store at −20°C. At this stage, the partials can be verified (optional) by subjecting ~100 to 200 ng to electrophoresis in 0.2 to 0.4% agarose, as described above in step 6. Before ligation, the samples should be reheated to 65°C for 10 min.

B. VELOCITY-GRADIENT SEPARATION OF DNA PARTIAL DIGESTS

1. A large-scale digestion of genomic DNA is carried out, using the amount of enzyme that previously was determined to give DNA fragments suitable for cloning. It is very difficult to scale-up and achieve the exact pattern of the small-scale digest; therefore, a limited kinetic sampling should be carried out.

* This is somewhat lower than the maximum insert size. Typically, higher ligation-packaging efficiencies are achieved if the partials are slightly below the 23-kb upper range, as this avoids introducing DNA that is too long to package. Since one will not get end-to-end (40-kb) DNA recombinants, one is better off.

Example (Based on Small-Scale Digest)

428 μl	DNA (300 μg)
60 μl	10X *Sau*3A buffer
11 μl	H_2O
1 μl	*Sau*3A (undiluted)
600 μl	total volume

A 5-μl aliquot is removed at time 0 and 190-μl aliquots are removed at 10, 20, and 30 min. To terminate, add 19 μl of 0.2 *M* EDTA and heat at 65°C for 10 min. Remove a 0.4 μl (~200 ng) aliquot from each time point for a 0.2 to 0.3% agarose gel run at 40 V for 2 h. Try to do the scale-up digestion of DNA and the electrophoretic analysis the same day the gradient is loaded.

2. Density-gradient ultracentrifugation is carried out in a Beckman SW28 rotor or equivalent. Form an ~36-ml gradient in a (1 × 3½ in.'') SW28 polyallomer tube (Beckman Co., #326823) using a 40 to 10% (w/v) sucrose density gradient* in 1 *M* NaCl, 20 m*M* Tris-HCl pH 7.5, 20 m*M* EDTA, 0.3% Sarkosyl (IBI).
3. Add 1 ml TE to the combined digests (total volume ~1.6 ml) and carefully layer over the gradient. Centrifuge in an SW 28 rotor at 25,000 rpm (85,000 × g) for 16 h at 10°C.
4. Gradient fractions (~40) should be 1 ml each. Collect directly into 15-ml polypropylene culture tubes.
5. Initially remove 50 μl from each tube and place into a 1.5-ml Eppendorf tube containing 200 μl of H_2O. Add 0.5 ml ethanol at −20°C, allow to precipitate on ice for 10 min, centrifuge at 12,000 × g for 10 min at 4°C, dry, and resuspend in 10 μl TE. A 2-μl aliquot is sufficient for analysis on a 0.3 to 0.4% agarose mini-gel using the standards described above in Section II.A, step 6. Overnight gel runs are critical for selecting appropriate fractions.
6. In a Beckman SW28 polyallomer tube, 3.6 ml of H_2O is added to each selected 0.9-ml fraction, 12.5 ml of −20°C ethanol is added, and the DNA is precipitated on ice for 10 to 20 min. The tube then is filled to the top with cold ethanol and centrifuged at 25,000 rpm (85,000 × g) at 4°C for 1 h. The ethanol is removed with care and the DNA pellet is dried with a stream of nitrogen. The pellet should be washed once with 70% ethanol to remove residual salt.
7. Resuspend the pellet in 0.5 ml of TE and determine the OD_{260}. One can expect ~10 μg per fraction (or more, depending on digestion kinetics). Always reconfirm sizes on a 0.3% agarose gel before proceeding.
8. Try to use the DNA as soon as possible, always heating to 65°C for 10 min before ligation. Fractions are stored best as ethanol precipitates, but also appear to be stable for up to 6 months when stored at 0.5 μg/μl in TE at −20°C.

V. TECHNIQUE III: VECTOR ARM PREPARATION

Phage arms which have been digested with restriction enzymes and are ready to use for library construction may be purchased commercially (Lambda DASH™ II, Stratagene). If only a few libraries are required, it is recommended that these be used. For arm preparation:

1. Phage DNA is prepared using a standard cesium chloride banding protocol.[7] Twenty-microgram aliquots of the vector DNA are stored as ethanol precipitates at −70°C.

* The gradient is formed by displacement. A 10% sucrose buffer is placed in the front chamber of a gradient maker, and a 40% sucrose buffer is placed in the rear chamber. The SW28 tube is loaded from the bottom. The gradient can be made 1 d ahead and held at 4°C.

Before use, the DNA is centrifuged at 12,000 × g for 10 min at 4°C in a microcentrifuge. After drying, the DNA is resuspended in 40 μl of TE (0.5 μg/μl).

2. For Lambda DASH(TM) II, digest 20 μg of DNA with *Hind*III, using 20 U (1 U of enzyme per microgram of DNA) in a volume of 50 μl at 37°C. The reaction should be stopped after 1 h.
3. Extract once with 50 μl of phenol-chloroform, then once with 150 μl of H_2O-saturated ether and blow off the excess ether with N_2. Ethanol precipitate by bringing the sample to 50 μl with water, add 5 μl of 3 *M* sodium acetate (pH 5.2), 100 μl of ethanol at −20°C, and precipitate on ice for 10 min. Resuspend the DNA in 40 μl of TE.
4. Digest next with *Bam*HI, again using 20 U (1 U of enzyme per μg of DNA) in a reaction volume of 50 μl. Incubate at 37°C for 1 h. Repeat the phenol-chloroform and ether extraction as in step 3.
5. Measure the volume recovered by pipetting and, if necessary, add water to 50 μl. Add an equal volume (50 μl) of 4 *M* ammonium acetate (made fresh, 3.02 g/10 ml). Vortex gently and add 0.6 volumes (60 μl) of isopropyl alcohol at room temperature. Ammonium acetate precipitation is more effective than conventional sodium acetate precipitation in preventing the precipitation of the small *Hind*III/*Bam*HI DNA fragments. Vortex and allow to precipitate at room temperature for 15 min. Centrifuge 10 min at 12,000 × g at 4°C and vacuum dry. Resuspend the DNA in 40 μl TE (0.5 μg/μl) and freeze at −20°C.

VI. TECHNIQUE IV: LIGATION OF GENOMIC DNA INTO VECTOR ARMS AND *IN VITRO* PACKAGING

If genomic DNA enzyme partial digests are available in sufficient amounts, it is advantageous to perform a number of small-scale ligations to determine those time points or gradient fractions that give optimal results. Each ligation then must be packaged and titered. Small aliquots of each time point or fraction are ligated with a constant amount of vector DNA. Those points/aliquots yielding the largest number of recombinants then can be pooled, and the pool tested to determine the optimum molar ratio of insert DNA to vector DNA, once more using varying amounts of insert and constant amounts of vector. The test ligations should bracket an equimolor insert-to-vector ratio, which is ~0.4 μg of 20-kb insert DNA: 1 μg of phage DNA (including the central fragment). When the best ratio is determined, a bulk ligation can be done. Although these procedures should optimize construction of the library, they consume both time and material, and an adequate library usually can be prepared based on agarose gel-selected aliquots.

1. Having identified the best time points by any one or a combination of the methods suggested, perform a test ligation using a small amount of vector and insert DNA which is packaged and evaluated as described in Technique V. If the test ligation and packaging is successful, a bulk ligation is performed using enough insert and vector DNA to produce a library of the desired number of recombinants, calculated as described in the introduction. For the example of human DNA, a successful library should have ~7.6×10^5 recombinants with 20-kb inserts. This would require the packaging of ~8 μg of total DNA, at 1×10^5 recombinants per microgram. In practice, this means three ligation reactions (8.4 μg DNA total, 6 μg vector, 2.4 μg insert) and eight packaging reactions (~8 μg), which should be plated on 24 150-mm plates at 3.3×10^4 plaques per plate.

Test Ligation

1 μl vector DNA (100 ng/μl)
2.5 μl insert DNA(20 ng/μl) heated to 65°C before ligation
1 μl 5X ligation buffer*
0.5 μl ligase (0.5 U)

---

5 μl total volume

Note: A vector control, using only vector DNA, should be ligated as well, to assess background nonrecombinants.

Bulk Ligation

4.0 μl (2 μg) Lambda DASH™ II/*Hind*III-*Bam*HI (0.5 μg/μl)
3.5 μl (0.8 μg) 20-kb insert (0.25 μg/μl), reheated to 65°C, quick-chilled before use
2.0 μl 5X ligase buffer (without PEG)
0.8 μl ligase (0.8 U)

10.0 μl (280 ng/μl) total volume

2. Incubate overnight in a 4-l beaker of room temperature (22°C) water that is allowed to equilibrate in a cold room (6 to 8°C). This provides a temperature gradient to optimize ligation.
3. Package the DNA with commercially available packaging mixes (we recommend Stratagene), following the manufacturer's recommendations and using ~3.6 μl (1 μg) for each packaging. Immediately freeze the remainder of the ligation at −20°C.** If several libraries are to be constructed, it is convenient (or at least cost effective) to have packaging mixes prepared[7] in the laboratory for the test ligation(s). These can be less efficient (typically two orders of magnitude) and still allow assessment of the best ligation reaction.

VII. TECHNIQUE V: TEST PLATING AND AMPLIFICATION

1. The day before the library is to be packaged, start overnight cultures of bacteria appropriate for the vector used. For Lambda DASH™ II, this is LE392 and P2392 (a P2 lysogen of LE392 that selects against nonrecombinant phage containing the central fragment). Both of these are supplied as standard cultures with the vector arms. P2392 should be stored as a glycerin stock and a frozen chip should be used as an inoculum. The bacteria are grown in 50 ml of NZCYM (Gibco Laboratories, M36550) broth with 0.2% maltose. Centrifuge 5 min at 3000 rpm (1500 × g) at room temperature and resuspend in 25 ml of 10 m*M* $MgSO_4$. These cells can be used for up to 1 week.
2. Test the packaged library on both LE392 and P2392 host cells by infecting 0.2 ml of the bacteria with 2- and 10-μl aliquots of the packaged library. Incubate 15 to 20 min at 37°C without shaking, add 3.0 ml of NZCYM-0.7% agarose at 45°C, and plate on 100 mm 1.1% NZCYM agar plates. If the packaging efficiency is 1×10^5 PFU/μg DNA, a 1-μg packaging would contain ~1×10^5 PFU/500 μl. A 10-μl aliquot would yield 2000 PFU; 2 μl would yield 400 PFU. The number and ratio of PFU observed on LE392 vs. P2392 indicate both the efficiency and percent recombinants in the library. The same number of PFU on both strains suggests that the library is 100% recombinant. A cell control using wild-type and recombinant phage controls is im-

* Ligation buffer should not contain polyethylene glycol (PEG), which may inhibit the packaging reaction.

** Try to time the experiment so that one can package the remainder within 24 to 48 h. Frozen ligations (−20°C) are by far more stable than those stored at 4 to 6°C, but are not as stable as packaged DNAs. Do not heat (65°C) ligations stored at −20°C.

perative, as wild-type phage should plaque only on LE392; a known recombinant should plaque equally well on both strains.

Two other methods can be used to determine the number of recombinants:

- Core ten random plaques, grow up and make DNA mini-preps, cut with *Eco*RI and/or *Eco*RI/*Hind*III, and examine their restriction patterns on a gel. These should differ from each other as well as from the intact vector.
- Lift (possible since top agarose was used) and hybridize with labeled genomic DNA from the same species used to construct the library. This effectively serves as a probe for high-copy-number (repeat) sequences and is quite specific. With Lambda DASH(TM) II arms, greater than 98% recombinants are possible.

3. For amplification, NZCYM plates should be poured and dried 1 or more days ahead. Warm the plates to 37°C for at least 2 h prior to plating. Since each ~0.5-ml package will contain $\sim 1 \times 10^5$ PFU, one will require three 150-mm plates (33,000 PFU per plate) for each packaging reaction. The infection ratio is <0.2 ml of packaged DNA plus 0.6 ml of bacterial suspension. Do not attempt to exceed this amount.

- Combine mixes (e.g., 8 × 0.5 ml = 4.0 ml) and P2392 bacterial cells (24 × 0.6 ml = 14.4 ml) in a 500-ml sterile glass bottle.
- Incubate 15 to 20 min at 37°C.
- Warm contents to 42°C and add 7 ml of NZCYM-0.7% top agar/plate (e.g., 24 × 7 ml = 168 ml), also warmed to 42 to 45°C. It is imperative that this temperature be maintained. A styrofoam container of water that has been prestabilized at 42 to 45°C is ideal.

4. For the purpose of quantitating the library, pipette a 0.5-ml aliquot of the "plating mixture" into 3 ml of top agarose held at 45°C and overlay an NZCYM plate. This should yield ~500 to 1500 PFU and provides an estimate of the number of plaques in the library. If a nonselective strain is used, the test plates could be lifted and screened or "test cored" for determining the recombinant efficiency.
5. Pipette 7.75 ml on each 150-mm plate and incubate 8 h to overnight at 37°C.
6. Do not allow the plaques to become too large or the incubation to proceed for an extended time period. It is advantageous to prevent overproduction of plaques, which can result in recombination between plaques and/or selective overrepresentation of a given phage.
7. Harvest the amplified library as described in Reference 7.
8. Determine the final titer (usually $>10^9$ PFU/ml) and store a reasonable stock, e.g., 10 ml plus 50 μl of $CHCl_3$ at 4°C. Store the remainder as 7% DMSO stocks at −70°C.

VIII. TECHNIQUE VI: TRANSFER AND SCREENING OF THE RECOMBINANT PHAGE LIBRARY

1. Plate, as described in Technique V, the appropriate number of phage, calculated as described in the Introduction, on 150-mm plates using 30,000 plaques on each plate.
2. Lift the library onto nitrocellulose filters (Schleicher & Schuell, #40-20570) in duplicate, as described.[7]
3. Screen the recombinant library for sequences of interest using conventionally labeled probes.

The conditions employed for hybridization are dependent on the relatedness of the probe to the DNA being selected. If a cDNA probe from the same species is employed, selection can be accomplished using very stringent conditions. If the probing is heterologous, i.e., employing a probe from one species to search for a similar gene in another species, much lower stringency is employed. It should be noted, however, that heterologous screening under conditions of extremely low stringency usually is not successful. When screening with oligonucleotide probes, 33 mers used at high stringency (60°C hybridization and wash) have proven to be very specific for closely related DNA sequences. It is important to screen duplicate lifts in the original library screening to reduce the number of false positives. Only those signals which are positive on both plates are picked for replating.

4. Core the positive plaques using the wide end of a Pasteur pipette.

Example of the Plating Sequence from a Typical Screening

- When picked from a 150-mm plate with nearly confluent plaques, the positive plaque represents 1 of about 50 to 100 in the core. Dilute 1/100 with 1X SM and plate 2 to 10 μl. Expect 100 to 500 PFU, of which 1 to 10% should give a positive signal.
- When the secondary cores are picked from a plate with 300 PFU, the core should have three to five plaques. Dilute 1:10 and plate 5 to 50 μl. One should expect 50 to 200 PFU with 20 to 30% positive.
- When the third cores are picked from a plate with 100 PFU, the core should have one or two plaques. Plate 2 to 20 μl undiluted and expect 10 to 50 isolated phage. Recover a positive isolated core for a stock lysate.[7]

IX. TECHNIQUE VII: CLONING GENOMIC DNA IN COSMIDS

The use of cosmid vectors allows the cloning of larger DNA fragments than can be achieved in λ vectors. This facilitates analysis of very large genes, characterization of extensive multigene families, and chromosome walking. Although in theory cosmid libraries may be more useful than phage libraries, they are somewhat more difficult to construct and to work with. The methods provided here that are based on well-established protocols[7,8,12] should alleviate problems in construction and screenings, but may not alleviate inherent problems in representation. These problems may be due to the selective unsuitability of certain large segments of eukaryotic DNA for replication in bacterial cells.

1. DNA is prepared, partial digests are made, and, if desired, DNA is size selected as described for phage libraries. DNA fragments from 30 to 42 kb are used for cosmid cloning.
2. Cosmid DNA can be purchased from commercial sources. The vector to be discussed is pWE15, available from Stratagene. Vector can also be grown using conventional plasmid protocols.
3. Before ligating in the insert DNA, it is necessary to restriction digest and dephosphorylate the vector arms.

- Digest 20 μg of the cosmid vector pWE15 with 20U *Bam*HI (1 U/μg of DNA) in a volume of 50 μl at 37°C for 1 h.
- Check for complete digestion by agarose gel electrophoresis. Complete digestion is marked by the appearance of a single band of linearized DNA.
- Extract the DNA once with an equal volume (50 μl) of phenol-chloroform and

once with an equal volume of chloroform/IAA. Adjust the aqueous phase to 0.3 *M* sodium acetate (pH 5.2) and ethanol precipitate. Wash the DNA pellet with 70% ethanol to remove residual salt.

- Resuspend the DNA at 1 mg/ml in water (20 μl).
- Dephosphorylate the DNA (0.2 μ*M*) with calf intestinal alkaline phosphatase (2.5 U/50 pmol of 5′ ends) in a 50 μl reaction for 1 h at 37°C. The reaction is stopped by incubation at 65°C for 10 min.
- Extract the DNA once with an equal volume of saturated phenol and once with chloroform/IAA. Adjust the aqueous phase to 2.5 *M* ammonium acetate, then ethanol precipitate and wash the precipitate with 70% ethanol.
- Resuspend the vector DNA in 20 μl of TE (1 μg/μl). Store at −20°C.

4. The efficiency of the dephosphorylation must be checked empirically using a test insert, which can be pBR322 linearized with *Bam*HI or any other DNA with *Bam*HI cohesive ends.

Vector/Vector Ligation

pWE15, *Bam*HI cut and dephosphorylated (1 μg/μl)	0.5 μl (0.5 μg)
H_2O	15.5μl
5X ligation buffer	4.0 μl
	20.0 μl total volume

Vector/Insert Ligation

pBR322, *Bam*HI cut (1 μg/μl)	3.0 μl (3 μg)
pWE15, *Bam*HI cut and dephosphorylated (1 μg/μl)	0.5 μl (0.5 μg)
5X ligation buffer	4.0 μl
H_2O	12.5 μl
	20.0 μl total volume

- Remove an aliquot (1 μl) from each reaction and add to 9 μl TE. Freeze these samples at −20°C for later analysis by gel electrophoresis.
- Add T4 DNA ligase (0.5 μl; 1 to 2 Weiss units) and incubate the reaction overnight, as described in Section VI, step 2.
- Remove another aliquot (1 μl) from each of the ligation reactions and analyze the samples before and after ligation by electrophoresis through a 0.4% agarose gel. The vector/vector ligation should have the same band before and after ligation. In the presence of the test insert, the DNA should be in high molecular weight concatamers after ligation.

5. For ligation of genomic DNA into the cosmid, the vector DNA is used in a tenfold molar excess. A ligation reaction is performed for each fraction which appears to contain DNA in the appropriate size range. The total concentration of DNA in the ligation reaction is 225 μg/ml.

Size-fractionated genomic DNA	6.0 μl (0.25 μg/μl)
Dephosphorylated vector DNA	3.0 μl (1 μg/μl)
5X ligation buffer (without PEG)	4.0 μl
H_2O	7.0 μl
	20.0 μl total volume

- Remove a 1-μl aliquot from each reaction and store at 4°C. Add 1 μl of T4 DNA ligase (1 to 2 Weiss units) to the reaction and incubate overnight, as described in Section VI, step 2.

- Remove another 1-μl aliquot from each reaction and analyze these samples, in addition to those in the preceding step, by electrophoresis through a 0.3% agarose gel. If ligation is successful, the eukaryotic DNA should be converted to high molecular weight concatamers.

6. Package the DNA following the manufacturer's recommendations. Use 5 μl (~1 μg) of the ligation mixture/packaging reaction.
7. Incubate the packaging reaction for 2 h at room temperature, dilute the reaction in 500 μl SM, and store at 4°C until ready for use.
8. Titering the Cosmid Library

- Inoculate 25 ml of LB media containing 0.4% maltose with a single colony of NM 554 bacteria and incubate at 37°C with vigorous shaking for about 16 h (overnight).
- Centrifuge the culture at 3000 rpm (1500 × g) at room temperature and resuspend the pellet in 12.5 ml of 10 m*M* $MgSO_4$. Store the cells at 4°C until needed. The cells are stable for at least 1 week.
- Dilute 5 μl of the packaged cosmid DNA to 50 μl and 250 μl with SM buffer.
- Add 25 μl of each dilution to 25 μl of cells in a sterile tube and incubate at room temperature for 30 min to allow adsorption of the packaged cosmid.
- Add 200 μl of LB media without antibiotic to each tube, invert several times, and incubate at 37°C for 1 h with occasional mixing. This allows expression of the ampicillin resistance gene.
- Centrifuge the culture at low speed and resuspend each pellet in 100 μl of fresh LB media.
- Plate the cells on LB agar plates (150 mm) containing 50 μg/ml of ampicillin. Incubate at 37°C until colonies are big enough to count (usually 18 to 24 h). Titers usually range from 10^5 to 10^6 colonies per microgram of chromosomal DNA.

9. Plating for Library Amplification

- In a sterile 50-ml tube, combine packaged, titered recombinants (3.4×10^5 are needed in the example of human DNA) with an equal volume of cells and incubate at room temperature for 30 min.
- Add four volumes of LB media to the tube and incubate the culture for 1 h at 37°C.
- Centrifuge the culture for 5 min at 3000 rpm (1500 × g) at room temperature. The cells are resuspended in LB media (0.5 ml per filter) and spread on 150-mm LB agar plates containing 50 μg/ml of ampicillin.[7] Using 20,000 colonies per plate, this would require 17 plates to represent the human genome.
- Incubate the plates at 37°C for 24 h.
- Add 2 ml of L-Broth + 15% sterile glycerol to each plate, swirl, and scrape off all the colonies with a sterile glass slide and pool.
- Mix carefully, pipette 1-ml aliquots into freezer vials, and store at −70°C.

X. TECHNIQUE VIII: PLATING COSMID RECOMBINANTS ONTO NITROCELLULOSE FILTERS

Effective plating of cosmids can be achieved in two ways. When a high density of colonies is desired, as in initial library plating, the Buchner funnel method (see below) should be used. For low-density platings (colony purification), the traditional spreader method is preferred.

1. Titer library by plating serial dilutions made in L-Broth using the Buchner funnel method. Attempt to put 1, 20, and 100 colonies on three filters (it is best to titer on filters because of possible inhibition).
2. Cut or purchase enough nitrocellulose circles (Schleicher & Schuell) with a diameter of ~117 mm for the initial library plating. This size fits easily into a Coors porcelain funnel (#60245).
3. When plating the library, calculate the appropriate amount of the bacterial suspension to give 20,000 colonies on each filter and dilute the suspension in LB medium such that the desired amount of bacteria is in 6 ml.
4. Label the 117-mm nitrocellulose filters with a ballpoint pen on the side opposite that where the bacteria will be plated. Place the filters on the surface of LB/ampicillin (50 μg/ml) plates to wet them.
5. Prepare a layer of three sterile 117-mm Whatman 3MM paper disks and place in the Buchner funnel. Pour 10 ml sterile LB medium over paper disks to make a level bed. The same disks can be used for many filters.
6. Remove a wet nitrocellulose filter from an LB/ampicillin plate and place on the filtration apparatus with the suction off. Carefully pipette the 6-ml bacterial suspension onto the surface of the nitrocellulose filter, leaving the outer 4 to 5 mm of the filter free of solution.
7. Slowly apply mild suction, taking care to evenly and completely draw the bacterial suspension through the filter. Transfer the filter (bacteria side up) back to the antibiotic plate on which it was wetted.
8. Plate the entire library in this way and incubate the plates at 37°C until the colonies are 0.5 mm in diameter (usually this is overnight). These filters are library master lifts.
9. To produce replicas, label and wet another set of nitrocellulose filters, as described in step 4, with the same numbers, followed by the letter "A". Replica lifts can be a larger (132 mm), precut size (Schleicher & Schuell, #40-20570).
10. Remove the master library filter from its plate and place on three sheets of 20 × 20-cm 3MM paper, colony side up. Using two forceps, carefully position the wetted replica filter above the master filter. Slowly lay the replica filter on the library filter, avoiding the formation of air bubbles between the filters.
11. Lay three sheets of 20 × 20-cm 3MM paper on the two filters (one may need to change these frequently to avoid contamination), followed by a 20 × 20-cm glass plate. Using the palms of the hands, press hard on the glass plate. The bacterial colonies from the library master filter will be transferred to the replica filter.
12. Remove the glass plate and the top three layers of filter papers. Invert the master-replica filters so the master is on the top. Using a 20-gauge needle dipped in India ink, punch 3 to 4 asymmetric holes through both of the filters. These holes will allow the orientation of the autoradiograph produced from the replica filter with the library master filter for isolating positive colonies.
13. Carefully peel the two filters apart, placing both filters, with the bacteria up, on their respective agar plates. Grow the replica colonies at 37°C overnight and allow the library master filters to incubate overnight at 25°C (room temperature). After overnight growth, store the library master filters on agar plates at 4°C, while screening the replica filters.
14. After the 37°C overnight incubation, the cosmids on the replica filter can be amplified by transferring them to an LB plate containing 50 μg/ml of chloramphenicol and

readily discriminate a single base change in genomic DNA, making them powerful tools for diagnosis of genetic diseases.[7] The specificity in oligo base pairing makes them useful as primers for DNA and/or RNA sequencing, as well as for primer extension analyses. Recently, perhaps the most important application of synthetic oligos has been in the polymerase chain-reaction assay (PCR).[8,9] In this assay, opposing synthetic primers complementary to each strand of the target sequence are utilized to prime multiple rounds of polymerization. The repetitive nature of this procedure results in up to a several millionfold amplification of the target sequences, thus providing large quantities of DNA for subsequent analyses. An important innovation in this methodology is the use of a heat-resistant DNA polymerase isolated from the bacterium *Thermus aquaticus*,[10] which has enabled the semiautomation of the entire process.

This chapter will provide detailed descriptions of the chemical synthesis of DNA, methods for purifying the chemically synthesized product, and protocols for the utilization of synthetic oligos in hybridization analyses, site-directed mutagenesis, and polymerase chain-reaction assays. Because of space limitatons, we will only cover the most popular applications of synthetic oligonucleotides.

II. CHEMISTRY

The synthesis of oligodeoxyribonucleotides (DNA) and oligoribonucleotides (RNA) is almost universally accomplished on machines designed for these specific purposes. The first decision to be made before starting a synthesis is to decide which chemistry to use. The decision as to which chemistry is chosen will affect which of the many possible postsynthetic steps are required to make the oligonucleotide useful for a biological experiment. It is to this end that we will attempt to provide information to make this task easier.

A. RNA VS. DNA

The practical synthesis of oligodeoxyribonucleotides was accomplished in the late 1970s whereas the practical (automated) synthesis of oligoribonucleotides has only recently been accomplished.[11] This slower development of the synthesis of RNA has been caused by the presence of the 2′-hydroxyl group in the ribose ring. This hydroxyl group must be protected during the synthesis and sterically retards the coupling reactions that occur at the 3′-position of the ribose. The selection of protecting groups for the 2′-position has been made difficult by all of the requirements that this protecting group must possess. The first criteria is that the 2′-protected ribo compound must be capable of being prepared in a pure state, free of any contamination from the 3′-protected compound. It must be stable during synthesis (coupling, capping, oxidation, and detritylation) and the post-synthetic steps (cleavage from the support, removal of base-protecting groups), and finally, it must not cause isomerization of the RNA backbone during its cleavage. These problems have finally been overcome, and we look forward to exciting developments in the synthesis and uses of synthetic ribonucleotides. Since the synthesis of RNA is a relatively new accomplishment, there are many studies still to be done to make the particular information about the synthesis of RNA comparable to the synthesis of DNA.

B. CHOICE OF CHEMISTRY

Two methods are currently being used for the synthesis of both DNA and RNA oligonucleotides. These two methods are known as the phosphoramidite and H-phosphonate method. The phosphoramidite method is currently the most popular technique and is depicted in Figure 1.

1. Phosphoramidite Method (DNA)

The phosphoramidite method consists of two variations, both of which are being used

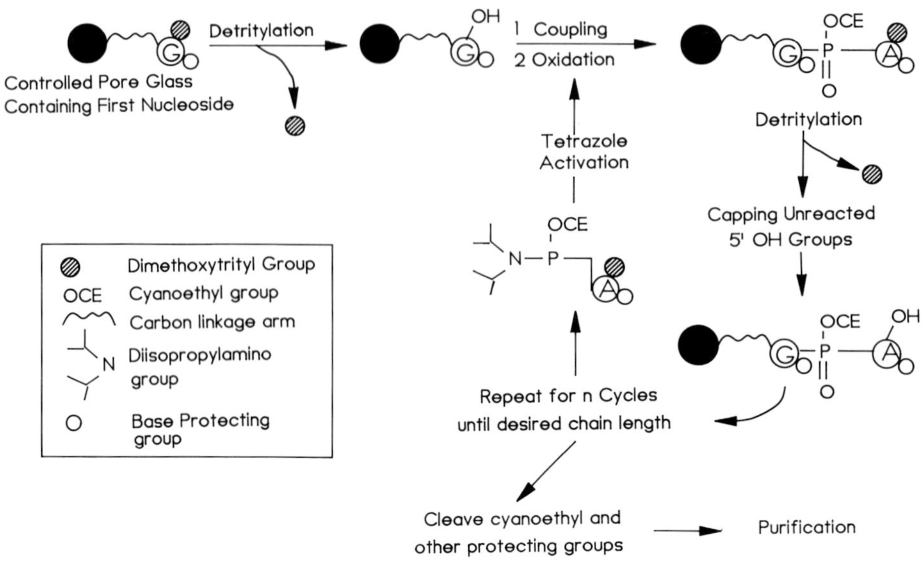

FIGURE 1. Solid-phase synthesis of DNA using phosphoramidite chemistry.

currently. For DNA, the *O*-methyl-phosphoramidite method of Beaurage and Caruthers[12] and the *O*-cyanoethyl-phosphoramidite method of Koster et al.[13] are commonly used. The method of Koster is preferred because there appear to be fewer side reactions and the post-synthetic steps are fewer, simpler, and do not require the use of the vile-smelling and toxic reagent thiophenol. One of the major side reactions that occurs during the synthesis of DNA using *O*-methyl protection of the phosphoramidite is the formation of 3-methyl-thymidine.[14] This base modification was shown to be present in up to 30% of the total thymidine in synthetic DNA produced using the *O*-methyl protection scheme. This base modification was used to explain the low cloning efficiency of DNA produced by this method.[15] Modifications of guanine bases during oligonucleotide synthesis have also been observed when the *O*-methyl phosphoramidite reagent is used and little or no modification when the *O*-cyanoethyl phosphoramidate is employed.[16] Another problem that manifests itself in the preparation of "mixed-sequence" oligonucleotides is the fact that the shelf life of the protected d-A and d-G is significantly different in the *O*-methyl series vs. the *O*-cyanoethyl series. When one uses fresh reagents, the ratio of d-A to d-G incorporated in a DNA fragment (at the same position) is roughly 1:1 using either protection scheme, but falls to 6:4 for 2-week-old *O*-cyanoethyl- and to 8:2 for 2-week-old *O*-methyl-protected phosphoramidites.[17] It would therefore be good laboratory practice to use fresh reagents when preparing oligonucleotides of "mixed sequence".

2. Hydrogen Phosphonate Method (DNA and RNA)

The other general method that is being used for the synthesis of DNA[18] and RNA[19] is called the hydrogen-phosphonate (H-phosphonate) method. A possible advantage of this method is that the cycle time is shorter than the phosphoramidite method (4 vs. 7 min per cycle), but it appears that it is difficult to make oligomers as long with the H-phosphonate method as one can with the phosphoramidite method. A recent study[20] compared the synthesis of a 107-bp oligo using H-phosphonate vs. *O*-methyl phosphoramidite. Ten clones were sequenced from plaques derived from each synthetic method. Seven out of ten clones were correct using the H-phosphonate method, whereas ten in ten were correct using the phosphoramidite method. (It should be noted that great care was exercised to reduce or eliminate the side reactions known to occur using standard protocols for the coupling of *O*-methyl phosphoramidites.) In spite of the slight difficulties with the H-phosphonate method, if a laboratory has the capability of preparing these reagents, then the cost of the H-phosphonate method is sufficiently less than the phosphoramidite method to make it a viable alternative method.

3. Phosphorothioate Oligodeoxynucleotides

Not only can oligomers of DNA and RNA be synthesized, but modified oligomers may also be synthesized. In the phosphoramidite and H-phosphonate methods, the phosphorous atom is in the +3 state and will finally end up in the +5 oxidation state, thereby allowing one to modify the phosphorous atom. The normal modification is, of course, to oxidize the phosphorous atom with the formal addition of an oxygen atom to generate the normal phosphodiester backbone of DNA and RNA. This oxidation is usually carried out by briefly treating the DNA with a solution of iodine in pyridine/water. If one wishes to incorporate a sulfur in place of an oxygen to form a phosphorothioate, then a solution of sulfur in carbon disulfide/pyridine may be utilized in place of the normal oxidizing reagent.[21] Phosphorothioate oligonucleotides have been utilized as anti-sense inhibitors of specific gene expression[22] *in vivo* and may be more effective as anti-sense agents since they are more resistant to certain nucleases than normal oligonucleotides.[22,23]

In an effort to eliminate the need for radioactive tagging of oligonucleotides, chemistries have been developed that allow nonradioactive reporter molecules to be covalently attached to synthetic oligonucleotides. Among the various types of reporter molecules are biotin, fluorophores, and proteins that have sensitive assays associated with their specific properties (such as alkaline phosphatase). These molecules can be attached anywhere in the oligonucleotide, but most commonly at the 5′-end and less frequently at the 3′-end. The attachment at the 5′-end allows the oligonucleotide to be used as both a probe and a primer. The methods for attaching molecules to DNA are many and it is not yet clear which of these methods is best. Therefore, several of the current techniques will be discussed.

4. 5′ Labeling

The fact that biotin bound to avidin forms a very stable complex and that biochemists were experienced in using these biotin-avidin complexes for the purification of biomolecules made the attachment of biotin to oligonucleotides an early target of nucleotide chemists.[24,25] Biotin was chemically attached to the 5′ terminus of oligonucleotides through a 5′ aminoalkyl linkage, and it was demonstrated that the "melting temperature" of the biotinylated oligo was approximately the same as the radiolabeled oligonucleotide. An alternative scheme for the attachment of a 5′ aminoalkyl group to an oligonucleotide has been developed, and it has been demonstrated that biotin is easily attached to the amino group.[26] All of the above reactions require a certain amount of manual chemistry that would not be easily accomplished on an automatic synthesizer. A method has been developed that allows the attachment of a 5′ aminoalkyl group directly to the oligonucleotide and can be accomplished automatically on any DNA synthesizer.[27] Reagents similar to these are now commercially available from a number of suppliers. Analogous to the 5′-aminoalkyl derivatives are 5′-thioalkyl derivatives of oligonucleotides which have been prepared and are now also commercially available. Both the thio and amino alkyl derivatives are suitably reactive for linking to various fluorescent molecules as well.[28]

The preparation of oligodeoxynucleotides conjugated to alkaline phosphatase has been reported to yield detection levels of target DNA of 2 attomoles.[29] In this method, an oligonucleotide containing an internal linker is first conjugated to the bifunctional reagent disuccinimidyl suberate, which is then conjugated to alkaline phosphatase. This method should also be applicable to 5′-aminoalkylated oligonucleotides.

A pertinent question to be asked is "which of the commonly used assay systems offers the highest sensitivity?'' An attempt has recently been made to answer this question by introducing fluorescent, chemiluminescent, or enzyme labels into a unique oligonucleotide. In a direct comparison, it was shown that enzymes commonly employed in immunoassays, alkaline phosphatase and horseradish peroxidase, are superior to fluorescein, Texas Red, rodamine, and isoluminol. The enzyme-based assays were as sensitive as, and sometimes even more sensitive than, ^{32}P detection.[30]

Many uses of oligonucleotides require that the 5′ terminus be phosphorylated. Although this is easily accomplished enzymatically, it is also easily accomplished using a reagent recently developed by Horn and Urdea.[31] This reagent is one of a family of reagents that can just be added to an automatic synthesizer, where it accomplishes its task as the final step of a synthetic program.

C. THE SOLID SUPPORT

Almost all DNA is currently synthesized on a material called controlled pore glass (CPG). This material is easily derivatized and a nucleoside (deoxy or ribo) can be easily attached. For the synthesis of oligonucleotides less than 50 bases in length, one can employ a CPG with a 500 pore size and a nucleoside loading of 20 to 30 μmol/g. For longer oligonucleotides (greater than 50 bases long), one should use a CPG with a 1000 pore size and a nucleoside loading of 10 to 12 μmol/g. The rationale behind these choices is that once the loading gets too high, the strands of polynucleotides begin to interact with each other, sterically hindering the coupling reactions. Thus, the longer the oligonucleotide, the lower the initial loading of the CPG should be.

D. REAGENTS FOR THE MODIFICATION OF OLIGONUCLEOTIDES

1. Amino Alkyl Phosphoramidites

One class of molecules which shows great promise is the amino alkyl phosphoramidities (AAP). Most commonly, these reagents are available with the amino group protected as the monomethoxy trityl derivative and the phosphoramidite as the cyanoethyl derivative. These molecules are currently available from several sources with alkyl groups of 3, 6, and 12 carbon atoms. These reagents provide an easy means of introducing a 5′ terminal primary aliphatic amine to an oligonucleotide. They are fully compatible with any commercial DNA synthesizer.

Protocol

1. Dissolve the AAP reagent in anhydrous acetonitrile according to the concentrations suggested by the manufacturer (about 50 mg/ml). These reagents are viscous oils and may take a few minutes to fully dissolve. Transfer the solution to the extra phosphoramidite port on your DNA synthesizer. It is recommended to make all transfers of anhydrous acetonitrile with a syringe for ease of handling and for minimum exposure to air. The AAP reagents are stable for 5 d after dissolving in acetonitrile.
2. Enter in the oligonucleotide sequence you wish to synthesize, remembering to program the AAP reagent as the last coupling step.
3. Prime the instrument and initiate the synthesis using the Trityul-OFF mode. Make sure the AAP reagent line is well primed.
4. The final detritylation reaction will yield the yellow color associated with the MMT group instead of the normal orange color associated with the DMT group.
5. Cleave and deprotect by incubating in ammonium hydroxide at 55 to 60°C for 6 to 18 h. Be sure to use a sealed reaction vessel to prevent ammonia from evaporating. A screw-cap microcentrifuge tube (1.5 ml) works well. Cool below room temperature and filter the ammonia solution through a cotton or glass wool plug in a Pasteur pipet.

Evaporate to dryness by vacuum centrifugation or rotary evaporation. Great care must be used when applying the vacuum. If the vacuum is applied too quickly, the ammonia solution may bump.

6. After the ammonia has evaporated, the residue may be dissolved and purified by PAGE.

2. Thio Alkyl Phosphoramidates (TAP)

There is currently only one TAP available, one with a six-carbon chain between the thio and the phosphate group. The thio group is protected by a trityl group. The thiol-protecting group is readily removed with silver nitrate/DTT to generate a free thiol on the 5′ terminus. The resulting 5′ thiol-alkyl oligonucleotide can be subsequently conjugated with any thiol-reactive molecule such as a maleimido- or iodoacetyl-derivatized adduct. Since the trityl thiol-protecting group is base stable, it can be retained on the 5′ thiol-modified oligonucleotide after ammonium hydroxide cleavage and deprotection. Thus, if desired, the trityl group can be used as a lipophilic tag to simplify the RP-HPLC purification.

Protocol

1. Same as step 1 for ALP
2. Same as step 2 for ALP
3. Prime the instrument and initiate the synthesis using the Trityl-ON mode, i.e., leave the final DMT group on. Make sure the TAP reagent line is well primed.
4. Cleave and deprotect by incubating in ammonium hydroxide at 55°C for 6 h. Be sure to use a sealed reaction vessel to prevent ammonia from evaporating. A screw-cap microcentrifuge tube (1.5 ml) works well. Evaporate to dryness by vacuum centrifugation or rotary evaporation. Remember, the trityl-protecting group on the 5′ thiol-modified oligonucleotide is still intact at this point! If desired, the 5′ thiol-modified oligonucleotide can be purified by RP-HPLC before proceeding to step 6 (see Section II.G).
6. Take up the crude and dried 5′ thiol-modified oligonucleotide of step 5 in distilled water and measure the UV absorbance at 260 nm. Aliquot five A260 units, evaporate to dryness, and resuspend in 50 μl of 0.1 *M* triethylammonium acetate, pH 7.5.
7. Add 7.5 μl of 1.0 *M* silver nitrate solution, vortex, and incubate at room temperature for 30 min.
8. Add 10 μl of 1.0 *M* dithiolthreitol (DTT), vortex, and incubate at room temperature for 15 min.
9. Centrifuge for 15 min to remove the precipitated silver salt of DTT and save the clear supernatant. Wash the precipitate with 50 μl of triethylammonium acetate buffer and centrifuge for 5 min. Combine the supernatants. This product can be stored for several months at −20°C with no decomposition.

E. LABELING AMINO-MODIFIED OLIGONUCLEOTIDES

The following protocol was designed to label an amino-modified oligonucleotide with either biotin or fluorescein. However, any amine-reactive reagent can be used with this basic protocol. Attachment of enzymes or proteins have been described in the literature.

Protocol

1. The following reaction is designed for a 1-μmol synthesis. In a 1.5-ml microcentrifuge tube, dissolve the total crude amino-modified oligonucleotide in 0.9 ml of deionized water.

2. Add 0.1 ml of 10X labeling buffer (1.0 *M* sodium bicarbonate/carbonate, pH 9). Vortex.
3. Biotin labeling: prepare fresh a 100 mg/ml solution of biotin-aminohexanoic-*N*-hydroxysuccinimide ester or biotin-di (aminohexanoic)-*N*-hydroxysuccinimide ester solution in *N,N*-dimethylformamide (DMF) by dissolving 25 mg in 0.25 ml of DMF. Fluoresceinisothiocyanate (FITC) labeling: prepare fresh a 100 mg/ml solution of FITC in DMF, i.e., dissolve 25 mg of FITC in 0.25 ml of DMF.
4. Add 0.25 ml of the freshly prepared solution of step 3 to the aqueous oligonucleotide solution.
5. Incubate at room temperature overnight. Sometimes longer oligonucleotides precipitate out of solution. If this occurs, add 0.25 ml 1X labeling buffer (0.1 *M* sodium bicarbonate/carbonate), vortex, and incubate at 42°C for 4 h. *Note:* If using FITC, perform the reaction in the dark.
6. To remove unreacted labeling reagent and buffer, pass the reaction mixture down a Sephadex G25 column (1 × 40 cm). The Sephadex column should be saturated first by passing 1 mg of sheared salmon sperm DNA through and washing with 100 ml of deionized water. Load the sample from step 5 and elute with deionized water. Collect 1-ml fractions and take absorbance readings at 260 nm. The product will be eluted in the void volume. Pool appropriate fractions. Concentrate and store at −20°C.

If desired, the labeled oligonucleotide can be purified further by HPLC or electrophoresis.

F. LABELING OF THIOL-MODIFIED OLIGONUCLEOTIDES

Any thiol-reactive reagent may be used to label a thiol-modified oligonucleotide. Below is a protocol for labeling a thiol-modified oligonucleotide with maleimidobutyryl biocytin.[32]

1. Remove excess DTT from the free thiol-modified oligonucleotide mixture by extracting three times with 100 μl of ethyl acetate. The DTT is removed in the ethyl acetate layer. You must immediately proceed to the next step since the free sulfhydryl group is unstable after the removal of DTT.
2. Prepare a 37 m*M* solution of maleimidobutyryl biocytin by dissolving 10 mg of maleimidobutyryl biocytin in 0.5 ml of 1.0 *M* sodium bicarbonate/carbonate (pH 9). Add to the thiol-modified oligonucleotide solution of step 1, vortex, and incubate overnight at room temperature.
3. Remove unreacted labeling reagent and buffer by procedures described in step 6 of Section II.E.

G. PURIFICATION

In many applications, further purification may not be necessary. However, to achieve optimum results, we recommend purifying by RP-HPLC or polyacrylamide gel electrophoresis.

1. Reverse-Phase HPLC

The lipophilic character of trityl, biotin, and fluorescein groups aids in oligonucleotide purification by RP-HPLC. A trityl-, biotin-, and fluorescein-labeled oligonucleotide is retained longer on the column, requiring a higher acetonitrile concentration to elute. Thus, as the last peak to elute, it is easily separated. For review, see Reference 33.

For HPLC purification, use a Hamilton PRP-1 or any RP-C18 analytical column (4.6 mm × 250 mm).[10] Set the detector wavelength to 260 nm. Use the following gradient system as a reference run: A = 0.1 *M* triethylammonium acetate (pH 7); B = 50% acetonitrile in 0.1 *M* triethylammonium acetate (pH 7); 20 to 80% B, 30 min; 1.0 to 1.5 ml/min. Since retention time will vary with the size of the oligonucleotide, you may wish to adjust the gradient conditions after the first run. *Note:* Amino-modified oligonucleotides with deprotected and free primary amine groups will not give good resolution using these HPLC conditions. It is recommended that polyacrylamide gel electrophoresis be used under such circumstances.

2. Polyacrylamide Gel Electrophoresis (PAGE)

Conventional PAGE methods can be used to separate modified oligonucleotides by employing a 1.5-mm-thick denaturing (urea) polyacrylamide gel.[34] The percentage of acrylamide and bis-acrylamide will depend on the size.

After electrophoresis, the labeled or modified oligonucleotide may be visualized as follows. Remove the glass plates from the gel and wrap the gel in transparent plastic wrap. In a darkroom, place the gel on a fluorescent TLC plate and use a UV lamp to observe fluorescent quenching. Avoid prolonged exposure to UV light, as cross-linking of the DNA may result. The labeled oligonucleotide runs the slowest (having the highest molecular weight) and is easily identified. Excise the labeled oligonucleotide with a razor blade. Crush the gel in a test tube and incubate in 0.1 *M* ammonium bicarbonate at room temperature overnight. Finally, pass the oligonucleotide through a Sephadex G25 column for desalting.

III. OLIGONUCLEOTIDE DESIGN

The design of an oligonucleotide should be based on a careful examination of the region of interest, preferably by thorough computer analysis to reduce the possibility of self-complementarity, which may compete for hybridization to the target region. The two most critical intrinsic factors affecting the stability of an oligonucleotide-DNA duplex are oligonucleotide length and the G + C content of the oligonucleotide.

Oligonucleotide length will primarily be determined by the genomic complexity of the organism being studied; however, oligonucleotides as short as 17 bases can be used to screen genomic libraries. However, cloned DNAs from higher eukaryotic organisms, when manipulated *in vitro,* can be screened and mutated readily using oligonucleotides much shorter than those needed to isolate or identify the gene in its normal genetic background, due to the overall decreased genetic complexity in the cloned system.

When screening genomic or cDNA libraries, one typically has two alternatives: use of a single, long synthetic oligonucleotide[35] or a mixture of oligonucleotides.[36] When using a single oligonucleotide probe for screening a library, one must take into account the codon frequencies of the organism being studied and the degeneracy of the third nucleotide (wobble position). When using a heterogenous mixture of shorter oligonucleotides containing all possible codon combinations, it is advisable to rescreen potential positively hybridizing clones with a second oligonucleotide mixture specific for another region of the protein. This eliminates many false positives due to the degeneracy of the probe.

In designing probes, it is useful to minimize the frequency of partially homologous sequences due to chance occurrences. Assuming a random distribution of nucleotides in a given DNA sequence, the frequency at which a given sequence is expected to occur in a sequence of defined length can be defined by *K*.[37] For probes that are designed to be completely homologous to their targets, the *K* value is equal to $(1/4)^I \times 2L$, where I is the oligonucleotide length and L is the number of nucleotides of unique sequence in the nucleic acid being screened (multiplied by 2, since DNA is double stranded). *K* values of unity define the minimum-length oligonucleotide that would be considered unique. Minimizing the *K* value reduces background signals due to imperfectly matched duplexes. For screening a cDNA or genomic library, the *K* value will be much higher than for a unique probe; therefore, a second screening with a mixed oligonucleotide probe specific for another region of the gene will decrease the overall *K* value and eliminate many false positives.

Base composition of the oligonucleotide should, if possible, contain a balance of G + C vs. A + T residues. Oligonucleotides that are too A + T rich may be unstable, while the converse may be true for G + C-rich oligonucleotides. In both cases, the ability of oligos to discriminate between completely and partially homologous sequences may be jeopardized.

IV. OLIGONUCLEOTIDE HYBRIDIZATION

It is important to determine optimum conditions for oligonucleotide hybridization in colony, plaque, and DNA/RNA blotting experiments. The filters are typically incubated with the ^{32}P-labeled oligonucleotide and then washed under conditions of increasing stringency.[38] Parameters affecting hybridization include oligonucleotide length, ionic strength, concentration of denaturants, temperature, and base composition. Stringency is routinely controlled by varying the temperature of the hybridization and subsequent washings. The T_d has been defined as the temperature at which half the oligonucleotide-DNA duplex is dissociated in 1 *M* Na^+. Suggs et al.[39] have developed a guideline for oligonucleotides between 12 and 20 bases long based on base composition: $T_d = 4(G + C) + 2(A + T)$ (°C), where G + C and A + T equal the number of G + C and A + T residues, respectively. Individual oligonucleotides may, however, vary from this rule, especially when larger than 20 bases. Hybridizations are typically done at 10 to 15°C below the calculated T_d, with subsequent washings approaching the T_d. It is advised that a mock experiment be performed to precisely determine the T_d for each oligonucleotide when the appropriate controls are available.[40]

The kinetics of oligonucleotide hybridization are much faster than their larger duplex counterparts due to their low sequence complexity and high concentrations. Thus, hybridization rates are much faster. Immobilized DNA or RNA on solid matrices need only be hybridized for a few hours with the oligonucleotide probe before washing the filters.

Protocol for Hybridization of Filters

1. Labeling of oligonucleotide

- 2.0 μ*M* oligonucleotide
- 67 μ*M* Tris-HCI, pH 8.3
- 10 m*M* $MgCI_2$
- 10 m*M* DTT
- 3.0 μ*M* $\gamma^{32}P$ ATP
- 2 U T4 polynucleotide kinase

For a 10 μl reaction, incubate at 37°C for 30 min. Heat denature the enzyme by incubating at 65°C for 5 min. Generally, the kinased oligonucleotide can be used directly in hybridizations or the unincorporated γ ATP can be removed by gel filtration or anion-exchange chromatography on DE52 resin.

2. Filter hybridization

For prehybridization, in a heat-sealable bag, incubate the filters, plaque, or colony at 65°C for 30 min to 1 h with:

- 6X SET (1X = 0.15 *M* NaCI, 0.015 *M* Tris-HCI, pH 8.0, 1 m*M* EDTA)
- 10X Denhardts (1X = 0.02% BSA, 0.02% PVP-40, 0.02% Ficoll)
- 0.1% SDS

- 20 μg/ml denatured herring sperm DNA (boil DNA for 5 min, then cool quickly on ice)

For hybridization, remove the prehybridization solution and replace it with the following solution:

- 6X SET
- 10X Denhardts
- 0.1% SDS
- 10^6 cpm/ml kinased probe

The bags are resealed and hybridized at a temperature (as explained in the text above) about 10 to 15°C below the T_d.

3. Washing filters

After hybridization (usually 2 h to overnight), the filters are removed and washed in 6X SSC and 0.1% SDS (1X SSC = 0.15 *M* NaCI, 0.015 *M* sodium citrate) (~50 ml per filter). If one has not determined the optimum wash temperature for the unique or mixed oligonucleotide(s), successive washings and autoradiographic cycles at progressively higher temperatures can be performed. If the T_d is known, one can immediately wash the filters at the predetermined temperature.

V. SITE-DIRECTED MUTAGENESIS

Many areas of modern research require specific alteration of DNA sequences for the purpose of assigning functional roles to the sequences themselves or the products which they encode. For studies concerning protein structure and function, the sequence of the gene encoding the protein of interest is altered usually at a single codon, such that a new amino acid is inserted at a position of the peptide being analyzed. The mutant protein is then expressed and either studied *in vivo* or purified and studied *in vitro*. It is sometimes desirable to alter DNA sequences which do not encode proteins. Usually, these sequences are involved in the regulation of a gene. Alteration of these sequences causes noticeable changes in the regulation or expression of the gene being analyzed, thus allowing the assignment of functional roles to the regulatory sequences. Recently, within the growing field of RNA enzyme analysis, site-specific mutagenesis has been useful for changing the nucleotides of "ribozymes" and assaying the transcribed products *in vitro*. The availability of *in vitro* transcription systems simplifies *in vitro* studies of small ribozymes. One only has to synthesize an oligonucleotide encoding the mutant ribozyme colinear with a T7 promoter sequence, anneal a 17-base oligonucleotide complementary to the promoter region, carry out a transcription reaction, and treat with DNAase. The result is microgram quantities of purified ribozyme. All site-specific mutagenesis procedures require the use of an oligonucleotide complementary to the region of the desired change. Thus, the nucleotide sequence of the molecule under study must be known. Nucleotide insertions, deletions, and substitutions are all possible, and, in general, the oligonucleotide should have 10 bases of complementarity on both sides of the change being made (we usually use 15 to 20 bases of complementarity on both sides if large deletions or insertions are being made). Finally, the entire gene, or a convenient fragment of the gene, needs to be cloned into a single-stranded vector for the procedure.

A. GENERAL DESCRIPTION OF THE PROCEDURE

Historically, the procedure of oligonucleotide-directed mutagenesis was developed using the single-stranded phage φX-174. The technique stemmed from the combination of a number

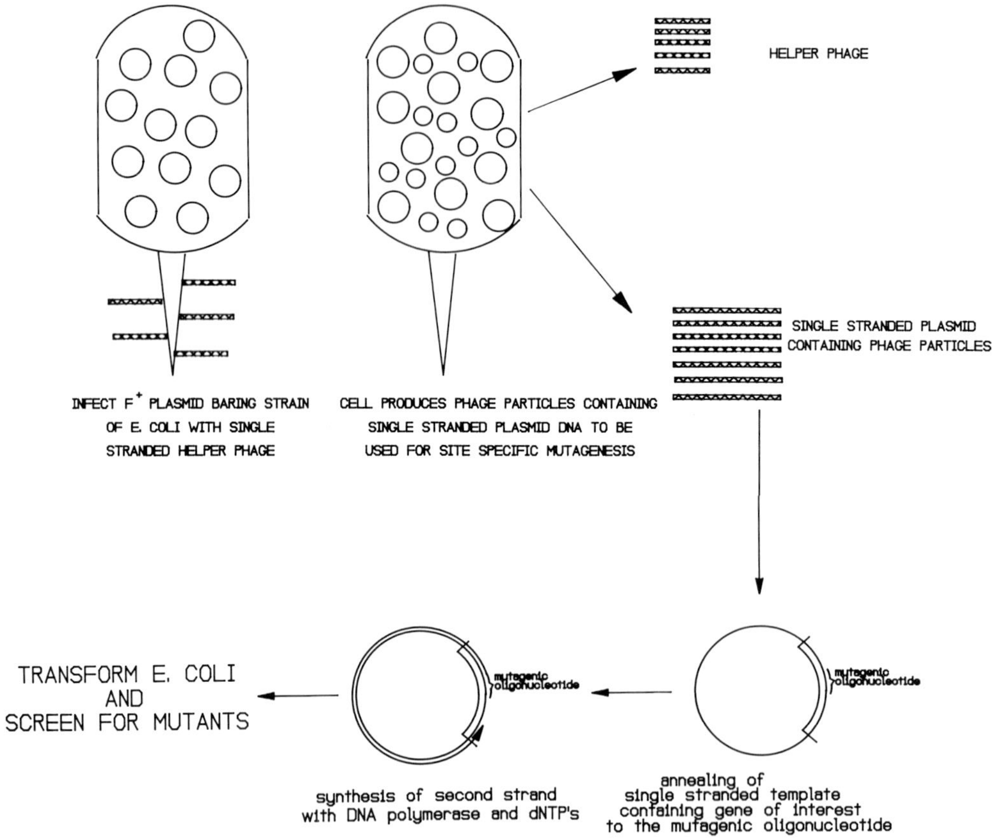

FIGURE 2. Generalized scheme for site-specific mutagenesis.

of observations about nucleic acids: (1) marker rescue of mutations in φX-174 by restriction fragments annealed to single stranded phage DNA, (2) the stability of DNA duplexes containing mismatches, and (3) the ability of *Escherichia coli* DNA polymerase I to extend oligonucleotide primers hybridized to single-stranded templates. In these pioneering studies, amber mutations were either corrected or created *in vitro* by using synthetically produced oligonucleotides (7 to 17 nucleotides in length) as primers for DNA polymerase I-mediated repair synthesis on the single-stranded templates. The success rate of the mutagenesis was monitored by phenotypic scoring of the resultant phage plaques.[41-44]

Today, commonly used procedures involve cloning the DNA to be mutagenized into a vector which will allow purification of single-stranded DNA. A short oligonucleotide contains the desired mutation and a mutant wild-type heteroduplex is formed. A complete second strand is formed *in vitro* using DNA polymerase, and the resultant mixture is used to transform the appropriate strain of *E. coli* (Figure 2). Wild-type and mutant clones are segregated *in vivo,* and a variety of screening procedures can be used to identify the desired mutation. Finally, the mutant clone is sequenced for verification.

B. CHOOSING A VECTOR

The choice of the vector you use for mutagenesis can be very important and depends on the sequence of the fragment you want to mutate and the mutagenesis protocol you use. Two of the protocols described below require specialized vectors, but the methods developed by Kunkel and Eckstein have no strict vector requirements except for the ability to produce purified single-stranded vector DNA in *E. coli.* M13 vectors usually work quite well when dealing with small fragments (500 bases or less). Larger fragments can be unstable, or even lethal to the phage, and therefore are incapable of being cloned. Sometimes there are even problems cloning small fragments into M13. These problems are rare and sequence dependent, but one must be aware of them while choosing the appropriate vector. In our lab, we generally use plasmid vectors containing a single-stranded filamentous phage-replication origin.[45-47] These plasmids use the plasmid-replication origin for propagation, but when transformed cells are superinfected with a helper phage, single-stranded plasmid molecules are packaged and secreted into the medium as uninfectious phage particles. The Bluescript vectors available from Stratagene, or the pTZ vectors from Bio-Rad, give good results with mutagenesis. We have created mutations in fragments as large as 3 kb and have had no problems with stability. When choosing the specific vector, the following two factors must be considered: (1) does the vector have the restriction sites you need to clone your fragment? and (2) will the correct strand be packaged into the phage particle upon superinfection with the helper phage? (This is usually unimportant if the mutagenic oligonucleotide has not been ordered yet.) Single-stranded DNA can be isolated from the "phagemids" mentioned above with the helper phage R408 (Stratagene).

C. MUTAGENESIS PROTOCOL

The overall scheme of mutagenesis described above is represented in Figure 2. Theoretically, the frequency of mutant to wild-type clones should be 50%, but in practice this ratio is much less. When this value is low, screening and isolation of the desired mutant can be laborious and time consuming. However, if a frequency of 50% or greater can be attained, one has only to sequence a few candidate clones for the isolation of the desired mutant. A good mutagenesis procedure must therefore result in a high ratio of mutant clones to wild-type clones.

Many factors influence the ratio of mutant to wild-type clones, resulting in the normally low ratio. We will describe some of these factors and discuss how they can be overcome. The initial *in vitro* mutagenesis procedures used *E. coli* DNA polymerase I holoenzyme for synthesizing the mutagenic strand. Certain characteristics of this enzyme made it an undesirable enzyme for mutagenesis. For example, this enzyme has high 5′ to 3′ exonuclease activity and can begin replication at nicks in double-stranded DNA.[48] Therefore, after completion of second-strand synthesis, this enzyme can hydrolyze the mutagenic primer from its 5′ end and replace it with the original template sequence. By using a high-quality preparation of purified Klenow fragment which does not contain the 5′ to 3′ exonuclease activity and ligase, the ratio of newly synthesized mutant DNA strands to wild type can be increased. However, strand displacement can still occur. Therefore, researchers commonly use T4 DNA polymerase, which is incapable of initiating replication at nicks in double-stranded DNA.[49]

Another potential problem is the presence of small RNA and DNA molecules which copurify with the single-stranded DNA template. These molecules act as random primers for the initiation of DNA synthesis. The presence of these small molecules will have no effect on the ratio of mutant to wild-type clones, however, if every template molecule is annealed to a mutagenic oligonucleotide also acting as a primer. In general, a primer/template ratio of approximately 10:1 is adequate, and an annealing time of 30 min is sufficient (20 m*M* Tris, 2 m*M* $MgCl_2$, 50 m*M* DTT), for complete saturation of the template strands with the mutagenic oligonucleotide.

Even when optimizing annealing conditions and using the correct DNA polymerase, the theoretical value of mutant to wild-type ratios is at most 50% and, in practice, difficult to achieve. This is probably due, at least in part, to the mismatch repair system of *E. coli*. Ideally, a 100% ratio of mutant to wild type is desirable. Many biochemical and genetic tricks have been developed which give 50 to 80% mutant to wild-type ratios. All are designed to select for the newly synthesized strand containing the mutagenic oligonucleotide and select against the original template strand containing the wild-type sequence. Four of the more successful methods will be described briefly below.

Kramer et al. used a "gapped duplex" approach whereby the template strand contains an amber nonsense mutation in an essential gene of the vector, and the newly synthesized mutagenic strand does not contain the amber mutation. By transforming a cell without suppressor activity, the mutagenized strand is more likely to survive than the template strand.[50] Carter et al. have selected against template strands by simultaneously removing a restriction site in the mutagenic strand and transforming a host which will cut the template strand containing the restriction site.[51] Kunkel engineered a strain of *E. coli* which inserts uracil residues into the template strand. The *in vitro* synthesized mutagenic strand does not contain uracil residues. When a wild-type *E. coli* strain is transformed with the heteroduplex, the uracil template strand does not survive as well as the mutagenic strand.[52] Finally, a method developed by Eckstein[53] incorporates phosphorothioate dNTPs into the mutagenic DNA. This results in a mutagenic strand insensitive to cleavage by certain restriction enzymes. The template strand is nicked with one of these enzymes and digested away with exonuclease. Then a new strand is synthesized complementary to the mutagenic strand. All of these methods have been used with fairly good success rates, and some are commercially available in kit form. Below is described a procedure based on Kunkel's method, which works well in our laboratory.

D. PREPARATION OF URACIL-CONTAINING SINGLE-STRANDED TEMPLATE

Transform Bio-Rad strain CJ236 (dut^- ung^-) with your phagemid, plate on LAmp Chl plates, and grow overnight at 37°C. The next day, pick a colony and grow in 5 ml of LB + Amp (100 μg/ml), chloramphenicol (25 μg/ml), and uridine (25 μg/ml) to saturation. Dilute the saturated culture 1/50 in 7 ml of LB + Amp, Chl, and uridine. Shake at 37°C for 1 h, add approximately 10^{10} to 10^{11} helper phage (we use strain R408 from Stratagene), and shake for approximately 12 h at 37°C.

Transfer 1.3 ml of the culture to four Eppendorf tubes. Spin out cells completely and transfer supernatant to new tubes. Add 2 μl of 10 mg/ml solution of RNAse A and incubate at 37°C for 1 h. Purify the ss DNA from the phage particles, beginning with the PEG step in the DNA sequencing section of this review. Combine the DNA from all four tubes by resuspending in 200 μl of TE total and store frozen. This is the single-stranded DNA you will use in all your subsequent mutagenesis experiments.

The template should be tested by transforming both wild-type *E. coli* and strain CJ236 with equal amounts of wild-type- and uracil-containing ss DNA templates. The uracil-containing template should give 1000 times less colonies than an equal amount of wild-type, single-stranded plasmid. The number of transformants should be the same in the mutant strain CJ236 with both DNAs.

Another way to test the template and oligonucleotide is to sequence the template to ensure that specific priming and good synthesis are occurring with your reagents. Once the uracil-containing, single-stranded DNA template has been made and tested, and the mutagenic oligonucleotide synthesized, the annealing and synthesis reactions described below may be performed.

E. PRACTICAL MUTAGENESIS PROTOCOL

Protocols for Annealing and Synthesis Reaction

1. Annealing — In a single tube, add the following components:

- Template — 2 μl (50 ng) ss DNA
- 5X annealing buffer — 2 μl (200 m*M* Tris, pH 7.5, 50 m*M* $MgCl_2$, 250 m*M* NaCl)
- Primer — 1 pmol kinased oligo
- H_2O — to 10 μl total volume

Put the tube in a 100-ml beaker half full with 70°C H_2O and let it cool over a 45-min period to 25°C.

2. Synthesis reaction — To the above reaction, add the following:

- 10X synthesis buffer — 1 μl (100 m*M* Tris, pH 7.6, 50 m*M* $MgCl_2$, 50 m*M* DTT)
- 10 m*M* ATP — 2 μl
- 5X dNTP — 4 μl (2.5 m*M* dATP, dCTP, dGTP, and dTTP)
- Ligase — 5 to 10 units
- T4 DNA polymerase — 1 unit
- H_2O — to 20-μl total volume

Hold the reaction on ice for 10 min, RT for 5 min, and then transfer to 37°C for 1 h.

At this point, transform a wild-type strain of *E. coli* known to have a high transformation efficiency (we use MC1061) with the entire mixture and plate all cells on LAmp plates. One should see anywhere from 30 to 200 colonies on each plate the next day, 75% of which will contain the desired mutation.

F. SCREENING FOR MUTANT CLONES

The protocol described above is designed to make screening mutants from wild-type clones quick and easy in the absence of a genetic selection. If all the reactions go well, and greater than 50% of the resultant clones contain the mutation of interest, then one needs only to pick three or four colonies or plaques (depending on the choice of vector) and sequence them. If this frequency cannot be attained, it is convenient to have another method for screening. If possible, the addition or removal of a restriction site in the sequence being made can facilitate screening. Then one can easily screen colonies by restriction analysis of double-stranded DNA. Another method is to transfer the colonies to nitrocellulose and probe with the mutagenic oligonucleotide which has been labeled with ^{32}P. Mutant clones are differentiated from wild type by intensities on autoradiographic exposures.[54]

G. MAKING MULTIPLE BASE CHANGES, INSERTIONS, AND DELETIONS

Many projects require that multiple changes be made in a particular region of the gene under investigation. A project analyzing a DNA operator sequence, for example, may require that a single nucleotide be substituted with the remaining three. This may be accomplished by using a mutagenic oligonucleotide degenerate at a single position. The same degenerate oligonucleotide can be used for colony hybridization, and sequence analysis would be used to identify each individual mutant. However, sequence analysis of a number of clones should be sufficient for mutant identification if an effective mutagenesis protocol has been used.

Multiple mutations using a single mutagenic oligonucleotide are sometimes impossible to create simultaneously, especially if the mutations are separated by 5 to 10 nucleotides. A hypothetical RNA structure has been tested by altering two to five of six different nucleotides in a small region of the gene with a single 56-nucleotide mutagenic oligomer. Since all six nucleotide changes could not be substituted simultaneously, plaque hybridizations and sequence analysis defined various combinations of the mutations in the resultant clonal population.[55] It may be possible to generate more mutations during one round of mutagenesis by either using the method of Eckstein or by transforming a mismatch repair mutant of *E. coli* when using the Kunkel method.

Finally, deletions and insertions are quite simple to make. We have attained 80% efficiency in the creation of an 800-nucleotide deletion of a yeast actin gene utilizing a 40-base oligonucleotide (20 bases of complementarity on each side of the deletion site).[56] To our knowledge, there is no size limit on deletions which can be generated using this approach. Insertions are, in principle, the same as deletions; however, the size is restricted by the cost and length of the oligonucleotide to be synthesized. In summary, mutagenic procedures and projects involving site-specific mutagenesis are subject to various modifications and improvizations depending on the investigator's needs. One has to remember that the ratio of mutant to wild-type clones needs to be maximized, the effort of attaining the mutants minimized, and the chance of creating unwanted mutations minimized. Once a fragment has been mutagenized, the entire fragment of interest should be sequenced and subcloned into the appropriate vector for further characterization.

VI. DNA AND RNA SEQUENCING

A. DNA SEQUENCING

The most predominantly used sequencing methods today are based on the method first developed by Sanger.[57] The method involved copying a single-stranded DNA template in the presence of chain-terminating analogues of the normal deoxynucleoside triphosphates. Originally, this protocol required a single-stranded DNA template, an oligonucleotide primer, DNA polymerase, α^{32}P-labeled ATP, and a ddNTP/dNTP mixture. The method was used to complete the sequence of the bacteriophage ϕX174, which contains a single-stranded DNA chromosome. Subsequently, the M13 cloning vectors developed by Gronenborn and Messing[58] allowed the isolation of any clonable DNA insert in the single-stranded form, allowing the Sanger sequencing method to be applied to the sequencing of cloned DNA fragments. Once a DNA fragment has been cloned, single-stranded template DNA can be purified by the following procedure.

Dilute an overnight culture of *E. coli* strain JM103 1/100 in 2X YT medium and grow at 37°C for 90 min with good aeration. Pick a plaque and inoculate 2 ml of the culture. Grow for 8 h at 37°C. Spin down the cells and add 1.5 ml of the supernatant to 0.3 ml of 20% PEG 6000, 2.5 *M* NaCl. Hold on ice for 15 min. Spin in a microfuge for 15 min. Discard the supernatant. Resuspend the pellet in 1.2 ml of TE and repeat the PEG precipitation ensuring that all of the supernatant is discarded. Resuspend the pellet in 200 μl of TE and transfer to a small (0.4-ml) Eppendorf tube. Extract one time with an equal volume of phenol, and once with phenol:chloroform. It is important to stay as far from the interface as possible while doing these extractions. Extract with 0.2 ml of CH_2Cl_2 and precipitate with ethanol. Resuspend in 25 μl of distilled H_2O and use 7 μl for each annealing reaction.

The original Sanger method has been improved by Tabor and Richardson.[59] Tabor and Richardson have modified the T7 phage DNA polymerase, which makes it more desirable for DNA sequencing. This procedure has been commercialized by U.S. Biochemicals and is available in the form of a DNA sequencing kit. We have used this kit to sequence double-stranded plasmids and linear double-stranded DNAs as well as single-stranded DNA.

For sequencing plasmids, we have found that small RNA contaminants are the biggest problem. A method for purifying plasmids suitable for DNA sequencing has been presented by Hattori and Sakaki.[60] Briefly, plasmid DNA is purified from a 1.5-ml overnight culture by the alkaline lysis procedure,[61] phenol extracted once, precipitated with ethanol, and resuspended in 50 μl of TE. The DNA is then RNAse-treated with 10 μg of RNAse A for 1 h at 37°C and precipitated on ice for 1 h with 30 μl of 20% PEG 6000, 2.5 *M* NaCl. The RNAse A must be highly active. The plasmid DNA is resuspended in 50 μl of TE after drying the precipitated pellet. For sequencing, 18 μl of the plasmid is denatured by adding 2 μl of 2 *N* NaOH, incubating at room temperature for 5 min, and adding 8 μl of filter-sterilized 5 *M* NH_4^+OAc (pH 7.4). Then 100 μl of ethanol is added and the mixture is iced for 5 min, spun in a microfuge for 5 min, decanted, and dried. One pmol of the sequencing primer is added to this, along with buffer and H_2O for the annealing reaction. The sequencing reactions are identical to that for single-stranded DNA.

An increasingly important application of double-stranded sequencing is that of sequence determination of products derived from the polymerase chain reaction (PCR) described later in this chapter. A method has been described for sequencing such double-stranded DNA products (U.S. Biochemicals Editorial Comments) with the U.S. Biochemicals Sequenase kit. They run the PCR products on a 10% polyacrylamide gel, excise the band of interest, and elute in 0.5 *M* NH_4Cl, 1 m*M* EDTA overnight. The PCR DNA is precipitated with two volumes of ice-cold ethanol and resuspended in a small volume of TE buffer. The Sequenase kit protocol was modified to accommodate sequencing of radiolabeled primer-PCR template complexes: 5 pmol of ^{32}P-labeled primer is boiled for 2 min in a 10-μl reaction containing 50 to 100 ng of purified PCR template and 2 μl of 5X Sequenase buffer. The reaction is then placed at room temperature for 15 to 30 min, after which 1.0 μl of 0.1 *M* DTT, 2.5 μl of water, and 2 μl of diluted (1:8 in TE) Sequenase (12.5 U/μl) are added. The labeling step is eliminated, and 3.5-μl aliquots of the reaction are incubated for 15 min at room temperature in 2.5 μl of the nucleotide-specific termination buffers supplied in the Sequenase kit. The reaction is stopped by the addition of 4.0 μl of formamide-dye mix and then heated to 100°C for 1 min. Then the samples are electrophoresed in a 10% polyacrylamide, 7 *M* urea-sequencing gel, and the gels dried and autoradiographed.

B. PRIMER EXTENSION ANALYSIS AND SEQUENCING OF RNAs USING OLIGONUCLEOTIDES

Oligonucleotide primers can be used in the analysis of RNAs by primer extension analysis using reverse transcriptase to identify the 5′ terminus of an RNA molecule and/or a branch point in partially spliced pre-mRNA intermediates.[62,63] Also, direct sequencing of RNA can be accomplished using the dideoxynucleoside triphosphate chain-termination method.[64] The appropriate oligonucleotide is first ^{32}P-kinase labeled, annealed with the denatured RNA sample, and reverse transcription is carried out in the presence of the four deoxynucleoside triphosphates and with dideoxynucleoside triphosphates for sequencing. The products are then analyzed on a DNA sequencing gel to size the reverse transcribed products or determine the RNA sequence. Below is a protocol used in our laboratory.

5′ ^{32}P Labeling of Oligonucleotide Primer

- 1 μ*M* oligonucleotide (10 pmol)
- 1 μl 10X kinase buffer (500 m*M* Tris-HCl, pH 8.0, 100 m*M* $MgCl_2$, 100 m*M* DTT)
- 1 μl τ ^{32}P dATP
- 2 U T4 polynucleotide kinase
- 10 μl total volume with H_2O

The reaction is incubated at 37°C for 30 min, then heated to 65°C for 5 min to denature the polynucleotide kinase. Clean the labeled oligonucleotide by adding 35 μl of H_2O and 5 μl of 5 *M* NaCl, mixing, and extracting two times with 50 μl of phenol and one time with 50 μl of CH_2Cl_2. Add 25 μl of 7.5 *M* NH_4^+ OAc and 300 μl of ethanol and precipitate by centrifuging 5 min. Discard the supernatant, dry the pellet, and resuspend in 20 μl of TE. Use 1 μl in the annealing reactions described below.

Annealing of Primer to RNA

- 1 to 10 μg total or poly A^+RNA
- 1 μl kinased primer (1 pmol)
- 1 μl 10X reverse transcriptase buffer (1X = 50 m*M* Tris, pH 8.3, 60 m*M* NaCl, 10 m*M* DTT)
- 10 μl total volume

The sample is heated to 80°C for 2 min, then slowly cooled to 45 to 50°C over a period of 2 h. Rapid cooling is not recommended since the likelihood of mispriming with partially homologous sequences is greatly increased. Then 1.2 μl of 36 m*M* $Mg(OAc)_2$ is added such that the final $Mg(OAc)_2$ is 6 m*M*.

Primer Extension

- 2 μl annealing solution
- 1μl 5X dNTP solution [5X = 0.375 μ*M* each dATP, dCTP, dGTP, and dTTP in 1X reverse transcriptase buffer with 6 m*M* $Mg(OAc_2)$]
- 4 U AMV reverse transcriptase (diluted in 1X RT buffer)
- 5 μl total volume

Incubate at 45 to 50°C for 30 min. Add 10 μl of loading dye mix (90% formamide, 1X TBE, 0.05% bromophenol blue and xylene cyanol). Heat the samples to 95°C for 3 min and load 5 μl of the reaction on a 6% polyacrylamide, 7 *M* urea-sequencing gel. Run the gel until the bromophenol blue is 1 to 2 in. from the bottom of the gel (oligonucleotides about 20 nucleotides long run approximately with the bromophenol blue).

Sequencing

After the annealing reaction is complete, add the following:

- 2 μl annealing solution
- 1 μl 5X dNTP solution
- 1 μl 5X ddNTP solution (four separate solutions in 1X reverse transcriptase buffer: 1 m*M* ddATP, 1 m*M* ddCTP, 1 m*M* ddGTP, and 1 m*M* ddTTP)
- 4 U reverse transcriptase
- 5 μl total volume

Prepare five reaction mixtures (one each for ddA, ddC, ddG, and ddT, and a control reaction with 1X RT buffer with 6 m*M* Mg/OAc_2). Incubate and run gel as described above in the primer extension protocols.

C. POLYMERASE CHAIN REACTION (PCR)

The polymerase chain reaction, or PCR, enables the exponential amplification and subsequent detection of rare sequences obtained from any biological sample.[8,9] The technique, conceived and developed by Mullis and colleagues at the Cetus Corporation, was originally shown to be useful for the rapid detection and discriminaton of the allelic differences in sickle cell vs. normal DNAs.[8,9] The ability to amplify sequences from minute quantities of biological samples has led to the widespread use of this technique in both basic research and diagnostic medicine.

Conceptually, the technique is simple and straightforward and is diagrammed in Figure

3. Synthetic oligonucleotides complementary to opposite strands of the target sequence and bracketing a defined region (usually 100 to several hundred nucleotides) are synthesized. These oligonucleotides are added in huge molar excess relative to the target DNA (or RNA) sample. The oligos serve as specific primers for repair replication catalyzed by a DNA polymerase, either the Klenow fragment of *E. coli* DNA polymerase or the heat-resistant DNA polymerase isolated from the thermophilic organism *Thermus aquaticus*.[10] Each round of repair replication is brief (1 to 3 min) and is followed immediately by a thermal denaturation step to separate both the original and newly created double-stranded DNA molecules. The next step utilizes a more moderate temperature to allow the oligos to anneal to the target sequences. The whole process, polymerization-denaturation-annealing, is usually repeated 20 to 30 times before the products of the amplification reaction are analyzed. The use of a thermoresistant polymerase has led to the semiautomation of the entire PCR process. Several different companies now market microprocessor-controlled devices which automatically carry out the cycling reactions. The investigator need only add the reagents and place the samples in the automatic thermal cycling devices. If the entire process proceeds as planned, the fold of amplification can be as high as several million. Theoretically, then, a sample containing only a few hundred molecules of target sequence can be amplified such that several hundred million copies of the target sequence are available for detection following PCR.

Detection is usually accomplished by designing and utilizing a third synthetic oligonucleotide complementary to sequences in between the two primer sites. An aliquot of the amplification reaction is electrophoresed in an agarose or polyacrylamide gel, blotted to an immobilizing support, and hybridized to the oligonucleotide probe. The probe may be tagged with either a radioactive nucleotide or a nonradioactive ligand such as biotin. Hybridization to an amplified product of the expected size, utilizing conditions of high stringency, provides rigorous proof of the fidelity of the amplification process. Alternatively, the amplified samples can be blotted directly to a nitrocellulose membrane without prior gel electrophoresis. The dot- or slot-blotted DNAs can then be probed as above. We have initiated the use of RNA templates for PCR and have been able to successfully amplify and detect HIV-1 sequences from peripheral blood samples of AIDS and ARC patients using this approach.[65] We have also demonstrated that it is possible to amplify and detect HIV-1 RNAs from as few as 100 original copies of template. The primary difference between using an RNA vs. a DNA template is that the first polymerization reaction utilizes reverse transcriptase primed by one of the two oligonucleotides which is complementary to the RNA. Subsequent rounds of amplification are idential to that described above for the DNA template. We present below a generalized laboratory protocol for the PCR assay. Variations in the cycling times, magnesium concentrations, primer, and dNTP concentrations may be necessary to optimize any individual sample.

1. DNA

1. Mix 0.5 μg of genomic DNA (at least 10 × less if you are amplifying from a plasmid) with 50 pmol of each primer in a 50-μl reaction containing the following: 50 *M* KCI, 10 m*M* Tris-HCI pH 8.3, 1.5 m*M* $MgCl_2$, 0.01% (w/v) gelatin, 200 μ*M* each dNTP. Finally, add 1.25 U of *Taq* DNA polymerase. Overlay the entire reaction with 50 μl of mineral oil.
2. Cycling times depend to a large extent on the sequence composition of the target and the length of the region to be amplified. Initial conditions to try are 95°C for 5 min, 42 to 45°C for 2 min, and 72°C for 3 min. Then the cycling should be repeated 30 times using 95°C for 2 min, 42°C for 2 min, and 72°C for 3 min.
3. The best source of *Taq* is U.S. Biochemicals or Amplitaq from Perkin-Elmer Cetus.
4. One tenth of the PCR is electrophoresed in 1.8% agarose and vacuum blotted in 0.4 *M* NaOH to Zeta-Probe (Bio-Rad) for at least 40 min.

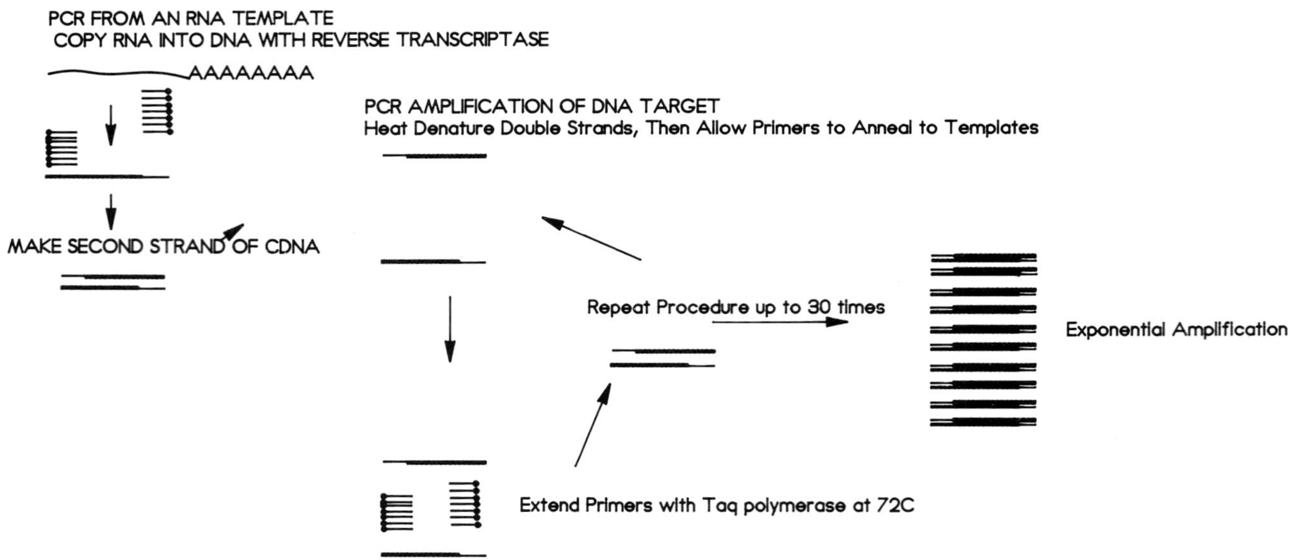

FIGURE 3. Schematic representation of RNA- and DNA-dependent gene amplification utilizing the polymerase chain reaction (PCR).

5. Hybridization is with 20 pmol of kinased oligo in 6XSSPE, 7% SDS, 0.5% Blotto (reconstituted powdered milk). Hybridization conditions are determined by probe length and composition.
6. Filters are washed in 6XSSC at hybridization temperature for at least three changes for 15 min each.

2. RNA

The reaction is set up as described above, except that the template is 0.5 μg of total RNA. The mixture is heated to 95°C for 5 min, allowed to cool to 42°C for 5 min, and then 2.0 U of AMV reverse transcriptase (diluted in above buffer) is added. Reverse transcription proceeds for no longer than 5 min at 42°C. The reaction is stopped by heating at 95°C for 3 min and primer annealing for 2 min at 42°C. The *Taq* DNA polymerase is added to 1.25 U final, and at 72°C for 3 min, followed by 30 cycles of 2 min at 95°C, 2 min at 42°C, and 3 min at 72°C.

Trouble Shooting:

If there is no product or, conversely, a large smear, check the oligonucleotide primers by kinase labeling and electrophoresing them in a denaturing polyacrylamide gel. Also, if a control template containing the target sequence is available, make sure that the oligos hybridize to this template. If the primers are okay, cut down their concentration (to 20 pmol) and/or titrate the Mg in the reaction from 1 to 5m*M*. Sometimes the nucleic acid template contains substances inhibitory to the reaction. If no product is obtained when all of the above have been tried, try diluting the template DNA or RNA 1:50 or 1:100 and then adding an aliquot of this diluton to the PCR mix.

VII. SUMMARY

This article has focused upon some of the methods for the chemical synthesis of DNA (and RNA) as well as some of the more common and important uses of synthetic oligos in biological and medical research. This is by no means an exhaustive treatment of synthetic DNA chemistry and uses, but is meant to serve as an introduction and survey of the uses, both real and potential, of this important tool. With the large number of companies offering synthetic DNA services, it is quite likely that the number of users of synthetic DNA and RNA will grow rapidly in the next few years. Synthetic oligonucleotides, once the toy of only organic chemists, are now one of the most commonly utilized tools in biology and medicine.

ACKNOWLEDGMENTS

This work was supported by the American Foundation For AIDS Research (Grant No. 000613R) and the IDEN Foundation.

REFERENCES

1. **Khorana, H. G.,** Total synthesis of a gene, *Science,* 203, 614, 1979.
2. **Riggs, A. and Itakura, K.,** Synthetic DNA and medicine, *Am. J. Hum. Genet.,* 31, 531, 1979.
3. **Rossi, J. J., Kierzek, R., Huang, T., Walker, P., and Itakura, K.,** An alternate method for synthesis of double-stranded DNA segments, *J. Biol. Chem.,* 256, 9229, 1982.
4. **Razin, A., Hirose, T., Itakura, K., and Riggs, A. D.,** Efficient correction of a mutation by use of chemically synthesized DNA, *Proc. Natl. Acad. Sci. U.S.A.,* 75, 4268, 1978.

5. **Itakura, K., Rossi, J. J., and Wallace, R. B.**, Synthesis and use of synthetic oligonucleotides, *Annu. Rev. Biochem.*, 53, 323, 1984.
6. **Rossi, J. and Zoller, M.**, Site-specific and regionally directed mutagenesis of protein-encoding sequences, in *Protein Engineering*, Oxender, D. and Fox, C. F., Eds., Alan R. Liss, New York, 1987, 51.
7. **Conner, B. J., Reyes, A. A., Morin, C., Itakura, K., Teplitz, R. L., and Wallace, R. B.**, Detection of sickle cell β^{s}-globin allele by hybridization with synthetic oligonucleotides, *Proc. Natl. Acad. Sci. U.S.A.*, 80, 278, 1983.
8. **Saiki, R. F., Scharf, S., Faloona, F., Mullis, K. B., Horn, G. T., Erlich, H. A., and Arnheim, N.**, Enzymatic amplification of β-globin sequences and restriction site analysis for diagnosis of sickle cell anemia, *Science*, 230, 487, 1985.
9. **Mullis, K. and Faloona, F. A.**, Specific synthesis of DNA *in vitro* via a polymerase catalyzed chain reaction, *Methods Enzymol.*, 155, 335, 1987.
10. **Saiki, R. K., Gelfand, D. H., Stoffel, S., Scharf, S. J., Higuchi, R., Horn, G. T., Mullis, K. B., and Erlich, H. A.**, Primer directed enzymatic amplification of DNA with a thermostable DNA polymerase, *Science*, 239, 487, 1988.
11. **Usman, N., Ogilvie, K. K., Jiang, M.-Y., and Cedergren, R. J.**, Automated chemical synthesis of long oligoribonucleotides using 2′-*O*-silylated ribonucleoside 3′-*O*-phosphoramidites on a controlled-pore glass support: synthesis of a 43-nucleotide sequence similar to the 3′-half molecule of an *Escherichia coli* formylmethionine tRNA, *J. Am. Chem. Soc.*, 109, 7845, 1987.
12. **Beaucage, S. and Caruthers, M.**, Deoxynucleoside phosphoramidites. A new class of key intermediates for deoxypolynucleotide synthesis, *Tetrahedron Lett.*, 22(20), 1859, 1981.
13. **Sinha, N. D., Biernat, J., McManus, J., and Koster, H.**, Polymer support oligonucleotide synthesis XVIII[1,2]: use of β-cyanoethyl-*N*, *N*-dialkylamin-/*N*-morpholino phosphoramidite of deoxynucleotides for the synthesis of DNA fragments simplifying deprotection and isolation of the final product, *Nucleic Acids Res.*, 12, 4539, 1984.
14. **Gao, X., Gaffney, B. L., Senior, M., Riddle, R. R., and Jones, R. A.**, Methylation of thymine residues during oligonucleotide synthesis, *Nucleic Acids Res.*, 13, 573, 1985.
15. **Urdea, M. S., Ku, L., Gee, Y. G., and Warner, B. D.**, *Base Modification and Cloning Efficiency of Oligodeoxyribonucleotides Synthesized by the Phosphoramidite Method: Methyl Versus Cyanoethyl Phosphorus Protection*, Nucleic Acids Research Symp. Ser. No. 16, IRL Press/Oxford University Press, Oxford, England, 1985, 257.
16. **Yeung, A. T., Dinehart, W. J., and Jones, B. K.**, Modifications of guanine bases during oligonucleotide synthesis, *Nucleic Acids Res.*, 16, 4539, 1988.
17. **Zon, G., Gallo, K. A., Samson, C. J., Summers, M. F., and Byrd, R. A.**, Analytical studies of 'mixed sequence' oligodeoxyribonucleotides synthesized by competitive coupling of either methyl- or β-cyanoethyl-*N,N*-diisopropylamino phosphoramidite reagents, including 2′-deoxyinosine *Nucleic Acids Res.*, 13, 8181, 1985.
18. **Froehler, B. C. and Matteucci, M. D.**, Nucleoside H-phosphonates: valuable intermediates in the synthesis of deoxyoligonucleotides, *Tetrahedron Lett.*, 27, 469, 1986.
19. **Stawinski, J., Stromberg, R., Thelin, M., and Westman, E.**, Studies on the t-butyldimethylsilyl group as a 2′-O-protection in oligoribonucleotide synthesis via the H-phosphonate approach, *Nucleic Acids Res.*, 16, 9285, 1988.
20. **Froehler, B. C., Ng, P. G., and Matteucci, M. D.**, Synthesis of DNA via deoxynucleoside H-phosphonate intermediates, *Nucleic Acids Res.*, 14, 5399, 1986.
21. **Ott, J. and Eckstein, F.**, Protection of oligonucleotide primers against degradation by DNA polymerase I, *Biochemistry*, 26, 8237, 1987.
22. **Stein, C. A., Subasinghe, C., Shinozuka, K., and Cohen, J. S.**, Physiochemical properties of phosphorothioate oligodeoxynucleotides, *Nucleic Acids Res.*, 16, 3209, 1988.
23. **Spitzer, S. and Eckstein, F.**, Inhibition of deoxyribonucleases by phosphorothioate groups in oligodeoxyribonucleotides, *Nucleic Acids Res.*, 16, 11691, 1988.
24. **Kempe, T., Sundquist, W. I., Chow, F., and Hu, S.**, Chemical and enzymatic biotin-labeling of oligodeoxyribonucleotides, *Nucleic Acids Res.*, 13, 45, 1985.
25. **Chollet, A. and Kawashima, E. H.**, Biotin labeled synthetic oligodeoxyribonucleotides: chemical synthesis and use as hybridization probes, *Nucleic Acids Res.*, 13, 1529, 1985.
26. **Wachter, L., Jablonski, J., and Ramachandran, K. L.**, A simple and efficient procedure for the synthesis of 5′ aminoalkyl oligodeoxynucleotides, *Nucleic Acids Res.*, 14, 7985, 1986.
27. **Connolly, B. A.**, The synthesis of oligonucleotides containing primary amino group at the 5′ terminus, *Nucleic Acids Res.*, 15, 3131, 1987.
28. **Connoly, B. A.**, Chemical synthesis of oligonucleotides containing a free sulfhydryl group and subsequent attachment of thiol specific probes, *Nucleic Acids Res.*, 13, 4485, 1985.
29. **Jablonski, E., Moomaw, E. W., Tullis R. H., and Ruth, J. L.**, Preparation of oligodeoxynucleotide-alkaline phosphatase conjugates and their use as hybridization probes, *Nucleic Acids Res.*, 14, 6115, 1986.

30. **Urdea, M. S., Warner, B. D., Running, J. A., Stempien, M., Clyne, J., and Horn, T.,** A comparison of nonradioisotopic hybridization methods using flourescent, chemilluminescent and enzyme labeled synthetic oligodeoxyribonucleotide probes, *Nucleic Acids Res.*, 16, 4937, 1988.
31. **Horn, T. and Urdea, M.,** A chemical 5′-phosphorylation of oligodeoxyribonucleotides that can be monitored by trityl cation release, *Tetrahedron Lett.*, 27, 4705, 1986.
32. **Bayer, E. A., Zalis, M. G., and Wilchek, M.,** 3-(*N*-Maleimido-propionyl) biocytin: a versatile thiol-specific biotinylating reagent, *Anal. Biochem.*, 149, 529, 1985.
33. **McLaughlin, L. and Piel, N.,** Chromatographic purification of synthetic oligonucleotides, in *Oligonucleotide Synthesis: A Practical Approach*, Gait, M., Ed., IRL Press, Oxford, 1984, 117.
34. **Maniatis, T., Fritsch, E., and Sambrook, J.,** *Molecular Cloning: A Laboratory Handbook*, Cold Spring Harbor Laboratory, Cold Spring Harbor, NY, 1982.
35. **Urlich, A., Berman, C., Dull, T., Gray, A., and Lee, J.,** Isolation of the human insulin-like growth factor I gene using synthetic DNA probe, *EMBO J.*, 3, 361, 1984.
36. **Goeddel, D., Yelverton, E., Urlich, A., Heyneker, H., Miozzari, G., Holmes, W., Seeburg, P., Dull, T., May, L., Stebbing, N., Crea, R., Maeda, S., McCandliss, R., Sloma, A., Tabor, J., Gross, M., Familletti, P., and Pestka, S.,** Human leukocyte interferon produced by *E. coli* is biologically active, *Nature*, 287, 411, 1981.
37. **Lathe, R.,** Synthetic oligonucleotide probes deduced from amino acid sequence data: theoretical and practical considerations, *J. Mol. Biol.*, 183, 1, 1985.
38. **Reyes, A., Studencki, A., and Wallace, B.,** Methods of oligonucleotide hybridization, in *Genetic Engineering, Principles and Methods*, Vol. 6, Setlow, J. and Hollaender, A., Eds., Plenum Press, New York, 1984, 157.
39. **Suggs, S., Wallace, B., Hirose, T., Kawashina, E., and Itakura, K.,** Use of synthetic oligonucleotides as hybridization probes: isolation of cloned cDNA sequences for human β_2-microglobulin, *Proc. Natl. Acad. Sci. U.S.A.*, 78, 6613, 1981.
40. **Colleaux, L., d'Auriol, L., Betermier, M., Cottarel, G., Jacquier, A., Galibert, F., and Dujon, B.,** Universal code equivalent of a yeast mitochondrial reading frame is expressed into *E. coli* as a specific double strand endonuclease, *Cell*, 44, 521, 1986.
41. **Gillam, S. and Smith, M.,** Site-specific mutagenesis using synthetic oligodeoxyribonucleotide primers. II. *In vitro* selection of mutant DNA, *Gene*, 8, 99, 1979.
42. **Gillam, S. and Smith, M.,** Site-specific mutagenesis using synthetic oligodeoxyribonucleotide primers. I., Optimum conditions and minimum oligodeoxyribonucleotide length, *Gene*, 8, 81, 1979.
43. **Hutchinson, C., Phillips, S., Edgell, M., Gillam, S., Jahnke, P., and Smith, M.,** Mutagenesis at a specific position in a DNA sequence, *J. Biol. Chem.*, 253, 6551, 1978.
44. **Razin, A., Hirose, T., Itakura, K., and Riggs, A.,** Efficient correction of a mutation by use of chemically synthesized DNA, *Proc. Natl. Acad. Sci. U.S.A.*, 75, 4268, 1978.
45. **Dente, L., Cesareni, G., and Cortese, R.,** pEMBL: a new family of single stranded plasmids, *Nucleic Acids Res.*, 11, 1645, 1983.
46. **Levinson, A., Silver, D., and Seed, B.,** Minimal size plasmids containing an M13 origin for production of single-stranded transducing particles, *J. Mol. Appl. Genet.*, 2, 507, 1984.
47. **Zagursky, R. and Berman, M.,** Cloning vectors that yield high levels of single-stranded DNA for rapid DNA sequencing, *Gene*, 27, 183, 1984.
48. **Kelly, R., Cozzarelli, N., Deutscher, M., Lehman, I., and Kornberg, A.,** Enzymatic synthesis of deoxyribonucleic acid, *J. Biol. Chem.*, 245, 39, 1970.
49. **Nossal, N.,** DNA synthesis on a double stranded DNA template by the T4 bacteriophage DNA polymerase and the T4 gene 32 protein, *J. Biol. Chem.*, 249, 5668, 1974.
50. **Kramer, W., Drutsa, V., Jansen, H., Kramer, B., Pflugfedlder, M., and Fritz, H.,** The gapped duplex DNA approach to oligonucleotide-directed mutation construction, *Nucleic Acids Res.*, 12, 9441, 1984.
51. **Carter, P., Bedoulle, H., and Winter, G.,** Improved oligonucleotide site-directed mutagenesis using M13 vectors, *Nucleic Acids Res.*, 13, 4431, 1985.
52. **Kunkel, T., Roberts, J., and Zakour, R.,** Rapid and efficient site-specific mutagenesis without phenotypic selection, *Methods Enzymol.*, 154, 367, 1987.
53. **Nakamaye, K. and Eckstein, F.,** Inhibition of restriction endonuclease *NciI* cleavage by phosphorothioate groups and its application to oligonucleotide-directed mutagenesis, *Nucleic Acids Res.*, 14, 9679, 1986.
54. **Wallace, B., Johnson, P., Tanaka, S., Schold, M., Itakura, K., and Abelson, J.,** Directed deletion of a yeast transfer RNA intervening sequence, *Science*, 209, 1396, 1980.
55. **Reyes, A. and Akeson, R.,** Generation of multiple independent substitution mutants by M13 *in vitro* mutagenesis using a single mutagenic oligonucleotide, *DNA*, 7, 579, 1988.
56. **Deshler, J., Larson, G., and Rossi, J.,** *Kluyveromyces lactis* maintains *Saccharomyces cerevisiae* intron encoded splicing signals, *Mol. Cell. Biol.*, 9, 2208, 1989.
57. **Sanger, F.,** Determination of nucleotide sequence in DNA, *Science*, 214, 1205, 1981.

58. **Gronenborn, B. and Messing, J.,** Methylation of single-stranded DNA *in vitro* introduces new restriction endonuclease cleavage sites, *Nature*, 272, 375, 1978.
59. **Tabor, S. and Richardson, C.,** DNA sequence analysis with a modified bacteriophage T7 DNA polymerase, *Proc. Natl. Acad. Sci. U.S.A.*, 84, 4767, 1987.
60. **Hattori, M. and Sakaki, Y.,** Dideoxy sequencing method using dematured plasmid templates, *Anal. Biochem.*, 152, 232, 1986.
61. **Birnboim, H. and Doly, J.,** A rapid alkaline extraction procedure for screening plasmid DNA, *Nucleic Acids Res.*, 7, 1513, 1979.
62. **Vijayraghavan, U., Parker, R., Tamm, J., Iimura, Y., Rossi, J., Abelson, J., and Guthrie, C.,** Mutations in conserved intron sequences affect multiple steps in the yeast splicing pathway, particularly assembly of the spliceosome, *EMBO J.*, 5, 1683, 1986.
63. **Padgett, R., Konarska, M., Grabowski, P., Hardy, S., and Sharp, P.,** Lariat RNA's as intermediates and products in the splicing of messenger RNA precursors, *Science*, 225, 898, 1984.
64. **Inoue, T. and Cech, T.,** Secondary structure of the circular form of the *Tetrahymena* rRNA intervening sequence: a technique for RNA structure analysis using chemical probes and reverse transcriptase, *Proc. Natl. Acad. Sci. U.S.A.*, 82, 648, 1985.
65. **Murakawa, G., Zaia, J. A., Spallone, P. A., Stephens, D. A., Kaplan, B. E., Wallace, R. B., and Rossi, J. J.,** Direct detection of HIV-1 RNA from AIDS and ARC patient samples, *DNA*, 7, 287, 1988.

Chapter 3

IN VITRO SITE-DIRECTED MUTAGENESIS

Donald R. Menick

TABLE OF CONTENTS

I. INTRODUCTION

Recent advances in genetic engineering have made it possible through site-directed mutagenesis to introduce single amino acid changes into proteins and provide a subtle approach to studying structure-function relationships in proteins. Application of this technique with both soluble and membrane proteins has already led to exciting new discoveries in the mechanism of catalysis[1-5] and protein folding.[6] In the past, the tools available to help determine what role a particular amino acid played in a protein's function were somewhat limited in application. Although characterization of mutant proteins provided a wealth of information, it was limited by the random nature of mutagenesis and was confined to the isolation of phenotypically selectable mutants. Chemical modification studies of amino acids have provided insight into the possible role of particular amino acids. However, again there are major drawbacks to the technique, such as the chemical specificity and bulkiness of the reagents. Site-directed mutagenesis allows the investigator to make precise changes at or near the active site, permitting a first-hand look at the interdependent roles of catalytic groups.[1]

Site-directed mutagenesis was first developed using the single-stranded bacteriophage, ϕX174. In these early studies, phenotypic screening was used to assess the efficiency of mutagenesis,[7-10] using radiolabeled mutagenic oligonucleotide as a hybridization probe to distinguish mutant clones. These early studies opened the way for the rapid screening of hundreds of putative mutants with silent, unknown, or unselectable phenotypes. The development of the M13mp series of vectors with (1) a polylinker region of unique restriction sites, (2) a phenotypic screen for insertion of fragments, and (3) easily purified single-stranded (ss) DNA in high yield made M13 the vector of choice in early protocols of site-directed mutagenesis.[11] M13 contains circular ss DNA packaged in a filamentous virion which binds to the F pilus of *Escherichia coli* and injects its DNA into the cell. The ss phage DNA is converted using host enzymes to double-stranded replicative form (RF) DNA. Early in the phage life cycle, the RF is replicated by a rolling circle mechanism that is terminated by a phage protein (gIIp), leaving a displaced, closed circular plus strand which is converted to the double-stranded RF. Later in the phage life cycle, the plus strand is packaged into viral coat proteins which are extruded from the cell without lysis.[12] Because of this, M13 can have very high titers of plaque-forming units (10^{12}/ml), thus giving excellent yields of ss DNA.

The overall strategy of mutagenesis as originally described by Zoller and Smith[11] is outlined in Figure 1. The DNA fragment of interest is ligated into the RF of M13 (M13mp-18/19), which is used to transfect *E. coli* JM101. Phage particles are isolated and their ss DNA containing the fragment is extracted. This serves as the template for mutagenesis. A mutagenic oligonucleotide complementary to the target region, except for one or more mismatches directing the desired change, is annealed to the ss DNA template containing the wild-type fragment. The mutagenic primer is extended *in vitro* by DNA polymerase I (large fragment) and ligated with T4 DNA ligase. *E. coli* JM101 is transfected with the closed circular heteroduplex and the resulting plaques screened for mutants by colony blot hybridization using the mutagenic oligonucleotide as probe. Putative mutants are plaque-purified and sequenced. The mutant RF DNA is isolated and the mutant insert cloned into an expression vector for analysis.

II. CRITICAL PARAMETERS OF MUTANT FREQUENCY

Although the theoretical yield of mutants should be 50%, the actual yield is often much lower (e.g., <1 to 15%). Mutation frequency depends on many factors, including the construction of the heteroduplex and the mismatch repair system of the host cell.

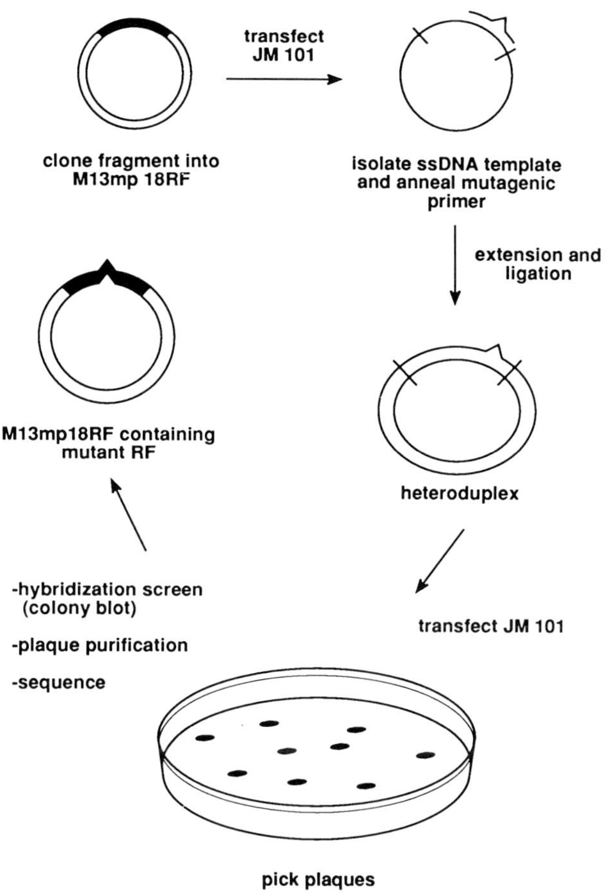

FIGURE 1. A simplified scheme for oligonucleotide-directed mutagenesis using M13 as the cloning vehicle. Details of each step are described in the text.

A. CONSTRUCTION OF THE HETERODUPLEX

Several factors can contribute to incomplete synthesis of the heteroduplex, including the purity of the ss template, possible secondary structures in the template (e.g., hairpin loops) that would cause the polymerase to pause or to fall off, inactive polymerase, and inefficient hybridization of the mutant primer to the template.[13] In any case, analysis of the extension-ligation product is important. Sufficient covalently closed circular DNA is critical for good mutant yields.

Another source for reduced mutant yields is the displacement of the mutant oligonucleotide by the polymerase I large fragment (Klenow fragment) during *in vitro* synthesis. The extent of displacement is dependent upon the ratio of polymerase I to DNA ligase present. Recently, investigators have replaced the Klenow fragment with T4 DNA polymerase because it will not cause "strand displacement" of the mutagenic primer.[13]

Trace amounts of dUTP in the dNTPs used for *in vitro* elongation will provide sites for uracil *N*-glycosylase. The release of uracil produces apyrimidine (AP) sites which can produce strand breaks (via AP endonuclease) and can block DNA synthesis.[13] This problem is overcome simply by using high-quality dNTP substrate.

B. MUTAGENIC STRATEGIES

The *E. coli* mismatch repair system reduces the number of mutants by preferentially

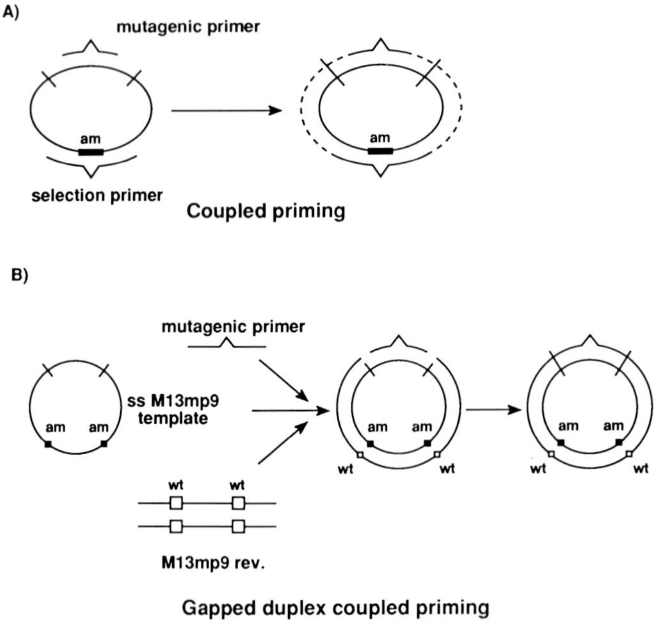

FIGURE 2. Schematic outlines of two strategies of site-directed mutagenesis using amber selection. (A) Coupled priming mutagenesis utilizing both a mutagenic primer and a second primer, which removes an amber (am) mutation in gene IV of the M13mp18 vector. (Adapted from Carter, P., Bedouelle, H., and Winter, G., *Nucleic Acids Res.*, 13, 4431, 1985.) (B) Gapped-duplex method of mutagenesis. Details of each step are described in the text. Symbols: am, amber codon, wt, corresponding wild-type codon. (Adapted from Kramer, W. and Fritz, H.-J., *Methods Enzymol.*, 154, 350, 1987.)

correcting the unmethylated minus strand using the methylated plus strand as template. The repair system is directed by methylation of GATC sites by *dam* methylase.[14] The replication of the daughter strand is transiently undermethylated and repair uses the methylated plus strand as template.[15] Therefore, upon transfection of *E. coli* with the M13 heteroduplex, the repair is directed toward the *in vitro* nonmethylated mutagenic strand. Several protocols have been devised to minimize mismatch repair. Early strategies included growing the viral plus strand in *E. coli dam* mutants so that it, too, was unmethylated.[16] Sarkar et al.[17] developed a protocol whereby mismatch repair is circumvented by transfection with mutant and wild-type ss circular DNA. The heteroduplex DNA is nicked with *Hind*III in the presence of ethidium bromide, then denatured. The host is transfected with the resulting closed circular ss DNA (frequency of transfection with the linear product is extremely low), which acts as a template for synthesis of the complementary strand.

Others have increased mutant frequency by genetically selecting against the progeny phage from the original "wild-type" template strand. Carter et al.[18] have devised a "coupled priming" technique which allows for the selection of the mutant strand. Two primers are used, one to generate the mutant of interest, the second to remove an amber mutation in gene IV of the M13mp18amIV vector (Figure 2A). After elongation and ligation, the heteroduplex DNA is transformed into a nonsuppressing strain in which only phage lacking the amber mutation will be viable.

Kramer et al.[19,20] have developed a gapped-duplex method in which the wild-type insert is cloned into the RF of M13mp9, which carries two amber mutations (Figure 2B). The M13mp9 ss DNA containing the insert is isolated for use as a template. This is hybridized

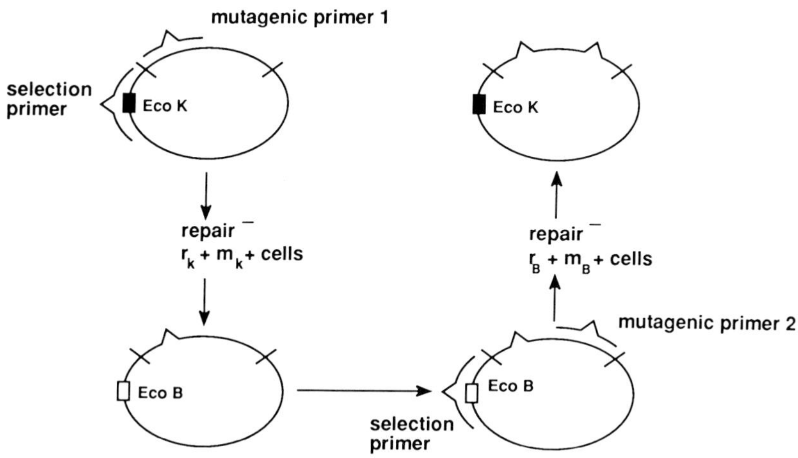

FIGURE 3. Schematic representation of the "cyclic selection" mutagenesis method. (From Carter, P., Bedouelle, H., and Winter, G., *Nucleic Acids Res.*, 13, 4431, 1985. With permission.)

to the linear denatured RF of an M13mp9 revertant which lacks amber mutations. This exposes a window of ss DNA, representing insert, to which a mutagenic primer is hybridized. The filled-in heteroduplex is transfected into nonsuppressor *E. coli* in which only progeny where the amber mutations have been removed are viable.

Carter et al.[18,21] have also developed a mutagenic strategy using a synthetic EcoK or EcoB site inserted 3′ to the cloned gene for selection of mutant progeny (Figure 3). Unlike the amber selection strategies described above, the EcoK/EcoB selectable marker can be retained after several rounds of mutagenesis[18] by cycling between the two markers (see Figure 3). Again, two primers are used, one mutagenic and one which changes the EcoK restriction site into an EcoB site. The heteroduplex is transfected into a repair-deficient K strain in which only phage with the EcoK site removed will be viable. A second mutation can be easily made by using one mutagenic primer and a second primer which converts the EcoB site to an EcoK site. The resulting heteroduplex is transfected into a repair-deficient B strain.

One of the most powerful strategies developed using selection against the original wild-type strand is the procedure developed by Kunkel.[22] It takes advantage of a mutant *E. coli* which is deficient in the enzyme dUTPase. When the M13 template ss DNA is grown in this host, as many as 30 dUTPs are incorporated in the place of dTTP (Figure 4). This is a nonmutagenic change, but when the heteroduplex (with U incorporated into the template) is transfected into an ung$^+$ strain, the U's are excised and the template strand is preferentially destroyed. The plus (template) strand is then repaired *in vivo* using the minus strand as template, which results in an enrichment of the mutant phenotype.

Eckstein and colleagues[23] have developed a method which avoids host-mediated repair altogether by *in vitro* elimination of the nonmutant template, which generates a pure homoduplex mutant DNA (Figure 5). Selective removal of the nonmutant template is based on the observation that some restriction enzymes will not cleave phosphorothioate DNA.[24] The mutagenic primer is annealed to the template, and extension and ligation are carried out in the presence of dCTPαS. The template (nonmutant, nonphosphorothioate) strand of the heteroduplex is nicked by NciI and the strand is partially or totally digested with exonuclease III. The mutant strand is then used as a template to reconstruct a ds homoduplex mutant molecule. This strategy is particularly useful in making insertions and deletions.

The mutagenic strategy presented in this chapter provides a simple, reliable, and straight-

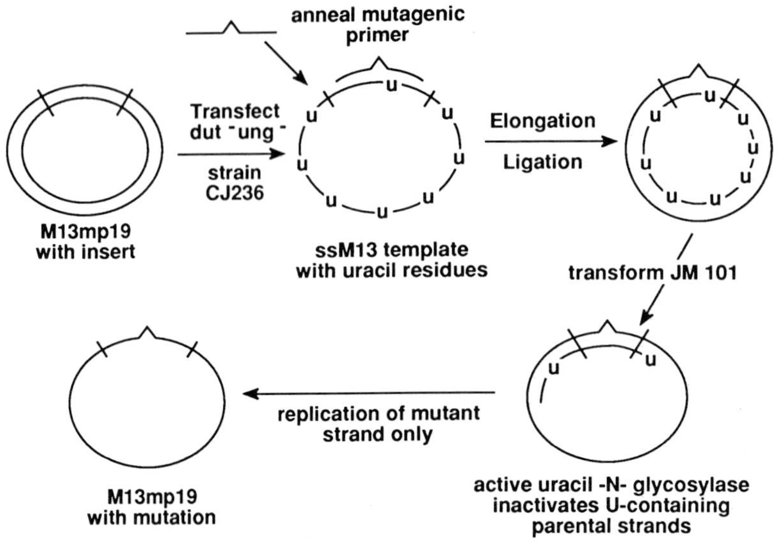

FIGURE 4. Simplified scheme for site-directed mutagenesis based on the method described by Kunkel.[13,22] See text for description.

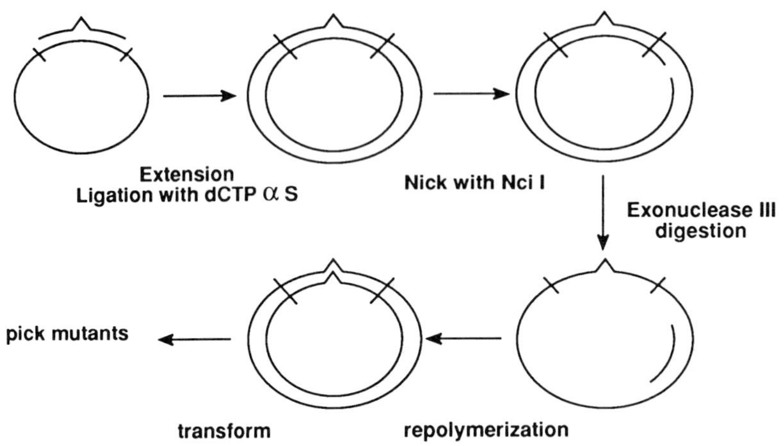

FIGURE 5. Scheme for phosphorothioate-based oligonucleotide-directed mutagenesis, as developed by Eckstein and colleagues.[23,24]

forward method for obtaining high yields of mutants in a very short time. First, mismatch repair is minimized by transforming the heteroduplex into an *E. coli* strain (*mut*L)[25] which is defective in mismatch repair.[15] Both amber selection strategies mentioned above (e.g., coupled priming[21] and gapped duplex[20]) have employed repair-deficient host strains to enhance mutant frequency. We have found that when *mut*L is used as a host, strand selection is not requisite for high mutant yield.[2] We consistently find mutation frequencies of 40 to 70% using only a single mutagenic primer when the repair *mut*L strain is used as host. *Mut*L strains have a higher error rate than wild-type strains (10^{-7} vs. 10^{-9} with wild type), but phage and plasmid replication normally has an error rate of 10^{-7}, so there is little added risk of making errors.[26] In addition, this method uses the phagemid pUC118 or pUC119 as a vector.[27] The Messing pUC118/119 vectors combine the advantages of a small, high-copy plasmid with the ability to generate ss DNA for use as a template for mutagenesis. Cloning and isolation of double-stranded pUC118/119 is much easier than isolation of M13 RF. Because pUC118 is much smaller than the M13mp vectors, larger fragments can be cloned into it without the instability problems experienced with M13mp vectors.[27] Third, this method uses T4 DNA polymerase, which, unlike the large fragment of DNA polymerase I (Klenow), does not cause strand displacement and therefore does not readily remove the hybridized mutagenic primer.

III. TYPICAL TIMETABLE FOR MUTAGENESIS EXPERIMENT

Prior to mutagenesis, design a mutagenic primer and clone fragment of interest into pUC118 or 119. Grow up phage and isolate ss DNA containing your fragment to use as a template.

- Day 1: Phosphorylate mutagenic oligonucleotides. Anneal the primer to the ss DNA template and begin extension-ligation. Prepare competent *E. coli mut*L. Check the extension-ligation mix on agarose gel. Transform competent *E. coli mut*L with the extension-ligation mix.
- Day 2: Pick 50 transformants onto the LB plate for the colony blot.
- Day 3: End-label the mutagenic oligonucleotide. Do the colony blot.
- Day 4: Grow the phagemid and isolate the ss DNA for sequencing.
- Day 5: Sequence the suspected mutants.

Since the mutant yield is usually greater than 50%, colony blot hybridization can be skipped and selected transformants can be sequenced on day 3 or 4.

IV. MATERIALS

A. STOCK SOLUTIONS

10X annealing buffer

200 m*M* Tris-HCl (pH 7.4)
20 m*M* $MgCl_2$
500 m*M* NaCl

10X kinase buffer

500 m*M* Tris-HCl (pH 8.0)
100 m*M* $MgCl_2$

10X T4 polymerase Rx buffer

5 m*M* dATP
5 m*M* dCTP
5 m*M* dGTP
5 m*M* dTTP
10 m*M* ATP
200 m*M* Tris-HCl (pH 8.8)
100 m*M* $MgCl_2$
20 m*M* DTT
(use HPLC-purified dNTPs)

20X SSC

3 *M* NaCl
0.3 *M* sodium citrate
10 m*M* EDTA (pH 7.2)

100X Denhardt's

2% bovine serum albumin
2% polyvinyl pyrrolidine
2% Ficoll

B. GROWTH MEDIA

LB (L-Broth)

10 g Bactotryptone
5 g yeast extract
10 g NaCl
Deionized H_2O to 1 l

2X YT

16 g Bactotryptone
10 g yeast extract
5 g NaCl
Deionized H_2O to 1 l

Glucose-Minimal Media

10.5 g KH_2PO_4
4.5 g K_2HPO_4
0.5 g NaCl
1 g NH_4Cl
Deionized H_2O to 1 l

After autoclaving, add the following filter-sterilized solutions:

- 1 ml of 1 *M* $MgSO_4 \cdot 7H_2O$
- 0.5 ml of 2% thiamine HCl in deionized H_2O
- 10 ml of 10% glucose

To prepare agar plates, add 15 g of Bacto-Agar to the mixture (1 l) before autoclaving, and cool to ~60°C before pouring. To prepare top agar, add 0.7 g of Bacto-Agar/100 ml and autoclave.

Ampicillin — The stock solution is 25 mg/ml of the sodium salt in water. Sterilize by filtration and store in aliquots at −20°C. The working concentration is 50 μg/ml.

Tetracycline — The stock solution is 12.5 mg/ml in 50% ethanol. Sterilize by filtration and store in aliquots at −20°C in the dark. The working concentration is 12.5 μg/ml. Because tetracycline is light sensitive, store the plates in the dark.

Kanamycin — The stock solution is 50 mg/ml in water. Sterilize by filtration and store in aliquots at −20°C. The working concentration is 70 μg/ml.

IPTG (isopropyl β-D-thiogalactopyranoside) — The stock solution is 100 m*M* in deionized water. Filter sterilize. Add 1 ml/l.

X-gal (5-bromo 4-chloro 3-indolyl β-D-galactoside) — The stock solution is 2% in dimethylformamide. Add 2.5 ml/l.

Note: Antibiotics, IPTG, and X-gal should be added from stock solutions just before pouring or after liquid medium has cooled below 60°C. Agar plates containing antibiotics should not be stored longer than 2 weeks before use.

C. STRAINS

E. coli MV1193: Δ*(lac-proAB), thi, rpsi*(strep^r), *endA, sbcB15, hspR4,* Δ*(srl-recA)* 306::Tn10(tet^r)[F′*:traD36, proAB, lacI*^{q}zΔM15]: MV1193 was used for all work with pUC118 and pUC119 vectors and was a generous gift of J. Messing (Waksman Institute, Piscataway, NJ).

E. coli BMH71-18 *mut*L: Δ*(lac-pro), supE,* [F′*:proA*$^+$*B*$^+$ *lacI*^{q}zΔM15] *mut*L::Tn10(tet^r): BMH71-18 *mut*L was used for transformation of the heteroduplex and was the generous gift of P. Carter, Genetech, Inc., San Francisco, CA.

Both MV1193 and BMH71-18 *mut*L are struck out on a glucose minimal plate with tetracycline at least every 3 weeks to maintain the episome. Well-isolated single colonies from these plates are used as inocula to start overnight cultures. Glycerols of both cultures are kept at −70°C. New plates are made by streaking from glycerols and not the old plate. This is especially important with BMH71-18 *mut*L, which has a higher frequency of mutation than repair$^+$ strains.[21]

D. PHAGE

The helper phage M13k07 has been constructed from an M13 phage that has the gene II of M13mp2. The origin of replication from p15A and the kanamycin resistance gene have been inserted between the A and B domains of the M13 origin of replication.[27] This creates an origin which is less efficient than the one carried by the pUC118/119 phage. Therefore, pUC118 with its high-copy number is preferentially packed. However, the M13k07 origin is efficient enough to produce a high titer of phage when pUC118/119 is not present.[27] M13k07 was generously supplied by J. Messing.

1. Growth of Helper Phage M13k07

Make 100-fold serial dilutions of M13k07 stock (10^{-2}, 10^{-4}, 10^{-6}, and 10^{-8}) and add 50 μl of 10^{-4}, 10^{-6}, and 10^{-8} dilutions to separate tubes containing 3 ml of LB top agar (maintained at 50°C). Immediately add 0.2 ml of overnight culture of MV1193, mix (vortex), and pour onto an LB plate. Allow the top agar to harden and incubate at 37°C overnight. Single plaques are picked and grown individually in 2 to 3 ml of YT containing kanamycin (70 μg/ml) overnight. The cells are pelleted, and the supernatant is titered and used as inoculum of M13k07. M13k07 will maintain titer for months when stored at 4°C.[27]

E. VECTORS

pUC118 and pUC119 are pUC18 and pUC19, respectively, with the M13 IG region inserted in the NdeI site.[27] pUC118 and pUC119 were supplied by J. Messing.

V. METHODS

A. PREPARATION FOR MUTAGENESIS

1. Design of Mutagenic Oligonucleotides

Oligonucleotides of 15 to 20 nucleotides are sufficient for making single- or double-point mutations as long as the mismatch(es) are positioned in the middle of the oligonucleotide. Longer oligonuleotides are necessary when making multiple mismatches to the template. A sequence having normal G + C content containing multiple mismatches requires flanking sequences of 6 to 10 nucleotides which are perfectly matched to the template.[21] The flanking sequences should be longer (10 to 15 nucleotides) if the oligonucleotide has high A + T content, or if there are large unpaired regions, as required in the construction of deletion or insertion mutants.

Competing priming sites are another point to consider when designing a mutant oligonucleotide. This problem can usually be avoided by computer analysis of the proposed oligonucleotide sequence with that of the template using one of the commercially available software programs. If a competing site is predicted, the oligonucleotide should be redesigned, either by lengthening it or shifting two or three bases to one side or the other of the target site. In any case, each mutagenic oligonucleotide should be tested for its priming by using it as a primer in dideoxy sequencing reactions using the same primer to template ratio that will be used in mutagenesis.

In addition, secondary structure of the oligonucleotide resulting from inverted repeats can interfere with its proper annealing to the template. Direct repeats near either end of the primer can lead to primer-template slippage, possibly resulting in an unwanted insertion.[20] Again, the solution may involve lengthening the primer or shifting two or three bases to one side or the other of the target site. Oligonucleotide synthesis can be done on any commercially available synthesizer. We use an Applied Biosystem Model 380B DNA synthesizer and purify the oligonucleotides as described in the Applied Biosystems manual.

2. Phosphorylation of Mutagenic Oligonucleotide

Prior to use, the mutagenic oligonucleotide is phosphorylated.

1. To a 1.5-ml Eppendorf tube, add the following:

   - 3 μl 10X kinase buffer
   - 3 μl 10 m*M* ATP (neutralized)

- 1.5 μl 100 m*M* DTT
- Oligonucleotide (200 pmol)
- H_2O to 30 μl
- 4 U of polynucleotide kinase

2. Incubate for 60 min at 37°C.
3. Stop the reaction by adding 5 μl of 100 m*M* EDTA and heat to 65°C for 10 min. (Mutagenic primers were tested as sequencing primers prior to use in heteroduplex synthesis in order to check for possible spurious priming from secondary hybridization of primer to template.)

3. Cloning Insert into pUC118/119

Clone the DNA insert to be mutated into the polylinker cloning site of the pUC118/119 vector using standard protocols such as those found in Maniatis et al.[28]

4. Transformation of *E. coli* MV1193

1. Competent cells are prepared by inoculating (1:100) 50 ml of LB media from an overnight culture of MV1193. The cells are grown with good aeration at 37°C to an A_{600} of 0.6 (approximately 2 h). The cells are harvested (spin 5 min at 3000 to 4000 × g) and gently resuspended in ice-cold 50 m*M* $CaCl_2$. (Do not vortex the cells. Resuspend them by swirling or, if necessary, gently pipetting up and down in a prechilled pipette. It is very important to keep the cells ice cold during and after this resuspension in 50 m*M* $CaCl_2$.)
2. The cells are left on ice for 30 min, harvested, and gently resuspended as described above in 2 ml of 50 m*M* $CaCl_2$. These cells are now competent and will remain so if kept on ice for approximately 25 h. (Glycerol can be added to 15% [final concentration] and the cells frozen in liquid nitrogen and stored at −70°C without a significant loss of transformation efficiency.)
3. Place 200 μl of competent cells in a cold 1.5-ml sterile Eppendorf tube on ice. (If frozen competent cells are used, thaw them on ice.)
4. To the competent cells, add 1 to 50 ng of DNA from the ligation reaction. Incubate on ice for 45 min.
5. Heat shock the cells by placing the tubes in a 42°C water bath for 1 min, after which 1 ml of LB is added and the cells are incubated at 37°C for 1 h without shaking.
6. Aliquots of 10 to 100 μl are spread onto LB-amp (100 μg ampicillin per milliliter) plates with IPTG and X-gal. The plates are incubated at 37°C for 12 to 16 h. The blue colonies are cells which carry the intact pUC118/119 vector, while white colonies are transformed with pUC118/119 containing an insert. (Do not leave the plates in the incubator longer than 16 h because small satellite colonies of untransformed cells will grow near the β-lactamase-secreting colonies and, if allowed to grow, can obscure true transformants.)
7. Pick the white colonies and restreak on LB-amp plates for single-colony isolates. Verify the presence of the insert by doing a plasmid mini-prep.[28]

5. Preparation of ss pUC118/119 DNA for Mutagenesis

1. Pick a single isolated colony of MV1193/pUC118/119 with the desired insert and place in 3 ml of LB-amp. Incubate overnight while shaking.
2. Inoculate 2 ml of 2X YT supplemented with 0.002% B_1 and 100 μg/ml of ampicillin and grow at 37°C, shaking until the culture reaches an A_{600} = 0.1.

3. The culture is infected with the helper phage M13k07 at a multiplicity of infection of 5 to 10 and incubated with shaking at 37°C for 1 h.
4. The infected cells are then diluted into 35 ml of 2X YT supplemented with 0.002% B_1 and 70 μg/ml kanamycin. Grow the phage for 10 to 14 h at 37°C with vigorous shaking.
5. Spin down the cells and any cell debris (10,000 × g for 10 min) and remove supernatant.
6. Add 1.25 ml of 5 *M* NaCl and 1.25 ml of 30% PEG for each 10 ml of supernatant, mix, and place on ice for 30 min.
7. Centrifuge at 30,000 × g for 15 min. Decant the supernatant and remove any remaining liquid with a sterile cotton swab. Resuspend the phage pellet in 500 μl TE.
8. Extract 1X with an equal volume of neutralized phenol, 1 to 2X with phenol/chloroform ([1:1:1/48] phenol/chloroform/isoamyl alcohol), vortexing each extraction for 0.5 to 1 min before spinning. (Repeat the chloroform/isoamyl/alcohol extraction if there is any material remaining at the interface.) Extract 1X with chloroform.
9. Add one tenth volume of 3 *M* sodium acetate. Vortex, then add 2.5 volume of 100% ethanol. Store at −70°C for 60 min.
10. Centrifuge for 10 min at 4°C, carefully remove the supernatant, and wash the pellet with 70% ethanol. Centrifuge, remove the supernatant, speed vac the pellet to dryness, and dissolve the pellet in 50 μl of TE.
11. Run a small aliquot on an agarose gel with a known amount of ss DNA to determine the DNA concentrations.

B. PROCEDURE FOR OLIGONUCLEOTIDE-DIRECTED MUTAGENESIS

1. Primer Annealing

To a 1.5-ml Eppendorf tube, add the following:

- 1 μl 10X annealing buffer
- 1 to 2 μl mutagenic oligonucleotide (6.6 pmol/μl)
- 1 μg ss DNA (corresponds to ~ 1.3 pmol, depending on size of insert)
- H_2O to a final volume of 10 μl (Molar ratios of primer to template usually vary from 2:1 to 10:1. Higher ratios of primer to template can result in a significant level of incorrect priming.)

Mix and place Eppendorf tubes containing the annealing reactions in a 500-ml beaker of H_2O at 80°C. Allow the water to cool to room temperature, then place the tube on ice.

2. Extension-Ligation

1. While the hybridization mixture is still on ice, add the following:

   - 1.5 μl 10X T4 polymerase Rx buffer
   - 2.5 U T4 DNA polymerase (diluted in 1X polymerase Rx buffer)
   - 3 U T4 DNA ligase

   Final volume <20 μl

2. Incubate on ice for 5 min. Bring to room temperature for 5 min, then place the tube in a 37°C water bath for 1.5 h. (Beginning the reaction at low temperatures allows a small number of bases to polymerize onto the 3′ end of the oligonucleotide, thus stabilizing the initial duplex between substrate and mutagenic primer.[13])
3. After 90 min, the reaction is terminated by the addition of 7 μl of 150 m*M* EDTA.
4. The reaction products are analyzed by agarose gel electrophoresis. Remove 2 μl from the mutagenesis reaction mix. Add 2 μl of H_2O and 1 μl of gel loading buffer (0.5%

SDS, 0.25% bromophenol blue, 25% glycerol), then apply to a 0.8% agarose gel in 1X TBE containing 0.5 μg/ml of ethidium bromide. In order to adequately evaluate the reaction products, the following standards must be run: (1) single-stranded phagemid DNA, (2) RF I, and (3) RF II forms of the phagemid being mutagenized. Successful elongation should result in the conversion of all the ss DNA into covalently closed circular (RF I) and nicked double-strand forms (RF II) of DNA.

3. Comments

Nearly all the ss DNA template should have been converted to ds closed-circular heteroduplex DNA. Incomplete synthesis of the heteroduplex often results in a very low mutant yield. Several factors can contribute to incomplete synthesis.

1. The ss template maybe contaminated. Therefore, repurification may be a requisite for efficient synthesis.
2. There may be inefficient hybridization of the mutant primer to the template. Try varying the hybridization conditions.
3. The reagent may be contaminated. Remake using only HPLC-purified dNTPs.
4. The DNA template may contain secondary structures which block polymerization. Single-stranded DNA binding proteins may be required to assist elongation or a Klenow fragment may be used instead of T4 polymerase.

4. Transformation of Heteroduplex into *E. coli mut*L

1. Prepare competent *E. coli mut*L as described in Section V.A.4. Add between 1 and 50 ng of double-stranded DNA (as estimated from gel) to 100 μl of competent cells.
2. Incubate on ice for 40 min, then heat shock at 42°C for 1 min. After the heat shock, add 0.5 ml of LB media to the cells and incubate at 37°C for 45 min.
3. Place 10 and 100 μl on LB-amp plates and incubate for 12 to 16 h at 37°C.

5. Screening for Mutants

1. Mutants can be rapidly screened by colony blot hybridization using [^{32}P] labeled mutagenic primer.[29] A total of 50 to 100 transformants are picked onto a LB-amp plate and incubated overnight. As a control, include one or two colonies containing a wild-type sequence. The colonies are transferred to a nitrocellulose filter and lysed by setting the nitrocellulose (colonies up) on Whatman 3-mm paper saturated with 0.5 *M* NaOH for 5 min.
2. Neutralize by transferring the filters (face up) to Whatman 3-mm paper saturated with 1 *M* Tris, pH 7.5 for 1 to 2 min.
3. Transfer the filters to Whatman 3-mm paper saturated with 0.5 *M* Tris (pH 7.5), 1.5 *M* NaCl for 2 min.
4. Air dry the filters on paper towels or other absorbent papers, then bake the filters in a vacuum at 80°C for 15 to 45 min.
5. Wet the filters in 6X SSC for 1 min, then prehybridize with shaking in 10X Denhardt's solution, 6X SSC, and 0.2% SDS at 67°C for 15 to 30 min.
6. Wash the filters in 6X SSC.
7. The mutagenic primer is phosphorylated using T4 polynucleotide kinase as described in Section V.A.2. Any unincorporated nucleotides are removed by desalting on a Sephadex G-25 spin column.
8. The labeled probe is diluted into 3 ml of 6X SSC and placed in a small petri dish. The probe can be reused several times over a period of 5 to 8 d (store at −20°C).

9. Remove the filter from the 6X SSC and place it on the probe solution face down so that no bubbles are trapped under the filter. Incubate for 1 to 4 h at room temperature. (Oligonucleotides longer than 20 bases may require hybridization conditions at temperatures of 37°C or higher.)
10. After hybridization, the filters are washed for 1 min in 6X SSC. Repeat three times. Blot off the excess 6X SSC (do not let the filters dry out). Wrap in plastic and autoradiograph at −70°C for 2 to 4 h. The mutagenic primer should hybridize to both mutant and wild-type DNAs. Subsequent washings at increasing temperatures will cause the labeled probe to dissociate from the nonmutant clones, but not from the mutant. The temperature, T_D, at which this occurs for oligonucleotides between 14 and 22 bases long can be estimated by using the "Wallace rule".[10] The T_D is predicted by the formula: $T_D = 4(\text{number of G} + \text{C}) + 2(\text{number of A} + \text{T})$. Filters should be washed at 5°C below the calculated T_D in 6X SSC for 3 min and then autoradiographed. If mutant and wild-type colonies cannot be discriminated, then the wash is repeated with a 2 to 3°C increase in temperature until the mutant colonies can be distinguished.

6. Expected Results

The expected mutant frequency should be at or above 50%. If the frequency is dramatically lower, check the *mut*L phenotype (see strains).

7. Purification of Mutant Clones

A positive colony may contain a small amount of wild-type DNA. Because of the high copy number of pUC vectors, it may not be segregated by simple single-colony isolation. Therefore, plasmid DNA is isolated from putative mutants as described[28] and then used to transform *E. coli* MV1193 in order to obtain a pure mutant clone.

8. Analysis of Mutant by Sequencing

To prepare ss DNA for sequencing, grow a 2-ml culture from a single isolated colony of MV1193 containing the putative mutant as described in Section V.A.5. The phagemid ss DNA can be sequenced using standard dideoxy chain termination carried out according to Sanger.[30,31] If the site of the mutation is too far from the site of the universal sequencing primer, a synthetic oligonucleotide (15- to 18-mer) is made that primes 50 to 150 nucleotides upstream from the mutation. Once positive mutants have been identified by sequence, the entire insert is sequenced to assure that no spurious mutations at other sites have occurred. The mutant insert can now be ligated into the appropriate expression vector and its phenotype can be assessed.

VI. CONCLUSIONS

Many believe that with the advent of site-directed mutagenesis, the study of protein structure, function, and folding has entered a golden age. This chapter has presented the reader with a systematic approach for introducing nonrandom single-stranded vectors. Space constrains a discussion of many other innovative approaches to mutagenesis, including the use of double-stranded vectors,[32] saturation mutagenesis,[33] and *in vitro* chemical mutagenesis.[34,35] One novel approach worth mentioning is taken by Miller and colleagues,[36,37] who are using nonsense suppressor strains of *E. coli* to rapidly introduce multiple amino acid substitutions at any target site. These suppressor strains can be used to effectively insert either lys, ser, glu, tyr, leu, phe, cys, his, ala, gly, or pro at an UAG site which has been engineered anywhere in the coding region of the gene. The technology for site-directed mutagenesis is rapidly evolving. One of the most exciting alternative approaches to site-directed mutagenesis is the use of the polymerase chain reaction (PCR). PCR offers a rapid and efficient method for the generation of site-directed mutants,[38] and soon its use may outstrip that of more conventional methods.

Protein engineering promises to provide us with insight into the fundamental questions of protein chemistry, and the methods presented in this chapter provide a powerful and simple way to rapidly produce any desired mutation. The difficult task ahead of us is the interpretation or overinterpretation[1] of a given alteration of the folding or catalytic activity of a protein. Our present understanding of the relationship between protein structure and function and how a protein will fold is too naive for us to predict why changing a given amino acid residue might lead to a change in catalytic activity. The mutagenized amino acid residue in question might play a direct, primary role in the reaction mechanism, or the replacement might lead to an alteration in activity via a secondary structural change.[4] Although it is impossible to answer this question satisfactorily without detailed structural information, it is possible to resolve it to some extent by further mutational analysis. Second-site suppression analysis,[35] double-mutant analysis,[29] and saturation mutagenesis have been used to complement site-directed mutagenesis studies in defining possible roles of amino acid residues and their interactions with other active-site residues during folding and enzyme catalysis. So although on its own site-directed mutagenesis will not uncover the rules of protein folding and the molecular basis of enzyme catalysis, when coupled to other biophysical, biochemical, and genetic techniques, it becomes a powerful and promising tool, and will continue to contribute to our understanding of protein chemistry.

REFERENCES

1. **Knowles, J. R.,** Tinkering with enzymes: what are we learning?, *Science,* 236, 1252, 1987.
2. **Menick, D. R., Lee, J. A., Brooker, R. J., Wilson, T. H., and Kaback, H. R.,** Role of cysteine residues in the *lac* permease of *Escherichia coli, Biochemistry,* 26, 1132, 1987.
3. **Menick, D. R., Carrasco, N., Antes, L., Patel, L., and Kaback, H. R.,** *Lac* permease of *Escherichia coli:* arginine-302 as a component of the postulated proton relay, *Biochemistry,* 26, 6638, 1987.
4. **Kaback, H. R.,** Use of site-directed mutagenesis to study the mechanism of a membrane transport protein, *Biochemistry,* 26, 2071, 1987.
5. **Fersht, A. R. and Leatherbarrow, R. J.,** Structure and activity of the tyrosyl-tRNA synthetase, in *Protein Engineering,* Oxender, D. L. and Fox, C. F., Eds., Alan R. Liss, New York, 1987, 269.
6. **Matouschek, A., Kellis, J. T., Jr., Serrano, L., and Fersht, A. R.,** Mapping the transition state and pathway of protein folding by protein engineering, *Nature,* 340, 122, 1989.
7. **Hutchison, C. A., Phillips, S., Edgell, M. H., Gillam, S., Jahnke, P., and Smith, M.,** Mutagenesis at a specific position in a DNA sequence, *J. Biol. Chem.,* 253, 6551, 1978.
8. **Gillam, S. and Smith, M.,** Site-specific mutagenesis using synthetic oligodeoxyribonucleotide primers. I. Optimum conditions and minimum oligodeoxyribonucleotide length, *Gene,* 8, 81, 1979.
9. **Wallace, R. B., Shaffer, J., Murphy, R. F., Bonner, J., Hirose, T., and Itakura, K.,** Hybridization of synthetic oligodeoxyribonucleotides to ϕX 174 DNA: the effect of single base pair mismatch, *Nucleic Acids Res.,* 6, 3543, 1979.
10. **Wallace, R. B., Schold, M., Johnson, M. J., Dembek, P., and Itakura, K.,** Oligonucleotide-directed mutagenesis of the human β-globin gene: a general method for producing specific point mutations in cloned DNA, *Nucleic Acids Res.,* 9, 3647, 1981.
11. **Zoller, M. J. and Smith, M.,** Oligonucleotide-directed mutagenesis of DNA fragments cloned into M13 vectors, *Methods Enzymol.,* 100, 468, 1983.
12. **Zinder, N. D. and Horiuchi, K.,** Multiregulatory element of filamentous bacteriophages, *Microbiol. Rev.,* 49, 101, 1985.
13. **Kunkel, T. A., Roberts, J. D., and Zakour, R. A.,** Rapid and efficient site-specific mutagenesis without phenotypic selection, *Methods Enzymol.,* 154, 367, 1987.
14. **Hattman, S., Brooks, J. E., and Masurekar, M.,** Sequence specificity of the P1 modification methylase (M · *Eco* P1) and the DNA methylase (M *Eco* dam) controlled by the *Escherichia coli dam* gene, *J. Mol. Biol.,* 126, 367, 1978.

15. **Kramer, B., Kramer, W., and Fritz, H.-J.,** Different base/base mismatches are corrected with different efficiencies by the methyl-directed DNA mismatch-repair system of *E. coli, Cell,* 38, 879, 1984.
16. **Kramer, W., Schughart, K., and Fritz, H.-J.,** Directed mutagenesis of DNA cloned in filamentous phage: influence of hemimethylated GATC sites on marker recovery from restriction fragments, *Nucleic Acids Res.,* 10, 6475, 1982.
17. **Sarkar, H. K., Viitanen, P. V., Padan, E., Trumble, W. R., Poonian, M. S., McComas, W., and Kaback, H. R.,** Oligonucleotide-directed site-specific mutagenesis of the *lac* permease of *Escherichia coli, Methods Enzymol.,* 125, 214, 1986.
18. **Carter, P., Bedouelle, H., and Winter, G.,** Improved oligonucleotide site-directed mutagenesis using M13 vectors, *Nucleic Acids Res.,* 13, 4431, 1985.
19. **Kramer, W., Drusta, V., Jansen, H.-W., Kramer, B., Pflugfelder, M., and Fritz, H. J.,** The gapped duplex DNA approach to oligonucleotide-directed mutation construction, *Nucleic Acids Res.,* 12, 9441, 1984.
20. **Kramer, W. and Fritz, H.-J.,** Oligonucleotide-directed construction of mutations via gapped duplex DNA, *Methods Enzymol.,* 154, 350, 1987.
21. **Carter, P.,** Improved oligonucleotide-directed mutagenesis using M13 vectors, *Methods Enzymol.,* 154, 382, 1987.
22. **Kunkel, T. A.,** Rapid and efficient site-specific mutagenesis without phenotypic selection, *Proc. Natl. Acad. Sci. U.S.A.,* 82, 488, 1985.
23. **Sayers, J. R., Schmidt, W., and Eckstein, F.,** 5′-3′ Exonucleases in phosphorothioate-based oligonucleotide-directed mutagenesis, *Nucleic Acids Res.,* 16, 791, 1988.
24. **Taylor, J., Schmidt, W., Gosstick, R., Okruszek, A., and Eckstein, F.,** The use of phosphorothioate-modified DNA in restriction enzyme reactions to prepare nicked DNA, *Nucleic Acids Res.,* 13, 8749, 1985.
25. **Glickman, B. W. and Radman, M.,** *Escherichia coli* mutator mutants deficient in methylation-instruction DNA mismatch correction, *Proc. Natl. Acad. Sci. U.S.A.,* 77, 1063, 1980.
26. **Carter, P., Bedouelle, H., Waye, M. M. Y., and Winter, G.,** *Oligonucleotide Site-directed Mutagenesis in M13: An Experimental Manual,* Anglian Biotechnology, Colchester, England, 1985.
27. **Vieira, J. and Messing, J.,** Production of single-stranded plasmid DNA, *Methods Enzymol.,* 153, 3, 1987.
28. **Maniatis, T., Fritsch, E. F., and Sambrook, J.,** *Molecular Cloning — A Laboratory Manual,* Cold Spring Harbor Laboratory, Cold Spring Harbor, NY, 1982.
29. **Carter, P. J., Winter, G., Wilkinson, A. J., and Fersht, A. R.,** The use of double mutants to detect structural changes in the active site of the tyrosyl-tRNA synthetase, *(Bacillus stearothermophilus), Cell,* 38, 835, 1984.
30. **Sanger, F., Nicklen, S., and Coulson, A. R.,** DNA sequencing with chain-terminating inhibitors, *Proc. Natl. Acad. Sci. U.S.A.,* 74, 5463, 1977.
31. **Sanger, F. and Coulson, A. R.,** The use of thin acrylamide gels for DNA sequencing, *FEBS Lett.,* 87, 107, 1978.
32. **Rossi, J. and Zoller, M.,** Site-specific and regionally directed mutagenesis of protein-encoding sequences, in *Protein Engineering,* Oxender, D. L. and Fox, C. F., Eds., Alan R. Liss, New York, 1987, 51.
33. **Hill, D. E.,** Mutagenesis with degenerate oligonucleotides: creating numerous mutations in a small DNA sequence, in *Current Protocols in Molecular Biology,* Ausubel, F. M., Brent, R., Kingston, R. E., Moore, D. D., Seidman, J. G., Smith, J. A., and Struhl, K., Eds., Greene Publishing Associates/Wiley-Interscience, New York, 1989, chap. 8, suppl. 6, sect. 8.2.1.
34. **Myers, R., Lerman, L., and Maniatis, T.,** A general method for saturation mutagenesis of cloned DNA fragments, *Science,* 229, 242, 1985.
35. **Shortle, D., Grisafi, P., Benkovic, S. J., and Bostein, D.,** Gap misrepair mutagenesis: efficient site-directed induction of transition, transversion, and frameshift mutations *in vitro, Proc. Natl. Acad. Sci. U.S.A.,* 79, 1588, 1982.
36. **Kim, C. W., Michaels, M. L., Markiewicz, P. G., and Miller, J. H,,** The use of *E. coli* nonsense suppressors to rapidly introduce amino acid substitutions in thymidylate synthetase, *J. Cell. Biochem.,* Suppl. 13A, 72, 1989.
37. **Kleina, L. G. and Miller, J. H.,** Structure-function studies of the *lac* repressor of *E. coli* K12, *J. Cell. Biochem.,* Suppl. 13A, 55, 1989.
38. **Ho, S. N., Hunt, H. D., Horton, R. M., Pullen, J. K., and Pease, L. R.,** Site-directed mutagenesis by overlap extension using the polymerase chain reaction, *Gene,* 77, 51, 1989.

produce antibodies, and purify the antibodies. This is the most convenient way to obtain anti-peptide antibodies, but considerations of the number of antisera required and costs may prevail. If one plans to make the antibodies one's self, then one needs to consider that the production of antisera requires the use of animals and that there are very strict laws and guidelines covering the use of animals in research. Primarily, one must have access to approved animal house facilities. One must also have approval from the appropriate "animal-use committee" for the use of animals in research covering the particular procedures used. The exact requirements vary somewhat between countries and institutions, but it is one's responsibility to properly comply. If one can't, then a commercial company will need to be used, at least for production of the antisera.

At this stage, one must also give serious consideration to the controls to be used in assays. Experiments using antibodies are very susceptible to misinterpretation due to irrelevant and/or nonspecific reactivities. It is quite possible, sometimes likely, that an animal used for immunization already possesses antibodies reactive to the test protein or to other antigens which may be present during an experiment. These reactivities could easily be misinterpreted to be due to the anti-peptide antibodies. At the very least, prebleeds to be used as controls should be taken before the animals are injected. The best approach is to also use as a control antibodies to an irrevelant peptide, raised at the same time, under the same conditions, and treated the same way as the test antibodies. These plans need to be made early so that both control and test antibodies will be available for the performance of correctly designed experiments.

III. PROCEDURES

The production of anti-peptide antibodies can be divided into four stages:

1. Selection and synthesis of peptide
2. Coupling of peptide to carrier protein
3. Immunization of rabbits and collection of antisera
4. Assay and purification of antibodies

A. SELECTION AND SYNTHESIS OF PEPTIDE

The demonstration a few years ago that anti-peptide antibodies commonly reacted with intact proteins surprised many immunologists, since the thinking at that time predicted that this generally wouldn't happen. Much research has since been done on the immunochemistry of antigen/antibody interactions. I would urge you to read recent reviews[12] covering this field since an understanding of antibody/antigen interactions will help you to better design and interpret your experiments. For instance, I'm sure some think of antibody binding as a specific all-or-none effect. This is not so. It can be said, with exaggeration, of course, that an antibody can bind to all antigens, with a range of affinities from very high to very low. Under the conditions normally used in assays, only the high-affinity (specific) binding is detected. However, even under these conditions the binding of the same antibody to different but similar antigens can and does occur. Indeed, in the work I have been involved in using anti-peptide antibodies, we rely on such cross-reactivity. We have shown that antibodies made to the J region (ANYGYTFGSGTRLTVV) of th β chain of the T-cell antigen receptor will react with the J region (ASGLSTFGRGTKLGVV) of shark immunoglobulin light chain[7] even though there is only 50% identity between these sequences. The binding area of an antibody covers only about five to eight amino acid residues. So, the J peptide comprises several antigenic determinants, and at least three binding specificities can be demonstrated in the antisera generated.[13] The critical segment for J region cross-reactivity comprises residues 7 to 13 and must be (F) (G) (9) (G) (T) (12) (L/V). Position 9 can be almost any residue, but position 12 must be one of the two positively charged residues, lysine (K) or arginine (R). The nature of the amino acid at position 12 appears to be critical to antigenicity overall. Even if the other 15 residues are identical, if position 12 comprises the neutral substituents glycine (G) or threonine (T), or the acidic substituent glutamic acid (E), then antibody binding is substantially reduced.[13]

This example illustrates several points. Cross-reactions can occur between nonidentical peptide regions, but a high degree of identity must occur in the critical segment comprising the antigenic determinant. Even in this segment some substitutions can occur, but certain amino acids have a much greater effect than others. Recently,[12] it has been suggested that replacement of the amino acids alanine, methionine, glutamine, arginine, serine, and threonine has the least effect on antibody binding and that the replacement of cysteine, glycine, histidine, asparagine, proline, and tyrosine has the most effect. Maximizing the presence of the latter in a peptide may promote greater specificity in the resulting antibody. Having selected a peptide sequence, it is very advisable to check the uniqueness of that sequence, and therefore the potential specificity of the antibody, by doing a computer search of protein data bases for matches with the selected sequence. Search programs are available for mainframes[14,15] and personal computers (e.g., the Pustell software package from International Biotechnologies, Inc., New Haven, CT). Look for matches of at least 50% identity, but bear in mind the points raised above. Identities must be clustered for there to be a potential for cross-reaction. Finding many such matches or matches with proteins likely to be present in your assays would indicate that the antibodies generated may not have the necessary discriminatory power, and a different peptide might need to be considered. Converse findings should increase confidence in your selection. Because the power of anti-peptide technology lies in the predetermined specificity of the antibodies generated, the possibility of cross-reactions needs to be borne in mind.

To be a useful reagent, the anti-peptide antibody must be able to bind to the intact protein for which it was designed. Much has been written[1-4,12,16,17] concerning the "rules" which should be applied when selecting the peptide in order to achieve this. My advice is not to be particularly concerned about these rules. My experience agrees with the advice of Lerner, who wrote in a recent review:[16] "The only rule is that there is no rule." Your primary consideration when selecting a peptide is the precision with which the peptide will answer the experimental question. Even if the rules say the peptide will not work, you should still do the experiment. There are enough exceptions to the rules in the literature to indicate a reasonable chance of success. The only condition which must be satisfied for binding to occur is that the antigenic determinant on the protein defined by the peptide be accessible by the antibody. This is influenced greatly by the type of assay you perform. In some assays, such as Western blot and ELISA, the protein is partially or completely denatured so that most determinants will be exposed. In others, such as immunoprecipitation, cell binding, tissue or cellular localization, or function studies, the protein is in its native folded configuration, and some determinants may be buried within the protein and inaccessible. However, even in these conditions, a surprisingly large proportion of the sequence of a protein seems to be accessible for binding by antibodies.[3]

In many cases, the particular antigenic determinant is of secondary importance to the development of a reagent specific for the intact protein. This would be the case, for instance, in the identification of gene products. To ensure success in these circumstances, one should probably select two or even three peptides, using the general guidelines discussed below. This approach has the additional advantage of providing internal controls. Thus, the binding of two or more different anti-peptide antibodies would unambiguously identify a protein. If Western blots will be used to detect the protein, then just about any peptide can be chosen. If reactivity against native protein is desired, then you may want to consider additional selection factors, discussed fully in recent reviews.[1-4,12,16-18,22] Computer programs are now available which claim to predict from sequence data which peptides will be antigenic in the native program. For instance, such programs are available in the GCG package[15,19] and the Pustell software for personal computers from IBI. Probably the most one needs to do is to select peptides from polar, hydrophilic regions of a protein based on the assumption that these are predominately located on the surface which makes contact with the aqueous environment, whereas hydrophobic residues make up the internal part of the protein. Again, there are computer programs that do this, but one can manually create a hydrophilicity profile by simply color coding the charged and uncharged residues.

Despite my comment above about ''no rules'', there are a few guidelines one should try to follow in all cases. First, use peptides that are 12 to 16 residues long. Success has been reported[20] using peptides of only five to seven residues, but it is generally considered best to use the longer peptides. Second, try to select peptides with a high proportion of polar, hydrophilic amino acids. These are best for reasons mentioned above and also because they are much more likely to be soluble in aqueous buffers. This makes them very much easier to work with. However, I do give procedures for working with hydrophobic peptides, since this is quite often necessary.

Although it is possible to manually synthesize the short peptides involved here,[21] I recommend using the automatic peptide synthesis machines that are now available. All the peptides I have used were synthesized with an Applied Biosystems 430A automatic peptide synthesizer, with excellent results. Machines from other manufacturers should be comparable. Most institutions have these machines available; otherwise, custom peptide synthesis services are available commercially.[11] Specify that the peptide be supplied in the C-terminal free acid and N-terminal free amine form, since this gives versatility for coupling and adds charged groups, which helps solubility. However, the charged groups may alter the antigenicity of internal peptides,[22] so you may want to consider getting the N-terminal amide and C-terminal acetyl or succinyl forms. Also, it is better if the peptide is lyophilized rather than rotary dried since lyophilized peptides are easier to dissolve. Small peptides should be about 90% pure ''as is'' from the synthesizer; for immunological work, no further purification is necessary. However, one does need an amino acid composition analysis. The composition of the peptide should be known exactly from the sequence, and any deviation from this would indicate a mistake or failure during synthesis.

B. COUPLING OF PEPTIDE TO CARRIER PROTEIN

Although some peptides are immunogenic (i.e., can illicit an antibody response) by themselves, it is generally necessary to couple peptides to carrier proteins to ensure an adequate antibody response. I couple via amino groups using glutaraldehyde. This is a quick, simple, and effective coupling procedure. However, glutaraldehyde can modify lysine-NH_2 groups and thus potentially alter the antigenicity of peptides containing lysine. This possibility is minimized by performing the coupling reaction at a pH well below the pKa of the side chains, thus keeping the amino groups in the NH_3 form which is unreactive with glutaraldehyde. Also, the concentration of glutaraldehyde is kept low to prevent overcoupling. Since coupling is through the N-terminal end, keep residues you think may be critical for antibody activity away from the N terminal by extending the N-terminal sequence three or four residues or by adding several spacer residues such as glutamic acid. An alternative procedure is to couple through C-terminal cysteine using *m*-maleimido-benzoyl-*N*-hydroxysuccinimide ester[23] (see References 1 and 22 for other procedures). With this procedure, it is necessary to add a cysteine which is not part of the original sequence to the C terminal and to avoid peptides with internal cysteines. Here I only describe the glutaraldehyde method.

First, it is necessary to dissolve the peptide. This can be easy or difficult, depending on the peptide. Generally, once in solution, peptides will stay in solution even if the solvent is changed. Thus, in a lot of cases, difficult peptides can be dissolved in extreme conditions and then returned to normal conditions for use. The routine is to first try to dissolve the peptide in aqueous buffer using a number of different procedures and then to try organic buffers if these fail. The coupling reaction is done in approximately 1 ml of 0.125 *M* sodium phosphate buffer (pH 7.2) containing 5 mg of peptide plus 5 mg of carrier protein, and so the aim is to try to reach this end condition when dissolving the peptide.

Bovine serum albumin (BSA) or keyhole limpet hemocyanin (KLH) are generally used as the carrier protein, although any protein is probably suitable. KLH is a very good immunogen, but has a high carbohydrate content which can cause cross-reactivity problems, and also it is not very soluble and may precipitate during the coupling procedure. I generally use BSA. It is advisable not to use BSA if you will be using systems in the presence of, or derived from, tissue-culture media containing bovine serum supplement. Ovalbumin (not related to or cross-reactive with BSA) is also a good carrier protein.[22] I suggest coupling to both BSA and ovalbumin, using one conjugate for immunization and the other for assays.

1. Dissolving the Peptide

Try these procedures successively until the peptide is dissolved. Hopefully, one will only need to use the first procedure.

1. Add 0.5 ml of water or 1 ml of phosphate buffer to 5 mg of peptide. Gently agitate overnight at room temperature. Some peptides dissolve better and faster in water than in salt solutions, so try both. It is necessary to incubate at least overnight since some peptides dissolve very slowly. Add 0.5 ml of 2X (0.25 *M*, pH 7.2) phosphate buffer to the water solution; the peptide will usually stay in solution even if it did not dissolve in the salt solution. Add 5 mg of carrier protein. Instead of adding the carrier protein separately, I find it more convenient to first dissolve the peptide and then add 2X phosphate buffer containing 10 mg/ml carrier protein. In the procedures outlined below, the 2X phosphate buffer contains 10 mg/ml carrier protein.
2. Try dissolving the peptide in 1 to 10 m*M* HCl or NaOH. After a few minutes, add the 2X phosphate buffer. This is usually sufficient to neutralize the peptide solution, but check the pH with indicator paper. You can use higher concentrations of acid or base, but you will have to adjust the pH of the final solution. If a precipitate forms after neutralization, check whether any peptide has remained in solution by centrifuging and taking an optical density reading of the supernatant at 220 nm (extinction coefficient is approximately 16). Sometimes a significant amount of peptide does remain in solution, especially, it seems, if a very fine, cloudy precipitate forms. Indeed, in such cases I have successfully performed conjugation reactions in the presence of the precipitate.
3. Add the peptide to 0.5 ml of water and then add glacial acetic acid or 80 to 90% formic acid until the peptide goes into solution. Add 3 to 4 ml of water and Speedvac (Savant Instruments, Inc., Farmingdale, NY) to 0.5 ml. Repeat this two times. Nearly all of the acid is removed by this procedure and the solution is easily neutralized with the 2X phosphate buffer. If a precipitate forms, again check whether any peptide has remained in solution as above.
4. Mix peptide and carrier protein in water and follow step 3, but Speedvac to dryness in the final step. Redissolve in water or phosphate buffer. The rationale for this procedure is that the peptide may bind to the protein and so be stabilized in solution.
5. When a peptide is so hydrophobic that none of the above procedures work, it is necessary to use organic solvents. Three solvents that have proven useful for this purpose are dimethylformamide (DMF), ethylmorpholine, and 1-methyl-2-pyrrolidinone. These solvents do not possess free primary or secondary amines and so are suitable for use with the glutaraldehyde conjugation method. Ethylmorpholine and

methylpyrrolidinone seem to be the most effective, but ethylmorpholine is very alkaline and dissolves plastic. Methylpyrrolidinone, on the other hand, has neutral pH, can be used with plastic (e.g., pipette tips), and is the solvent of first choice. However, it may be necessary to experiment with all three solvents to find the one best suited for a particular peptide. Make 10, 20, 30, 40, and maybe 50% solutions of solvents in water and do preliminary experiments to see which will completely dissolve the peptide. Use 0.5 ml of this to dissolve 5 mg of peptide and add 0.5 ml of 2X phosphate buffer. Check the pH with indicator paper and adjust if necessary. If the peptide precipitates, it may be necessary to use a higher concentration of solvent, a larger volume, or a different solvent.

2. Coupling Reaction

Once the peptide is dissolved in the coupling buffer, add 11 μl of 12.5% glutaraldehyde (use the highest grade available). The final concentration should be 0.1 to 0.15%, so adjust accordingly if your final volume is different from 1 ml. Incubate at room temperature for 30 to 45 min. The solution will turn yellow. The carrier protein may precipitate if it becomes extensively cross-linked by the glutaraldehyde. If this happens, reduce the incubation time or the glutaraldehyde concentration. Dialyze the solution extensively against three changes of phosphate or Tris-buffered saline containing 0.05% sodium azide (an antimicrobial agent).

I don't attempt to quantitate the degree of peptide coupling. If the peptide was properly solubilized, I have never experienced a failed conjugation reaction, at least as judged by the ability of the conjugate to stimulate anti-peptide antibodies. Some workers maintain that it is not wise to spend time immunizing an animal if you don't know whether the peptide has been successfully coupled. The best way to assess this is by using radiolabeled peptide and determining how much of the label becomes associated with the protein.[1,22] Peptides can be labeled by incorporating ^{3}H- or ^{14}C-labeled amino acids during peptide synthesis or by labeling the peptide after synthesis. Peptides containing tyrosine and histidine can be labeled directly with ^{125}I.[24,25] If amino groups are not used for coupling to carrier protein, the Bolton and Hunter reagent can be used.[26]

C. IMMUNIZATION AND COLLECTION OF ANTISERA

Several types of animals, such as goats, sheep, mice, and chickens, can be used for immunization, but rabbits are almost universally used since they are readily available, large enough to yield sufficient quantities of sera, and are very well suited for boarding in animal-house facilities. In some very infrequent cases, the peptide may be sufficiently similar to a self-component of the rabbit so that no response will be generated, even after repeated immunizations. If such is the case, one should consider using a more phylogenetically different species, such as chickens.[27-29] Another remote possibility in these cases is that the rabbit will develop autoimmune disease and be of limited usefulness for antiserum production.

Always use adjuvants when immunizing with protein-peptide conjugates. Adjuvants are immunomodulators that greatly enhance the immune response over that generated by antigen alone. Freund's adjuvant has been the classical adjuvant used, but can sometimes produce toxic and allergenic side effects. I strongly recommended the RIBI (RIBI Immunochem Research, Inc., Hamilton, MT)[30] adjuvant system, which is nontoxic, nonallergenic, and routinely gives as good or better results than Freund's adjuvant. For rabbits, use the MPL + TDM ready-to-use emulsion. Reconstitute the emulsion with 1 ml sterile PBS as per the manufacturer's instructions. Dilute the peptide conjugate to approximately 1 mg/ml in PBS and mix 0.5 ml (approximately 500 μg of protein-peptide conjugate) with 0.5 ml of the emulsion. Store the rest of the emulsion and peptide solutions at 4°C. Pull the emulsion into a 3-ml syringe using an 18-gauge needle. Make sure there are no air bubbles by moving the plunger up and down. Replace the 18-gauge with a 22-gauge needle and push the plunger until the emulsion just extrudes from the needle. Immunize the rabbit by giving several subcutaneous injections around the neck and rump areas and one intramuscular injection. Pinch the skin between your thumb and forefinger and lift it up. At the base of your thumb and forefinger insert the needle into the space formed between the skin and body tissue. Move the needle from side to side to make sure it is free and inject about 200 μl of emulsion. To give the intramuscular injection, quickly jab the needle about 5 mm into the muscle at the rump and pull on the plunger a little. If the plunger moves relatively easily and blood is drawn up, remove the needle and try another spot; otherwise, inject 200 μl. After 14 d, repeat the immunization with the remaining peptide-conjugate and emulsion solution stored earlier.

The rabbit is bled 1 month after the first injection. Several methods can be used, but the best technique involves bleeding the rabbit from the central artery of the ear using a hubless 18-gauge hypodermic needle. The procedure is described in detail by Gordon.[31] Collect the blood in a 50-ml polypropylene centrifuge tube. Shake the tube vigorously one or two times to induce clotting and incubate at 37°C for 60 min to allow the clot to form. Loosen the clot from the tube by running a Pasteur pipette completely around the sides of the tube. This is critical to allow the clot to shrink correctly. Place the tube on ice for 1 h or incubate overnight at 4°C. The clot should shrink to nearly half its original size. Collect the serum using a Pasteur pipette and centrifuge to remove any remaining blood cells. Add 0.05% sodium azide and store at 4°C; serum is stable for years in these conditions.

Assay the serum for anti-peptide activity (see below). Generally, this protocol produces serum with good antibody titers. If the titer is very low, repeat the immunization protocol described above. In some cases, it has been reported[2] that it can take several months to achieve reasonable titers. Once good titers are obtained, the rabbit can be bled at weekly intervals to collect sufficient quantities of serum. After a time, the titer will drop, but can be restored with an additional immunization, although you should keep in mind that the nature and affinities of the antibodies can change over a prolonged immune response.

D. PURIFICATION OF ANTIBODIES

In most cases, whole serum can be used in experiments, but sometimes unacceptable background can occur. Occasionally this is due to non-antibody contaminants in the serum and can be simply eliminated by purifying the whole immunoglobulin fraction of serum. A good procedure for doing this is the octanoic acid method of Steinbuch and Audran.[32]

1. Add 200 ml of acetate buffer (0.06 *M*, pH 4.0) to 100 ml of serum. The pH should be 4.8.
2. Stirring vigorously, add dropwise 6.8 g (7.5 ml) of octanoic acid. Stir vigorously for 30 to 60 min. A dense precipitate will form. The immunoglobulin remains in solution.
3. Filter through a conical filter paper (Whatman Ltd., Maidstone, England).
4. To each 100 ml of filtrate, add 4.0 ml of 1 *M* Tris-HCl (pH 9.0) and then 26 g of solid ammonium sulfate. Steps 1 to 4 must be performed at room temperature.
5. Stir vigorously for 1 to 2 h at 4°C and then centrifuge at 16,000 × g for 30 min.
6. Dissolve the precipitate in approximately 25% of the original serum volume in phosphate- or Tris-buffered saline. Dialyze three times against buffered saline.
7. Centrifuge at 16,000 × g for 30 min. The supernatant should contain greater than 95% pure immunoglobulin. Store at 4°C with 0.05% sodium azide or in aliquots at −20°C.

In many cases, however, the background activity is not due to nonspecific binding, but rather, to the activity of irrelevant antibodies. For instance, *Escherichia coli* bacteria are part of the natural intestinal flora of rabbits and, therefore, the rabbit possesses high background antibody levels against bacterial components. This makes it practically impossible to use untreated antisera when using bacterial systems. Fortunately, it is usually easy to eliminate this activity by mixing the sera with a lysate extract of bacteria not expressing the experimental protein. The bacterial components bind with the offending antibodies and so mask their activity in the assay. Extracts can be prepared by sonication or by freeze/thawing. I find the freeze/thaw procedure of Huynh et al.[33] to be simple and effective.

Unfortunately, in many instances it is not so easy to eliminate this type of background activity and the only alternative is to purify the anti-peptide antibodies. This is best done by affinity chromatography on peptide-conjugated Sepharose (Pharmacia Fine Chemicals, Uppsala, Sweden). However, there can be a price to pay for using this technique, and that is the possible loss of high-affinity antibodies. This can occur because the very high affinity antibodies may bind so tightly that they cannot be recovered from the column. For this reason, it is probably best not to use affinity-purified antibodies if it is unnecessary. However, in my own work it is necessary to routinely use affinity-purified antibodies and I have not experienced any problems. I routinely use two types of affinity supports, epoxy-activated Sepharose (Pharmacia) and activated CH-Sepharose (Pharmacia). Both of these couple through amino groups, but use different chemistries. Some peptides couple better to one or the other of these Sepharoses, but it is rare for both not to work. AH-Sepharose (Pharmacia) with carbodiimide (*N*-ethyl-*N*-[3-dimethylaminopropyl] carbodiimide hydrochloride)[34] as the cross-linking agent can be used if coupling via carboxyl groups is desired.

1. Coupling Peptide to Epoxy-Activated Sepharose[34]

1. Dissolve 5 mg of peptide using the procedure established above in a final volume of 1.5 ml. Organic solvents can be used. Add 1.5 ml of 0.2 *M* sodium carbonate buffer (pH 10.7). Check the final pH with indicator paper. Measure the OD at 280 or 250 nm or, if these are not adequate, use 220 nm. Remember to use the appropriate blank.
2. Swell 1 gm of Sepharose beads in 10 ml of water for 15 min. Wash with 100 ml of water on a sintered glass funnel. Be careful not to dry the gel; it should remain quite moist.
3. Add the gel to the peptide solution and transfer to a small beaker. Incubate overnight at 37°C with gentle shaking. (Do not use a magnetic stirrer.)
4. Transfer to a centrifuge tube and centrifuge at 1000 × g for 2 min. Save the supernatant. Resuspend the gel in 10 ml of water, centrifuge, and save the supernatant. Resuspend the gel in 10 ml of 1 *M* sodium chloride, centrifuge, and save the supernatant. Resuspend the gel in 1 *M* ethanolamine (pH 8.0) and incubate overnight on an end-over-end shaker at room temperature. This treatment blocks any remaining active groups. Centrifuge and resuspend the gel in 3 ml of buffered saline.
5. Measure the OD of the wash solutions and note the volumes. Determine the percentage coupling by comparing the starting number of OD units (OD reading multiplied by volume) with the total number of OD units recovered. Coupling efficiencies will range from about 1% (i.e., not discernible) to 50%, but in my experience are usually about 10%. Don't be too concerned if you get a low coupling efficiency. For example, 1% coupling of a 16-mer represents approximately 0.03 μmol, which would give a theoretical maximum antibody binding capacity of 4.5 mg. This is usually more than adequate.

2. Coupling Peptide to Activated CH-Sepharose[34]

1. Dissolve the peptide as in step 1 above. Add 1.5 ml of 0.2 *M* sodium bicarbonate buffer (pH 8). Check the pH of the final solution with indicator paper.

2. Swell 1 g of Sepharose beads in 10 ml of 1 m*M* hydrochloric acid for 15 min. Wash on a sintered glass filter with 200 ml of 1 m*M* hydrochloric acid.
3. Add the gel to the peptide solution and incubate at room temperature for 1 h on an end-over-end shaker.
4. Do steps 4 and 5 as above.

3. Affinity Purification of Anti-Peptide Antibodies

1. Pour the gel suspension (there should be approximately 3 ml of packed gel) into a small glass column (internal diameter, 1 cm; length, 5 to 10 cm) (Kontes, Vineland, NJ; Bio-Rad, Richmond, CA). Wash with 15 ml of buffered saline and then with 10 ml of elution buffer. There are a number of different procedures that can be used to elute antibodies from affinity columns. I find that changing the pH is a particularly effective method. Antibodies can be eluted in alkaline (0.1 *M* sodium carbonate or diethanolamine, pH 11.5) or acid conditions (1 *M* propionic acid). Propionic acid also lowers the polarity of the eluent and so reduces hydrophobic interactions. Finally, wash the column with 15 ml of buffered saline.
2. Filter the serum with a 0.45-μm filter or centrifuge at 15,000 × g to remove aggregates. Apply the serum to the column at a flow rate of 5 to 10 $ml/cm^{-2}/h^{-1}$. A slow flow rate is best. The volume to apply will depend on the binding capacity of the column. Collect the flowthrough and assay the fractions to determine when the column becomes saturated. Wash with buffered saline. This is usually sufficient to remove nonspecifically adsorbed components, but for high-stringency washing, use 10 ml of 1 *M* sodium chloride, 25 m*M* Tris (pH 8.0), followed by 10 ml of 0.5% triton X-100, 0.1% sodium deoxycholate, 1 m*M* EDTA, 25 m*M* Tris (pH 8.0), and, finally, buffered saline.
3. Allow the washing buffer to just completely enter the gel and then stop the flow. Using a Pasteur pipette, wash the sides of the column with approximately 0.5 ml of elution buffer and allow this to just enter the gel. This procedure ensures a rapid changeover between wash buffer and elution buffer, which promotes maximal recovery of antibody. Add approximately 1 ml of elution buffer to the top of the gel and elute with 10 ml of elution buffer. Collect the elution fractions in tubes containing enough 3 *M* Tris (for acid elution, use pH 8.5; for alkaline elution, use pH 7.5) to neutralize the solution. Monitor the OD 280 nm readings and pool the peak fractions. Wash the column with buffered saline containing 0.05% sodium azide and store at 4°C.
4. Dialyze the pooled fractions exhaustively against three changes of buffered saline containing 0.05% sodium azide (omit the azide if the antibodies will be used with cells and viability is important). Store frozen in aliquots.

E. ASSAYS FOR ANTI-PEPTIDE ANTIBODIES

1. ELISA

The best assay for directly measuring anti-peptide activity is the enzyme-linked immunosorbent assay or ELISA.[1] Briefly, peptide is absorbed to a solid phase, usually a well of a 96-well plastic microtiter plate (e.g., Flow Laboratories, Inc., McLean, VA), and then incubated with the antibody solution. Unbound antibody is washed away and the bound antibodies are detected with anti-immunoglobulin antibodies conjugated with an enzyme (either alkaline phosphatase or horseradish peroxidase). Wells containing anti-peptide antibodies develop color when the enzyme substrate is added.

1. Dilute the peptide from a stock solution to 1 μg/ml in either phosphate-buffered saline or 0.2 *M* sodium carbonate (pH 9.6) and add 100 μl to the wells of a 96-well microtiter plate. Incubate **uncovered** overnight at 37°C to allow the buffer to evaporate. The

wells must be completely dry before proceeding. Which buffer to use is a matter of trail and error since, with some peptides, there is a large difference in signals between the buffers. Some workers add methanol to the dried wells for 10 min to fix the peptides.[23] The stock solution is prepared as described above. Dilute peptides usually stay in solution even if the stock solution was made with organic solvent. However, if this is not the case and you don't get a usable signal, you will have to use a protein-peptide conjugate in the ELISA. The same conjugate used to raise the antibody can be used with unconjugated protein (1%) in the diluent to inhibit anti-protein activity, but it is best to use the peptide conjugated to a protein unrelated to the one used for the immunization. Dilute the peptide conjugate to 10 μg/ml in the carbonate buffer, add 100 μl to the wells, and incubate **covered** overnight at 4°C. In this case, the wells are not dried.

2. Wash the plate with washing solution (20 m*M* sodium phosphate [pH 7.5], 0.15 *M* sodium chloride, 0.05% Tween 20 detergent) two times. There are automatic machines to do this,[35] but an adequate job can be done using a laboratory squirt bottle. Fill the wells with washing solution using the squirt bottle and then, holding firmly, rapidly flip the plate over a sink to empty the wells. Pat upside down on paper towels to remove any remaining solution.
3. Add 100 μl of diluent (washing solution containing 1% fish gelatin [Norland Products, New Brunswick, NJ]) and incubate at room temperature for 1 h. This step is to block any remaining sites on the plastic. Some wells should be left uncoated (i.e., incubated with buffer not containing peptide) and reacted with antibody the same way as coated wells. This is a control for the blocking step to check the level of nonspecific binding to the plastic. Wash two times with washing solution. Add 100 μl diluent to each well.
4. Prepare 1/10, 1/100, 1/1000, and 1/10,000 dilutions of the antibody solution in diluent. Add 100 μl of each dilution to the first well of separate rows. Then do eight serial twofold dilutions (100 μl) along each row. Incubate at room temperature for 1 h. Wash three times. Be careful not to allow spillover between wells, which can cause cross-contamination.
5. For the second reagent, use an enzyme-antibody conjugate against the immunoglobulin of the species used for the immunization.[11] Thus, for rabbits, use goat or sheep anti-rabbit IgG (heavy- plus light-chain) conjugate. If chickens were used for immunization, use sheep, goat, or rabbit anti-chicken immunoglobulin. For ELISA, I prefer to use peroxidase conjugates.[11] Dilute the conjugate in diluent according to the manufacturer's instructions or as previously determined as optimal. Add 100 μl to each well, incubate for 1 h, and then wash three times.
6. Wash once with phosphate-buffered saline. Prepare a 1.5% stock solution of 2,2'-azinobis(3-ethylbenz-thiazoline sulfonic acid) (ABTS) in water. This can be stored for several days at 4°C in the dark. ABTS gives a dark green color reaction with horseradish peroxide. Dilute the ABTS to 0.03% in 0.1 *M* citrate buffer (pH 4.0) containing 0.01% hydrogen peroxide. Add 100 μl to the microtiter plate wells. Incubate for 15 to 30 min. Read the OD at 420 nm in an ELISA plate reader.[35] The titer is usually defined as the dilution of the antibody which gives 50% of the maximum OD reading. A good estimate of the titer can also be made by determining the color endpoint visually.

2. Western Blots

The Western blot technique is a very good method for determining antibody activity and for detecting cross-reactive determinants on proteins.[36] Thus, it is an excellent way to assay for the presence of gene products using anti-peptide antibodies and to do preliminary characterization of these products. Proteins are first separated in SDS-PAGE and then transferred to a solid-phase support, such as nitrocellulose or polyvinylidene difluoride membranes, where they can be reacted with antibodies in much the same manner as in ELISA.

1. Run SDS-PAGE slab gel. I use the new mini-gel systems (Hoefer Scientific Instruments, San Francisco, CA; Bio-Rad Laboratories, Richmond, CA), which give faster analysis time and very good resolution, with the Laemmli buffer system[37] under standard conditions[1] and as recommended by the gel system manufacturer. Gels are run under reducing conditions (i.e., in the presence of 2-mercaptoethanol) to ensure complete denaturation of the proteins. The amount of sample to load is variable and is influenced by the conditions used. As a general rule, a band that can be seen in a normal Coomassie brilliant blue[1] protein stain should be easily visualized in a Western blot. With a good antibody, amounts down to 1 ng can be detected. For checking anti-peptide activity, use 10 to 100 ng of protein-peptide conjugate. It is good practice to use prestained molecular-weight markers (Bethesda Research Laboratories, Gaithersburg, MD) so that you can visually monitor both gel electrophoresis and the transfer process. (*Note:* Wear plastic gloves at all times when handling the PAGE gels and the transfer membranes.)
2. Remove the gel from the electrophoresis apparatus and soak in blotting medium (0.025 *M* Tris, 0.193 *M* glycine, 20% reagent grade methanol) for 30 min.
3. The transfer membrane can be either nitrocellulose (BA-85; Schleicher and Schuell, Keene, NH) or polyvinylidene difluoride membranes (Immobilon PVDF; Millipore Corp., Bedford, MA). I prefer Immobilon since it is easier to handle and has greater binding capacity.[38] Soak the Immobilon membrane, which has been precut slightly bigger than the gel, for 30 sec in methanol and then for 30 min in transfer buffer.
4. Slide a sheet of Whatman 3MM paper (Whatman Ltd., Maidstone, England) under the gel and lift both from the blotting medium. Lay paper side down onto a Scotch pad (3M Company, Minneapolis, MN) prewetted with transfer buffer. Using a test tube or a pipet as a "rolling pin", apply firm pressure to remove air bubbles and excess buffer from beneath the membrane. This step is critical for good transfer of proteins. Next, overlay with another sheet of wetted 3MM paper and then a Scotch pad.
5. Enclose in a gel holder and place in the blot transfer apparatus. Place the gel on the negative side and the Immobilon membrane on the positive side. Transfer conditions will vary, depending on the transfer apparatus and whether cooling is used.[36] Use conditions recommended by the manufacturer of the transfer apparatus. For example, using the Hoefer Mighty Small Transphor unit with cooling, I transfer at 250 V for 45 min.
6. Block reactive sites by soaking the Immobilon membrane in diluent (1% fish gelatin, 0.05% Tween 20, 25 m*M* Tris (pH 8.0), 0.15 *M* sodium chloride) with shaking for 15 min. Use about 20 to 50 ml per membrane. Repeat two times. The diluent should be made fresh each time using stock gelatin and Tween solutions and 10X Tris/sodium chloride buffer. Store the stock solutions in the cold. High backgrounds can result from solutions stored at room temperature.
7. Dilute the anti-peptide antibody in diluent and incubate 20 ml with the membrane for 1 to 2 h with gentle shaking. The dilution to use will depend on the titer of the antibody. It is best to use the lowest concentration of antibody possible to minimize background and nonspecific binding. For the antibody controls (see earlier discussion), use the same dilution of sera or concentration of affinity-purified immunoglobulin as used for the test antibodies. The best control is to show that the reaction can be competitively inhibited by the specific peptide. This is easily done by including free peptide in the

antibody solution; 10 to 100 μg/ml is usually sufficient. If the peptide is not soluble, one might try using a protein-peptide conjugate, although the amount to use will be variable and dependent on the degree of substitution.

8. Wash three times with diluent for 10 min each.
9. Incubate for 1 to 2 h with the second reagent antibody/enzyme conjugate (see ELISA step 5). For Western blots, I find that alkaline phosphatase conjugates[11] are the most sensitive, being equal to or better than radiolabeled reagents. Alkaline phosphatase is not inhibited by azide, as is horseradish peroxidase, so antibody solutions can be stored in 0.1% sodium azide in the cold for several weeks and reused several times. Sometimes, though, such storage can result in high backgrounds.
10. Wash the membrane three times, 5 min each, with Tris-buffered saline. Don't use phosphate since it is a weak chelator of divalent metal ions such as magnesium and zinc, which are needed by alkaline phosphatase for activity. Make 50 mg/ml stock solutions of 5-bromo-4-chloro-3-indoylphosphate disodium salt (BCIP) and nitroblue tetrazolium chloride (NBT) in water and 70% dimethylformamide, respectively. Use 100% dimethylformamide if you use the p-toluidine salt of BCIP. Store both stock solutions at −20°C. To 10 ml of substrate buffer (0.1 *M* diethanolamine [pH 9.8], 1 m*M* magnesium chloride), add 66 μl of NBT and 33 μl of BCIP, and incubate with the membrane with gentle shaking. Incubate until the background color starts to appear, usually in 5 to 30 min. Stop the reaction by incubating with 10 m*M* EDTA for 15 min. Wash several times with water and then dry the membrane.

ACKNOWLEDGMENTS

I thank Erika Bullesbach for her expert advice on using organic solvents and Andrew Diamanduros for excellent technical help. This work was supported in part by Grant #DCB 88 02353 from the National Science Foundation and by Grant #GM 42437 from the National Institutes of Health.

REFERENCES

1. **Harlow, E. and Lane, D., Eds.,** *Antibodies: A Laboratory Manual,* Cold Spring Harbor Laboratory, Cold Spring Harbor, NY, 1988.
2. **Sutcliffe, J. G., Schinnick, T. M., Green, N., and Lerner, R. A.,** Antibodies that react with predetermined sites on proteins, *Science,* 219, 660, 1983.
3. **Walter, G.,** Production and use of antibodies against synthetic peptides, *J. Immunol. Methods,* 88, 149, 1986.
4. **Lerner, R. A.,** Tapping the immunological repertoire to produce antibodies of predetermined specificity, *Nature,* 299, 592, 1982.
5. **Sutcliffe, J. G., Milner, T. M., and Bloom, F. E.,** Identifying the protein products of brain-specific genes with antibodies to chemically synthesized peptides, *Cell,* 33, 671, 1983.
6. **Schluter, S. F. and Marchalonis, J. J.,** Antibodies to synthetic joining segment peptide of the T-cell receptor β chain: serological cross-reaction between products of T-cell receptor genes, antigen binding T-cell receptors and immunoglobulins, *Proc. Natl. Acad. Sci. U.S.A.,* 83, 1872, 1986.
7. **Schluter, S. F., Rosenshein, I. L., Hubbard, R. A., and Marchalonis, J. J.,** Conservation among vertebrate immunoglobulin chains detected by antibodies to a synthetic joining segment peptide, *Biochem. Bipohys. Res. Commun.,* 145, 699, 1987.
8. **Yoneda, Y., Imamoto-Sonobe, N., Matsuoba, Y., Iwanoto, R., Kiho, Y., and Uchida, T.,** Antibodies to Asp-Asp-Glu-Asp can inhibit transport of nuclear proteins into the nucleus, *Science,* 242, 275, 1988.
9. **Marchalonis, J. J. and Warr, G. W., Eds.,** *Antibodies as A Tool: The Application of Immunochemistry,* Interscience, New York, 1982.
10. **Weir, D. M., Herzenberg, L. A., Blackwell, C., and Herzenberg, L. A., Eds.,** *Handbook of Experimental Immunology,* 4th ed., Mosby, Saint Louis, MO, 1986.

11. List of commercial suppliers, *Science,* 243(2), G23, 1989.
12. **Getzoff, E. D., Tainer, J. A., Lerner, R. A., and Geysen, H. M.,** The chemistry and mechanism of antibody binding to protein antigens, *Adv. Immunol.,* 43, 1, 1989.
13. **Marchalonis, J. J., Schluter, S. F., Hubbard, R. A., McCabe, C., and Allen, R. C.,** Immunoglobulin epitopes defined by synthetic peptides corresponding to joining region sequence: conservation of determinants and dependence upon the presence of an arginyl or lysyl residue for cross-reaction between light chains and T-cell receptor chains, *Mol. Immunol.,* 25, 771, 1988.
14. **George, D. G., Barker, W. C., and Hunt, L. T.,** The protein identification resource (PIR), *Nucleic Acids Res.,* 14, 11, 1986.
15. **Devereux, J., Haeberli, P., and Smithies, O.,** A comprehensive set of sequence analysis programs for the VAX, *Nucleic Acids Res.,* 12, 387, 1984.
16. **Lerner, R. A.,** Antibodies of predetermined specificity in biology and medicine, *Adv. Immunol.,* 36, 1, 1984.
17. **Berzofsky, J. A.,** Intrinsic and extrinsic factors in protein antigenic structure, *Science,* 229, 932, 1985.
18. **Westhof, E., Altschuh, D., Moras, D., Bloomer, A. C., Mondragon, A., Klug, A., and Van Regenmortel, M. H. V.,** Correlation between segmental mobility and the location of antigenic determinants in proteins, *Nature,* 311, 123, 1984.
19. **Jameson, B. A. and Wolf, H.,** The antigenic index: a novel algorithm for predicting antigenic determinants, *CABIOS,* 4, 181, 1988.
20. **Walter, G., Hutchinson, M. A., Hunter, T., and Eckart, W.,** Purification of polyoma virus medium-size tumor antigen by immunoaffinity chromatography, *Proc. Natl. Acad. Sci. U.S.A.,* 79, 4025, 1982.
21. **Tan, J. P.,** Synthetic peptide vaccine design; synthesis and properties of a high-density multiple antigenic peptide system, *Proc. Natl. Acad. Sci. U.S.A.,* 85, 5409, 1988.
22. **Doolittle, R. F.,** *Of URFS and ORFS: A Primer on How to Analyze Derived Amino Acid Sequences,* University Science Books, Midvalley, CA, 1986, 63.
23. **Green, N., Alexander, H., Olson, A., Alexander, S., Shinnick, T. M., Sutcliffe, J. G., and Lerner, R. A.,** Immunogenic structures of the influenza virus hemagglutinin, *Cell,* 28, 447, 1982.
24. **Hunter, W. M. and Greenwood, F. C.,** Preparation of iodine-131 labeled human growth hormone of high specific activity, *Nature,* 194, 495, 1962.
25. **Sonada, S. and Schlamowitz, M.,** Studies of I-125 trace labeling of immunoglobulin G by chloroamine-T, *Immunochemistry,* 7, 885, 1970.
26. **Bolton, A. E. and Hunter, W. M.,** The labeling of proteins to high specific radioactivities by conjugation to a I-125 containing acylating agent, *Biochem. J.,* 133, 529, 1973.
27. **Warr, G. W.,** Preparation of antigens and principles of immunization, in *Antibody as A Tool: The Applications of Immunochemistry,* Marchalonis, J. J. and Warr, G. W., Eds., Interscience, New York, 1982, 21.
28. **Chase, M. W.,** Animal handling, in *Preparation of Antigens and Antibodies,* Williams, C. A. and Chase, M. W., Eds., Academic Press, New York, 1967, 254.
29. **Herbert, W. J.,** Laboratory animal techniques for immunology, in *Handbook of Experimental Immunology, Vol. 3, Applications of Immunological Methods,* Weir, P. M., Ed., Blackwell Scientific, Oxford, 1978, A4.1.
30. **Rudback, J. A., Cantrell, J. L., and Ulrich, J. T.,** Molecularly engineered microbial immunostimulators, in *Technological Advances in Vaccine Development. Proceeding of UCLA Symposia on Molecular and Cellular Biology,* Alan R. Liss, New York, 1988, 443.
31. **Gordon, L. K.,** A reliable method for repetitively bleeding rabbits from the central artery of the ear, *J. Immunol. Methods,* 44, 241, 1981.
32. **Steinbuch, M. and Audran, R.,** The isolation of IgG from mammalian sera with the aid of caprylic acid, *Arch. Biochem. Biophys.,* 134, 279, 1969.
33. **Huynh, T. V., Young, R. A., and Davis, R. W.,** Constructing and screening cDNA libraries in λgt10 and λgt11, in *DNA Cloning: A Practical Approach,* Vol. 1, Glover, D. M., Ed., IRL Press, Oxford, 1985, 49.
34. *Affinity Chromatography: Principles and Methods,* Pharmacia Fine Chemicals, Uppsala, Sweden, 1983.
35. *Science,* 243(2), G26, G80, G148, 1989.
36. **Beisiegel, U.,** Protein blotting, *Electrophoresis,* 7, 1, 1986.
37. **Laemmli, E. K.,** Cleavage of structural proteins during the assembly of the head of bacteriophage T4, *Nature,* 227, 680, 1970.
38. **Pluskal, M. G., Przekop, M. B., Kavonian, M. R., Vecoli, C., and Hicks, D. A.,** Immobilon PVDF transfer membrane: a new membrane substrate for Western blotting of proteins, *BioTechniques,* 4, 272, 1986.

Chapter 5

POLYMERASE CHAIN REACTION TO AMPLIFY DNA FROM PHAGE PLAQUES

Eric S. Miller and Catherine Jozwik

TABLE OF CONTENTS

I. INTRODUCTION

The polymerase chain reaction[1,2] has found applications in nearly every life science endeavor, including forensics, diagnostic medicine, molecular archeology, and fundamental gene stucture and function studies, to name a few.[3] More recently, use of the heat-resistant Taq DNA polymerase purified from *Thermus aquaticus* has significantly simplified the amplification procedure,[4] to the extent that PCR is often the technique of choice for what previously involved tedious DNA (or RNA) cloning and sequencing strategies. Although larger eukaryotic genomes are primarily considered in PCR applications, prokaryotic genomes, or DNA already contained on plasmid or phage cloning vectors, are also readily amenable to manipulation using PCR.

A brief review of the literature reveals that the uses of PCR are limited only by the need for short oligonucleotides complementary to sequences flanking the region to be amplified. Sequential cycles of denaturing, annealing, and extension reactions, usually between 25 to 35 cycles in a programmable heat block (e.g., Perkin Elmer-Cetus Instruments[4]), can produce millions of copies of the DNA between the opposing oligonucleotide primers. Because the primers need not hybridize throughout their entire length, degenerate primers derived from consensus sequences or amino acid sequences can be used to amplify DNA of unknown sequence.[4-6] Similarly, primers on either or both ends of the amplified DNA can be synthesized to contain desired restriction enzyme recognition sites, promoters (e.g., T7, SP6), ribosome binding sites, or any other combination of gene control or protein recognition sites.[4,7,8] By the completion of the third extension cycle the primers hybridize to essentially all newly synthesized templates and are fully complementary. PCR has also been used in *in vivo* footprinting assays to demonstrate sequence recognition by DNA binding proteins.[9] A compendium of several PCR applications has recently appeared.[10]

Because the DNA templates used for PCR amplification need not be especially pure, almost any source of DNA will suffice. Bacterial chromosomal preparations,[11] boiled plasmid preparations,[12] intact plasmid-containing cells,[13] M13 plaques,[12] λ gt11 plaques,[13,14] and T-even phage plaques[15] are examples of unfractionated substrates suitable for PCR. This often obviates the need for constructing gene libraries since the amplified genomic DNA of interest, purportedly to at least 10-kilobase (kb) pairs, can be directly cloned or sequenced. Sequencing strategies usually involve Sanger dideoxynucleotide reactions using one of the amplifying primers, or an internal primer, on PCR material purified from agarose gels[16] or through microconcentrators (Centricon-100; Amicon,, Danvers, MA). Asymmetric amplification, where one primer is in 20- to 100-fold molar excess over the other,[17-19] can also yield single-stranded DNA suitable for sequencing by the Sanger method using a variety of DNA polymerases (Taq, T7, Klenow, etc.).

A simple protocol for PCR DNA amplification using unfractionated bacteriophage T4 plaques is described. As a general procedure, the method should be applicable to other double- or single-stranded DNA phages, including plaque-forming eukaryotic DNA viruses. It should be noted that T-even bacteriophages have extensively modified genomes comprised of glucosylated, hydroxymethylated cytosine-containing DNA of 166 kb pairs. The DNA is refractory to digestion by many commonly used restriction endonucleases. The products of PCR amplification, however, contain unmodified cytosines and can be readily cleaved with restriction enzymes.

II. PROTOCOLS

A. TEMPLATE PREPARATION

Bacteriophage lysates are prepared, diluted, and used to infect an appropriate host to obtain well-isolated plaques on a bacterial top agar overlay. For the T-even bacteriophages,

T2, T4, and T6, and the RB phages,[20] *Escherichia coli* B is routinely used as the propagating and plating host, with incubations carried out at 30°C.

Single plaques are removed with the tip of a sterile Pasteur pipet, placed into 40 μl of sterile distilled H_2O, and the agar plugs are briefly disrupted.

B. PRIMERS

Synthetic oligonucleotide primers are conveniently 15 to 25 nucleotides long, should be complementary to opposite strands flanking the region to be amplified, and of as equally distributed nucleotide composition as possible. Inter- or intra-primer complementarity should be avoided in the design of primers. The oligonucleotides are suspended in dH_2O at 25 pmol/μl and stored at −20°C.

C. PCR AMPLIFICATION

10X PCR Buffer

100 m*M* Tris-Cl, pH 8.4
500 m*M* KCl
15 m*M* $MgCl_2$
200 μg/ml gelatin

dNTP Mixture

Deoxynucleotides (Boehringer Mannheim) are separately suspended in dH_2O to a concentration of 8 m*M* and the pH is adjusted to 7.0. The four dNTPs are combined in equal portions and stored in convenient aliquots at −20°C.

Reaction Mixture

20 μl plaque suspension
5 μl dNTP mix
5 μl 1% Triton X-100
5 μl 10X PCR buffer
1 μl each primer (0.5 μ*M* final concentration)
1 μl Taq DNA polymerase (1 to 2 U)
12 μl dH_2O

Thoroughly mix all components in a 0.5-ml microtube, overlay with 50 μl of mineral oil, and place in a programmable heat block. Sequential incubations at 95°C (1 min), 52°C (1 min), and 72°C (1 min) in a Perkin-Elmer Cetus Thermal Cycler (Norwalk, CT) have proven satisfactory; increasing the time of the extension reaction (72°C) by 10 sec in successive cycles, or incorporating a 10-min extension after the final cycle, has in some cases enhanced the yield of amplified DNA. A total of 25 cycles is routinely used in our laboratory. These incubation parameters generally work well, but they, and other parameters for PCR, may need to be optimized[21] for different templates or primer sequences.

Following removal of the mineral oil from the top of the sample, 5 μl of the 50-μl reaction volume can be analyzed by agarose gel electrophoresis to assess yield and product homogeneity. The remaining amplified DNA is removed to a clean microtube, precipitated by standard procedures, and can then be cloned, sequenced, or used as desired.

Taq DNA polymerase as supplied by Promega (Madison, WI) and USB/Perkin-Elmer Cetus have been successfully used in our laboratory for PCR amplification of phage plaque

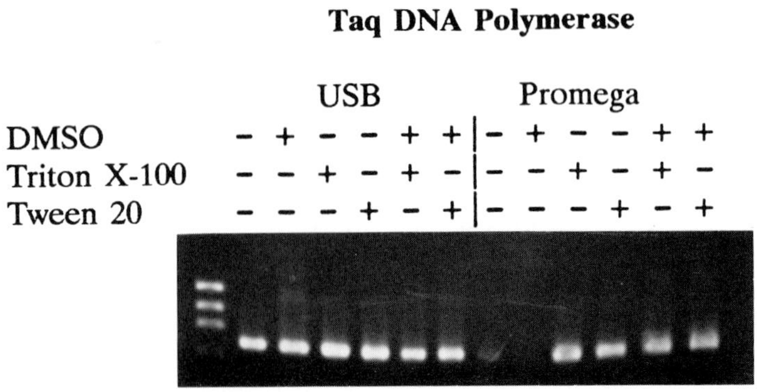

FIGURE 1. Conditions for maximizing PCR amplification of DNA from T-even phage plaques. Single bacteriophage T4 plaques were treated as described in the standard protocol, with or without the addition of DMSO, Triton X-100, or Tween 20. Where omitted, reagents were replaced with 5 μl of dH_2O. Conditions were evaluated for amplification grade Taq DNA polymerase obtained from USB (Perkin-Elmer Cetus; lot no. 63874) and Promega (lot no. M617A). The primers used flank a region of 699 base pairs that includes the T4 *reg*A gene[15] and were 20 (P5; 5′- CCCGAGTTATTAGGCTCTTG-3′) and 23 (P62; 5′-CTAATGGCTGCAGTTCCTCGTGG-3′) nucleotides long. Standard reaction conditions were used, followed by agarose (0.8%) gel electrophoresis and ethidium bromide staining of 5 μl of the amplification mixture. Molecular weight standards in the left-most lane were 1.35, 1.08, 0.87, and 0.60 kb pairs. The standard reaction mixture includes 0.1% Triton X-100 to accommodate various Taq DNA polymerase preparations and more refractory DNA templates.

DNA. Amplification-grade Taq polymerase from Promega required the addition of a nonionic detergent (Figure 1) to the reaction mix: AmpliTaq (Perkin-Elmer Cetus) distributed by U.S. Biochemical Corporation (Cleveland, OH) amplified the template under all conditions tested, although slightly more DNA was amplified in the presence of Triton X-100. It is therefore advisable to use nonionic detergent routinely in plaque amplification protocols.

Maximum amplification was observed with primer concentrations at 0.5 μ*M* in the reaction mix, although lower concentrations do work (Figure 2). Nonspecific amplification may be prevented in some situations by using reduced amounts of primer.[21]

Single plaques need not be resuspended in water, but can be simply picked with the end of a sterile toothpick and the toothpick swirled in the standard 50-μl reaction volume (adjusted with additional dH_2O). However, when screening phage mutants or newly isolated phage, the remaining volume of the plaque suspension can be stored and used later to isolate phage after the PCR analysis. The standard protocol is also suitable for amplifying DNA from plasmid-containing cells that are lifted from colonies with toothpicks and swirled in the reaction mixture.

Programmable temperature blocks are available in a variety of formats from several manufacturers, including Coy Instruments (Ann Arbor, MI), Ericomp (San Diego, CA), Perkin-Elmer Cetus (Norwalk, CT), Techne (Princeton, NJ), and USA Scientific (Ocala, FL).

ACKNOWLEDGMENTS

This work was supported by National Institutes of Health (NIH) Grant GM38659. We thank B. Singer and C. Tuerk for initial conversations on PCR. The use of trade names in this publication does not imply endorsement by the North Carolina Agriculture Research

Primer Concentration

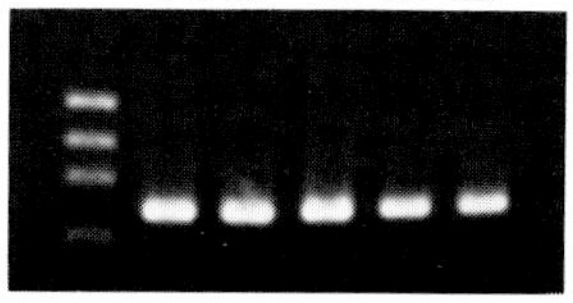

FIGURE 2. Effect of primer concentration on PCR amplification of DNA from bacteriophage plaques. Single T4 plaques were treated and subjected to PCR amplification as described in the standard protocol using two units of Taq DNA polymerase (USB). Primers P62 and P5 (see legend to Figure 1) were adjusted to final concentrations in the 50-μl reaction volume as follows: A, 1.0 *μM*; B, 0.5 *μM*; C, 0.25 *μM*; D, 0.13 *μM*; E, 0.06 *μM*., 5 μl of each reaction were subjected to electrophoresis through 0.8% agarose and stained with ethidium bromide. Markers were as in Figure 1. The standard protocol uses 0.5 *μM* of each primer, although significant amplification is obtained at lower concentrations. Lower amounts of primer may reduce primer mismatching and prevent nonspecific amplification in some situations.

Service, nor criticism of similar ones not mentioned. This is Paper No. 12398 of the Journal Series of the North Carolina Agriculture Research Service, Raleigh, NC 27695-7601.

REFERENCES

1. **Saiki, R. K., Scarf, S., Faloona, F., Mullis, K. B., Horn, G. T., Erlich, H. A., and Arnheim, N.,** Enzymatic amplification of B-globin genomic sequences and restriction site analysis for diagnosis of sickle cell anemia, *Science,* 230, 1350, 1985.
2. **Mullis, K. B. and Faloona, F. A.,** Specific synthesis of DNA *in vitro* via a polymerase catalyzed chain reaction, *Methods Enzymol.,* 155, 335, 1987.
3. **White, T. J., Arnheim, N., and Erlich, H. A.,** The polymerase chain reaction, *Trends Genet.,* 5, 185, 1989.
4. **Saiki, R. K., Gelfand, D. H., Stoffel, S., Scharf, S. J., Higuchi, R., Horn, G. T., Mullis, K. B., and Erlich, H. A.,** Primer-directed enzymatic amplification of DNA with a thermostable DNA polymerase, *Science,* 239, 487, 1988.
5. **Mack, D. H. and Sninsky, J. J.,** A sensitive method for the identification of uncharacterized viruses related to known virus groups: hepadnavirus model system, *Proc. Natl. Acad. Sci. U.S.A.,* 85, 6977, 1988.
6. **Lee, C. C., Wu, X., Gibbs, R. A., Cook, R. G., Munzy, D. M., and Caskey, C. T.,** Generation of cDNA probes directed by amino acid sequence: cloning of urate oxidase, *Science,* 239, 1288, 1988.
7. **Scharf, S. J., Horn, G. T., and Erlich, H. A.,** Direct cloning and sequence analysis of enzymatically amplified genomic sequences, *Science,* 233, 1076, 1986.
8. **Stoflet, E. S., Koeberl, D. D., Sarkar, G., and Sommer, S. S.,** Genomic amplification with transcript sequencing, *Science,* 239, 491, 1988.
9. **Axelrod, J. D. and Majors, J.,** An improved method for photofootprinting yeast genes *in vivo* using Taq polymerase, *Nucleic Acids Res.,* 17, 171, 1989.
10. **Erlich, H., Ed.,** *PCR Technology: Principles and Applications for DNA Amplification,* Stockton Press, New York, 1989.
11. **Shyamala, V. and Ames, G. F.-L.,** Amplification of bacterial genomic DNA by the polymerase chain reaction and direct sequencing after asymmetric amplification: application to the study of periplasmic permeases, *J. Bacteriol.,* 171, 1602, 1989.
12. **Gussow, D. and Clackson, T.,** Direct clone characterization from plaques and colonies by the polymerase chain reaction, *Nucleic Acids Res.,* 17, 4000, 1989.

13. **Sandhu, G. S., Precup, J. W., and Kline, B. C.,** Rapid one step characterization of recombinant vectors by direct analysis of transformed *Escherichia coli* colonies, *BioTechniques,* 7, 689, 1989.
14. **Nishikawa, B. K., Fowlkes, D. M., and Kay, B. K.,** Convenient uses of polymerase chain reaction in analyzing recombinant cDNA clones, *BioTechniques,* 7, 730, 1989.
15. **Miller, E. and Jozwik, C.,** Sequence analysis of conserved *regA* and variable *orf43.1* genes in T4-like bacteriophages, *J. Bacteriol.,* 172, 1990.
16. **Kretz, K. A., Carson, G. S., and O'Brien, J. S.,** Direct sequencing from low-melt agarose with sequenase, *Nucleic Acids Res.,* 17, 5864, 1989.
17. **Innis, M. A., Myambo, K. B., Gelfand, D. H., and Brow, M. A. D.,** DNA sequencing with *Thermus aquaticus* DNA polymerase and direct sequencing of polymerase chain reaction amplified DNA, *Proc. Natl. Acad. Sci. U.S.A.,* 85, 9436, 1988.
18. **Mihovilovic, M. and Lee, J. E.,** An efficient method for sequencing PCR amplified DNA, *BioTechniques,* 7, 14, 1989.
19. **Gyllensten, U. B.,** PCR and DNA sequencing, *BioTechniques,* 7, 700, 1989.
20. **Russel, R. L. and Huskey, R. J.,** Partial exclusion between T-even bacteriophages; an incipient genetic isolation mechanism, *Genetics,* 78, 989, 1974.
21. **Williams, F. J.,** Optimization strategies for the polymerase chain reaction, *BioTechniques,* 7, 762, 1989.

II. DNA AND GENOMIC ANALYSIS (ELECTROPHORETIC METHODS)

Chapter 6

PROTOCOLS FOR MANUAL DIDEOXY DNA SEQUENCING

Barton E. Slatko

TABLE OF CONTENTS

I. INTRODUCTION

As DNA sequencing has rapidly become standard practice in many laboratories, a large variety of new cloning vectors and sequencing strategies and techniques have been developed. The need for methods allowing the more rapid and efficient sequencing of large DNA templates has provided an additional impetus for change. In most instances, these advances evolved independently of the current focus on automating various procedures in DNA sequencing protocols. These developments are also applicable to current automated technology, although some modifications in sequencing reaction protocols and reagents have been required. Until such time as automated DNA sequencing is in general use, established methods are still required.[1,2]

It is the purpose of this chapter to describe current approaches and to provide specific protocols for manual dideoxy DNA sequencing. These protocols are only a subset of a large variety of procedures currently in use and in the literature. As such, they provide a current overview of basic DNA sequencing technology. This chapter will concentrate on dideoxynucleotide chain termination methods[3-5] in *Escherichia coli*.

The overall strategy of dideoxy DNA sequencing usually involves: (1) cloning the appropriate DNA fragments to be sequenced into a suitable vector, (2) purifying them as templates, (3) performing the sequencing reactions by the dideoxynucleotide termination method[1-3] or by chemical degradation methods,[6-14] (4) electrophoresing the reaction products on appropriate denaturing polyacrylamide gels, (5) processing the gel for autoradiography, and (6) analyzing the resultant data to complete a final contiguous DNA sequence.

Most DNA sequencing projects begin with the insertion of experimental DNA into a suitable *E. coli* cloning vector, and although it is also possible to sequence directly from RNA (e.g., genomic or transcribed from T3, T7, or SP6 RNA polymerase promoter vectors),[15,16] from genomic DNA,[17,18] or from asymmetric, symmetric, or a two-stage polymerase chain reaction (PCR™),[19-36] these methods will not be described in this chapter.

Protocols used to prepare DNA fragments for dideoxy sequencing in *E. coli* are commercially available in a variety of DNA sequencing and cloning manuals and references[37-43] (see Appendix: Sources of Reagents and Supplies for Dideoxy DNA Sequencing and Other Applications). Many of these manuals and publications describe M13 and related single-stranded, male-specific phage biology and cloning techniques[43-47] and will not be described here. The general overall strategy employed involves subcloning the DNA to be sequenced, as appropriately sized fragments, into plasmid, phage, or phagemid vectors. Procedures should be followed as steps in any generalized cloning strategy: preparation of competent bacterial cells, digestion of vector DNA with restriction endonuclease(s), ligation of vector and insert, transformation into *E. coli* cells, and preparation of a single-stranded or double-stranded template for sequencing. Methods for purification of templates for dideoxy DNA sequencing are presented below.

When determining which *E. coli* vector system to use for DNA sequencing, the following might be considered:

1. The ability to identify, by positive screening, vectors containing DNA inserts
2. The presence of restriction enzyme sites which allow ligation of appropriate DNA subfragments to the vector
3. Whether single-stranded or double-stranded DNA will be used as the template for the synthesis reaction, since dideoxy DNA sequencing can be efficiently performed from single-stranded or from double-stranded DNA templates and many sequencing vectors contain f1, fd, or M13 origins of replication which can be activated for the production of single-stranded DNA[13-15,43,48-61] (see Appendix: Sources of Reagents and Supplies for Dideoxy DNA Sequencing and Other Applications)

4. The presence of a ''high-copy-number'' plasmid origin of replication for increased yield of vector DNA, allowing direct sequencing from mini-preparations of the double-stranded vector template
5. The availability of primer molecules to initiate the dideoxy sequencing reactions, since many ''sequencing vectors'' utilize readily available specific (universal) primers to initiate the reactions
6. The strategy to be used for efficient sequencing of a large DNA fragment (e.g., overlapping deletion subcloning, restriction digest ''shotgun'' subcloning, ''primer walking'', etc.)

When a subfragment of less than 500 nucleotides is cloned into a vector, determination of its sequence is usually straightforward. If the experimental DNA fragment can be ligated to the vector in two orientations, individual clones oriented in opposite directions can be subsequently identified by restriction analysis or if the vector produces single-stranded DNA, by annealing the single-stranded DNA of independent clones to one another and analyzing their mobilities on agarose gels (the ''C'' test).[43] A primer, homologous to a priming site on the vector adjacent to the cloning site, can then be used to sequence clones in both orientations (in effect, reading from both ends toward the middle). Since one can usually determine 250 to 500 nucleotides of sequence information following the primer, this strategy will yield the entire sequence of that fragment. If a second primer is available, complementary to the DNA sequence on the opposite side of the cloning site from the position of the first primer, then double-stranded DNA sequencing of only one clone with the two primers will effect the same result.

Sequencing a large DNA molecule can be approached in a random ''shotgun'' fashion by the subcloning of small fragments generated by DNAse I digestion, by physical shearing, or by restriction digesting with endonucleases which make frequent cuts in the fragment (e.g., those with four base recognition specificity). These fragments are pooled and then the entire pool is ligated into a vector (''shotgun cloning'').[3,43,51,57,62-65] Sequence data from independent clones are generated using customized primers which anneal to either side of the polylinker cloning site in the vector. The final sequence is compiled by hand or computer from overlaps in the individual fragments.[51,66-69] However, with more complex (longer-length) sequences, this approach becomes tedious since it requires the purifying, cataloging, ligating, and cloning of numerous individual fragments. In addition, as the project nears completion, finding the final 1% of the unknown sequence can consume a disproportionately large amount of time.

Less random (more ordered) methods of sequencing a large DNA molecule can be performed by making directed overlapping (nested) sets of deletions from a clone containing the entire DNA fragment to be sequenced. The nested deletions can be generated from single-stranded DNA using T4 or T7 polymerase,[70,71] or from double-stranded DNA using DNAse I, Bal31, exonuclease III, transposon-mediated (Tn5, Tn9) deletions, or other methods.[46,51,54,58,59,63-65,70-101]

A third method for DNA sequencing involves the making of synthesized oligonucleotide primers from the 3′ end of a sequence previously ascertained (for instance, sequence generated using a universal primer homologous to the vector). Currently, ''primer walking'' or the ''look and leap'' method[1,51,82,102] is perhaps the simplest method available. It requires only an initial cloning of the entire DNA fragment and a single large template purification step since the same template is utilized throughout the project.

Whichever method is utilized, both strands of the DNA fragment should be sequenced since independent complementary confirmation leads to higher accuracy, especially when sequencing artifacts such as ''compressions'', generate sequence ambiguity. Sequencing the opposite strand is often required to resolve such ambiguities.

II. DIDEOXY DNA SEQUENCING METHODS

A. SEQUENCING PROTOCOLS

The DNA sequencing protocols described below are based upon the dideoxynucleotide chain termination method of Sanger et al.[3-5] First, an appropriate primer is annealed to the complementary region of single-stranded or double-stranded DNA template. The primer-template complex is then incubated with a DNA polymerase enzyme in the presence of all four deoxynucleotide triphosphates and dideoxynucleotide triphosphates. *E. coli* DNA polymerase I large (Klenow) fragment,[3-5,103-105] reverse transcriptase,[15,23,106-110] T7 DNA polymerase,[25,26,35,105,109,111-115,159] *Thermus aquaticus* DNA polymerase,[19,22,24,27,29-34,36,116-118,159] or other thermophylic polymerases (i.e., *Bacillus stearothermophilus* polymerase,[159] Vent™ polymerase) may be used to catalyze the reaction in which DNA molecules of different chain lengths are synthesized (due to the occasional incorporation of a dideoxynucleotide lacking a 3′ OH group, which blocks further chain elongation). The relative concentration of the deoxynucleotide triphosphate (dNTP) to dideoxynucleotide triphosphate (ddNTP) is adjusted so that, with a given low probability, an occasional incorporation of dideoxynucleotide occurs. Four separate reaction mixes, each containing all four dNTPs and one of the four ddNTPs, generates four different sets of fragments, each set corresponding to termination at specific nucleotide residues. Electrophoresis of the four sets of fragments on denaturing polyacrylamide gels separates these fragments on the basis of their size. Since synthesis is initiated at only one site, where an oligonucleotide primer anneals to the template, loading each of the four reactions in a separate adjacent lane produces a pattern in which the sequence of the newly synthesized complementary strand can be directly read from the gel. Reading the gel from the fastest to the slowest migrating band corresponds to reading the complement of the template in the 5′ to 3′ direction.

In order to visualize the dideoxy-terminated chains on the gel, various means of labeling have been developed. Conventionally, ^{32}P or ^{35}S have been used for incorporation into the nascent chain by using α-labeled deoxynucleotide triphosphates in the reaction mixture.[3-5,119-121] Certain criteria influence the choice of label. Sequencing with ^{32}P allows for very short exposure times without post-electrophoresis fixing or drying of the gels. A disad vantage to using this higher-energy-emission isotope is the presence of broader bands which may obscure data at the top of the subsequent gel autoradiogram. Another disadvantage is that the reaction products must be immediately electrophoresed after performing the sequencing reactions because the high-energy emission degrades the synthesized strand.[122] Sequencing with ^{35}S has several significant advantages over ^{32}P dideoxy DNA sequencing.[119-123] First, the low-energy β-emissions of ^{35}S result in very narrow, sharply defined bands on the autoradiograms. This allows more sequence information to be read from the upper portion of a gel than is possible when using ^{32}P. Second, unlike ^{32}P, the lower-energy emissions of ^{35}S do not appear to cause breaks in the sugar phosphate backbone of DNA. In practice, this means that ^{35}S dideoxy sequence reaction can be stored at −20°C for as long as 1 week prior to electrophoresis with no loss in clarity. Third, the weak β-emissions of ^{35}S result in a lower radiation dose for researchers performing DNA sequence analysis on a regular basis. The disadvantages of using ^{35}S include the necessity for slightly longer exposure times and for drying a fixed-sequencing gel, since urea and water quench the ^{35}S emissions. Exposure of 2 h is sufficient to read a sequence when using 2 μl of 500 to 3000 Ci/mmol [α-^{32}P] dATP, while overnight exposure may be necessary when utilizing 400 to 1000 Ci/mmol [α-^{35}S] dATP.

A variant technique involves labeling the primer itself by using T4 polynucleotide kinase to transfer a [γ-^{32}P] or [γ-^{35}S] from rATP to the 5′ end of the primer.[6,9] End-labeling is recommended for sequencing templates where standard labeling gives less than optimum results or for large templates, such as λgt11. Several automated methods for sequencing involve oligonucleotide primers labeled at the 5′ end with fluorescent dyes[111,124-129] or biotin for chemiluminescent techniques,[98] and one method uses fluorescent-tagged dideoxy nucleotide triphosphates.[130,131]

Currently, there are two basic methods for dideoxy DNA sequencing. "Standard" dideoxy methods involve a simultaneous enzymatic chain extension and a dideoxy chain termination to produce a set of fragments ranging in a size of 1 to over 500 bases on standard polyacrylamide DNA sequencing gels. A second method, "extension" sequencing, initiates enzyme-catalyzed primer extension in the presence of trace amounts of deoxynucleotide triphosphates, including α-labeled deoxynucleotide triphosphate. This is followed by a brief chain elongation and termination step in the presence of small amounts of dNTPs and a relatively high concentration of ddNTPs.[25,35,105,109,111-115,131,159] This protocol produces a nested set of fragments, ranging in size from 10 to over 750 bases, as determined on standard polyacrylamide DNA sequencing gels.

Appendix B: Composition of DNA Sequencing Reagents outlines the nucleotide/dideoxynucleotide mixtures in micromolar concentrations used for all of these approaches. The appropriate mixture must be used for each specific radioactive label and method (standard or extension sequencing). The protocols assume that radioactive label derives from α-dATP (or γ-rATP for end-labeling), although the nucleotide concentrations within the reaction mixtures can be easily adjusted to utilize label incorporated from any nucleotide.

Choice of polymerase may depend on the secondary structure of template due to extended palindromic regions and G + C content. Sequencing reactions are commonly done at 25 to 37°C, and higher temperatures can be used with templates containing potential secondary structure. Incubation temperatures of DNA polymerase I large (Klenow) fragment may be 25 to 50°C. *Thermus aquaticus* Bst on Vent™ DNA polymerases may be incubated at 37 to 75°C, with the initial 30 s of the reaction no higher than 55°C to allow for the primer to be annealed and extended. Improved sequencing of regions with high secondary structure may be accomplished with any polymerase by substituting either 2′-deoxyinosine-5′-triphosphate, or 7-deaza-2′-deoxyguanosine-5′-triphosphate[51,132-134] for the deoxyguanosine triphosphate or 7-deaza-2′-deoxyadenosine for the deoxyadenosine triphosphate. Since the optimum temperature for T7 DNA polymerase is 25 to 42°C,[111] these substitutions may be the only methods for improving the sequencing data from T7 DNA polymerase reactions. *T. aquaticus* DNA polymerase should not be used with 2′-deoxyinosine-5′-triphosphate.[116]

Both the "standard" and "extension" method can be utilized with DNA polymerase I large (Klenow) fragment, T7 DNA polymerase, reverse transcriptase, *B. stearothermophilus* polymerase, Vent™ polymerase, and *T. aquaticus* polymerase. Both methods can be used for [α-^{32}P] dNTP or [α-^{35}S] dNTP label incorporation. Both methods, and all of the above mentioned enzymes, can be used for end-labeled primers kinased with [γ-^{32}P] ATP or [γ-^{35}S] ATP. The sequencing reactions can be performed in nonsiliconized microcentrifuge tubes, as described below, which should be sterilized before use. For more efficient sequencing of a large number of templates simultaneously, the reaction can be performed in sterile, boiling-resistant, 64-well microtiter trays.[78,135-137] The round- or "v"-bottom wells are preferable to the flat-bottom wells.

For shorter sequencing (e.g., closer to the priming site), it is recommended to use the "standard" sequencing protocol. For longer (e.g., farther from the priming site) sequencing, the "extension" protocol works best. Whichever method, whichever label, whichever labeling method, and whichever polymerase is utilized for the sequencing reactions, the ultimate accuracy of DNA sequence determination is dependent upon electrophoretic resolution in appropriate percentages any types of denaturing acrylamide gels.

1. Standard Sequencing Reactions

a. Single-Stranded DNA Sequencing

Step A: Preparation of Annealed Template and Primer

1. In a 0.5-ml microcentrifuge tube, add:

   - 1.5 to 2 μg of single-stranded template DNA (~0.5 pmol template)
   - 1 μl of a stock solution of an appropriate primer (stock = 2.5 μg/ml) (~0.5 pmol primer)
   - 1 μl of the appropriate 10X sequencing buffer (see Appendix B: Composition of DNA Sequencing Reagents)
   - Distilled water to a total volume of 10 μl

Note: There should be a primer:template ratio of 1:1 to 5:1 (see Appendix E: Molar Ratio Calculations).

2. Mix gently by drawing the solution up and down with a pipette.
3. Incubate at 60°C for 6 min.
4. Incubate at room temperature for 20 min.

Note: More stringent annealing may be accomplished by incubating (after the 60°C step) at 37°C for 16 to 20-mers and 42°C for primers larger than 20-mers. Highly A-T-rich primers must be annealed at lower temperatures. Maximum temperature for a specific primer (16- to 25-mers) may be calculated by adding 2°C for every A-T and 4°C for every G-C and then subtracting 15% from the total amount, since the 10X buffer is low ionic strength. If longer annealing times are performed (e.g., 45 to 60 min), they should be done at the higher temperatures to prevent annealing of the primers at potential secondary priming sites.

Step B: Preparation of Deoxy/Dideoxy Reaction Tubes

1. Label four 0.5-ml microcentrifuge tubes A, C, G, and T, respectively. Choose the appropriate set of four standard sequencing mixes (outlined in Appendix B: Composition of DNA Sequencing Reagents) for the radioactive label and polymerase you are using. These mixtures can be made with standardized deoxy/dideoxy stock solutions (as outlined in Appendix C: Dilution Protocols from Stock Solutions of Nucleotides).
2. Add 3 μl of the A mixture to the bottom of the A tube, and 3 μl of the C, G, and T mixtures to the bottom of the C, G, and T tubes, respectively.

Step C: Addition of Label and Polymerase

1. To the annealed primer and template mixture, add 2 μl of either [α-^{32}P] dATP or [α-^{35}S] dATP (500 to 3000 Ci/mmol, 10 mCi/ml water).
2. Add 1 μl (5 U) of polymerase (3000 to 5000 U/ml) (DNA polymerase I large, Klenow) fragment, *T. aquaticus* DNA polymerase, or T7 DNA polymerase; for reverse transcriptase, follow the manufacturer's suggestions; this is usually 5 to 20 U of avian reverse transcriptase or 200 U of murine reverse transcriptase).
3. Mix by pipetting the solution up and down, followed by brief microcentrifugation. This reaction is now referred to as the TPK mixture. The total volume is 13 μl.

Step D: Addition of Deoxy/Dideoxy Nucleotides

1. Add 3.0 μl of the TPK mixture into the tube labeled A. Changing pipette tips each time, repeat this addition to the C, G, and T tubes.
2. After each addition, mix the solutions by gently pipetting the solution up and down.
3. Incubate from 25 to 50°C for DNA polymerase I (large) Klenow fragment, 37 to 75°C for *T. aquaticus* DNA polymerase, 37°C for reverse transcriptase, or 25 to 42°C for T7 DNA polymerase. Allow the reaction to proceed for 10 to 15 min.
4. Add 1 μl of dNTP Chase mixture (see Appendix B: Composition of DNA Sequencing Reagents) to each microcentrifuge tube.
5. For reactions at temperatures greater than 37°C, add 0.5 μl more of the polymerase to each tube.
6. Mix by gently pipetting the solution up and down.
7. Incubate at the appropriate polymerase temperature (25 to 75°C) for 10 to 15 min.
8. Add 6 μl of stop/loading dye solution (see Appendix B: Composition of DNA Sequencing Components).
9. Before loading the gel, incubate the samples in a boiling water bath for 2 min.
10. Load 2 to 3 μl of each reaction into separate adjacent wells in the sequencing gel (see Polyacrylamide Gel Electrophoresis, Section II.B).

b. Double-Stranded DNA Sequencing

The dideoxy sequencing method[3-5] can be used to sequence double-stranded templates, such as plasmid DNA, as well as large linear double-stranded DNA, such as bacteriophage λ DNA.[35,100,104,138-158,160-163] Although this protocol recommends alkaline denaturation, the plasmid can be denatured by boiling.[80,163-167] Linearization of certain double-stranded templates may avoid the appearance of artifact bands[15,85,100,143,158,169] or be required for certain applications.[36,170-173,195-198]

Step A: Denaturation of the Template

1. Dissolve 2 μg or purified double-stranded template DNA (~0.5-pmol template) in 20 μl of distilled water or 10 m*M* Tris, 1 m*M* EDTA pH 7.5.
2. Add 2 μl of 2 *M* NaOH, 2 m*M* EDTA. Gently mix by drawing up and down with a pipette. Incubate at room temperature for 5 min.
3. Place on ice and quickly add the following in the order listed:

   - 7 μl distilled water
   - 7 μl 3 *M* NaOAc, pH 6.0*
   - 75 μl 95% EtOH (4°C)

   Mix thoroughly between each addition by drawing the solution up and down with the pipette.

4. Place the microcentrifuge tube on dry ice for 10 to 15 min to precipitate the DNA. Spin in a microcentrifuge at 4°C for 10 min.
5. Carefully remove the EtOH layer with a Pasteur pipette. Fill the microcentrifuge tube with 70% EtOH and vortex briefly. Microcentrifuge the pellet for 10 min at 4°C and carefully remove the EtOH layer as previously described.

* The pH of the template solution should be checked at this step by spotting 1 μl on pH paper. Excess NaOAc buffer may be added in 1-μl increments until the pH is 7.0 or lower.

6. Completely dry the pellet in a vacuum centrifuge for 10 min. The pellet may be stored at −20°C for several weeks before sequencing.

Step B: Preparation of Annealed Template and Primer

1. To the lyophilized denatured template, add 8 μl of water, followed by 1.0 μl of the appropriate 10X sequencing buffer (see Appendix B: Composition of DNA Sequencing Reagents). Mix and wash the sides of the tube thoroughly by drawing the solution up and down with the pipette.
2. Add 1 μl (~1 pmol) of an appropriate primer solution (approximately 2.5 μg of primer resuspended in 500 μl of water). Mix gently by drawing the solution up and down with a pipette.

Note: There should be approximately 1 to 5 primer molecules per template molecule (see Appendix E: Molar Ratio Calculations).

3. Incubate for 30 min at 37°C for 16- to 20-mers and 42°C for primers larger than 20-mers. Highly A-T-rich primers must be annealed at lower temperatures. Maximum temperature for a specific primer may be calculated by adding 2°C for every A-T and 4°C for every G-C, and then subtracting 15% from the total amount, since the 10X buffer is low ionic strength. If longer annealing times are performed (e.g., 45 to 60 min), they should be done at the higher temperatures to prevent annealing of the primers at potential secondary priming sites. Leave the priming reaction at the specific annealing temperature until ready to add polymerase.

Step C: Preparation of Deoxy/Dideoxy Reaction Tubes

1. Label four 0.5-ml microcentrifuge tubes A, C, G, and T, respectively. Choose the appropriate set of four standard sequencing mixes (outlined in Appendix B: Composition of DNA Sequencing Reagents) for the radioactive label and polymerase being used. These mixtures can be made from standardized deoxy/dideoxy stock solutions (see Appendix C: Dilution Protocols from Stock Solutions of Nucleotides).
2. Add 3 μl of the A mixture to the bottom of the A tube, and 3 μl of the C, G, and T mixtures to the bottom of the C, G, and T tubes, respectively.

Step D: Addition of Label and Polymerase

1. To the annealing reaction, add 2 μl of either [α-^{32}P] dATP or [α-^{35}S] dATP (500 to 3000 Ci/mmol, 10 mCi/ml water).
2. Add 1.0 μl of polymerase (5 U) (3000 to 5000 U/ml) (DNA polymerase I large, Klenow) fragment, *T. aquaticus* DNA polymerase, or T7 DNA polymerase; for reverse trancriptase, follow the manufacturer's suggestions; this is usually 5 to 20 U of avian reverse transcriptase or 200 U of murine reverse transcriptase).
3. Mix by pipetting the solution up and down, followed by brief microcentrifugation. This reaction is now referred to as the TPK mixture. The total volume is now 13 μl.

Step E: Addition of the Deoxy/Dideoxy Nucleotides

1. Add 3.0 μl of the TPK mixture to the tube labeled A. Changing pipette tips each time, repeat this addition to the C, G, and T tubes.
2. After each addition, mix the solutions by gently pipetting the solution up and down.

3. Incubate from 25 to 50°C for DNA polymerase I large (Klenow) fragment, 37°C for reverse transcriptase, 37 to 75°C for *T. aquaticus* DNA polymerase, or 25 to 42°C for T7 DNA polymerase. Allow the reaction to proceed for 10 to 15 min.
4. Add 1 μl of dNTP Chase mixture (see Appendix B: Composition of DNA Sequencing Reagents) to each microcentrifuge tube.
5. For reactions run at temperatures greater than 37°C, add 0.5 μl more of the polymerase to each tube.
6. Mix by gently pipetting the solution up and down, followed by brief microcentrifugation.
7. Incubate at the appropriate polymerase temperature (25 to 75°C) for 10 to 15 min.
8. Add 6 μl of stop/loading dye solution (see Appendix B: DNA Sequencing Reagents).
9. Before loading the gel, incubate the samples in a boiling water bath for 2 min.
10. Load 2 to 3 μl of each reaction into separate adjacent wells in the sequencing gel (see Polyacrylamide Gel Electrophoresis, Section II.B).

c. End-Labeled Sequencing the Single- or Double-Stranded Template

End-labeled primer sequencing is primarily used for double-stranded template sequencing of very large templates (such as λgt11) or for those templates which give less than superior sequencing results. Primers may be end-labeled with ^{32}P or ^{35}S. ^{32}P end-labeled primers (not the sequencing reactions) may be stored up to 1 week before use. End-labeled primers may already be available in laboratories set up for custom primer synthesis used in "primer walking" ("look and leap") approaches[1,51,82,102] since synthesized primers may be quality-control tested by end-labeling and sizing on polyacrylamide gels. The end-labeled sequencing protocol described below (except for the labeling procedure) is also compatible with primers labeled on the 5′ end with fluorescent dyes.[111,124-129] *T. aquaticus* DNA polymerase, T7 polymerase, reverse transcriptase, or DNA polymerase I large (Klenow) fragment can be used in this method.

Step 1A: Kinasing the Primer with [γ-^{32}P] rATP*

1. In a 0.5-ml microcentrifuge tube, mix together the following:

- 2.5 μl 10X T4 polynucleotide kinase buffer (see Appendix B: Composition of DNA Sequencing Reagents)
- 7 μl [$\gamma^{32}P$] rATP, 3000 Ci/mmol, 0.00165 μmol/ml water (11.5 pmol)
- 1 μl T4 polynucleotide kinase (10,000 U/ml)
- 13.5 μl water
- 1 μl (10.5 pmol of 24-mer) of stock primer solution (stock = 2.5 μg primer resuspended in 30 μl of water)

2. Allow the reaction to proceed at 37°C for 30 to 60 min.
3. Heat kill the reaction at 95°C for 5 min. Briefly microcentrifuge at room temperature.

Step 1B: Kinasing the Primer with [γ-^{35}S] rATP*

1. In a 0.5-ml microcentrifuge tube, mix together the following:

- 2.5 μl 10X T4 polynucleotide kinase buffer (see Appendix B: Composition of DNA Sequencing Reagents)

* The above kinase reactions are enough for sequencing 20 templates using 0.50 pmol or primer. These reactions may be stored for 1 week at −20°C.

- 2 μl [γ-^{35}S] rATP, 1300 Ci/mmol, 0.0078 μmol/ml (13 pmol)
- 18.5 μl water
- 0.5 μl (5 U) T4 polynucleotide kinase (10,000 U/ml) *per hour* (total amount = 20 U)
- 1 μl (10.5 pmol or 24-mer) of stock primer solution (stock = 2.5 μg primer resuspended in 30 μl water)

2. Allow the reaction to proceed at 37°C for 4 h.
3. Heat kill the reaction at 95°C for 5 min. Briefly microcentrifuge at room temperature.

Step 2: Sequencing the Single- or Double-Stranded Template

The end-labeled primer can now be annealed to a single-stranded or double-stranded template, as described in the previous standard DNA sequencing protocol sections. There is no need to remove the unincorporated γ-rATP from the end-labeling reaction, as it will not incorporate in the sequencing reaction.

Similarly, the sequencing reactions follow the previous standard DNA sequencing protocol, with the following exceptions:

1. The labeled nucleotide should be omitted from the reactions (and replaced with an equal volume of water.
2. End-labeled standard sequencing mixes must be substituted for the standard sequencing mixes (see Appendix B: Composition of DNA Sequencing Reagents).

2. "Extension" Sequencing Reactions

Standard sequencing reactions usually resolve less than 500 nucleotides from a priming site using standard acrylamide gels run for 2 to 6 h. Obtaining further sequence information usually requires subcloning (such as "shotgun" cloning,[3,43,51,62-65,57] directional *in vitro* or *in vivo* deletions,[46,51,54,58,59,63-65,70-101] or synthesis of new primers from a newly derived DNA sequence downstream from a previous sequencing reaction[1,51,82,102]). Often, coordinate usage of these techniques is required to obtain the necessary sequence information. The time involved in a large sequencing project can be greatly reduced by increasing the number of nucleotides which can be accurately determined from a single priming step. The extension protocol allows accurate sequence information to be obtained in excess of 750 nucleotides from a single priming site.[25,35,103,105,109,111-115,131,159]

The procedure involves primed DNA synthesis, first in the presence of limiting concentrations of all four dNTPs, followed by a second step in the presence of relatively high concentrations of ddNTPs and additional low concentrations of dNTPs. In the first step, primed DNA synthesis is initiated and DNA chains are extended until the dNTP pool is depleted. In the second step, synthesis continues (due to the presence of an influx of dNTPs), but terminates after the incorporation of a ddNTP. By regulating the relative concentration of dNTPs at either step, various sized chain-length extensions can be generated from a given set of four reactions.

a. Extension Sequencing of Single-Stranded DNA Templates

Step A: Preparation of Annealed Template and Primer

1. Prime the template, as previously described in the standard single-stranded template sequencing protocol.

Step B: Addition of Label and Polymerase (Extension Reaction)

1. To the annealed primer and template, add 2 μl of extension mix NK (see Appendix B: Composition of DNA Sequencing Reagents).
2. Add 2 μl of either [α-^{32}P] dATP or [α-^{35}S] dATP (500 to 3000 Ci/mmol, 10 mCi/ml).
3. Add 1 μl (5 U) of DNA polymerase I large (Klenow) fragment, T7 DNA polymerase, or *T. aquaticus* DNA polymerase. For reverse transcriptase, refer to the manufacturer's instructions; this is usually 5 to 20 U of avian reverse transcriptase or 200 U of murine reverse transcriptase. Mix thoroughly by drawing the solution up and down with a pipette.
4. Incubate for 5 min at 25 to 50°C for DNA polymerase I large (Klenow) fragment, 37°C for reverse transcriptase, 37 to 75°C for *T. aquaticus* DNA polymerase, or 25 to 42°C for T7 DNA polymerase. The initial 30 s of any given reaction should not be greater than 55°C or greater than the maximum annealing temperature for the specific primer. This mixture is referred to as the TPK mixture. The total volume is 15 μl.

Step C: Termination Reaction

1. While the extension reaction proceeds, label four microcentrifuge tubes A, C, G, and T.
2. Choose the appropriate sequencing mixes. Use either the set of four short extension-termination mixes for sequencing 10 to 500 bases or the set of four long extension-termination mixes for sequencing greater than 500 bases (see Appendix B: Composition of DNA Sequencing Reagents).

Note: These mixtures can be made from standardized deoxy/dideoxy stock solutions (see Appendix C: Dilution Protocols from Stock Solutions of Nucleotides).

3. Add 3 μl of the A termination mixture to the bottom of the A tube and 3 μl of the C, G, and T termination mixtures to the bottom of the C, G, and T tubes, respectively.
4. Add 3.2 μl of the TPK mixture to the tube labeled A. Changing pipette tips each time, repeat this addition to the C, G, and T tubes.
5. After each addition, mix the solution by gently pipetting the solution up and down.
6. Incubate for 5 min from 25 to 50°C for DNA polymerase large I (Klenow) fragment, 37°C for reverse transcriptase, 37 to 75°C for *T. aquaticus* DNA polymerase, or 37°C for T7 DNA polymerase.
7. Add 6 μl of stop/loading dye solution (see Appendix B: Composition of DNA Sequencing Reagents).
8. Before loading the gel, incubate the samples in a boiling water bath for 2 min.
9. Load 2 to 3 μl of each reaction into separate adjacent wells in the gel (see Polyacrylamide Gel Electrophoresis, Section II.B).

b. Extension Sequencing of Double-Stranded DNA Templates

The extension-termination protocol can be used to sequence double-stranded templates, such as plasmid DNA, as well as large linear double-stranded DNA, such as bacteriophage λ DNA.[35,100,104,138-162] Although this protocol recommends alkaline denaturation, the plasmid can be denaturated by boiling.[80,163-167] Linearization of certain double-stranded templates may avoid the appearance of artifact bands,[15,100,143,168,169] or may be required for certain applications.[37,170-173]

Step A: Denaturation of the Template

1. Denature the double-stranded template, as described in the standard double-stranded DNA sequencing protocol.

Step B: Preparation of Annealed Template and Primer

1. Anneal the primer, as described in the standard double-stranded DNA sequencing protocol.

Step C: Addition of Label and Polymerase (Extension Reaction)

1. To the annealed primer and template, add 2 μl of extension mix NK (see Appendix B: Composition of DNA Sequencing Reagents).
2. Add 2 μl of either [α-^{32}P] dATP or [α-^{35}S] dATP (500 to 3000 Ci/mmol, 10 mCi/ml.
3. Add 1 μl (5 U) of DNA polymerase I large (Klenow) fragment, T7 DNA polymerase, or *T. aquaticus* DNA polymerase. For reverse transcriptase, refer to the manufacturer's instructions; this is usually 5 to 20 U of avian reverse transcriptase or 200 U of murine reverse transcriptase. Mix thoroughly by drawing the solution up and down with a pipette.
4. Incubate for 5 min at 25 to 50°C for DNA polymerase I large (Klenow) fragment, 37°C for reverse transcriptase, 37 to 75°C for *T. aquaticus* DNA polymerase, or 25 to 42°C for T7 DNA polymerase. The initial 30 s of any given reaction should not be greater than 55°C or greater than the maximum annealing temperature for the specific primer. This mixture is referred to as the TPK mixture. The total volume is 15 μl.

Step D: Termination Reaction

1. While the extension reactions are proceeding, label four microcentrifuge tubes A, C, G, and T.
2. Choose the appropriate sequencing mixes. Use either the set of four short extension-termination mixes for sequencing 10 to 500 bases or the set of four long extension-termination mixes for sequencing greater than 500 bases (see Appendix B: Composition of DNA Sequencing Reagents). These mixtures can be made from standardized deoxy/dideoxy stock solutions, as described in Appendix C: Dilution Protocols from Stock Solutions of Nucleotides).
3. Add 3 μl of the A termination mixture to the bottom of the A tube and 3 μl of the C, G, and T termination mixtures to the bottom of the C, G, and T tubes, respectively.
4. Add 3.5 μl of the TPK mixture to the tube labeled A. Changing pipette tips each time, repeat this addition to the C, G, and T tubes.
5. After each addition, mix the solution by gently pipetting the solution up and down.
6. Incubate for 5 min at 25 to 50°C for DNA polymerase I large (Klenow) fragment, 37°C for reverse transcriptase, 37 to 75°C for *T. aquaticus* DNA polymerase, or 37°C for T7 DNA polymerase. The initial 30 s of any given reaction should not be greater than 55°C or greater than the maximum annealing temperature for the specific primer.
7. Add 6 μl of stop/loading dye solution (see Appendix B: Composition of DNA Sequencing Reagents).
8. Before loading the gel, incubate the samples in a boiling water bath for 2 min.
9. Load 2 to 3 μl of each reaction into separate adjacent wells in the gel (see Polyacrylamide Gel Electrophoresis, Section II.B).

c. End-Labeled Extension Sequencing of Single- and Double-Stranded Templates

End-labeled primers can be used with the "extension" sequencing protocol in exactly the same manner as they are used with standard sequencing protocols. Extension sequencing protocols may be modified to include end-labeled primers by transferring a 5′ label to the primer with T4 polynucleotide kinase and replacing the labeled nucleotide from the deoxy/dideoxy nucleotide reaction mix with an equal volume of water. End-labeled sequencing may be used on either single- or double-stranded templates. Primers may be end-labeled with either [γ-^{32}P] rATP or [γ-^{35}S] rATP. The sequencing protocol (except for the labeling procedure) is also compatible with primers labeled on the 5′ end with fluorescent dyes[111,124-129] or biotin.[98]

1. The procedure for end labeling the primer is identical to that described for the end-labeled standard sequencing.
2. The end-labeled primer can be annealed, exactly as described in the end-labeled standard sequencing, to a single- or double-stranded template. If a double-stranded template is used, the templates must be denatured before annealing to the primer.
3. Extension and termination sequencing reactions can be conducted exactly as in the extension-termination protocols described above, with the only changes being the omission of labeled nucleotide (replacing with an equal volume of water) and the use of extension mix K in lieu of extension mix NK, in the extension reaction step (see Appendix B: Composition of DNA Sequencing Reagents).

B. POLYACRYLAMIDE GEL ELECTROPHORESIS

There are a wide variety of commercially available DNA sequencing gel electrophoresis boxes, and each comes with detailed instructions concerning its specific set-up. General methods of gel handling and processing are described in various publications.[121,174-177] A few important considerations, with respect to sequencing gel solutions, and some important helpful hints are provided below.

1. Preparing and Pouring the DNA Sequencing Gel

Numerous techniques are used to prepare the sequencing gel, depending upon the apparatus used and the individual sequencer's experience. To begin, the glass plates should be coated with silicon, following the manufacturer's instructions. This enables easy removal of the polyacrylamide gel from the glass plate sandwich, which is necessary for autoradiography. Immediately before use, meticulously clean the plates using a soapy sponge, followed by thorough rinses in distilled water and then alcohol. Carefully wipe the plates dry to remove any residual particles.

Form the gel sandwich by using the two glass plates and two 0.15 to 0.8-mm-thick × 1-cm-wide polystyrene or Teflon side spacers (cleaned as above, but not siliconized) and tape the plates together (sides and bottom) using 3.1-cm-wide yellow vinyl electrical tape. Alternatively, the plates may be clamped together with "book-binder" clamps. To prevent leakage, tape the bottom of the sandwich a second time, forming "hospital bed corners", or insert a bottom spacer and clamp, as above. A 25 to 64-well, 0.15- to 0.8-mm-thick polystyrene or Teflon comb of thickness corresponding to the side spacers being used ("shark's tooth" or regular), cleaned as above, should be tested for a tight fit by inserting it at the top of the sandwich and then should be removed before pouring the gel. The comb and the top of the spacers should correspond in thickness to one another.

Sequencing experiments utilizing different acrylamide gel compositions are useful under some circumstances. For example, 4 to 6% gels give a tighter spacing of bottom gel bands, than do 8% gels, resulting in the ability to obtain more sequence data per gel. This may be somewhat of a compromise, however, since the bands may be slightly fuzzier than observed in 8% gels.

Buffer-gradient gels,[120] electrolyte-gradient gels,[178] or "wedge" gels using tapered spacers[77,179] in combination with ultrathin gels[3,120,180,181] also increase the amount of sequence obtained from a single gel. Combinations of different acrylamide concentrations and gel types may also be useful.[133,134,171,182,183] For example, gels in which the bottom half is 6% acrylamide and the top half is 8% acrylamide, in which the entire gel is a buffer gradient[121] or where three sections of the gel are poured with different acrylamide concentrations,[184] have been utilized. Creative gel combinations may greatly minimize the number of gels necessary for sequencing projects. These gels are generally as readable and as easily prepared (with practice) as those previously described.

The use of the novel gel matrix Hydrolink™ (AT Biochem) has currently not been personally investigated in sufficient detail to describe all of its electrophoretic properties; nevertheless, the pouring and processing of the gels is reported to be similar to standard acrylamide gels, with the notable exceptions that no urea is necessary (the buffer contains a sufficient denaturant), the gels are less easily torn during handling, postelectrophoresis fixation is not necessary for ^{35}S reactions, and the gel percentage required to achieve the same electrophoretic resolution differs from that of a standard acrylamide gel.

The following protocol is for the preparation of a 60-ml 8% acrylamide gel solution. Use Appendix D: Polyacrylamide Sequencing Gel Mixtures, as a guide for the preparation of other gel types and concentrations.

1. Make a 38:2 (w/w) stock solution of acrylamide/bis-acrylamide by dissolving 38 g of ultrapure acrylamide and 2 g of bis-acrylamide in a final volume of 100 ml of deionized water, stirring with low heat. The temperature must be kept below 55°C or the acrylamide may be degraded. Store the solution in a brown bottle at 4°C. There are commercially available "premixed" or "preweighed" ultrapure acrylamide/bis-acrylamide solutions and powders. These eliminate the necessity of making up stock solutions and the potential exposure to the two neurotoxins. The final concentration of acrylamide in these solutions should be 40 or 50%, although 50% solutions often require resuspension of crystalized material before use when stored at 4°C.
2. Add 25.2 g of ultrapure urea, 12 ml of 5X TBE buffer (1X TBE = 0.1 *M* Tris, 0.08 *M* boric acid, 0.001 *M* disodium EDTA, pH 8.4) and 12 ml of a 40% 38:2 acrylamide/bis-acrylamide stock solution to a final volume of 60 ml in deionized water.
3. Gently mix and heat the solution (below 55°C), and when all solids have dissolved, cool the solution to 20°C by chilling on ice. Gel mixes, prepared as described above, normally need not be filtered or degassed. If solutions retain particulate matter after mixing, filtering (and degassing to prevent small bubbles from forming during electrophoresis) may be necessary.
4. After the solution reaches 20°C, transfer 0.6 ml to a 1.5-ml tube to test polymerization time. Add 0.6 μl of TEMED and 6.5 μl of 10% ammonium persulfate (not more than 1 week old and stored at 4°C) and mix the solution by inverting it several times. Polymerization should occur within 5 min. If the polymerization time differs substantially from 5 min, test another 0.6-ml sample using more or less TEMED and/or ammonium persulfate.
5. If the first test conditions were correct, add 650 μl of 10% ammonium persulfate and 60 μl of TEMED to the 60-ml acrylamide mixture. Swirl the solution for thorough mixing and immediately pour the gel using a 25-ml pipette to deliver the solution between the glass plates of the glass plate sandwich previously prepared. Alternatively, the acrylamide solution can be mixed in a 250-ml sidearm flask and the solution poured between the glass plate sandwich through the sidearm.
6. Hold the plates at a low angle during pouring, tilting them slightly so that the gel solution moves slowly down one side. Air bubbles can be prevented by pouring the

gel solution at a steady pace while maintaining constant contact between the fluid being poured and the fluid already between the plates. Any air bubbles that form can be removed by tapping the glass plates behind the bubbles or by rocking the plates until the bubbles move to the top of the gel solution.

7. Insert the comb, beginning at one notched corner of the sandwich, and slowly position it to prevent trapping air bubbles. The comb should be inserted far enough into the gel solution so that the final depth of the wells is about 2 mm. If a ''shark's tooth'' comb is being used, position it in the gel, flat side down, about 2 mm into the gel solution. Position the sandwich horizontally on the counter top with the top of the gel elevated with a rubber stopper (about 5 cm high). Pour extra acrylamide gel solution onto the comb to ensure full coverage of the comb/plate interface.
8. Clamp the sides of the gel together with two metal binder clips (one on each side) positioned about 4 cm from the top of the gel. The clips should be clamped over the side spacers and not over the gel itself.
9. Gels should be allowed to polymerize at least 2 h before use, with clamps placed in the same relative positions in which they will be in the sequencing apparatus. They can be stored for 48 h at room temperature in sealed plastic bags containing a small volume of 1X TBE gel buffer, with the top of the gel sandwich resting about 5 cm above the bottom.
10. Immediately before electrophoresis, remove the tape or spacer from the bottom of the sandwich, remove the topmost side clamps, and flood the top of the gel around the comb with 1X TBE. Remove extraneous polyacrylamide from around the comb with a razor blade and remove the comb, beginning from one side and slowly working across to the other side so that buffer replaces the comb teeth and prevents the wells from collapsing. Take care not to damage the wells when removing the comb. If using a ''shark's-tooth'' comb, clean and reinsert it ''teeth'' down, until the points of the teeth barely touch the polymerized gel.
11. Rinse the glass surfaces free of excess acrylamide, dry the plates, and clamp them into the electrophoresis apparatus. Fill the upper buffer reservoir to 1 cm above the gel wells. Providing there are no upper buffer chamber leaks, fill the lower buffer reservoir and check to ensure that no air bubbles are trapped below the glass plates. Rinse the wells thoroughly with 1X TBE using a Pasteur pipette or a 10-ml syringe to remove urea leached from the gel and any residue acrylamide material.

Note: Level, ''non-smiling'' electrophoretic bands require careful parallel alignment of the buffer levels with the bottom of the gel sandwich and the bottom of the gel wells. The alignment of the bottom buffer can be a particular problem due to capillary action, since the bottom buffer tends to climb up between the back glass plate and the lower buffer reservoir when the gel sandwich is clamped to the gel box. This ''smiling'' can be prevented in a number of ways. First, a small piece of folded filter paper can be inserted on each side between the gel sandwich and the electrophoresis box (about 5 cm above the level of the bottom buffer). This forces the bottom of the sandwich approximately 2 mm away from the gel box, allowing the bottom buffer to form a level surface. Second, the gel sandwich should be attached to the gel box with only two metal binder clips located in the same positions in which they were previously placed while the gel was polymerizing. Another important factor in preventing ''smiling'' is to ensure even conduction of the joule heat generated during electrophoresis. An aluminum plate (0.4 cm thick, 34 $\times$ 22 cm) should be clamped onto the front glass plate with the same two metal clips used to hold the gel sandwich to the gel box. The aluminum plate helps to prevent the formation of ''hot'' and ''cold'' spots in the gel (and helps to prevent the glass plates from cracking) by evenly distributing the heat generated during electrophoresis. The aluminum plate must be carefully positioned so that it does not touch any buffer during electrophoresis.

2. Loading and Running the DNA Sequencing Gel

The most inexpensive tool for loading DNA sequencing reactions into the thin wells of a sequencing gel is to use a 20-μl micropipettor with a modified micropipette tip, made by heating its end over a flame until it can be drawn out with a pair of forceps to form a very small diameter bore. Once the plastic is cool, trim the tip with a razor blade at a point where the diameter of the tip is about 0.15 mm. Alternatively, several "DNA sequencing pipettors", utilizing disposable ultrathin tips, are now commercially available. These tips do not break, rarely clog, and can be reused without "carryover" radioactivity between samples when rinsed as described below. Because these loading tips are so slender, DNA sequencing reactions can be easily layered onto the bottom of the wells.

1. Heat the DNA sequence samples (in stop/loading dye solution) in their microcentrifuge tubes (or in the microtiter tray), with the caps closed, in a boiling water bath for 2 min. If multiple sequence loadings are to be performed, it is suggested to remove and denature 2 to 3 μl from each the reactions to prevent excessive heat degradation of the rest of the sample when it is reheated, due to the presence of the DTT and formamide in the reactions.
2. Just prior to loading the gel, rinse the wells again with 1X TBE to remove excess urea.
3. Remove the reaction tubes from the water bath and load 2 to 3 μl of each sample per well. Between samples, rinse the loading tip twice with 1X TBE from the lower reservoir. With practice, 25 to 64 samples can be loaded in less than 5 min, thereby eliminating any need for electrophoresing one set of samples into the gel before loading the next set.
4. After the samples are loaded, the gels are run at 47 to 60 W constant power (65°C) (about 1300 V, depending upon the volume of buffer used, the percentage of acrylamide, and the gel box configuration). If electrophoresis is stopped just before the bromophenol blue dye runs off an 8% polyacrylamide gel, sequences just downstream from the priming site can be read at the bottom of the autoradiograph. Longer runs allow more sequence to be read from the top of the gel. Reloading gels in adjacent lanes, in conjunction with monitoring the positions of the bromophenol blue and xylene cyanol running dyes, allows a large contiguous sequence to be determined. Long gel electrophoresis runs (up to 12 h), buffer-gradient gels, 4 to 6% acrylamide, or "wedge" gels can extend the resolution to as many as 1000 nucleotides from the priming site.

3. Processing the DNA Sequencing Gel

DNA sequencing employing [α-^{35}S] dATP requires fixing and drying the acrylamide gel following electrophoresis in order to remove the urea and water which quench the weak β emissions. This procedure also improves clarity for ^{32}P sequencing reactions. Many mishaps occur during this part of the protocol because the gel is very thin and fragile and can easily be torn. Simple procedures for processing the gel, requiring a minimum of handling, have been described.[121,173,176,177] Hydrolink™ gels containing ^{35}S do not require soaking before drying, since the gels do not contain urea; in addition, these gels have been reported to have increased strength to resist tearing during handling.

C. HELPFUL HINTS

Several reports[51,82,119,121,123,174,175,185] discuss a variety of potential difficulties which may be encountered when performing dideoxy DNA sequencing reactions. These problems include dark-lane smearing, poor gel resolution, and nonexistent, faint, short, or artifact bands. In addition to the remedies suggested in previous reports, the following suggestions may prove to be helpful:

1. Autoradiograph resolution can be improved by ensuring the X-ray film and gel are tightly abutted during exposure in the X-ray cassette. This is achieved by layering several sheets of paper towels in the cassette before enclosing the dried gel and X-ray film. Care should be taken not to overpack the cassette, which may cause light leaks around the edges of the cassette.
2. Dideoxy sequencing is sensitive to primer/template molar ratios. The best ratio appears to be about 1:1. Even changing the ratio to above 5:1 may result in increased shadow bands and difficulty in reading some portions of the sequence.
3. Electrophoresis at 47 to 60 W constant power (55 to 65°C) for long runs sometimes results in a loss of resolution at the top of the gel. This phenomenon does not result from a pH gradient within the gel since recirculation of top and bottom buffers does not improve the resolution. The "loss of resolution" begins at the top of the gel and migrates towards the bottom, as a function of electrophoretic running time. Long-run buffers have been reported to correct this problem,[63,75] although none, in our hands, appear to totally ameliorate it. This appears to be due to the glycerol present in the sequencing DNA polymerase. Diluting the polymerase, when possible, into 1X sequencing buffer, before addition to the sequencing reactions, may eliminate this loss of resolution.
4. To read DNA sequences unambiguously beyond 300 nucleotides, it is essential to prevent "smiling" or "frowning" bands. The two most important factors are even heating of the gel and horizontal buffer levels. If the gel setup described previously is followed carefully, this should not be a problem. When the gel is processed after electrophoresis, care must be taken not to stretch the gel, since this causes distortion of the bands on the autoradiograph.
5. Loading duplicate lanes for each set of four-base reactions may, in theory, facilitate the correct ordering of bands, especially in higher regions of the gels (in order to unambiguously read a sequence beyond 300 to 350 nucleotides). Each base-specific reaction can be located next to the other three when loaded in the order: GATCGTAC. In practice, however, it is seldom necessary to "duplicate load" and it uses eight gel lanes rather then four for the same sequence. It is more beneficial to reload the same sequence after a time period of electrophoresis and run the gel longer so that overlap of the sequence occurs (a short and long run of the same four lanes, as opposed to a "duplicate loaded" short or long run).
6. When "shotgun cloning" or when using restriction endonuclease digests to clone fragments into a sequencing vector, multiple occurrences of the same insert/vector combination will likely be recovered. To prevent sequencing and analysis of the same (informational) template, sequencing reactions involving only one of the four nucleotide chain extension reactions (for example, the T-reaction, "T-tracking"[12,43] from each template) can be performed and electrophoresed side-by-side. Identical banding patterns will be observed among identical templates, and duplicate or weak templates can be eliminated from further analysis.
7. Stock solutions of 38:2 acrylamide/bis-acrylamide are stored at 4°C in brown or foil-wrapped bottles. Stock solutions should be good for several months. If an ammonia smell is detected, the solution should be discarded. When dissolving the acrylamide and bis-acrylamide, the solution should not be heated to greater than 55°C and the solution should be stirred constantly to maintain a uniform temperature. Filtering is unnecessary when using ultrapure reagents.
8. Should recurrent problems of gels "sticking" to glass plates occur, resiliconization of the plates may be necessary. This usually occurs after 20 to 25 electrophoresis runs. A more thorough cleaning of the glass plates, as previously described, may also alleviate such difficulties.

9. Excessive heating of the template-primer annealing reaction in the presence of DTT (present in the 10X sequencing buffer) often leads to smeary sequencing resolution. The annealing reaction previously described is recommended unless otherwise counter-indicated, such as for very high "G + C" or palindromic sequences. Similarly, excessive boiling of the completed reactions in the presence of formamide (in the stop/loading dye) leads to DNA chain breakage and smeared sequences.
10. Pre-electrophoresis of sequencing gels, to remove the salt front, is usually not necessary. If the salt front does interfere with DNA sequence interpretation on short electrophoresis runs, prerun the gels for 30 min before loading the samples. The pre-electrophoresis can also serve to prewarm the gel if hotter gels are required to resolve compressed sequence regions.

D. INTERPRETATION OF GEL ELECTROPHORESIS PATTERNS

1. If one more sequencing lanes stops or fades out before the others, the wrong dNTP/ddNTP ratio was likely used in that sequencing reaction. This may be due to depletion of one nucleotide due to the frequency at which it occurs in that specific sequence. Similarly, a sequence may not begin until fairly high up in the ladder, relative to the others. In either case, readjust the dNTP/ddNTP ratio for that lane. Adjustment may be an addition of ddNTP (in the appropriate buffer) for ascertainment of sequencing information lower in the gel (closer to the primer) or may require dilution of the ddNTP concentration, while maintaining the concentration of all other components, by diluting the reaction mixture with a solution containing all components in the dNTP/ddNTP mixture except the ddNTP.
2. A heavy dark background (often with a "ladder-like" effect) in all lanes with an unreadable sequence is usually due to template impurity or nuclease contamination of the reactions. This is more often observed with double-stranded templates than with single-stranded ones and is likely due to polymerase synthesis initiating at nicks in the template (nick translation) instead of at the sequencing primer. Alternatively, this effect may be due to contamination of the template with high salt, PEG, or alcohol, or may result from an insufficiently purified template.
3. Dark background smears are sometimes observed in the "A" track that appear to be associated with different lots of α-labeled dATP, especially when the lot is nearing the end of one half-life. This problem can be mitigated by lowering the amount of the "A" TPK mix in the "A" tube (i.e., add 2 μl of TPK mix to 3 μl of "A" dideoxy/deoxy mix).
4. Dark shadow bands occur infrequently and may become more of a problem with long exposure times. Keeping exposure times short often makes autoradiographs easier to read. The finer bands obtained with short exposures allow more sequence data to be read from the top of the autroradiograph. "T" and "A" bands can leave faint shadow bands in the "G" and "C" tracks, and "A" bands can also give faint shadow bands in the "T" track. Occasionally, the "A" track can give dark shadow bands in the other three lanes. By paying careful attention to spacing between nucleotides and consideration of points 5, 6, 7, and 8 which follow, it is usually easy to determine what is a shadow band and what is not.
5. Using DNA polymerase I (Klenow fragment), the first "A" in a run of "A"s is usually the darkest; the second "G" in run of "G"s is always very dark when the run of "G"s is preceded by a "T", and the second "C" in a run of "C"s is always the most intense, while the first "C" is very light[3] and the third "C" is frequently dark as well. Other polymerases may show varying amounts and intensities of these patterns or different patterns altogether. The use of a manganese buffer, with adjusted

dNTP/ddNTP ratios can also affect these patterns, reducing the banding to nearly equal intensities.[115]

6. Stretches of single-stranded DNA which are rich in ''G'' + ''C'' or ''A'' + ''T'', or which are extensively palindromic, may be particularly difficult to sequence accurately due to formation of stable secondary structures. Occasionally, the banding pattern abruptly terminates at or shortly before such sequences and is very faint after them. In such cases, the template secondary structure in the sequencing (TPK) reaction prevents the polymerase from advancing past the ''foldback'' position. Performing sequencing reactions at 42 to 75°C, the utilization of additional DNA polymerase I large (Klenow) fragment, of using reverse transcriptase, T7 DNA polymerase, or *T. aquaticus* DNA polymerase may help alleviate such problems.
7. One other consequence of sequencing regions with secondary structure may be autoradiographic bands that are so tightly spaced (''compressed'')[4,5,132] that the sequence cannot be determined. These compressions occur as electrophoretic artifacts rather than actual sequencing reaction artifacts.[3] For example, M13mp7 contains a very unusual 42-base palindromic sequence which is very difficult to sequence due to the compression (pile up) of bands. The first 21 bases of the insertion sites are complementary to the second 21 bases and, hence, a very stable base-paired loop can be formed. When electrophoresed at 1300 V, the sequence can be read from the primer up to the *Pst*I site. However, the second half of the palindrome, from the *Pst*I site to the second *Eco*RI site, is compressed on the gel and is uninterpretable. The sequence picks up just after the second *Eco*RI site. When the same set of reactions is run on a gel at 1650 V (higher temperature), the sequence can be read up to the second *Bam*HI site, although bands are still compressed between the *Bam*HI and *Eco*RI sites. Compressions are often manifest as dark shadow bands or missing, extraneous, or displaced bands, as well as compressed sequences. Such regions are easily recognized by the lack of uniform spacing with respect to adjacent banding patterns. While varying the sequencing reaction conditions may not help alleviate these problems, running ''hotter'' (i.e., 65°C) or ''preheated'' electrophoretic gels may help to eliminate secondary structure in the newly synthesized DNA single strand. The replacement of dGTP by dITP (inosine), 7-deaza-A or 7-deaza-G[51,132-134] in sequencing reactions can also help prevent formation of secondary structure in nascent strands, thereby eliminating band compressions by allowing molecules to run in their proper molecular-weight positions. Formamide denaturing gels may also be useful in the resolution of electrophoretic compression.[12,133,134]
8. Sequencing resolution, when using dITP in lieu of dGTP, can lead to loss of ''weak'' C bands, banding in all ITP lanes in positions near secondary structure regions, and mobility shifts in dITP reaction lanes relative to non-dITP lanes. The use of 7-deaza-G may help resolve these problems, but it appears not able to resolve all compressions. *T. aquaticus* DNA polymerase should not be used in conjunction with dITP.[116] T7 DNA polymerase often shows banding across all four lanes in the presence of dITP.
9. Often, sequencing the complementary strand provides the correctly spaced and accurate sequence. For example, the observed autoradiographic sequence, 5′ dCCGGC[G]G 3′, often shows an extra ''shadow'' ''G'' band running slightly behind the 3rd ''C'', while the complementary autoradiographic sequence, 5′ dCGCCGG 3′, gives the proper spacing and correct sequence.

APPENDICES

APPENDIX A: ISOLATION AND PURIFICATION OF TEMPLATES

SINGLE-STRANDED TEMPLATES FROM M13 AND RELATED PHAGES

The following protocol, a modification of Schreier and Cortese,[186] was developed for M13 single-stranded DNA isolation. It is applicable to any similar system where the end product is high-quality single-stranded DNA, such as is required for efficient DNA sequencing. Many other protocols reported in the literature[43,135,150,159,187] have also been used successfully for this process. However, the following protocol has been optimized for yield and quality (see Appendix F: Materials and Media, for relevant recipes and stock solutions). Protocols for generation of *in vitro* asymmetric (single-stranded) PCR products are not described in this review.[36,170-172,195-198]

1. Inoculate 500 μl of YT media with a single JM101 (or similar) colony grown on a fresh minimal media plate. If the host strain contains an F factor with a selectable drug resistance marker, grow the strain on medium containing the appropriate antibiotic.
2. Shake at 37°C for 1.5 to 2 h.
3. Inoculate 2 ml of YT, in a 50-ml plastic-capped centrifuge tube, with 200 μl of the exponentially growing cells.
4. Pick a single, isolated, colorless plaque, with a sterile wire or toothpick, from a fresh plaque purification plate and mix into the 2 ml of YT.
5. Shake the tube at 37°C for 5 to 8 h. It is important not to incubate longer than 8 h to decrease the chance of defective phage production. Because of the biology of M13, defective phage that arise spontaneously are secreted into the media and can reinfect cells in which the recombinant clone is growing. Eventually, every cell in the culture may contain the defective phage, causing a mixed population of template DNA.
6. Divide the infected culture into two 1.5-ml microcentrifuge tubes (1 ml each) and spin for 15 min at 4°C to pellet the cells.
7. Carefully decant the supernatant (containing the phage) into two new tubes, making sure not to disturb the cell pellets.
8. Recentrifuge, as above, and again pour off the supernatants into two new tubes.
9. Add 200 μl of 2.5 *M* NaCl, 20% polyethylene glycol 6000.[60,188,189] Mix the solutions and put on ice for 30 min or overnight at 4°C.
10. Spin down the precipitated phages in the microcentrifuge, for 10 min at 4°C. If the phage yield is good, a visible precipitate will be present.
11. Pour off and discard the supernatants and allow the tubes to drain upside down for 10 min onto a paper towel.
12. Add 100 μl of 1X TE buffer, pH 8.0, and dissolve the pellet by vortexing the tube.
13. Pool the two tubes from the same original infection into a single 1.5-ml tube.
14. Add 100 μl of phenol (equilibrated in 0.5 *M* Tris-HCl, pH 7.5) and vortex the tube.
15. Microcentrifuge the tube 5 min in order to separate the phases.
16. Remove the top (aqueous) phase (200 μl), avoiding the phenol, and transfer it to another microcentrifuge tube.
17. Add 20 μl of 3 *M* NaOAc (pH 6.0) and 200 μl of ethanol and mix the contents.
18. Precipitate the DNA at −20°C for 1 h.
19. Collect the DNA by centrifugation in a microcentrifuge for 15 min at 4°C. It is useful to orient all the microcentrifuge tubes being spun in the same way in order to more easily keep track of the position of the pellets.

20. Carefully pour off the supernatants so as not to dislodge the pellets. The pellets may or may not be visible at this stage.
21. Rinse the pellets with 500 μl of 70% ethanol, mix gently, and recentrifuge as above.
22. Carefully pour off the alcohol, drain the tube, and allow the pellet to completely air dry.
23. Dissolve the pellet in 15 μl of 1X TE buffer, pH 8.0.
24. Phage DNA may be stored at −20°C indefinitely. The final yield of template DNA should be at least 7.5 μg. The concentration can be verified by running a 0.7% agarose gel with 2 μl of the sample alongside single-stranded vector template DNA as a standard.

DOUBLE-STRANDED PLASMID TEMPLATES

Many different protocols have been used successfully for the isolation of plasmid DNA as templates for DNA sequencing.[43,100,144,150-158,160-167,190] The protocols presented below have been found to reliably and consistently yield DNA suitable for DNA sequencing. Refer to Appendix F: Materials and Media for relevant recipes and stock solutions. Protocols for generation of *in vitro* asymmetric (single-stranded) or double-stranded PCR products are not described in this review.[36,170-173,195-198]

Preparation of M13 RF DNA[43]

1. Pick a plaque from a fresh plaque purification plate with a sterile toothpick or sterile platinum wire and inoculate 2 ml of YT medium containing 10 μl of a host cell culture growing in minimal medium for 2 to 3 h. If the host strain contains an F factor with a selectable drug resistance marker, grow the strain on medium containing the appropriate antibiotic.
2. Incubate, with shaking, at 37°C, for 2 to 3 h.
3. At the same time, start a culture of uninfected cells by inoculating 10 ml of YT medium with a single colony of host cell from an M9 minimal plate. (It is important to maintain the host strain on minimal medium to select for the presence of the F episome unless the F factor is selectable by drug resistance. Increasing incubation time and/or successive passing of these strains in rich medium may result in a loss of the episome). Cells grown below 34°C may not produce the F factor-encoded sex pilus necessary for M13 phage infection. Cells that do not have the episome cannot be infected by M13 phage and result in lower yields of RF DNA.
4. After 2 to 3 h, the uninfected cells should have an OD_{590} = 0.3, and the infected cells will have a phage titer of 10^{10} to 10^{11} pfu/ml. About 500 μl of the infected 2-ml culture should be added to the 10-ml host cell culture so that a multiplicity of infection of between 1 and 10 is achieved.
5. Incubate at 37°C for an additional 4 to 5 h. To reduce the risk of defective phage formation, do not exceed a total incubation time of 8 h.
6. Pellet the cells at 2500 × g for 5 min. The supernatant, which contains the phage, should have a titer of about 10^{11} to 5×10^{12} pfu/ml. The supernatant can be stored at 4°C (in the optional presence of 0.01% sodium azide) or can be used to make single-stranded template DNA (see Single-Stranded Templates from M13 and Related Phages, step 9, this appendix).
7. Drain the cell pellet by inverting the centrifuge tube on a paper towel.
8. Directly proceed with plasmid purification to isolate the RF DNA or store the pellet at −20°C either as a frozen pellet or resuspended in 200 μl of 10% sucrose, 50 m*M* Tris, pH 8.0. RF DNA can be isolated from infected cells by standard plasmid purification methods. A 100-ml culture yields 10 to 50 μl of RF DNA. Two recommended protocols are supplied below.

Cesium Chloride/Ethidium Bromide Plasmid Preparation[152]

1. Inoculate 10 ml of L-Broth with antibiotics in an 18-mm tube with a single bacterial colony and incubate overnight at 37°C. It is not necessary to shake the culture.
2. Inoculate the miniculture into 500 ml of fresh medium plus appropriate antibiotics. Incubate the 500-ml culture overnight in a shaker at 37°C.
3. Collect the cells by centrifugation at 2500 × g for 5 min.
4. Resuspend the cell pellet in 10 ml of 25% sucrose, 50 m*M* Tris, pH 8.0, at room temperature or at 37°C.
5. Add 5.0 ml of 250 m*M* EDTA, pH 8.0. Swirl and then add 3.0 ml of freshly prepared 250 m*M* Tris, pH 8.0, containing 10 mg/ml of lysozyme.
6. Leave the suspension on ice for 1 h, swirling occasionally.
7. Forcefully pipet 12.0 ml of 50 m*M* Tris, 62.5 m*M* EDTA, 1% Triton X-100, pH 8.0, into the cell suspension and swirl gently to facilitate mixing. Leave the mixture on ice for several minutes to ensure complete lysis. The solution should become translucent and viscous.
8. Transfer the lysed mixture to a 50-ml plastic centrifuge tube and centrifuge it at 31,000 × g at 4°C for 45 min to pellet the chromosomal DNA and cell debris.
9. Remove the nonviscous portion of the supernatant with a pipet. Up to 30 ml can be collected. Discard the viscous portion of the supernatant, if any, together with the pellet. Store the supernatant frozen at −20°C or proceed directly to the next step.
10. Weigh 20.0 g of solid CsCl into a 50-ml plastic screw-cap tube. Pipet in 22.0 g of the cleared lysate and mix the solution gently until the CsCl has dissolved. Add 1.0 ml of 5 mg/ml ethidium bromide in 10 m*M* Tris, 1 m*M* EDTA, 100 m*M* NaCl, pH 8.0, and again mix gently.
11. Transfer the solution to two 5/8 × 3 in. polyallomer centrifuge tubes and seal them. If necessary, the tubes can be stored in the dark at room temperature for several days.
12. Spin the tubes in a Ti70.1 (or similar) rotor for 42 h at 190,000 × g at 17°C or 240,000 × g for 12 h followed by 4 h at 190,000 × g.
13. Under ultraviolet (UV) illumination, pierce the tops of the tubes with a scalpel and collect only the supercoiled plasmid (lower) band by syringe. The upper band is nicked plasmid and chromosomal DNA, and the lower band contains the covalently closed circular plasmid DNA. Transfer the plasmid solution to a sterile, glass screw-cap tube.
14. Remove the ethidium bromide by gently extracting three to four times with equal volumes of ice-cold *n*-butanol (equilibrated with 1/20th volume sterile water to prevent dehydration of the plasmid/CsCl solution) until the aqueous phase is colorless. After each extraction, remove and discard the upper butanol layer.

Note: This step should be performed under a hood. Alternatively, the ethidium bromide may be extracted, at room temperature, with isopropanol previously equilibrated with a 0.5 volume of CsCl saturated water.

15. Transfer the extracted plasmid/CsCl solution to dialysis tubing. The solution might have a pale, yellowish color, but should have no fluorescence under UV illumination.
16. Dialyze for 24 h against 1 l of 1X TE buffer, pH 8.0, changing the dialysis buffer three to four times.
17. Transfer the dialyzed DNA solution to a preweighed 50-ml sterile plastic polypropylene centrifuge tube. Determine the weight of the transferred solution, and, hence, its volume (density = 1 g/ml).

Note: Be careful of cross-contamination among several different preparations; change gloves each time after squeezing out the tubing.

18. Add 5 *M* NaCl to achieve a 0.4 *M* final concentration. Mix gently, then add two volumes of 100% isopropanol. Gently vortex, mix, and precipitate the plasmid overnight at −20°C.
19. Centrifuge the tube at 30,000 × g at 0°C for 15 min and discard the supernatant. Recentrifuge the tube at 2500 × g for 30 s and aspirate off the residual liquid. Note where the pellet should be; it may not be visible.
20. Leave the tube on the bench to air dry (15 min) and then add 0.75 ml of dH_2O. Angle the tube so that the pellet is immersed. Let the pellet dissolve for 30 min. Add 10 μl of 1 *M* Tris, pH 7.5, 0.1 *M* EDTA and mix. When the pellet is totally dissolved (the solution wets the tube equally in all positions), transfer it to a labeled microcentrifuge tube.
21. The usual yield is around 400 μg of plasmid (from pBR322) and about 1 mg of plasmid (from pUC and other "high-copy-number plasmid) per liter of saturated cell culture.
22. Electrophorese 10 μl (about 1 μg) on an agarose gel together with samples of DNA vector of known concentration and visually confirm the concentration of the plasmid preparation.
23. Measure the OD_{260} and OD_{280}. The usual OD_{260}/OD_{280} ratio is 1.8 to 1.9. One OD_{260} = 50 μg/ml double-stranded DNA.

Plasmid Mini-Preparation (Alkaline SDS Lysis)[151]

1. Inoculate 10 ml of medium containing appropriate antibiotics in an 18-mm tube with a single bacterial colony. Incubate the culture overnight at 37°C in a shaking incubator. The culture should grow to saturation.
2. Pour the saturated culture into a 50-ml polypropylene centrifuge tube. Pellet the cells by centrifugation at 2500 × g for 5 min. Pour off the supernatant and discard it. Retain the cell pellet. (At this point, the cell pellet may be stored frozen at −20°C.)
3. Add 1 ml of freshly prepared 50 m*M* glucose, 25 m*M* Tris, 10 m*M* EDTA, pH 8.0, containing 10 mg/ml of lysozyme, to the pellet and immediately resuspend it by gentle vortex mixing. Leave the cells in the lysozyme at room temperature for 5 min.
4. Swirl the tube to ensure that the cell suspension is homogeneous, then add 2.0 ml of freshly prepared 0.2 *M* NaOH, 1% SDS. Rock the tube back and forth until the solution clears to completely lyse the cells.
5. Immediately add 1.5 ml of freshly prepared 3 *M* NaOAc, pH 4.8. Rock the tube back and forth until a flocculent, white precipitate forms and the solutions mix completely.
6. Immediately spin down the precipitate at 31,000 × g at 4°C for 10 min. Pour the nonviscous supernatant, containing the plasmid DNA, into a fresh 50-ml centrifuge tube containing 3 ml of ice-cold isopropanol and discard the pellet.
7. Gently mix the supernatant and the alcohol, then centrifuge at 31,000 × g for 10 min and discard the supernatant. Briefly recentrifuge to collect any residual liquid and withdraw it by aspiration. Allow the nucleic acid pellet to air dry at room temperature (15 min).
8. Dissolve the pellet in 850 μl of 1X TE buffer, pH 8.0. Add 75 μl of 5 *M* NaCl and transfer the solution to a 1.5-ml microcentrifuge tube containing 575 μl of 95% isopropanol. Mix and precipitate the nucleic acid at room temperature for 10 min. (If necessary, the tube can be stored overnight at −20°C.)
9. Collect the precipitated nucleic acid by microcentrifuging the tube for 10 min at 4°C. Discard the supernatant. Recentrifuge the tube briefly and remove the residual liquid by aspiration with a drawn-out Pasteur pipet. Air dry the pellet at room temperature (15 min).
10. Add 500 μl of 200 μg/ml RNAse in 1X TE, pH 8.0. Incubate the tube at 37°C for 30 min to digest the RNA; flick the tube occasionally during the incubation to help dissolve the pellet.

11. Add 50 μl of 5 *M* NaCl and 350 μl of 95% isopropanol. Mix and leave at room temperature for 10 min, then spin for 2 min in the microcentrifuge and discard the supernatant.
12. Recentrifuge briefly and aspirate away the residual liquid. Air dry the pellet thoroughly, until the pellet becomes transparent; then resuspend it in 100 μl of 1X TE buffer, pH 8.0. Store the mini-prep frozen at −20°C. The yield of plasmid DNA is 10 to 20 μg.
13. Phenol/chloroform extract and reprecipitate the DNA as follows:

- Add 400 μl of 1X TE, pH 8.0, plus 50 μl of 5 *M* NaCl to the mini-preps, and mix.
- Add 100 μl of 1:1 phenol/chloroform and flick the tube rapidly until the solution appears milky.
- Spin the tube for 1 min in the microcentrifuge.
- Transfer the top (aqueous) phase to a fresh tube, discard the lower phase, and interface.
- Add 350 μl of 95% isopropanol to the aqueous layer, mix, and allow the nucleic acid to precipitate at room temperature for 15 min. Spin the tube in the microcentrifuge for 3 min to pellet the nucleic acid and discard the supernatant. Briefly recentrifuge the tube and aspirate away the residual liquid.
- Air dry the pellet thoroughly. Resuspend it in 100 μl of 1X TE.
- Store the mini-prep at −20°C.

ISOLATION OF DNA FROM PHAGE λ

CsCl Method[40,191]

This procedure is used to isolate DNA from phage λ and involves a plate lysate, a CsCl step gradient, and phenol extraction. The yield is approximately 50 μg from a single 150 mm plate of λ-infected *E. coli.* Plate lysis with λ is preferred to lysis in liquid culture with larger numbers of phage DNA preparations because correct calculation of the multiplicity of infection is not as critical. Other preparative methods have been reported which also provide adequate templates for sequencing.[145-150,159,160,162-164,172] Refer to Appendix F: Materials and Media for relevant stock solutions.

1. Inoculate the host bacterium, *E. coli* ER1578 (or other compatible strain), into 10 ml of L-Broth and incubate at 37°C overnight with shaking.
2. Transfer the overnight culture to a 50-ml polypropylene centrifuge tube and collect the cells by centrifugation at 3000 × g for 5 min. Discard the supernatant.
3. Resuspend the pellet in 10 m*M* $Mg_2SO_4 \cdot 7H_2O$ to an OD_{590} of 2.5.
4. Transfer 300 μl of the above suspension to a sterile 1.5-ml microcentrifuge tube and infect the culture with about 10^5 pfu of the appropriate phage in SM buffer. (Phage stored in SM buffer with a drop of chloroform are stable at 4°C).
5. Incubate at 37°C without agitation for 10 min to adsorb the phage to *E. coli.*
6. Add 10 ml of 0.3% melted Bacto-agar (50°C) to the infected culture. Mix gently and plate onto two 150-mm L agar plates containing 100 μg/ml of ampicillin (or other required antibiotic).
7. After the agar solidifies, incubate the plates at 37°C overnight.
8. Add 10 ml of SM buffer to the surface of each plate, followed by a few drops of chloroform.
9. Allow the phage to diffuse into the buffer for 2 to 4 h at room temperature.
10. Collect the buffer in a 30-ml plastic screw-cap tube and rinse the plate with an additional 2.5 ml of SM buffer.

11. Centrifuge the phage solution at 20,000 × g for 30 min at 4°C to pellet the debris.
12. With a sterile 10-ml plastic pipette, gently layer 6.5 ml of a density (p) = 1.6 g/ml CsCl solution (in 10 m*M* Tris-HCl, 20 m*M* $MgCl_2$, 10 m*M* NaCl, pH 7.5) below 5 ml of a density (p) = 1.4 g/ml CsCl solution in a 1 × 3.5-in. centrifuge tube.
13. With a sterile plastic pipette, layer 20 ml of the cleared phage solution over the CsCl step gradient.
14. Centrifuge at 65,000 × g for 90 min at 15°C in a swinging bucket rotor to band the phage.
15. Recover the phage band between the CsCl phases by placing a 10-μl glass micropipette, attached to plastic tubing (ID,1/32 in., OD, 3/32 in., wall, 1/32 in.), at the bottom of the phage band and slowly syphon the phage (1 to 3 ml) into a 30-ml Corex tube. Remove 100 μl of the CsCl/phage solution and store at 4°C in a sterile microcentrifuge tube.
16. Add EDTA to the remaining phage/CsCl solution to a final concentration of 20 m*M* and incubate the solution at 65°C for 30 min.
17. Dialyze the solution for 4 h against two changes (1 l each) of 1X TE buffer, pH 7.5, at 4°C.
18. Add RNAse A to a final concentration of 1 mg/ml and incubate the solution at 37°C for 1 h.
19. Phenol/chloroform extract the phage solution several times until the interphase is clear.
20. Extract the solution with 100% *n*-butanol to reduce the volume to approximately 0.5 ml.
21. Transfer the DNA solution to a microcentrifuge tube and precipitate it by adding NH_4OAc to a final concentration of 1.5 *M*. Add 1.1 volume of isopropanol. Place the tube at −20°C overnight.
22. Pellet the DNA by microcentrifugation for 10 min.
23. Pour off the supernatant and rinse the pellet in 70% ethanol.
24. Recentrifuge, aspirate off the liquid, and air dry the pellet.
25. Redissolve the pellet in 100 μl of 1X TE buffer, pH 8.0.

Notes:

1. ER1578 is Δ(*lac*)U169 Δ(*lon*) *araD*139 *rpsL* (Str^R) *supF mcrB1 hsdR2 zjj*202::Tn*10* (Tet^R) *serB*28 (pMC9: Amp^R *lac*I^q).[192,193]
2. To check for insertions into λgt11, IPTG and X-Gal are added to the top agar at a final concentration of 0.67 μ*M* and 0.0067%, respectively. Phage containing insert DNA give rise to white plaques while wild-type phage produce blue plaques on this medium.
3. When it is necessary to isolate phage from individual plaques, it is important that the plates are dry so that phage do not diffuse from one plaque into another. To dry the plates after pouring on the top agar, let the plates remain on the bench top with the lids ajar for approximately 1 h, then close and incubate at 37°C overnight.
4. When infecting from a plaque (in lieu of steps 1 to 4 above), remove the plaque from the plate with a sterile Pasteur pipette and resuspend the phage in 1 ml of SM buffer over a drop of $CHCl_3$. Dilute 1 to 10 and use 20 μl for infection.

Isolation of DNA from Phage λ (Quick Mini-Preparation)[40,191]

This procedure is used to isolate DNA from phage λ. It involves a plate lysate and phenol extraction, but no CsCl step gradient. The yield is approximately 20 μg from an 85-mm plate of λ-infected *E. coli*. Plate lysis with λ is easier with larger numbers of phage DNA preparations because correct calculation of the multiplicity of infection is not as critical with plate lysis as with culture lysis. There seems to be no difference in the purity of the λ DNA isolated in this manner, compared to the previous CsCl procedure. Refer to Appendix F: Materials and Media for relevant stock solutions.

1. Inoculate the host bacterium, *E. coli* ER1578 (or other compatible strain), into 10 ml of L-Broth, and shake the culture overnight at 37°C.
2. Transfer 100 μl of the above overnight to a sterile 5-ml glass tube and infect the culture with about 10^4 to 10^5 pfu of the appropriate phage in SM buffer. (Phage stored in SM buffer with a drop of chloroform are stable at 4°C for several years.)
3. Incubate at 37°C for 10 min to adsorb the phage to *E. coli*.
4. Add 4 ml of 0.3% melted top agarose (50°C) to the infected culture. Mix gently and plate onto a freshly prepared 85-mm 1% agarose plate containing 50 μg/ml ampicillin.
5. After the agarose solidifies, incubate the plates at 37°C for approximately 8 to 15 h. Overnight incubations may slightly reduce titers; shorter incubations also work well.
6. Using a sterile glass spreading rod, gently scrape the top agarose into a 35-ml plastic centrifuge tube. Add 3 ml of SM buffer and 10 drops of chloroform.
7. Vortex vigorously for 15 s and let stand for 10 min at room temperature.
8. Centrifuge the solution at 5000 × g for 10 min.
9. Collect the supernatant (the phage lysate), add two drops of chloroform, and briefly vortex mix.
10. Remove 700 μl of phage lysate stock and place in a sterile 1.5-ml microcentrifuge tube. The titer of the phage should be greater than 10^9 pfu/ml.
11. Add 1 μl of 1 mg/ml DNAse, followed by 1 μl of 1 mg/ml RNAse.
12. Incubate the solution at 37°C for 30 min.
13. Add an equal volume of 20% PEG/2.5 *M* NaOAc to the tube and incubate for 1 h on ice.

Note: Do not allow this step to exceed 4 h or much of the chromosomal DNA may also be precipitated in the subsequent steps.

14. Microcentrifuge the solution for 10 min at 4°C. Pour off the supernatant, recentrifuge for 2 min and aspirate the remaining liquid in the tube with a drawn-out capillary pipette.
15. Resuspend the pellet in 500 μl of SM buffer. (If the pellet remains undissolved, it will dissolve in the next step.)
16. Add 5 μl 10% SDS and 0.5 μl 0.5 *M* EDTA to the solution and incubate at 65°C for 15 min. Mix gently to resuspend the pellet.
17. Extract the solution with an equal volume of buffer-equilibrated phenol. Mix the contents of the tube by vigorous vortexing. Separate the phases by microcentrifugation for 2 min at room temperature.
18. Transfer the top (aqueous) layer to a new microcentrifuge tube and repeat step 17, except use an equal volume of a 1:1 mixture of phenol:chloroform.
19. Transfer the aqueous top layer to a new microcentrifuge tube and extract once more with an equal volume of chloroform.
20. Transfer the aqueous top layer to a new microcentrifuge tube and add 0.1 volume of 3 *M* NaOAc, followed by 1.1 volume isopropanol.
21. Incubate at −20°C for 30 min.
22. Microcentrifuge at 4°C for 15 min. Aspirate away the liquid supernatant and air dry the (invisible) pellet.
23. Resuspend the pellet in 50 μl of 1X TE buffer, pH 7.5. The yield should be approximately 20 μg of pure λ DNA.

Notes:

1. ER1578 is Δ(*lac*)U169 Δ(*lon*) *araD*139 *rpsL* (Str^R) *supF mcrB1 hsdR2 zjj*202::Tn*10* (Tet^R) *serB*28 (pMC9: Amp^R, *lacI*^{q}).[192,193]
2. To check for insertions into λgt11, IPTG and X-Gal are added to the top agar at a final concentration of 0.67 μM and 0.0067%, respectively.
3. When infecting for individual plaques, it is important that the plates are dry so that phage from plaques do not contaminate other plaques. To dry plates after pouring on top agar, let the plates remain on the benchtop with the lids ajar for approximately 1 h, then close and incubate at 37°C overnight.
4. Agar may be used in place of agarose in step 4. However, unless the final preparation is extensively re-extracted (e.g., repeating steps 17 to 19) and drop dialyzed[194] against 1X TE buffer, pH 8.0, after step 23, it will likely not be able to be digested with restriction endonuleases and will not serve as a clean template for DNA sequencing reactions.
5. When infecting from a plaque (in lieu of steps 1 and 2 above), remove the plaque from the plate with a sterile Pasteur pipette and resuspend the phage in 1 ml of SM buffer over one drop of $CHCl_3$. Dilute 1:10 and use 20 μl for infection.
6. If freshly prepared (same day) agarose plates are unavailable or inconvenient, older plates may be substituted. Add 5 ml of sterile SM buffer to the plate and allow the buffer to soak into the plate for 15 min. Remove the excess buffer with a sterile pipette. Take care after pouring the top agarose, since it is only loosely attached to the bottom agarose; incubate the plates right side up.

APPENDIX B: COMPOSITION OF DNA SEQUENCING REAGENTS

Standard ^{32}P Sequencing Mixes (DNA Polymerase I Large, Klenow, Fragment)[a] (μM concentrations)

	A mix	C mix	G mix	T mix
ddATP	61.0	—	—	—
ddCTP	—	150.0	—	—
ddGTP	—	—	150.0	—
ddTTP	—	—	—	150.0
dATP	1.6	1.6	1.6	1.6
dCTP	33.0	4.4	44.0	44.0
dGTP	33.0	44.0	4.4	44.0
dTTP	33.0	44.0	44.0	1.6

Standard ^{35}S Sequencing Mixes (DNA Polymerase I Large, Klenow, Fragment)[a] (μM concentrations)

	A mix	C mix	G mix	T mix
ddATP	22.0	—	—	—
ddCTP	—	300.0	—	—
ddGTP	—	—	300.0	—
ddTTP	—	—	—	450.0
dATP	—	—	—	—
dCTP	33.0	5.6	56.0	56.0
dGTP	33.0	56.0	5.6	56.0
dTTP	33.0	56.0	56.0	5.6

Standard $^{32}P/^{35}S$ Sequencing Mixes (*Thermus aquaticus* Polymerase)[a] (μM concentrations)

	A mix	C mix	G mix	T mix
ddATP	450.0	—	—	—
ddCTP	—	500.0	—	—
ddGTP	—	—	222	—
ddTTP	—	—	—	520
dATP	10	10	10	8
dCTP	20	20	20	16
dGTP	20	20	20	16
dTTP	20	20	20	16

Standard $^{32}P/^{35}S$ Sequencing Mixes (Reverse Transcriptase)[b] (μM concentrations)

	A mix	C mix	G mix	T mix
ddATP	5.0	—	—	—
ddCTP	—	25.0	—	—
ddGTP	—	—	10.0	—

Standard $^{32}P/^{35}S$ Sequencing Mixes (continued)
(Reverse Transcriptase)[b]
(μ*M* concentrations)

ddTTP	—	—	—	40.0
dATP	2.0	2.0	2.0	2.0
dCTP	125.0	100.0	125.0	125.0
dGTP	125.0	125.0	100.0	125.0
dTTP	125.0	125.0	125.0	100.0

Standard End-Label Sequencing Mixes[c]
[$\gamma^{32}P$]rATP, [$\gamma^{35}S$]rATP
(μ*M* concentrations)

	A mix	C mix	G mix	T mix
ddATP	150.0	—	—	—
ddCTP	—	150.0	—	—
ddGTP	—	—	150.0	—
ddTTP	—	—	—	150.0
dATP	4.4	44.0	44.0	44.0
dCTP	44.0	4.4	44.0	44.0
dGTP	44.0	44.0	4.4	44.0
dTTP	44.0	44.0	44.0	4.4

Long Extension
Sequencing Termination Mixes[c]
(μ*M* concentrations)

	A mix	C mix	G mix	T mix
ddATP	300.0	—	—	—
ddCTP	—	100.0	—	—
ddGTP	—	—	150.0	—
ddTTP	—	—	—	500.0
dATP	22.0	244.0	244.0	244.0
dCTP	244.0	22.0	244.0	244.0
dGTP	244.0	244.0	22.0	244.0
dTTP	244.0	244.0	244.0	22.0

Short Extension
Sequencing Termination Mixes[c]
(μ*M* concentrations)

	A mix	C mix	G mix	T mix
ddATP	150.0	—	—	—
ddCTP	—	150.0	—	—
ddGTP	—	—	150.0	—
ddTTP	—	—	—	150.0
dATP	4.4	44.0	44.0	44.0
dCTP	44.0	4.4	44.0	44.0
dGTP	44.0	44.0	4.4	44.0
dTTP	44.0	44.0	44.0	4.4

[a] In 5 m*M* $MgCl_2$, 10 m*M* Tris, pH 7.5, 5 m*M* DTT.

[b] In 5 m*M* $MgCl_2$, 10 m*M* Tris, pH 8.3 , 5m*M* DTT, 50 m*M* NaCl.

[c] Should be made up in a or b above, depending upon the enzyme to be utilized.

Thermus aquaticus DNA Polymerase, DNA Polymerase I (Klenow), or T7 DNA Polymerase 10X Sequencing Buffer

100 m*M* Tris-HCl, pH 7.5
50 m*M* $MgCl_2$
75 m*M* DTT

Reverse Transcriptase 10X Sequencing Buffer

100 m*M* Tris, pH 8.3
50 m*M* $MgCl_2$
500 m*M* NaCl
75 m*M* DTT

Stop/Loading Dye Solution

Deionized formamide containing:
0.3% bromophenol blue
0.3% xylene cyanol FF
0.37% disodium EDTA, pH 7.0

10X Kinase Buffer

500 to 700 m*M* Tris, pH 7.6
100 m*M* $MgCl_2$
50 m*M* DTT

Extension Mix K

7.5 μ*M* dATP
7.5 μ*M* dCTP
7.5 μ*M* dGTP
7.5 μ*M* dTTP

Extension Mix NK

7.5 μ*M* dCTP
7.5 μ*M* dGTP
7.5 μ*M* dTTP

dNTP Chase

0.25 m*M* dATP
0.25 m*M* dCTP
0.25 m*M* dGTP
0.25 m*M* dTTP

APPENDIX C: DILUTION PROTOCOLS FROM STOCK SOLUTIONS OF NUCLEOTIDES

PART A: PROTOCOL FOR MAKING dNTP AND ddNTP STOCK SOLUTIONS

If individual powdered dNTP and ddNTP reagents are used for making DNA sequencing reagents, it is recommended that the following protocols be followed (Steps 1 to 3) for making stock solutions. If stock solutions of dNTPs and ddNTPs of predetermined concentrations are purchased, it is recommended to verify the concentrations by the procedure in Step 3. The following are recommended final (working) stock concentrations:

Powdered reagent	Stock concentration (m*M*)
dATP (2′-deoxyadenosine 5′-triphosphate) (disodium salt)	2
dCTP (2′-deoxycytosine 5′-triphosphate) (trisodium salt)	2
dGTP (2′-deoxyguanosine 5′-triphosphate) (trisodium salt)	2
dTTP (2′-deoxythymidine 5′-triphosphate) (trisodium salt)	2
ddATP (2′,3′-dideoxyadenosine 5′-triphosphate) (trisodium salt)	1
ddCTP (2′,3′-dideoxycytosine 5′-triphosphate (tetrasodium salt)	1
ddGTP (2′,3′-dideoxyguanosine 5′-triphosphate) (trisodium salt)	1
ddTTP (2′,3′-dideoxythymidine 5′-triphosphate) (tetrasodium salt)	1

Step 1: Resuspension of the Powdered Nucleotide to the Proper Concentration

To determine the volume of ice-cold, sterile water required to dissolve a given weight of powdered nucleotide (to achieve the proper final concentration), use the following equations (1 and 2a or 2b):

$$\frac{\text{g of nucleotide powder}}{\text{mol wt of nucleotide*}} \times 10^6 = \mu\text{mol of nucleotide} \quad (1)$$

$$\frac{\mu\text{mol of nucleotide}}{\text{Y ml of water}} = 2\ \text{m}M\ \text{(final concentration for dNTPs)} \quad (2a)$$

or

$$\frac{\mu\text{mol of nucleotide}}{\text{Y ml of water}} = 1\ \text{m}M\ \text{(final concentration for ddNTPs)} \quad (2b)$$

Example: 5 mg of powdered dATP is purchased: $5 \times 10^3/589.2 \times 10^6 = 8.48$ μmol of dATP; 8.48 μmol/Y = 2 m*M* final concentration; Y = 4.24 ml of water.

Step 2: Checking the Actual Nucleotide Concentrations

All nucleotide solution concentrations should be verified by optical density. Formula 3 is used in conjunction with the information presented in Table 1 to dilute the stock nucleotide solutions for optical density (OD) measurements and to independently calculate stock concentrations.

1. Dilute the dNTP solutions 1:120 in the appropriate diluent (see Table 1).
2. Dilute the ddNTP solutions 1:30 in the appropriate diluent (see Table 1).

* Molecular weights are given in Table 1.

3. Read the OD at the λ_{max} for each nucleotide, blanking the spectrophotometer against the correct diluent.
4. Use Equation 3 to calculate the µmolar concentrations of the original stock solutions.

$$\frac{\text{OD } \lambda_{max} \times \text{the dilution factor}}{\text{extinction coefficient}} = \text{m}M \text{ nucleotide concentration} \tag{3}$$

5. Adjust the stock solutions with sterile water, as necessary, and remeasure the OD. Alternatively, use the concentrations determined by the OD and recalculate the amounts required for each mixture in Part B below.

Step 3: Aliquot the Stock Solutions into Convenient Volumes and Store at −20°C

TABLE 1

Nucleotide	F.W.	λ_{max}	Extinction coefficient[a]	Dilution factor[b]	Diluent
dATP	589.2	260	15.4	120	0.01 *M* P.B., pH 7.0[c]
dCTP	551.1	280	13.0	120	0.01 *N* HCl, pH 2.0
dGTP	609.2	252	13.7	120	0.01 *M* P.B., pH 7.0[c]
dTTP	583.1	267	10.0	120	0.01 *M* P.B., pH 7.0[c]
ddATP	594	260	22.2	30	0.01 *M* P.B., pH 7.0[c]
ddCTP	575.2	280	18.1	30	0.01 *N* HCl, pH 2.0
ddGTP	639.2	252	19.3	30	0.01 *M* P.B., pH 7.0[c]
ddTTP	608.2	267	15.5	30	0.01 *M* P.B., pH 7.0[c]

Note: F.W., formula weight; P.B., phosphate buffer.

[a] $A_m \times 10^{-3}$, lot of nucleotide specific, see accompanying product sheets.
[b] To be used in Equation 3.
[c] 0.01 *M* phosphate buffer, pH 7.0, is made by mixing four parts of 1 *M* K_2HPO_4, pH 9.5, with three parts of 1 *M* KH_2PO_4, pH 2.7 (one part); verify the pH.

PART B: DILUTION PROTOCOLS

Extension Sequencing Mix NK ($\alpha^{32}P$ or $\alpha^{35}S$)
(30 reactions; units in µl)

Stock soln	Stock conc (m*M*)	
dATP	2	—
dCTP	2	2.5 (of 1/10 dil)
dGTP	2	2.5 (of 1/10 dil)
dTTP	2	2.5 (of 1/10 dil)
ddATP	1	—
ddCTP	1	—
ddTTP	1	—
$MgCl_2$	100	3.0
Tris-HCl	100	6.0
DTT	100	3.0
H_2O		40.5
Total		60.0

Extension Sequencing Mix K ($\gamma^{32}P$ or $\gamma^{35}S$) (30 reactions; units in μl)

Stock soln	Stock conc (m*M*)	
dATP	2	2.5 (of 1/10 dil)
dCTP	2	2.5 (of 1/10 dil)
dGTP	2	2.5 (of 1/10 dil)
dTTP	2	2.5 (or 1/10 dil)
ddATP	1	—
ddCTP	1	—
ddGTP	1	—
ddTTP	1	—
$MgCl_2$	100	3.0
Tris-HCl	100	6.0
DTT	100	3.0
H_2O		38.0
Total		60.0

Long Extension-Sequencing Termination Mixes[a] (30 reactions; units in μl)

Stock soln	Stock conc (m*M*)	A Mix	C mix	G mix	T mix
dATP	2	1.0	11.0	11.0	11.0
dCTP	2	11.0	1.0	11.0	11.0
dGTP	2	11.0	11.0	1.0	11.0
dTTP	2	11.0	11.0	11.0	1.0
ddATP	1	27.0	—	—	—
ddCTP	1	—	9.0	—	—
ddGTP	1	—	—	14.0	—
ddTTP	1	—	—	—	45.0
$MgCl_2$	100	4.5	4.5	4.5	4.5
Tris-HCl	100	9.0	9.0	9.0	9.0
DTT	100	4.5	4.5	4.5	4.5
H_2O		11.0	29.0	24.0	0.0
Total		90.0	90.0	90.0	97.0

Short Extension-Sequencing Termination Mixes[a] (30 reactions; units in μl)

Stock soln	Stock conc (m*M*)	A mix	C mix	G mix	T mix
dATP	2	2.0 (1/10 dil)	2.0	2.0	2.0
dCTP	2	2.0	2.0 (1/10 dil)	2.0	2.0
dGTP	2	2.0	2.0	2.0 (1/10 dil)	2.0
dTTP	2	2.0	2.0	2.0	2.0 (1/10 dil)
ddATP	1	13.5	—	—	—
ddCTP	1	—	13.5	—	—
ddGTP	1	—	—	13.5	—
ddTTP	1	—	—	—	13.5
$MgCl_2$	100	4.5	4.5	4.5	4.5
Tris-HCl	100	9.0	9.0	9.0	9.0
DTT	100	4.5	4.5	4.5	4.5
H_2O		50.5	50.5	50.5	50.5
Total		90.0	90.0	90.0	90.0

Standard End-Label Sequencing Mixes[a]
[γ^{32}P]rATP or [γ^{35}S]rATP
(30 reactions; units in μl)

Stock soln	Stock conc (m*M*)	A mix	C mix	G mix	T mix
dATP	2	2.0 (1/10 dil)	2.0	2.0	2.0
dCTP	2	2.0	2.0 (1/10 dil)	2.0	2.0
dGTP	2	2.0	2.0	2.0 (1/10 dil)	2.0
dTTP	2	2.0	2.0	2.0	2.0 (1/10 dil)
ddATP	1	13.5	—	—	—
ddCTP	1	—	13.5	—	—
ddGTP	1	—	—	13.5	—
ddTTP	1	—	—	—	13.5
$MgCl_2$	100	4.5	4.5	4.5	4.5
Tris-HCl	100	9.0	9.0	9.0	9.0
DTT	100	4.5	4.5	4.5	4.5
H_2O		50.5	50.5	50.5	50.5
Total		90.0	90.0	90.0	90.0

Standard ^{32}P Sequencing Mixes
(DNA Polymerase I Large, Klenow, Fragment)[b]
(30 reactions; units in μl)

Stock soln	Stock conc (m*M*)	A mix	C mix	G mix	T mix
dATP	2	7.5 (1/100 dil)	7.5 (1/100 dil)	7.5 (1/100 dil)	7.5 (1/100 dil)
dCTP	2	1.5	2.0 (1/10 dil)	2.0	2.0
dGTP	2	1.5	2.0	2.0 (1/10 dil)	2.0
dTTP	2	1.5	2.0	2.0	7.5 (1/100 dil)
ddATP	1	5.5	—	—	—
ddCTP	1	—	13.5	—	—
ddGTP	1	—	—	13.5	—
ddTTP	1	—	—	—	13.5
$MgCl_2$	100	4.5	4.5	4.5	4.5
Tris-HCl	100	9.0	9.0	9.0	9.0
DTT	100	4.5	4.5	4.5	4.5
H_2O		54.5	45.0	45.0	39.5
Total		90.0	90.0	90.0	90.0

Standard ^{35}S Sequencing Mixes
(DNA Polymerase I Large, Klenow, Fragment)[b]
(30 reactions; units in μl)

Stock soln	Stock conc (m*M*)	A mix	C mix	G mix	T mix
dCTP	2	1.5	2.5 (1/10 dil)	2.5	2.5
dGTP	2	1.5	2.5	2.5 (1/10 dil)	2.5
dTTP	2	1.5	2.5	2.5	2.5 (1/10 dil)
ddATP	1	2.0	—	—	—
ddCTP	1	—	27.0	—	—
ddGTP	1	—	—	27.0	—
ddTTP	1	—	—	—	40.5
$MgCl_2$	100	4.5	4.5	4.5	4.5

Standard ^{35}S Sequencing Mixes (continued)

Tris-HCl	100	9.0	9.0	9.0	9.0
DTT	100	4.5	4.5	4.5	4.5
H_2O		65.5	37.5	37.5	24.0
Total		90.0	90.0	90.0	90.0

Standard ^{32}P/^{35}S Sequencing Mixes (Reverse Transcriptase)[c] (150 reactions; units in μl)

Stock soln	Stock conc (n*M*)	A mix	C mix	G mix	T mix
dATP	2 m*M*	5.0 (1/10 dil)	5.0 (1/10 dil)	5.0 (1/10 dil)	5.0 (1/10 dil)
dCTP	2 m*M*	31.5	20	31.5	31.5
dGTP	2 m*M*	31.5	31.5	20	31.5
dTTP	2 m*M*	31.5	31.5	31.5	20
ddATP	1 m*M*	2.5	—	—	—
ddCTP	1 m*M*	—	12.5	—	—
ddGTP	1 m*M*	—	—	5	—
ddTTP	1 m*M*	—	—	—	13.5
$MgCl_2$	1 *M*	2.5	2.5	2.5	2.5
Tris-HCl	1 *M*	5.0	5.0	5.0	5.0
DTT	1 *M*	2.5	2.5	2.5	2.5
NaCl	1 *M*	25.0	25.0	25.0	25.0
H_2O		363.0	364.5	372.0	363.5
Total		500.0	500.0	500.0	500.0

Standard ^{32}P/^{35}S Sequencing Mixes (*Thermus aquaticus* Polymerase[b] (30 reactions; units in μl)

Stock soln	Stock conc (m*M*)	A mix	C mix	G mix	T mix
dATP	2	4.5 (1/10 dil)	4.5 (1/10 dil)	4.5 (1/10 dil)	4.5 (1/10 dil)
dCTP	2	9 (1/10 dil)	9 (1/10 dil)	9 (1/10 dil)	8 (1/10 dil)
dGTP	2	9 (1/10 dil)	9 (1/10 dil)	9 (1/10 dil)	8 (1/10 dil)
dTTP	2	9 (1/10 dil)	9 (1/10 dil)	9 (1/10 dil)	8 (1/10 dil)
ddATP	1	40.5	—	—	—
ddCTP	1	—	45.0	—	—
ddGTP	1	—	—	20	—
ddTTP	1	—	—	—	52
$MgCl_2$	100	4.5	4.5	4.5	5
Tris-HCl	100	9.0	9.0	9.0	10
DTT	100	4.5	4.5	4.5	5
H_2O		—	16.0	20.5	—
Total		90	94.5	90.0	100

[a] Table presents calculations for solutions in 5 m*M* $MgCl_2$, 10 m*M* Tris-HCl, pH 7.5, 5 m*M* DTT. If reverse transcriptase is to be utilized, NaCl should be added to a final concentration of 50 m*M* (add 18 μl of 1 *M* NaCl to each mixture, reducing the water in each by 18 μl).

[b] In 5 m*M* $MgCl_2$, 10 m*M* Tris-HCl, pH 7.5, 5 m*M* DTT.

[c] In 5 m*M* $MgCl_2$, 10 m*M* Tris-HCl, pH 8.3, 50 m*M* NaCl, 5 m*M* DTT.

APPENDIX D: POLYACRYLAMIDE SEQUENCING GEL MIXTURES

Standard Nonbuffer-Gradient Gels

% Gel	4%	6%	8%
1. Ultrapure urea	25.2 g	25.2 g	25.2 g
2. 38:2 Acrylamide/bis-acrylamide (40%)	6 ml	9 ml	12 ml
3. 5X TBE	12 ml	12 ml	12 ml
4. H_2O	21 ml	18 ml	15 ml
Total volume	60 ml	60 ml	60 ml

Buffer-Gradient Gels

Part A (0.5X TBE)

% Gel	4%	6%	8%
1. Ultrapure urea	42 g	42 g	42 g
2. 38:2 Acrylamide/bis-acrylamide (40%)	10 ml	15 ml	20 ml
3. 10X TBE[a]	5 ml	5 ml	5 ml
4. H_2O	53 ml	48 ml	43 ml
Total volume	100 ml	100 ml	100 ml

Part B (2.5X TBE)

% Gel	4%	6%	8%
1. Ultrapure urea	21 g	21 g	21 g
2. 38:2 Acrylamide/bis-acrylamide (40%)	5 μl	7.5 μl	10 μl
3. 10X TBE[a]	12.5 μl	12.5 μl	12.5 μl
4. Sucrose	5 g	5 g	5 g
5. 1% Bromophenol Blue	500 μl	500 μl	500 μl
6. H_2O	12.5 μl	10 μl	7.5 μl
Total volume	50 μl	50 μl	50 μl

Note: Pouring of buffer-gradient gels is described in Reference 120.

[a] 10X TBE is required, in lieu of 5X TBE, for 6 and 8% buffer-gradient gels, Part B, to maintain appropriate volumes.

APPENDIX E: MOLAR RATIO CALCULATIONS

1. Plasmid, λ, cosmid, restriction fragment, linear DNA, etc.

   a. Single-stranded DNA

$$\# \text{ pmoles} = \frac{\#\mu\text{g DNA} \times 10^6 \text{ pg}/\mu\text{g}}{\# \text{ bases} \times 330 \text{ pg/pmol base}}$$

   b. Double-stranded DNA

$$\# \text{ pmoles} = \frac{\#\mu\text{g DNA} \times 10^6 \text{ pg}/\mu\text{g}}{\# \text{ base pairs} \times 660 \text{ pg/pmol base pair}}$$

2. Oligonucleotides (linkers, adaptors, probes, and primers)

   a. Single-stranded: 40 μg = 1 OD_{260}

$$\# \text{ pmoles} = \frac{\#\mu\text{g oligonucleotide} \times 10^6 \text{ pg}/\mu\text{g}}{\# \text{ bases} \times 330 \text{ pg/pmol base}}$$

$$\# \text{ pmoles} = \frac{[\# \text{ OD}_{260} \times 40\ \mu\text{g/OD}_{260}] \times 10^6 \text{ pg}/\mu\text{g}}{\# \text{ bases} \times 330 \text{ pg/pmol base}}$$

   b. Double-stranded: 50 μg = 1 OD_{260}

$$\# \text{ pmoles} = \frac{\#\mu\text{g oligonucleotide} \times 10^6 \text{ pg}/\mu\text{g}}{\# \text{ base pairs} \times 660 \text{ pg/pmol base pair}}$$

$$\# \text{ pmoles} = \frac{[\# \text{ OD}_{260} \times 50\ \mu\text{g/OD}_{260}] \times 10^6 \text{ pg}/\mu\text{g}}{\# \text{ base pairs} \times 660 \text{ pg/pmol base pair}}$$

APPENDIX F: MATERIALS AND MEDIA

The items listed below should be on hand before these protocols are initiated. This list does not include items that would routinely be a part of a laboratory involved in nucleic acid research, such as centrifuges, autoclaves, etc.

Electrophoresis Reagents and Supplies

Electrophoresis power supply (1000 to 2000 V)
Electrophoresis gel apparatus
Acrylamide (ultrapure)
Bis-acrylamide (ultrapure)
Urea (ultrapure)
10% Ammonium persulfate
TEMED
1X TBE (0.1 *M* Tris-HCl, 0.08 *M* boric acid, 1 m*M* EDTA, pH 8.4) (see Appendix B: Composition of DNA Sequencing Reagents)
5% Methanol, 5% acetic acid
Glass plates
Yellow vinyl electric tape (3M Corp.)
Spacers and combs
Silicone or silane
Gel dryer
Autoradiography cassette
Darkroom developing access
Shallow "fixer" tray (see References 112 and 145)

DNA Sequencing Reagents

5X TBE (0.5 *M* Tris-HCl, 0.4 *M* boric acid, 5 m*M* EDTA, pH 8.4) (see Appendix B: Composition of DNA Sequencing Reagents)
[α-^{32}P] dATP (500 to 1500 Ci/mmol) or [α-^{35}S] dATP (500 to 1500 Ci/mmol) Stock nucleotide reagents (ddNTP and dNTP premixes or stock solutions of nucleotides) (see Appendix C: Stock Solutions of Deoxy and Dideoxy Nucleotides)
Formamide stop/loading dye solution solution (see Appendix B: Composition of DNA Sequencing Reagents)
10X sequencing buffer (see Appendix B: Composition of DNA Sequencing Reagents)
Template
Primer(s)
Sequencing enzyme(s)

General Reagents

Phenol (equilibrated in 0.5 *M* Tris, pH 7.8)
Chloroform
Isoamyl alcohol
Sterile test tubes (13 × 100 mm)
100 m*M* IPTG (isopropyl-β-D-thiogalactopyranoside)
2% X-gal (5-bromo-4-chloro-3-indolyl-β-D-galactoside) in *N-N*-dimethylformamide
Sterile 27-gauge platinum wire inoculating loop (or toothpicks)
Restriction enzymes, T4 DNA ligase, DNA polymerase

Recipes for Media

YT Medium[41]

Casein hydrolysate (Tryptone)	8 g/l
Yeast extract	5 g/l
NaCl	5 g/l

Adjust pH to 7.2 to 7.4 with NaOH.

YT Agar

YT medium which contains 15 g of agar per liter

Plating Agar

YT medium which contains 6 g of agar per liter, autoclaved and cooled to 45 to 50°C

Add (per 3 ml):

1.0 ml 2% X-gal (5-bromo-4-chloro-3-indolyl-β-D-galactoside) in dimethylformamide (avoid plastics with concentrated dimethylformamide)

0.02 ml 100 m*M* IPTG (isopropyl-β-D-thiogalactopyranoside)

Minimal Medium (1 l)[41]

M9 salts

Na_2HPO_4	6 g
KH_2PO_4	3 g
NaCl	0.5 g
NH_4Cl	1 g

Adjust to pH 7.4 and autoclave.

When cool (45°C), add supplements, as follows:

1 ml 1.0 *M* $MgSO_4$

1 ml 0.1 *M* $CaCl_2$

1 ml 1% thiamine-HCl

5 ml 40% glucose

Minimal Agar Plates

Minimal medium which contains 15 g of agar per liter. Add agar after adjusting pH, autoclave, and allow to cool to 45°C before adding supplements, as above.

Luria Broth (L Broth)[41]

Bacto-tryptone	10 g
Bacto-yeast extract	5 g
NaCl	10 g
H_2O to 1 liter	

Adjust pH to 7.5 and autoclave.

Luria Agar Plates

Luria medium which contains 15 g of agar per liter. Add agar after adjusting pH, autoclave, and allow to cool to 45°C before adding antibiotics.

Single-Stranded Template DNA Isolation Stock Solutions
Precipitation Solution
2.5 *M* NaCl, 20% PEG 6000
Phage Resuspension Buffer (1X TE Buffer)
0.01 *M* tris, 0.001 *M* EDTA, pH 8.0

Alkaline Lysis Mini-Prep Stock Solutions
10X Tris/EDTA/Glucose
250 ml Tris, 100 m*M* EDTA, 500 m*M* glucose, pH 8.0
Dilute the stock solution tenfold into dH_2O before use. Add solid lysozyme to a final concentration of 10 mg/ml; place the mixture on ice and allow the lysozyme to dissolve. Swirl the solution before use to ensure that it is homogeneous.
10X Lysis Solutions
2 *M* NaOH
10% SDS
Dilute and mix the stock solutions tenfold into dH_2O for use.
1X Precipitation Mix
3 *M* NaOAc, pH 4.8
1X TE Buffer
0.01 *M* Tris, 0.001 *M* EDTA, pH 8.0

CsCl Plasmid Preparation Solution:
1X Sucrose/Tris
25% sucrose, 50 m*M* Tris, pH 8.0
1X EDTA
250 m*M* EDTA, pH 8.0
1X Tris/Lysozyme
0.25 *M* Tris, pH 8.0
Add solid lysozyme to a final concentration of 10 mg/ml and allow to dissolve while incubating on ice. Gently swirl the solution to mix. Prepare fresh each time.
1X Lysis Mix
50 m*M* Tris, pH 8.0, 62.5 m*M* EDTA, 1% Triton X-100
1X TE Buffer
0.01 *M* Tris, 0.001 *M* EDTA, pH 8.0

λ DNA Preparation solutions
Cell Resuspension Buffer
10 m*M* $Mg_2SO_4{\cdot}7H_2O$
SM Buffer
50 m*M* Tris-HCl, pH 7.5, 8 m*M* $Mg_2SO_4{\cdot}7H_2O$, 100 m*M* NaCl, 0.01% gelatin
CsCl Solutions
ρ = 1.4 g/ml (w/v): 53 g of CsCl plus 87 ml of 10 m*M* Tris-HCl, 20 m*M* $MgCl_2$, 10 m*M* NaCl, pH 7.5
ρ = 1.6 g/ml (w/v): 81 g of CsCl plus 79 ml of 10 m*M* Tris-HCl, 20 m*M* $MgCl_2$, 10 m*M* NaCl, pH 7.5
1X TE Buffer
0.01 *M* Tris, 0.001 *M* EDTA, pH 8.0
20% polyethylene glycol 6000 (PEG)
10% SDS
0.5 *M* EDTA
3 *M* NH_4OAc

REFERENCES

1. **Moores, J.**, Current approaches to DNA sequencing, *Anal. Biochem.*, 163, 1, 1987.
2. **Smith, L.**, DNA sequence analysis: past, present and future, *Am. Biotechnol. Lab.*, May 1989, p. 10.
3. **Sanger, F., Nicklen, S., and Coulson, A. R.**, DNA sequencing with chain-terminating inhibitors, *Proc. Natl. Acad. Sci. U.S.A.*, 74, 5463, 1977.
4. **Sanger, F. and Coulson, A. R.**, The use of thin acrylamide gels for DNA sequencing, *FEBS Lett.*, 87, 107, 1978.
5. **Sanger, F., Coulson, A. R., Barrell, B. G., Smith, A. J. H., and Roe, B. A.**, Cloning in single-stranded bacteriophage as an aid to rapid DNA sequencing, *J. Mol. Biol.*, 143, 161, 1980.
6. **Maxam, A. and Gilbert, W.**, A new méthod for sequencing DNA, *Proc. Natl. Acad. Sci. U.S.A.*, 74, 560, 1977.
7. **Gish, G. and Eckstein, F.**, DNA and RNA seqeunce determination based on phosphorothioate chemistry, *Science*, 240, 1520, 1988.
8. **Ambrose, B. J. and Pless, R. C.**, DNA sequencing: chemical methods, *Methods Enzymol.*, 152, 522, 1987.
9. **Maxam, A. and Gilbert, W.**, Sequencing end-labeled DNA with base-specific chemical cleavages, *Methods Enzymol.*, 65, 499, 1980.
10. **Rubin, J. M. and Schmid, C. W.**, Pyrimidine-specific chemical reactions useful for DNA sequencing, *Nucleic Acids Res.*, 8, 4613, 1980.
11. **Hindley, J.**, *Lab. Tech. Biochem. Mol. Biol.*, 10, 230, 1983.
12. **Brown, N. L.**, DNA sequencing, *Methods Microbiol.*, 17, 259, 1984.
13. **Kieny, M., Lathe, R., and Lecoq, J.**, New versatile cloning and sequencing vectors based upon bacteriophage M13, *Gene*, 26, 91, 1983.
14. **Arnold, W. and Puhler, A.**, A family of high copy-number plasmid vectors with single end-label sites for rapid nucleotide sequencing, *Gene*, 70, 171, 1988.
15. **Mierendorf, R. C. and Pfeffer, D.**, Direct DNA sequencing of denatured plasmid DNA, *Methods Enzymol.*, 152, 556, 1987.
16. **Geliebter, J.**, Dideoxy nucleotide sequencing of RNA and uncloned cDNA, *Focus*, 9, 5, 1987.
17. **Church, G. and Gilbert, W.**, Genomic sequencing, *Proc. Natl. Acad. Sci. U.S.A.*, 81, 1991, 1984.
18. **Little, P.**, A close look at the genome, *Nature*, 310, 369, 1984.
19. **McMahon, G., Davis, E., and Wogan, G.**, Characterization of c-Ki-*ras* oncogene alleles by direct sequencing and enzymatically amplified DNA from carcinogen-induced tumors, *Proc. Natl. Acad. Sci. U.S.A.*, 84, 4974, 1987.
20. **Wong, C., Dowling, C., Saiki, R., Higuchi, R., Erlich, H., and Kazazian, H.**, Direct genomic sequencing of amplified single-copy DNA: rapid characterization of unknown β-thalassemia mutations, *Nature*, 330, 384, 1987.
21. **Wirschnik, L., Higuchi, R., Stoneking, M., Erlich, H., Arnheim, N., and Watson, A.**, Length mutations in human mitochondrial DNA: direct sequencing of enzymatically amplified DNA, *Nucleic Acids Res.*, 15, 529, 1987.
22. **Oste, C.**, Polymerase chain reaction, *BioTechniques*, 6, 162, 1988.
23. **Stoflet, E., Koeber, L., Sarkar, G., and Sommer, S.**, Genomic amplification with transcript sequencing, *Science*, 239, 491, 1988.
24. **Innis, M., Myambo, K., Gelfand, D., and Brown, M.**, DNA sequencing with *Thermus aquaticus* DNA polymerase and direct sequencing of polymerase chain reaction-amplified DNA, *Proc. Natl. Acad. Sci. U.S.A.*, 85, 9436, 1988.
25. **Mihovilovic, M. and Lee, J.**, An efficient method for sequencing PCR amplified DNA, *BioTechniques*, 7, 14, 1989.
26. **Frye, R., Cogswell, P., and Liu, E.**, The analysis of mutations in the *ras* gene family using the polymerase chain reaction (PCR), *Clontech Cruiser*, 1, 1, 1989.
27. **Engelke, P. R., Hoener, P., and Collins, F.**, Direct sequencing of enzymatically amplified human genomic DNA, *Proc. Natl. Acad. Sci. U.S.A.*, 84, 4974, 1988.
28. **Gorman, K. and Steinberg, R.**, Simplified method for selective amplification and direct sequencing of cDNAs, *BioTechniques*, 7, 326, 1989.
29. **Carothers, A. M., Urlab, G., Mucha, J., Grunburger, D., and Chasin, L. A.**, Point mutation analysis in a mammalian gene: rapid preparation of total RNA, PCR amplification of cDNA and *Taq* sequencing by a novel method, *BioTechniques*, 7, 494, 1989.
30. **Levedakou, E., Landegren, U., and Hood, L. E.**, A strategy to study gene polymorphism by direct sequence analysis of cosmid clones and amplified genomic DNA, *BioTechniques*, 7, 438, 1989.
31. **Gorman, K. and Steinberg, R.**, Simplified method for selective amplification and direct sequencing of cDNAs, *BioTechniques*, 7, 326, 1989.

32. **Gibbs, R., Nguyen, P. N., McBride, L. J., Koepf, S., and Casky, C. T.,** Identification of mutations leading to the Lesch-Nyhan syndrome by automated direct sequencing in *in vitro* amplified cDNA, *Proc. Natl. Acad. Sci. U.S.A.,* 86, 1919, 1989.
33. **Gyllensten, U. B.,** PCR and DNA sequencing, *BioTechniques,* 7, 700, 1989.
34. **Mitchell, L. and Merrill, C. R.,** Affinity generation of single-stranded DNA for dideoxy sequencing, following the polymerase chain reaction, *Anal. Biochem.,* 178, 239, 1989.
35. **Roth, M. E., Lacy, M. J., McNeil, L., and Kranz, D.,** Analysis of T-cell receptor transcripts using the polymerase chain reaction, *BioTechniques,* 7, 746, 1989.
36. **Brow, M. A. D.,** Sequencing with Taq polymerases in *PCR Protocols: A Guide To Methods and Applications,* Innis, M., Gelfand, D. H., Snisky, J. J., and White, T. J., Eds., Academic Press, New York, 1989.
37. **Hackett, P., Fuchs, J., and Messing, J.,** *An Introduction to Recombinant DNA Techniques,* Benjamin Cummins, Menlo Park, CA, 1986.
38. **Perbal, B.,** *A Practical Guide To Molecular Cloning,* 2nd ed., Academic Press, New York, 1988.
39. **Ausubel, F., Brent, R., Kingston, R., Moore, D., Smith, J., Seidman, J., and Struhl, K., Eds.,** *Current Protocols in Molecular Biology,* John Wiley & Sons, New York, 1987.
40. **Maniatis, T., Frisch, E. F., and Sambrook, J.,** *Molecular Cloning: A Laboratory Manual,* 2nd ed., Cold Spring Harbor Laboratory, Cold Spring Harbor, NY, 1989.
41. **Miller, J.,** *Experiments in Molecular Genetics,* Cold Spring Harbor Laboratory, Cold Spring Harbor, NY, 1972.
42. **Rodriquez, R. and Tait, R.,** *Recombinant DNA Techniques: An Introduction,* Benjamin Cummings, Menlo Park, CA, 1983.
43. **Messing, J.,** New M13 vector for cloning, *Methods Enzymol.,* 101, 20, 1983.
44. **Denhardt, D. T., Dressler, D., and Ray, D. S.,** *Single-Stranded DNA Phages,* Cold Spring Harbor Laboratory, Cold Spring Harbor, NY, 1978, 379.
45. **Ray, D. S.,** *Comprehensive Virology,* Vol. 7, Fraenkel-Conrat, H. and Wagner, R. R., Eds., Plenum Press, New York, 1977, 105.
46. **Barnes, W.,** DNA sequencing with single-stranded phage vectors, in *Genetic Engineering,* Vol. 2, Setlow, J. K. and Hollander, A., Eds., Plenum Press, New York, 1980, 185.
47. **Van Wezenbeek, P., Hulsebos, T., and Schoenmakers, J.,** Nucleotide sequencing of the filamentous bacteriophage M13 DNA genome: comparison with phage fd, *Gene,* 11, 129, 1980.
48. **Short, J., Fernandez, J., Sorge, J., and Huse, W.,** Lambda ZAP, a bacteriophage lambda expression vector with *in vivo* excission properties, *Nucleic Acids Res.,* 16, 7583, 1988.
49. **Zagursky, R. J. and Berman, M. L.,** Cloning vectors that yield high levels of single-stranded DNA for rapid DNA sequencing, *Gene,* 27, 183, 1984.
50. **Felton, J.,** M13 host strain JM103 contains 2 restriction systems, *BioTechniques,* 1, 42, 1983.
51. **Barnes, W. M.,** Sequencing DNA with dideoxyribonucleotides as chain terminating inhibitors: hints and strategies for big projects, *Methods Enzymol.,* 152, 538, 1987.
52. **Zagursky, R.,** pBR322 plasmids that yield single-stranded DNA, *Focus,* 7, 4, 1985.
53. **Dente, L., Cesareni, G., and Cortese, R.,** pEMBL: a new family of single-stranded plasmids, *Nucleic Acids Res.,* 11, 1645, 1983.
54. **Peng, Z. and Wu, R.,** A simple and rapid nucleotide sequencing strategy and its application in analyzing a rice histone3 gene, *Gene,* 45, 135, 1986.
55. **Zinder, N. D. and Boeke, J. D.,** The filamentous phage (Ff) as vectors for recombinant DNA — a review, *Gene,* 19, 1, 1982.
56. **Levinson, A., Silver, D., and Seed, B.,** Minimal size plasmids containing an M13 origin for production of single-stranded transducing particles, *J. Mol. Appl. Genet.,* 2, 507, 1984.
57. **Messing, J., Crea, R., and Seeburg, P. H.,** A system for shotgun DNA sequencing, *Nucleic Acids Res.,* 9, 309, 1981.
58. **Hong, G. F.,** A systematic DNA sequencing method, *J. Mol. Biol.,* 158, 539, 1982.
59. **Misra, T. K.,** A new strategy to create ordered deletions for rapid nucleotide sequencing, *Gene,* 34, 263, 1985.
60. **Vieira, J. and Messing, J.,** Production of single-stranded phage DNA, *Methods Enzymol.,* 153, 3, 1987.
61. **Heidecker, G., Messing, J., and Gronenborn, B.,** A versatile primer to DNA sequencing in the M13mp2 cloning system, *Gene,* 10, 69, 1980.
62. **Lin, H., Lei, S., and Wilcox, G.,** An improved DNA sequencing strategy, *Anal. Biochem.,* 147, 114, 1985.
63. **Anderson, S.,** A shotgun DNA sequencing strategy using cloned DNAseI-generated fragments, *Nucleic Acids Res.,* 9, 3015, 1981.
64. **Deininger, P.,** Random subcloning of sonicated DNA: application to shotgun DNA sequence analysis, *Anal. Biochem.,* 129, 216, 1983.
65. **Deininger, P.,** Approaches to rapid DNA sequence analysis, *Anal. Biochem.,* 135, 247, 1983.

66. **Roe, B.,** Computer programs for molecular biology: an overview of DNA sequencing and protein analysis packages, *BioTechniques,* 6, 560, 1988.
67. **Devereux, J., Haeberli, P., and Smithes, O.,** A comprehensive set of sequence analysis programs for the VAX, *Nucleic Acids Res.,* 12, 387, 1984.
68. **Keller, C., Corcoran, N., and Roberts, R.,** Computer programs for handling nucleic acid sequences, *Nucleic Acids Res.,* 12, 379, 1984.
69. **Staden, R.,** A computer program to enter DNA gel reading data into a computer, *Nucleic Acids Res.,* 12, 499, 1984.
70. **Dale, R., McClure, B., and Houchins, J.,** A rapid-stranded cloning strategy for producing a sequential series of overlapping clones for use in DNA sequencing: application to sequencing the corn mitochondrial 18S rDNA, *Plasmid,* 13, 31, 1985.
71. **Liu, Z. and Hackett, P. B.,** Rapid generation of subclones for DNA sequencing using the reverse cloning procedure, *BioTechniques,* 7, 722, 1989.
72. **Ozkaynak O. and Putney, S.,** A undirectional deletion technique for the generation of clones for sequencing, *BioTechniques,* 5, 770, 1987.
73. **Frischauf, A. M., Garoff, H., and Lerach, H.,,** A subcloning strategy for DNA sequence analysis, *Nucleic Aids Res.,* 8, 5541, 1980.
74. **Poncz, M., Solowiejczyk, D., Ballantine, M., Schwartz, E., and Surrey, S.,** "Non-random" DNA sequence analysis in bacteriophage M13 by the dideoxy chain-termination method. *Proc. Natl. Acad. Sci. U.S.A.,* 79, 4289, 1982.
75. **Barnes, W. M., Bevan, M., and Son, P. H.,** Kilo-sequencing: creation of an ordered nest of asymmetric deletions across a large target sequence carried on phage M13, *Methods Enzymol.,* 101, 98, 1983.
76. **Henikoff, S.,** Undirectional digestion with exonucleaseIII creates targeted breakpoints for DNA sequencing, *Gene,* 28, 351, 1984.
77. **Ansorge, W. and Labeit, S.,** Field-gradients improve resolution in DNA sequencing gels, *J. Biochem. Biophys. Methods,* 10, 237, 1984.
78. **Bankier, A. T.,** Advances in dideoxy sequencing, *BioTechniques,* 2, 72, 1984.
79. **Barnes, W. M. and Bevan, M.,** Kilo-sequencing: an ordered strategy for rapid DNA sequence data acquisition, *Nucleic Acids Res.,* 11, 349, 1983.
80. **Hong, G. F.,** A method for sequencing single-stranded cloned DNA in both directions. *BioSci. Rep.,* 1, 243, 1981.
82. **Smith, A.,** DNA sequence analysis by primed synthesis, *Methods Enzymol.,* 65, 560, 1980.
83. **Smith, A. J. H.,** The use of exonucleaseIII for preparing single-stranded DNA for use as a template in the chain termination sequencing method, *Nucleic Acids Res.,* 6, 831, 1979.
84. **Guo, L.-H. and Wu, R.,** New rapid methods for DNA sequencing based on exonucleaseIII digestion followed by repair synthesis, *Nucleic Acids Res.,* 10, 2065, 1982.
85. **Guo, L.-H., Yang, R. C. A., and Wu, R.,** An improved strategy for rapid sequencing of both strands of long DNA molecules cloned in a plasmid, *Nucleic Acids Res.,* 11 5521, 1983.
86. **Crouse, G. F., Frischauf, A. M., and Lehrach, H.,** An integrated and simplified approach to cloning into plasmids and single-stranded phages, *Methods Enzymol.,* 101, 78, 1983.
87. **Ahmed, A.,** Use of transposon-promoted deletions in DNA sequencing analysis, *J. Mol. Biol.,* 178, 941, 1984.
88. **Ahmed, A.,** A rapid procedure for nucleotide sequencing using transposon-promoted deletions, *Gene,* 39, 305, 1985.
89. **Ahmed, A.,** Use of transposon-promoted deletions in DNA sequence analysis, *Methods Enzymol.,* 155, 177, 1987.
90. **Vieira, J. and Messing, J.,** The pUC plasmids, and M13mp7 derived system for insertion mutagenesis and sequencing with synthetic universal primers, *Gene,* 19, 259, 1982.
91. **Okita, T.,** Non-random sequencing of exonucleaseIII-deleted complementary DNA, *Anal. Biochem.,* 144, 207, 1985.
92. **Burton, F., Loeb, D., McGraw, R., Edgell, M., and Hutchison, C.,** A directed nucleotide sequencing approach for single-stranded vectors based upon recloning intermediates of a progressive DNA synthesis reaction, *Gene,* 67, 159, 1988.
93. **Nag, D., Huang, H., and Berg, D.,** Bidirectional chain termination nucleotide sequencing: transposon Tn5seq1 as a mobile source of primer sites, *Gene,* 64, 135, 1988.
94. **Ahmed, A.,** A simple and rapid procedure for sequencing long (40 kb) DNA fragments, *Gene,* 61, 363, 1987.
95. **Shen, W. and Waye, M.,** A novel method for generating a nested set of unidirectional deletion mutants using mixed oligodeoxynucleotides, *Gene,* 70, 205, 1988.
96. **Hasan, N., Kim, S., Podhajska, A., and Szybalski, W.,** A novel multistep method for generating precise unidirectional deletions using *Bsp*MI, a class II restriction enzyme, *Gene,* 50, 55, 1986.

97. **Lee, Y.-M. and Lee, S.-C.,** A DNA sequencing strategy, *Anal. Biochem.*, 175, 521, 1989.
98. **Oshima, R.,** Rapid identification of M13 phage deletions for sequence analysis, *BioTechniques*, 6, 510, 1988.
99. **Aegerter, E. and Trachsel, H.,** Determination of size and orientation of DNA fragments cloned in phage M13 by S1 nuclease mapping, *Nucleic Acids Res.*, 15, 372, 1987.
100. **Steggles, A.,** A rapid procedure for creating nested sets of deletions using mini-prep plasmid DNA samples, *BioTechniques*, 7, 241, 1989.
101. **Sorge, J. and Blinderman, L. A.,** ExoMeth sequencing of DNA: eliminating the need for subcloning and oligonucleotide primers, *Proc. Natl. Acad. Sci. U.S.A.*, 86, 9208, 1989.
102. **Strauss, E., Kobori, J., Siu, G., and Hood, L.,** Specific-primer directed DNA sequencing, *Anal. Biochem.*, 154, 353, 1986.
103. **Stambaugh, K. and Blakesley, R.,** Extended DNA sequencing with Klenow fragment: the Kilobase sequencing™ system, *Focus*, 10, 29, 1988.
104. **Lim, H. and Pene, J.,** Optimal conditions for supercoil DNA sequencing with the *Escherichia coli* DNA polymerase I large fragment, *Gene Anal. Tech.*, 5, 32, 1988.
105. **Richterich, P., Heller, C., Wurst, H., and Pohl, F.,** DNA sequencing with direct blotting electrophoresis and colorimetric detection, *BioTechniques*, 7, 52, 1989.
106. **Zagursky, R. J., Baumeister, K., Lomax, N., and Berman, M. L.,** Rapid and easy sequencing of large double-stranded DNA and supercoiled plasmid DNA, *Gene Anal. Tech.*, 2, 89, 1985.
107. **Bartlett, J., Gaillard, P., and Joklik, W.,** Sequencing of supercoiled plasmid DNA, *BioTechniques*, 4, 208, 1986.
108. **Graham, A., Steven, J., McKechnie, D., and Harris, W.,** Direct DNA sequencing using avian myeloblastosis virus and Maloney-murine virus reverse transcriptase, *Focus*, 8, 4, 1978.
109. **Krawetz, S.,** DNA sequencing with AMV (avian myoblastosis virus) reverse transcriptase and a modified T7 polymerase (Sequenase), *BioTechniques*, 5, 620, 1987.
110. **Karathanasis, S.,** M13 DNA sequencing using reverse transcriptase, *Focus*, 4, 6, 1982.
111. **Johnston-Dow, E., Mardis, E., Heiner, C., and Roe, B. A.,** Optimized methods for fluorescent and radiolabeled DNA sequencing, *BioTechniques*, 5, 754, 1987.
112. **Tabor, S. and Richardson, C. C.** DNA sequence analysis with a modified bacteriophage T7 DNA polymerase, *Proc. Natl. Acad. Sci. U.S.A.*, 84, 4767, 1987.
113. **Toneguzzo, F., Glynn, S., Levi, E., Mjolsness, S., and Hayday, A.,** Use of chemically modified T7 DNA polymerase for manual and automated sequencing of supercoiled DNA, *BioTechniques*, 6, 460, 1988.
114. **Tabor, S. and Richardson, C. C.,** Selective inactivation of the exonuclease activity of bacteriophage T7 DNA polymerase by *in vitro* mutagenesis, *J. Biol. Chem.*, 264, 4767, 1989.
115. **Tabor, S. and Richardson, C. C.,** Effect of manganese ions on the incorporation of dideoxynucleotides by bacteriophage T7 DNA polymerase and *Escherichia coli* DNA polymerase I, *Proc. Natl. Acad. Sci. U.S.A.*, 86, 4076, 1989.
116. **Peterson, M.,** DNA sequencing using Taq polymerase, *Nucleic Acids Res.*, 16, 10915, 1988.
117. **D'Alessio, J. M., Kullman, J., Salyards, A. J., and Campbell, J. H.,** Highly purified DNA polymerase from *Thermus aquaticus*, *Focus*, 11, 30, 1989.
118. **Gross, L.,** Recent developments in Thermal Base™ Taq sequencing, *Strategies*, 3, 28, 1989.
119. **Ornstein, D. L. and Kashdan, M.,** Sequencing DNA using ^{35}S labeling: a troubleshooting guide, *BioTechniques*, 3, 476, 1985.
120. **Biggin, M. D., Gibson, T. J., and Hong, G. F.,** Buffer-gradient gels and ^{35}S-label as an aid to rapid DNA sequence determination, *Proc. Natl. Acad. Sci. U.S.A.*, 80, 3963, 1983.
121. **Williams, S. A., Slatko, B. E., Moran, L. S., and DiSimone, S. M.,** Sequencing in the fast lane: a rapid protocol for [α-^{35}S]-dATP dideoxy DNA sequencing, *BioTechniques*, 4, 138, 1986.
122. **Hodgson, C., Fisk, R., and Willett, L.,** Radiochemical degradation of nucleic acids by 32-phosphorus, *BioTechniques*, 6, 208, 1988.
123. **Reed, A., Kost, T., and Miller, T.,** Simple improvements in ^{35}S dideoxy DNA sequencing, *BioTechniques*, 4, 306, 1986.
124. **Connel, C., Fung, S., Heiner, C., Bridgeham, J., Chakerian, V., Heron, E., Jones, B., Menchen, S., Mordan, W., Raff, M., Racknor, M., Smith, L., Springer, J., Woo, S., and Hunkapillar, M.,** Automated DNA sequence analysis, *BioTechniques*, 5, 342, 1987.
125. **Brumbaugh, J., Middendorf, L., Grone, D., and Ruth, J.,** Continuous on-line DNA sequencing using oligodeoxynucleotide primers with multiple fluorophores, *Proc. Natl. Acad. Sci. U.S.A.* 85, 5610, 1988.
126. **Smith, L., Sanders, J., Kaiser, R., Hughes, P., Dodd, C., Heiner, C., Kent, S., and Hood, L.,** Fluorescence detection in automated DNA sequence analysis, *Nature*, 321, 674, 1986.
127. **Smith, L., Fung, S., Hunkapiller, M., Hunkapiller, T., and Hood, L.,** The synthesis of oligonucleotides containing an aliphatic amino group at the 5′ terminus: synthesis of fluorescent DNA primers for use in DNA sequence analysis, *Nucleic Acids Res.*, 13, 2999, 1985.

128. **Ansorge, W., Sproat, B., Stegemann, J., Schwager, C., and Zenke, M.,** Automated DNA sequencing: ultrasensitive detection of fluorescent bands during electrophoresis, *Nucleic Acids Res.,* 15, 4593, 1987.
129. **Middendorf, L., Brumbaugh, J., Grone, D., Morgan, C., and Ruth, J.,** Large-scale DNA sequencing, *Am. Biol. Lab.,* August 1988, p. 14.
130. **Prober, J., Trainor, G., Dam, R., Hobbs, F., Robertson, C., Zagursky, R., Cocuzza, R., Jenson, M., and Baumeister, K.,** A system for rapid DNA sequencing with fluorescent chain-terminating dideoxynucleotides, *Science,* 238, 336, 1987.
131. **Amorese, D. and Hochberg, A.,** A fluorescence labeled DNA analysis system for sequencing and mapping, *Am. Biotech. Lab.,* March 1989, p. 38.
132. **Mills, D. R. and Kramer, F. R.,** Structure-independent nucleotide analysis, *Proc. Natl. Acad. Sci. U.S.A.,* 76, 2232, 1979.
133. **Barr, P., Thayer, R., Najarian, R., Seela, F., Laybourn, P., and Tolan, D.,** 7-deaza-2′-triphosphate: enhanced resolution in M13 dideoxy sequencing, *BioTechniques,* 4, 428, 1986.
134. **Mizusawa, S., Nishimura, S., and Seela, F.,** Improvement of the dideoxy chain-termination method of DNA sequencing by the use of deoxy-7-deazaguanosine triphosphate in place of dGTP, *Nucleic Acids Res.,* 14, 1319, 1986.
135. **Ner, S., Goodin, D., Dielak, G., and Smith, M.,** A rapid droplet method for Sanger dideoxy sequencing, *BioTechniques,* 6, 408, 1988.
136. **Wilson, R., Yuen, A., Clark, S., Spence, C., Arakelian, P., and Hood, L.,** Automation of dideoxynucleotide DNA sequencing reaction using a robotic workstation, *BioTechniques,* 6, 776, 1988.
137. **Ortlepp, S. and McKay, I.,** Performing nucleic acid reaction using predispensed lyophilized reaction mixtures, *BioTechniques,* 7, 1110, 1989.
138. **Hattori, M. and Sakaki, Y.,** Dideoxy DNA sequencing method using denatured plasmid templates, *Anal. Biochem.,* 152, 232, 1986.
139. **Zhang, H., Scholl, R., Browse, J., and Somerville, C.,** Double-stranded DNA sequencing as a choice for DNA sequencing, *Nucleic Acids Res.,* 16, 1220, 1988.
140. **Chen, E. Y. and Seeburg, P. H.,** Supercoil sequencing: a fast simple method for sequencing plasmid DNA, *DNA,* 4, 165, 1985.
141. **Haltiner, M., Kempe, T., and Tijian, R.,** A novel strategy for constructing clustered point mutations, *Nucleic Aids Res.,* 13, 1015, 1985.
142. **Wallace, R. B., Johnson, M. J., Suggs, S. V., Miyoshi, K., Bhatt, R., and Itakura, K.,** A set of synthetic oligodeoxyribonucleotide primers for DNA sequencing in the plasmid vector pBR322, *Gene,* 16, 21, 1981.
143. **Gravel, R. A., Quan, F., and Korneluk, R. G.,** Rapid and reliable dideoxy sequencing of double-stranded DNA, *Gene,* 40, 317, 1985.
144. **Riley, D.,** Very rapid nucleotide sequence analysis of improved double-stranded minipreps, *Gene,* 75, 193, 1989.
145. **Steffens, D. L. and Gross, R. W.,** Sequencing of cloned DNA using bacteriophage λgt11 templates, *BioTechniques,* 7, 674, 1989.
146. **Manfioletti, G. and Schneider, C.,** A new and fast method for preparing high quality λ DNA suitable for sequencing, *Nucleic Acids Res.,* 16, 2873, 1988.
147. **Reddy, K. J., Kuwabara, T., and Sherman, L. A.,** A simple and efficient procedure for the isolation of high-quality phage λ DNA using a DEAE-cellulose column, *Anal. Biochem.,* 168, 324, 1988.
148. **White, B. A. and Rosenzweig, S.,** A reliable method for the purification of λ DNA, *BioTechniques,* 7, 694, 1989.
149. **Chisholm, D.,** A convenient moderate-scale procedure for obtaining DNA from bacteriophage lambda, *BioTechniques,* 7, 21, 1989.
150. **Del Sal, G., Manfioletti, G., and Schneider, C.,** The CTAB-DNA precipitation method: a common miniscale preparation of template DNA from phagemids, phages or plasmids suitable for sequencing, *BioTechniques,* 7, 514, 1989.
151. **Birnboim, H. L. and Doly, J.,** Rapid alkaline extraction procedure for screening recombinant plasmid DNA, *Nucleic Acids Res.,* 7, 1513, 1979.
152. **Clewell, D. and Helinski, D.,** Supercoiled circular DNA-protein complex in *Escherichia coli:* purification and induced conversion to an open-circular form. *Proc. Natl. Acad. Sci. U.S.A.,* 62, 1159, 1969.
153. **Windle, B.,** A simple technique for preparing pure λ DNA, *BioTechniques,* 6, 402, 1988.
154. **Just, L., Frankis, R., Lowery, W., Meyer, R., and Paddock, G.,** Rapid high-yield purification of plasmids by alkaline extraction, PEG precipitation and agarose gel chromatography, *BioTechniques,* 1, 136, 1983.
155. **Sadhu, C. and Gedamu, L.,** A procedure for the preparation of RNA-free plasmid DNA, *BioTechniques,* 6, 20, 1988.
156. **Kado, C. and Liu, S.,** Rapid procedure for detection and isolation of large and small plasmids, *J. Bacteriol.,* 145, 1365, 1981.

157. **Griffith, O.,** Large-scale isolation of plasmid DNA using high-speed centrifugation methods, *BioTechniques,* 6, 725, 1988.
158. **Gaterman, K., Rosenberg, G., and Kaufer, N.,** Double-stranded sequencing, using mini-prep plasmids, in eleven hours, *BioTechniques,* 6, 950, 1988.
159. **Mardis, E. R. and Roe, B. A.,** Automated methods for single-stranded DNA sequencing reactions on a robotic workstation, *BioTechniques,* 7, 840, 1989.
160. **Attal, J., Puissant, C., and Houdebine, L.-M.,** An improvement of a rapid method using Qiagen columns to purify plasmids, *BioTechniques,* 8, 269, 1990.
161. **Kim, B. and Jue, C.,** Direct sequencing of lambda-gt11 recombinant clones, *BioTechniques,* 8, 156, 1990.
162. **Liszewski, M., Kumar, V., and Atkinson, J.,** "Midi-Prep" isolation of plasmid DNA in less than two hours for sequencing, subcloning and hybridizations, *BioTechniques,* 7, 1079, 1988.
163. **Holmes, D. and Quigley, M.,** A rapid boiling method for the preparation of bacterial plasmids, *Anal. Biochem.,* 114, 193, 1981.
164. **Holmes, D.,** Improved heating technique for screening recombinant DNA plasmids in *E. coli, BioTechniques,* 2, 68, 1984.
165. **Kraft, R., Tardiff, J., Krautner, K., and Leinwand, L.,** Using mini-prep plasmid DNA for sequencing double-stranded templates with Sequenase™, *BioTechniques,* 6, 554, 1988.
166. **Wang, Y.,** Double-stranded DNA sequencing using unpurified templates produced by rapid boiling, *BioTechniques,* 6, 843, 1988.
167. **Wilmizig, M.,** LiCl-boiling method for plasmid mini-preps, *Trends Genet.,* 1, 158, 1985.
168. **Smith, M., Leung, D., Gillam, S., Astell, C., Montomery, D., and Hall, B.,** Sequence of the gene for iso-1-cytochrome c in *Saccharomyces cereviseae, Cell,* 16, 753, 1979.
169. **Wallace, R. B., Schold, M., Johnson, M., Dembek, P., and Itakura, K.,** Oligonucleotide directed mutagenesis of the human B-globin gene: a general method for producing specific mutations in cloned DNA, *Nucleic Acids Res.,* 9, 3647, 1981.
170. **Winship, P. R.,** An improved method for directly sequencing PCR amplified material using dimethyl sulfoxide, *Nucleic Acids Res.,* 17, 1266, 1989.
171. **James, R. and Bradshaw, R. A.,** Strand-separation of DNA fragments and their isolation from denaturing polyacrylamide gels, *Anal. Biochem.,* 140, 456, 1984.
172. **Melzer, S., Mane, S., Wood, P., Johnson, L., and Needleman, S.,** Sequencing products of the polymerase chain reaction directly, without purification, *BioTechniques,* 8, 142, 1990.
173. **Page, G.,** Some tips on sequencing with double-stranded plasmids, *Baseline,* 1, 2, 1990.
174. **Wigboldus, J.,** Handling ^{35}S gels, *BioTechniques,* 5, 630, 1987.
175. **Kobayashi, Y.,** A method to cast thin sequencing gels, *Focus,* 10, 73, 1988.
176. **Horseman, N. D. and Hankovich, S.,** A chamber and method which simplify fixing ultrathin sequencing gels, *BioTechniques,* 7, 340, 1989.
177. **Black, M., Wilson, E., and Hruby, D.,** A simple gel-saving tip, *BioTechniques,* 7, 1083, 1989.
178. **Sheen, J.-Y. and Seed, B.,** Electrolyte gradient gels for DNA sequencing, *BioTechniques,* 6, 725, 1988.
179. **Olsson, A., Moks, T., Muhlen, M., and Gaal, A. B.,** Uniformly spaced banding patterns in DNA sequencing gels by the use of a field-strength gradient, *J. Biochem. Biophys. Methods,* 10, 83, 1984.
180. **Garoff, H. and Ansorge, W.,** Improvements of DNA sequencing gels, *Anal. Biochem.,* 115, 450, 1981.
181. **Ansorge, W. and Barker, R.,** System for DNA sequencing with resolution up to 600 base pairs, *J. Biochem. Biophys. Methods,* 9, 33, 1984.
182. **Naylor, L. and van de Sande, J.,** Improved sequence resolution of highly repetitive DNA fragments, *Nucleic Acids Res.,* 14, 5939, 1986.
183. **Hampsey, D. M.,** A simple method to increase the resilience of DNA sequencing gels, *BioTechniques,* 7, 150, 1989.
184. **Isfort, R. and Ihle, J.,** The 4-6-8 method of sequence analysis, *BioTechniques,* 6, 138, 1988.
185. **Blakesley, R.,** Diagnostic dideoxy DNA sequencing, *Focus,* 5, 1, 1983.
186. **Schreier, P. and Cortese, R.,** A fast and simple method for sequencing DNA cloned in the single-stranded bacteriophage M13, *J. Mol. Biol.,* 129, 169, 1979.
187. **Kristensen, T., Voss, H., and Ansorge, W.,** A simple and rapid preparation of M13 sequencing templates for manual and automated dideoxy sequencing, *Nucleic Acids Res.,* 15, 5507, 1987.
188. **Yamamoto, K., Alberts, B., Benzinger, R., Lawhorne, L., and Treiber, G.,** Rapid bacteriophage sedimentation in the presence of polyethylene glycol and its application to large scale virus production, *Virology,* 40, 734, 1970.
189. **Lis, J.,** Fractionation of DNA fragments by polyethylene glycol induced precipitation, *Methods Enzymol.,* 65, 347, 1980.
190. **Wang, L., Weber, D., Johnson, T., and Sakaguchi, A.,** Supercoil sequencing using unpurified templates produced by rapid boiling, *BioTechniques,* 6, 839, 1988.
191. **Maina, C., Limberger, R., and McReynolds, L.,** personal communication, 1988.

192. **Backmann, B.,** *Escherichia coli and Salmonella typhimurium: Cellular and Molecular Biology,* Neidhardt, F. C., Ingraham, J. L., Low, K. B., Schaechter, M., and Umbarger, H., Eds., American Society of Microbiology, Washington, D.C., 1987, 1190.
193. **Raleigh, L.,** personal communication, 1989.
194. **Gorisch, H.,** Drop dialysis: time course of salt and protein exchange, *Anal. Biochem.,* 173, 393, 1988.
195. **Dicker, A., Volkenandt, M., Adamo, A., Barrenda C., and Bertino, J.,** Sequence analysis of a human gene responsible for drug resistance: a rapid method for manual and automated direct sequencing of products generated by the polymerase chain reaction, *BioTechniques,* 7, 830, 1989.
196. **McBride, L., Koepf, S., Gibbs, R., Salser, W., Mayrand, P., Hunkapiller, M., and Kronick, M.,** Automated DNA sequencing methods involving polymerase chain reaction, *Clin. Chem.,* 35, 2196, 1989.
197. **McBride, L., Koepf, S., Gibbs, R., Nguyen, P., Salser, W., Mayrand, P., Hunkapiller, M., and Kronick, M.,** Interfacing in vitro DNA amplification with DNA sequencing strategies, in *Current Communications in Molecular Biology,* Erlich, H., Gibbs, R., and Kazazian, H. H., Eds., Cold Spring Harbor Laboratory, Cold Spring Harbor, NY, 1989, 211.
198. **Gyllensten, U. and Erlich, H.,** Generation of single-stranded DNA by the polymerase chain reaction and its amplification to direct sequencing of the HLA-DQA locus, *Proc. Natl. Acad. Sci. U.S.A.,* 83 7652, 1988.

Note: A recent volume of *BioTechniques,* Volume 9(1), 1990, edited by B. Roe and G. Church, describes several new DNA sequencing methods.

Chapter 7

PULSED-FIELD ELECTROPHORESIS: PROCEDURES, PROBLEMS, AND PRACTICALITIES

Kevin L. Bentley, Charles J. Bieberich, and Frank H. Ruddle

TABLE OF CONTENTS

I. INTRODUCTION

The most common method for separating DNA molecules is by gel electrophoresis. The versatility of this separation technique, using an agarose or acrylamide matrix, can be illustrated by its use in DNA purification, restriction mapping, sequencing, mobility shift assays, gene mapping, and a host of other applications. Conventional agarose gel electrophoresis has been used to demonstrate the separation of DNA molecules up to 50 kilobase pairs (kbp) in size, although the effective resolution is generally considered to be less than 30 kbp. Thus, the major drawback of this method is the steady decrease in resolving power as DNA molecules approach 50 kbp. The way to achieve resolution in the 20- to 50-kbp range is normally limited to lowering the concentration of the gel matrix, a fairly unsatisfactory modification resulting in diffuse bands and low mechanical gel strength.

With the advent of pulsed-field gel electrophoresis (PFE), a breakthrough has been achieved, resulting in the resolution of DNA molecules of megabase proportions.[1,2] Commercially available instruments, prepared markers, and specialized reagents have allowed PFE to approach the status of a routine laboratory technique. Physical mapping of mammalian chromosomes has especially benefited from PFE technology. The genes or the regional location of the genes for a number of hereditary diseases, including Duchenne muscular dystrophy,[3-5] Huntington's chorea,[6] and cystic fibrosis,[7,8] have been mapped. The location and order of multigene families, such as the histocompatibility locus,[9,10] and immunoglobulin genes[11] have been determined. Preparative pulsed-field gels have been developed to accumulate megabase DNA for enriched libraries.[5] PFE has been used to study chromosomal rearrangements,[12,13] translocations,[4,14-16] deletions,[3,11] and fragile sites.[17] Double minute chromosomes have been separated on pulsed-field gels.[18] In lower organisms, PFE has been used to produce electrophoretic karyotypes,[19-21] and provides the means for analysis of clones from yeast artificial chromosome (YAC) libraries.[22]

A number of excellent reviews covering various aspects of this subject have appeared in the literature.[1,23-30] The purpose of this review is not to examine exhaustively the literature of PFE, but, rather, to provide researchers with practical information about the capabilities and procedures of pulsed-field technology. To this end, we will provide (1) an overview of the theory of pulsed-field electrophoresis, (2) brief considerations of the various instruments available, (3) a discussion of the electrophoretic conditions involved in DNA separation by PFE, and (4) step-by-step protocols used by our laboratory for the preparation of DNA samples, blotting, and the hybridization of pulsed-field gels. Readers will also be directed to more detailed information available in the literature.

II. THEORY OF PULSED-FIELD ELECTROPHORESIS

In conventional agarose electrophoresis, separation is achieved by the sieving properties of a gel matrix composed of pores of various dimensions. Small molecules have access to a greater number of pores, while larger molecules take longer to find pores of accommodating size. The mobility of DNA molecules up to 20 to 30 kbp is therefore size dependent. Molecules larger than this exhibit a size-independent mobility in which all molecules migrate through the gel at roughly the same rate. The model for describing the electrophoretic behavior of large molecules, termed biased reptation, states that very long molecules travel in a snake-like manner through tubes formed by the pores of the agarose matrix.[31-33] The molecule is represented as a series of charged beads separated by freely orienting connections. The tube is discontinuous, or segmented, with each pore representing a new segment of the tube. The head of the molecule semirandomly chooses a pore, influenced or biased by the electric field and thermal forces, and the trailing portions follow the same pathway.

Schwartz and Cantor reestablished size dependence on large DNA molecules by using nonuniform, alternating electric fields deployed at nearly right angles to each other.[1,2] Separation was explained by the need for large DNA molecules to reorient themselves with respect to the changing electric fields. Nonhomogeneous fields were thought to be necessary for reorientation, and obtuse angles between the fields resulted in the best resolution. Consider a stretched linear molecule, migrating parallel to the direction of an electric field. When subjected to a new field placed perpendicular to the first electric field, the molecule must change its direction of migration. To "turn the corner", the molecule must gather itself, reorient with the new field, and penetrate a pore at right angles to the original direction of travel. The time required for reorientation, a period of molecular relaxation, is a function of the molecule's size.

An alternative explanation was suggested by Southern et al., based on evidence gathered with the crossed-field apparatus.[34] Large DNA molecules did not reorient themselves, but remained in an extended conformation between field alternations. When field orientations were changed, the molecule migrated in the new direction, leading with the former trailing end of the molecule. The new trailing end then follows the new leading end through the pore, producing size-dependent separation. In this model, obtuse angles, but not heterogeneous fields, were required since the crossed-field apparatus consists of a gel rotating between two orientations across a uniform voltage gradient.

To this point, direct evidence for the electrophoretic movement of large DNA fragments through agarose matrices had not been observed on the molecular level. Refinements in the model were made using computer simulations based on the agarose matrix as a series of obstacles through which the molecule must navigate, rather than as a tube.[35] Smith et al.[36] used agarose-embedded DNA cast on microscope slides and observed the motion of the ethidium bromide-stained molecules by fluorescence microscopy as conventional and pulsed electric fields were applied. The observations agreed in many respects with the computer simulations of Deutsch.[35] DNA molecules assume an extended conformation during migration through a gel, but alternatively contract and expand in length as obstacles are met. When obstacles are encountered, the ends of the molecules extend around the obstacles in a U shape, a configuration lasting until one of the arms slips free of the obstruction and follows the path of the longer arm. The molecules also display elasticity. In an electric field, the molecules extend, but contract to globular conformations in the absence of the field. Extension appears greatest during the U shape.

The evidence of Smith et al. also most nearly follows the Southern model describing DNA mobility in a gel rotating through a uniform field at an obtuse angle.[36] Molecules proceeding in one direction reverse their direction when the field is rotated through 120°, and the end that had been the tail becomes the new head. Both ends of the molecule actually proceed parallel to the new field direction, forming the U shape, with one finally slipping into the path of the longer arm. The leading part of the molecule bunches up and may actually loop upon itself, forming a fork. This suggests that the segments of the molecule immediately trailing the head can follow the path of the head through the pores faster than the head can find a new pore to penetrate. Loops of the molecule can actually penetrate pores not entered by the leading end, but eventually the true end pulls the loops through the correct pore, straightening out the molecule. These observations were predicted by Southern, with the exception that the nonrelaxing, constantly extended molecule was not observed.

Lai et al. tested the Schwartz and Cantor model predicting conformational changes (relaxation) prior to migration in a new direction.[37] Since time required for conformational change is proportional to length, a certain amount of time from each new field orientation is used by the molecule to relax before again moving parallel to the new field. By allowing molecules to relax or reorient in zero field intervals between active, reoriented fields, an increased mobility of DNA molecules was observed. A period of zero electric field during which the molecule can relax appears to reduce the time required for the molecule to begin moving in the direction of the new electric field. However, the mobility of all molecules was not affected in the same manner by the zero interval. The size of the molecule with the increased mobility was dependent on the duration of the active field. This appears to be consistent with the fluorescence microscopy observations of molecules assuming a globular resting state in zero fields.[36,38]

More recent evidence from fluorescence microscopy studies indicates that while elongation and relaxation occur, the speed at which a molecule relaxes is also dependent on the angle of orientation of the fields.[38] Through the use of a novel electrophoretic method, pulse-oriented electrophoresis, field angles can be emulated by a series of very short alternating perpendicular pulses. The pulse times (3 s) are shorter than the time required for molecular reorientation. While relaxation time in zero fields took about 60 s, relaxation in active 120°-emulated fields took 6 s and relaxation took 20 s in active 90°-emulated fields. In active fields, molecular reorientation to a new field was faster than in zero. This apparently contradicts the results of Lai et al. and may result from differences in the methods used, pulse-oriented electrophoresis (POE) vs. programmable, autonomously controlled electrodes (PACE), or may point to the need for caution in extrapolating observations on the behavior of individual molecules to populations. Alternatively, these experiments may be addressing different aspects of electrophoresis with pulsed fields, mobility, and resolution.

Collectively, these results confirm many of the major predictions of both models, although not all observations have been satisfactorily explained. While a clearer picture of the pulsed-field electrophoretic behavior of large DNA molecules is emerging, it is also clear that the interrelationships of factors involved in the phenomenon may be more complex than originally thought.

III. INSTRUMENT DESIGN

A number of variations in the design of pulsed-field equipment have appeared since the description of the original electrode geometry[1,2] (Figure 1) and many of the instrument designs are commercially available.[39] Modifications and design innovations have been made to widen the range of separation, to improve resolution, and to straighten the lanes for easier comparison. Each design has its own advantages and disadvantages, making the choice of apparatus difficult and important. Trade-offs (cost, resolution, etc.) and the primary application of the instrument (restriction mapping, chromosome separation, YAC technology) should be kept in mind when making the choice. In this section, brief descriptions of the various designs are given along with some advantages and disadvantages. Most of our expertise has been with three types of instruments, the OFAGE apparatus of Carle and Olson,[40] the CHEF instrument of Chu, Volrath, and Davis,[41] and the crossed-field unit, or "Waltzer", developed by Southern et al.[34] Note that we adhere to the emerging convention of referring to the technology of megabase DNA separation as pulsed-field electrophoresis (PFE). We will refer to the nonhomogeneous field design of Schwartz and Cantor as pulsed-field gradient gel electrophoresis (PFGE) and will use terms such as OFAGE, FIGE, and PFGE to describe only the instrument design.

A. PULSED-FIELD GRADIENT GEL ELECTROPHORESIS (PFGE)

By applying two alternating electric fields, one homogeneous and the other nonhomogeneous, Schwartz and Cantor demonstrated the separation of yeast chromosomes up to 2000 kbp[2] (Figure 1, part 3). A point-source electrode, positioned at the corner of the gel box, generated the continuously varying field strength in one dimension. Double inhomogeneous fields were also tested and found to give better resolution. This electrode design was later modified by placing electrodes entirely around the square gel box and altering the

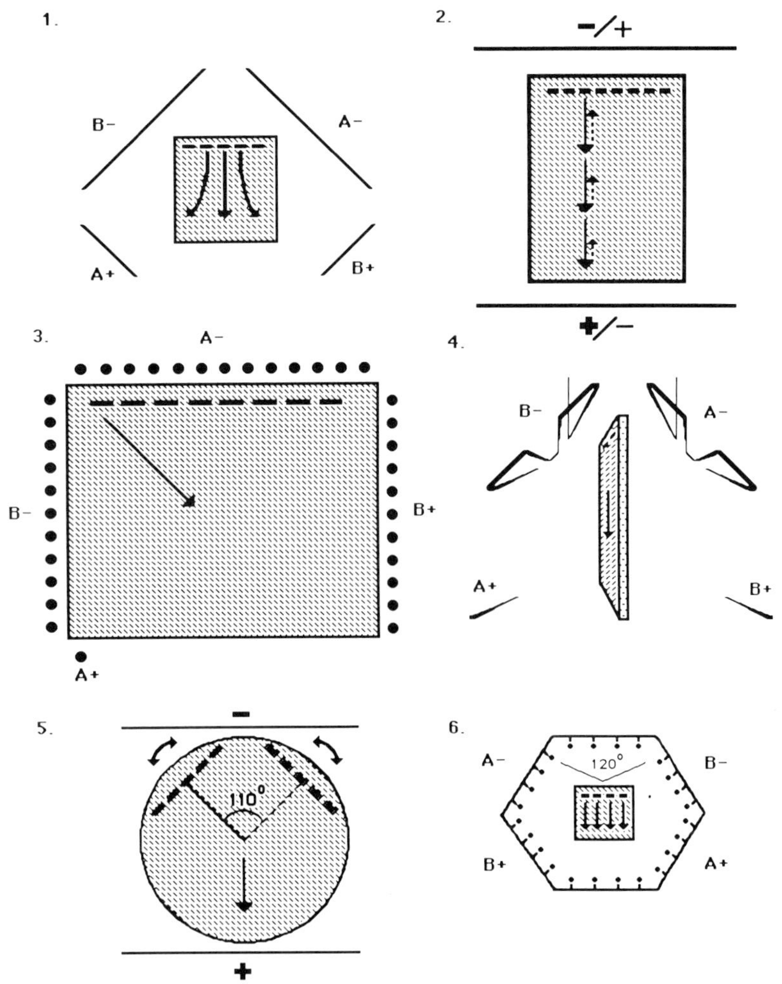

FIGURE 1. Schematic diagrams of electrode and gel configurations used in pulsed-field electrophoresis instruments: (1) OFAGE, (2) FIGE, (3) PFGE, (4) TAFE, (5) crossed field, (6) CHEF. The PACE electrode arrangment is similar to the CHEF apparatus. (Figures are not drawn to the same scale.)

switching pattern to produce nonuniform fields that circled the gel,[42] and by ramping the voltages at clamped electrodes.[43] With these modifications, resolution was improved, but lane distortion, while also improved, was not entirely eliminated. An updated version is commercially available from LKB.

B. ORTHOGONAL-FIELD-ALTERNATION GEL ELECTROPHORESIS (OFAGE)

Developed by Carle and Olson,[40] the OFAGE system together with the PFGE apparatus represent the first generation of pulsed-field electrophoresis units. OFAGE units use two nonhomogeneous electric fields (Figure 1, part 1) generated from pairs of electrodes deployed at 45° angles relative to the gel. Voltage is switched between each pair of electrodes at regular intervals. Because the positive and negative electrodes of each pair differ in size, the smaller electrode acts more like a point source, an effect which produces the field heterogeneity. Thus, while the angle between the electrode pairs is 90°, the angles between the fields within the gel vary from near 90° at the top to about 150° at the bottom of the gel.

While this design can achieve separation of megabase DNA molecules, as with the original PFGE apparatus, the nonuniformity of the electric fields and the variation in the angle of the fields across the gel are responsible for its major disadvantages. DNA molecules tend to stack up as they migrate into different areas of the gel, resulting in a series of compressions and expansions of DNA molecules. Since lane geometry is also distorted, especially in outer lanes, size estimations and comparisons are apt to be difficult. Coupled with the small gel size (10 × 10 cm), the lane distortion also reduces the number of samples that can be satisfactorily run on a single gel.

The major advantage of this apparatus is the simplicity of the gel box design and electrical circuitry. Using simple schematic drawings (helpfully provided by Maynard Olson), our biology instrumentation facility was able to construct, at very low cost, a serviceable apparatus, providing our initiation into the complexities of PFE technology.

C. FIELD-INVERSION GEL ELECTROPHORESIS (FIGE)

Separation of large DNA molecules can also be achieved using a single electrode pair and periodically inverting a uniform field through 180° [44] (Figure 1, part 2). By altering the duration of the forward and reverse fields, or by differences in the voltage, forward movement of DNA molecules is achieved. An additional modification, switch-time ramping, is used to control a size independence displayed by some DNA molecules subjected to field inversion.[45] The duration of the switching interval is progressively changed during electrophoresis.

In general, the range of molecules best resolved by FIGE is less than 1000 kbp, smaller than that achieved with other methods. However, one advantage is the ability to resolve molecules less than 50 kbp, generally considered to be the lower limit of most other PFE instruments. Another advantage is that FIGE is one of the less costly of the marketed instruments (DNA-Star), and plans for simple, inexpensive "lab-built" instruments are available in the literature.[46]

D. TRANSVERSE ALTERNATING-FIELD ELECTROPHORESIS (TAFE)

TAFE instruments are a vertical alternative to the previous horizontal electrode arrangements and result from attempts to correct lane distortion (Figure 1, part 4).[47] All gel lanes across the gel width are exposed equally to homogeneous electric fields directed through the thickness of the gel, thus producing straight lanes. However, because the field angle varies considerably from the top to the bottom of the gel, the stacking effect observed in previous instruments is not eliminated. While gels are still fairly small, limiting the sample number per run, the TAFE design has the advantage of simplicity in design and circuitry and, in addition, affords a unique benefit. Due to its design, the gel box is readily adapted for use as an electrotransfer apparatus. Complete TAFE systems are available from Beckman Instruments.[48]

E. CONTOUR-CLAMPED HOMOGENEOUS ELECTRICAL FIELD (CHEF)

The CHEF apparatus uses a novel arrangement of electrodes to produce homogeneous electric fields in two orientations a uniform 120° apart (Figure 1, part 6).[41] The electric fields are generated by a hexagonal array of electrodes, each clamped to a predetermined voltage. While the field is switched between the A and B sets of electrodes, a ramp of intermediate electric potentials is presented at the other faces of the contour. Thus, all portions of the gel are subjected to more nearly uniform fields than in any previous instrument. The effect of this tight control on the electric fields is straight lanes and good resolution over a broad range of sizes. *Schizosaccharomyces pombe* chromosomes, approximately 3000 to 7000 kbp, are readily resolved.[49] In addition, this instrument also easily resolves molecules in the 1- to 50-kbp range. For this reason, the CHEF is our method of choice for resolution of fragments less than 200 kbp.

CHEF systems are commercially available from Bio-Rad[50] and BRL. An equipment option is also available from LKB to modify the Pulsaphor unit for CHEF capability.

F. CROSSED-FIELD OR ROTARY PULSED-FIELD GEL ELECTROPHORESIS

The crossed-field[34] or rotating gel[29,51] design incorporates a single homogeneous electric field, but achieves separation by periodically reorienting the gel itself with respect to the field at an angle of 110° (Figure 1, part 5). Sharply resolved DNA molecules, in straight lanes, over a broad range of sizes can be produced without the need for complex arrays of clamped electrodes (Figures 2A and 3A). Coupled with the large gel size, 20 × 20 cm, each gel run can handle at least 20 to 30 samples. This design also makes it much easier to predict effects of altered pulse time, voltage, and run time, without running numerous gels. We use the crossed-field apparatus of Southern (received through the Oxford University Biology Department) extensively for long-range physical mapping of mammalian genomes and are very pleased with the instrument's performance. The only apparent disadvantage with the crossed-field or rotary gel in its present design is that molecules less than 50 kbp are not easily resolved because of the rotational time requirements. Immunetics also markets a rotary instrument, the Rhotaphor.

G. PROGRAMMABLE, AUTONOMOUSLY CONTROLLED ELECTRODES (PACE)

The latest entry into PFE instrumentation is the PACE system, incorporating a closed-contoured (hexagonal) array of independently regulated electrodes.[23,52] All electrophoretic parameters, i.e., uniformity of field, voltage gradient, pulse duration, and field orientation, can be precisely controlled at each electrode, at any time during the gel run. This creates the potential to generate any desired electric field and perform the switching operations of all the instruments described previously. By manipulation of the parameters, the range of DNA fragments resolvable extends upward from 100 bp and uniform migration of molecules is achieved throughout the gel. Run times for megabase molecules (*S. pombe* chromosomes), taking up to 1 week with other instruments, can be reduced to 24 h.

Recently, Bio-Rad introduced their newest version of the CHEF pulsed-field instrument incorporating PACE architecture.

IV. ELECTROPHORETIC PARAMETERS AFFECTING PFE

Electrophoretic mobility of DNA molecules through a gel is influenced by a number of interacting factors. Changing only one variable can greatly alter the mobility of a molecule, in turn altering the resolution. For example, to retain the same resolution between molecules when the voltage gradient is increased, a compensatory decrease in the switch or pulse time is required. Decreasing the voltage to resolve very large molecules requires longer pulse times.[13,37,49] With an understanding of how individual parameters affect the behavior of molecules and a little experience, one can fairly closely predict the range of resolution of molecules for a given set of conditions.

By modulating the pulse interval, DNA molecules of different size can be resolved. Longer switch times are required to resolve molecules of larger size. As switch times are reduced, the area of resolution shifts to smaller molecules. For example, chromosomes of *S. pombe,* ranging from 3000 to 7000 kbp, require pulse times of 30 to 60 min, while the smallest chromosomes of *Saccharomyces cerevisiae* (200 to 300 kbp) can be resolved with pulses of about 25 s. For any given switch interval, a range of sizes will be resolved to varying degrees, making the choice of the appropriate time dependent on the size range of the molecules of interest. Molecules larger and smaller than this size range will not be well resolved. Broader ranges of resolution can be achieved by using more than one switch

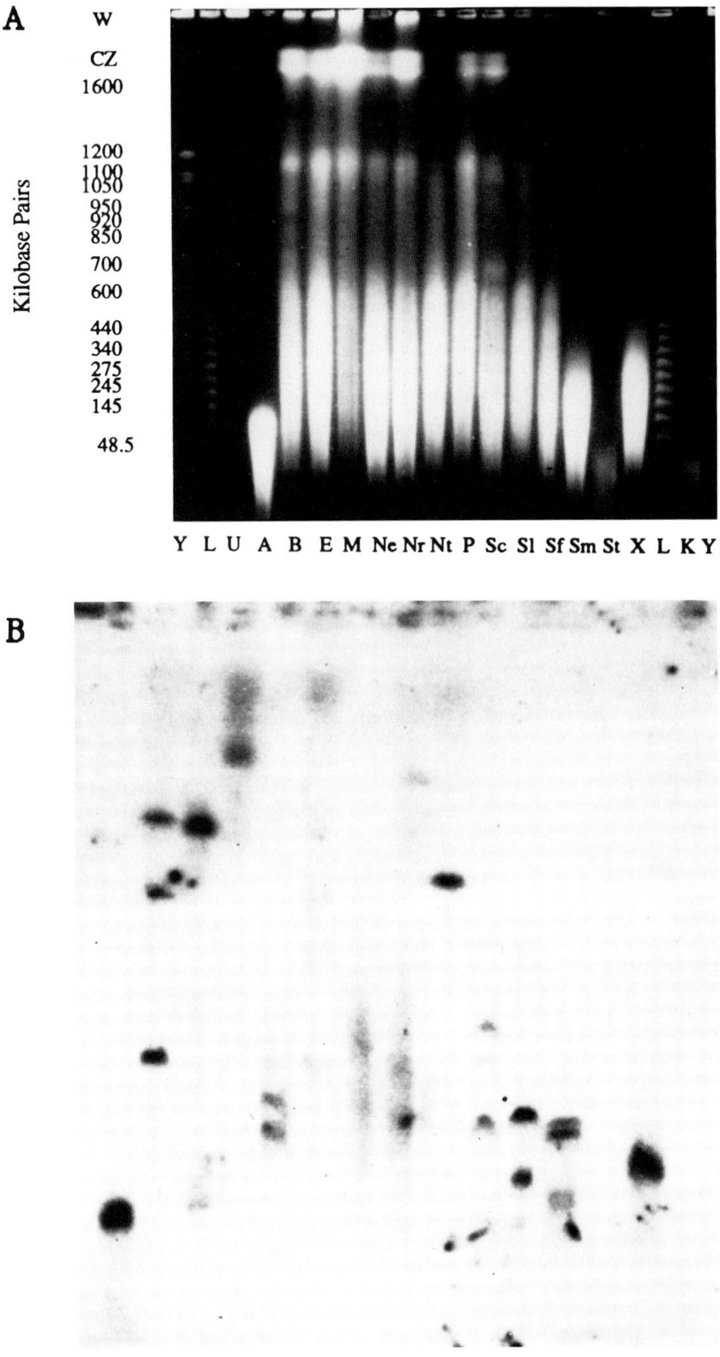

FIGURE 2. Ethidium bromide-stained gel (A) and autoradiograph of the nitrocellulose blot (B). Human DNA (1 μg per lane) digested with rare-cutter enzymes was electrophoresed on a 1.5% agarose gel at 10°C, 0.5X TBE, using the "Waltzer" or crossed-field unit of Southern. Conditions used were 6 V/cm^2, 80-s switch for 20 h, and 5 V/cm^2, 150-s switch for 26 h. Gels were blotted, hybridized, and washed as described in Table 2. A single-stranded M13 probe (human cytokeratin K10) was prepared as described in Table 3. The blot was washed to a final stringency of 65°C. Film exposure was for 2 d (−80°C), as this was the fourth reuse of this blot. Abbreviations: Y, yeast chromosome marker; L, lambda concatemer marker; U, uncut DNA control; A, Apa I; B, Bss HII; E, Eag I; M, Mlu I; Ne, Nae I; Nr, Nar I; Nt, Not I; P, Pvu I; Sc, Sac II; Sl, Sal I; Sf, Sfi I; Sm, Sma I; St, Sty I; X, Xho I; K, 1-kb marker.

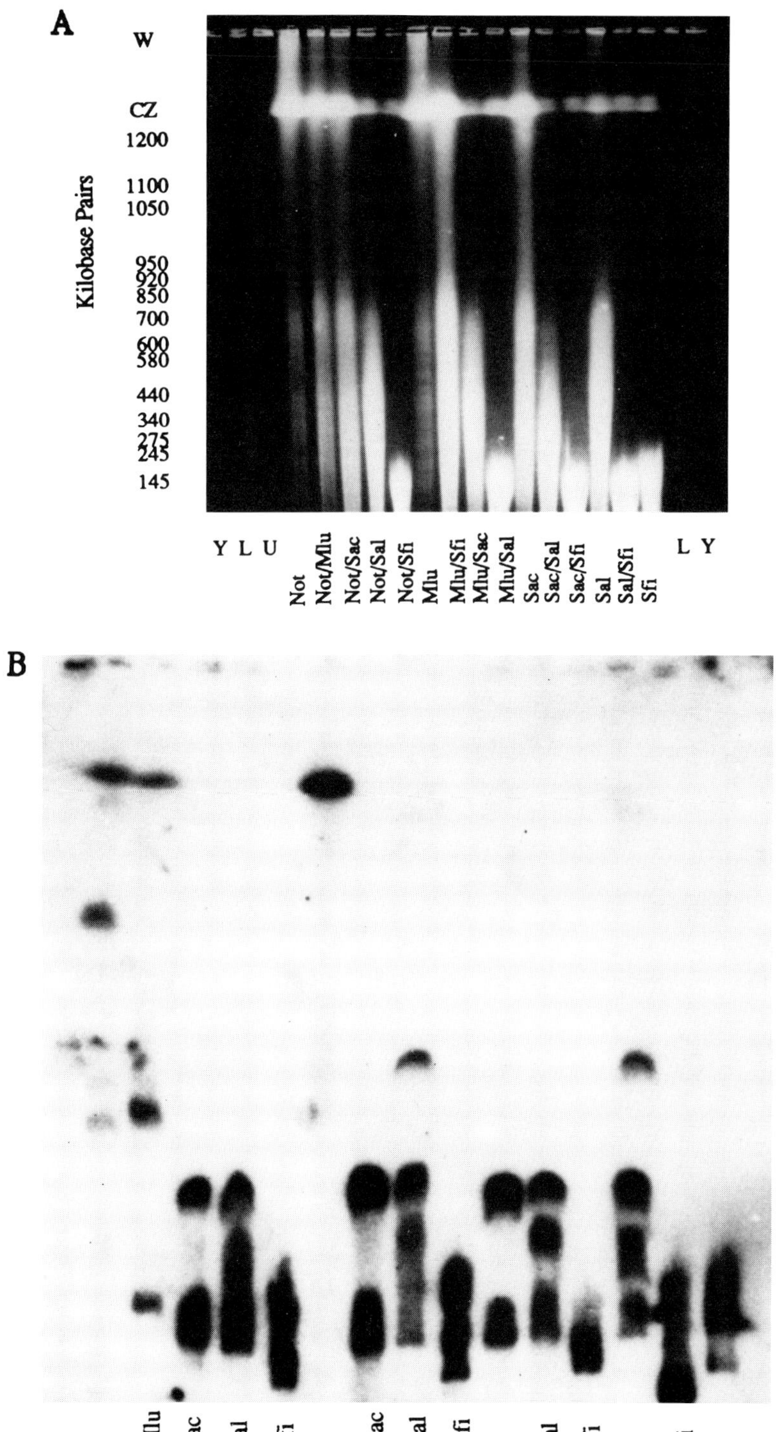

FIGURE 3. Ethidium bromide-stained gel (A) and autoradiograph of the Nytran blot (B). Double enzyme digests of mouse DNA (2.5 μg per lane) were electrophoresed on 1.5% agarose gels at 14°C in 0.5X TBE using the "Waltzer" unit. Conditions used were 6 V/cm^2, 80-s switch time for 40 h. All blotting, hybridization, and probe preparation conditions were as described in Tables 2 and 3. The blot was probed for the first time with a sequence from the mouse Hox 3.3 gene and washed to a stringency of 65°C (Table 2). Film exposure was for 6 h (−80°C).

interval during the same gel run (see Figure 2). Intervals can be changed either continuously (ramping) or discontinuously (stepping), depending on whether the instrument has programmable or manual control of this function.

The voltage used also affects the resolution of molecules in a size-dependent manner. Generally, higher voltages can be used to separate smaller molecules, while very large molecules require a reduction in the voltage gradient. *Schizosaccharomyces pombe* chromosomes fail to resolve at the higher voltages used to separate λ concatemers. Why this is so is not known. While pulse times do not affect the length of the gel run, the size of the voltage gradient certainly does. Smaller λ concatemers can be separated in a few hours; *S. pombe* chromosomes take up to 5 d.

The angle of reorientation (120 or 110°) is generally fixed with most instruments and therefore remains a constant. With instruments such as PACE that are capable of varying the reorientation angle, changing the angle during a gel run can greatly affect the mobility of DNA molecules. Progressively increasing the angle slows the molecules, enhancing band sharpness, but compressing the interband spacing. Decreasing the angle greatly increases the mobility of large molecules. *S. pombe* chromosomes were well resolved in 24 h using a 96° reorientation angle.[23]

DNA mobility is highly influenced by temperature. Mobility increases with increasing temperature, but there is a concomitant loss of resolution and band sharpness. We normally run gels at 10°C, colder than usually seen in the literature and resulting in longer run times, but we feel that resolution and band sharpness are better at this temperature (see Figures 2 and 3). Buffer cooling is generally advised to compensate for the heat generated by the voltage gradients, and fairly tight temperature maintenance is needed since fluctuations can cause difficulties when comparing different gel runs.

As with conventional agarose electrophoresis, the agarose concentration and the type of agarose used can also alter the mobility and resolution of DNA molecules in pulsed-field gels. Mobility rates increase and band sharpness decreases as the agarose concentration is reduced. Megabase-sized molecules are generally separated on gels with concentrations of less than 1%.

The type of agarose used can also affect DNA mobility. Faster runs are produced in low electroendosmosis (EEO) agarose.[53] Different grades of agarose can have a pronounced affect on the mobility rate and resolution of large DNA molecules.[54] Gels of the same concentration, but different grade, vary in their ability to resolve large molecules. This phenomenon, in which DNA seems to bind to the agarose, is known as "running inhibition".[53] There is a correlation with the metal ion content, although this may not be the only cause. Running inhibition also affects transfer of DNA from agarose gels to nitrocellulose and nylon filters, so the choice of agarose may be critical to different aspects of PFE technology.

The type of electrophoresis buffer used and its concentration also produce effects in PFE. Buffers with low ionic strength, such as 0.5X TBE (Table 1), result in slightly faster runs.[13,24,34] The low ionic strength also reduces the current load and makes temperature control easier. At least in the case of small DNA fragments, better resolution is achieved using TBE buffer, when compared to TAE buffer.[55] Whether this holds true for very large DNA molecules is not known.

V. PROTOCOLS FOR PULSED-FIELD ELECTROPHORESIS

This section contains detailed procedures used in our laboratory for processing DNA plugs, digesting samples, electrophoresing, blotting, and hybridizing pulsed-field gels. We have tried many different protocols, borrowing widely, combining, modifying, and optimizing until methods have produced consistent results in our hands. However, as with any

TABLE 1
Reagents for Preparing and Digesting DNA Embedded in Agarose

Reagent	Composition	Preparation/storage
NDS, pH 9.5	500 m*M* Na_2EDTA 10 m*M* Tris-HCl 1% Lauroyl sarcosine	Filter sterilize, store at 4°C
Proteinase K (stock)	10 mg/ml in NDS	Autodigest 1 h at 50°C, store at −20°C
Dulbecco's phosphate-buffered saline (PBS), pH 7.2 (1X)	Solution A (without divalent cations) 2.7 m*M* KCl 1.5 m*M* KH_2PO_4 137.0 m*M* NaCl 15.2 m*M* Na_2HPO_4	Autoclave, store at 4°C
Hank's balanced salts solution (HBSS), pH 7.4 (1X)	1.3 m*M* $CaCl_2$ 5.4 m*M* KCl 0.4 m*M* KH_2PO_4 0.5 m*M* $MgCl_2·6H_2O$ 0.4 m*M* $MgSO_4·7H_2O$ 137.0 m*M* NaCl 0.4 m*M* NaH_2PO_4 4.2 m*M* $NaHCO_3$ 5.6 m*M* D-Glucose 0.001% Phenol red	Filter sterilize, store at 4°C
0.25% trypsin in CMF (calcium, magnesium free)	HBSS without calcium and magnesium 0.0025 mg/ml trypsin	Filter sterilize, store at −20°C
SEM buffer	1000 m*M* sorbitol 50 m*M* Na_2EDTA 7.5% 2-mercaptoethanol	Filter sterilize, store at 4°C
STEM buffer	SEM buffer 10 m*M* Tris-HCl, pH 7.5	Filter sterilize, store at 4°C
SSC (1X)	150 m*M* NaCl 15 m*M* Na citrate, pH 7.0	
Multimer buffer	2X SSC 3% ficoll 10 m*M* Na_2EDTA 10 m*M* Tris-HCl, pH 7.5 0.05% orange-G	Filter sterilize, store at 4°C
YPD medium	2% glucose 2% bacto-peptone 1% yeast extract (2% agar for plates)	Autoclave
TE, pH 7.5	10 m*M* Tris-HCl, pH 7.5 1 m*M* Na_2 EDTA	Autoclave

TABLE 1 (continued)
Reagents for Preparing and Digesting DNA Embedded in Agarose

10X restriction buffer (incomplete)	Prepare as per manufacturer's instructions without BSA or reducing agent; aliquot in 1-ml volumes	Autoclave, store at −20°C
TBE (1X), pH 8.3	89 m*M* Tris-HCl 89 m*M* boric acid 1.25 m*M* Na_2EDTA	
Loading buffer	0.5X TBE 10 m*M* Na_2EDTA 0.05% orange-G	Autoclave, store at 4°C
Spermidine	Prepare 100 m*M* stocks in sterile, distilled water; aliquot in 1-ml volumes	Autoclave, store at −20°C
DTT		
2-Mercaptoethanol		
BSA, 1 mg/ml	Pentax fraction V; prepare in sterile, distilled water; aliquot in 1-ml volumes	Filter sterilize, store at −20°C
PMSF	100 m*M* solution in dimethyl sulfoxide	Prepare fresh

methodology, procedures and their modifications vary from laboratory to laboratory, and what is successfully applied in one laboratory may produce inconsistent results in another's hands. For those introducing PFE technology to their laboratories, we recommend that you try these protocols, but feel free to modify them to suit your needs.

A. PREPARATION OF DNA EMBEDDED IN AGAROSE PLUGS

DNA purified by conventional methods is unsuitable for use in pulsed-field analysis. Due to their large axial ratio, DNA molecules are highly sensitive to shearing forces exerted during normal preparative procedures. For this reason, techniques have been developed to extract DNA from cells and tissues in a manner that eliminates shearing and renders the DNA useful not only for pulsed-field electrophoresis, but conventional electrophoresis and cloning as well. The cells or tissues are initially immobilized in blocks of low-gelling-temperature agarose, called plugs or inserts[1,2] or, alternatively, in beads formed by vortexing a molten agarose-cell mixture in paraffin oil.[56] Proteins, salts, and detergents freely diffuse through the agarose, while DNA molecules remain trapped in the matrix. DNA purification and subsequent restriction-enzyme digestions are therefore relatively simple to perform and the agarose plugs are easily manipulated. DNA from many sources can be extracted using these techniques, modified to adapt them to the needs of various types of organisms.

Several aspects of the embedding process are important. First, the DNA in the plugs is highly susceptible to nuclease digestion. We autoclave or filter sterilize all solutions that come in contact with the plugs and aliquot them in volumes adequate for one round of processing. The remaining solution in the aliquot is discarded. Plugs are manipulated with glass cover slips and flame-bent Pasteur pipettes that have been rinsed in alcohol and flamed. Gloves are worn at all times. Plugs are stored in NDS (see Table 1) which contains 0.5 *M* EDTA, protecting the DNA from divalent cation-dependent nuclease degradation.

Second, the type of agarose used is extremely important for subsequent restriction digestion of the embedded DNA. Many batches of agarose contain impurities which inhibit restriction enzymes. The cause for this is not known, but experiments have shown no correlation between endonuclease inhibition and sulfate or moisture content of the agarose, or with the EEO properties.[53] We use a low-gelling-temperature agarose from FMC Bio-Products, called InCert Agarose, pretested and competent for restriction-enzyme digestion, and eliminating the time and effort required to screen batches of agarose.

Third, since pulsed-field electrophoresis is sensitive to overloading, the amount of DNA contained in the plugs must be controlled. For a particular source of DNA, the amount loaded in a gel lane must be determined experimentally; however, for mammalian cells, we typically use 1 to 2 μg per lane. We routinely embed approximately 1.5×10^6 cells per plug, using a concentration of 0.66×10^{-11} g of DNA per cell. Hence, each plug having a volume of 100 μl contains 10 μg of DNA. Although cell concentrations are given in the protocols, they are not set in stone. Adjustments should be made when conditions warrant.

The plug mold used in our laboratory was made in the university machine shop from Perspex; however, commercially manufactured instruments are generally supplied with molds and combs. Our mold consists of a number of identical bars with the dimensions, 170 × 19 × 10 mm. One edge of each bar has been cut with slots to the outside dimensions of the comb we use in our pulsed-field gels. When taped together with plastic electrical tape, the wells formed hold 100 μl of the agarose-cell suspension. Plug pieces cut across the width of the plug fit exactly into wells made in the gel with the comb. Before using the molds, the bars are washed by boiling in 0.25 *M* HCl (5 to 10 min), rinsed thoroughly with distilled water, and dried at 80°C.

The protocols given in this section are routinely used in our laboratory for DNA purification from four major sources: mammalian tissue-culture cells, yeast, bacteriophage, and dissected mammalian tissues and organs. Formulas, preparation and storage of buffers, and solutions are given in Table 1. Unless otherwise noted, all chemicals are available from Sigma.

1. DNA Extraction from Tissue-Culture Cells

1. Tissue-culture cells are harvested by scraping or by trypsinization. If trypsin is used, the cells should be washed once in tissue-culture medium containing fetal calf serum to inactivate residual trypsin. Wash the cells twice in ice-cold Dulbecco's phosphate-buffered saline (PBS) without divalent cations, count an aliquot, and resuspend at 3×10^7 cells per ml. Be sure the cell suspension is homogeneous by disrupting cell clumps with gentle pipetting. Warm the cell suspension to 37°C. Centrifugations are performed at $1000 \times g$ for 3 min.
2. Prepare a 1% low-gelling-temperature agarose solution in PBS and cool to 37°C. Mix equal volumes of the agarose solution with the cell suspension and quickly dispense into a prechilled plug mold. Hold the mold on ice for 5 to 10 min until the plugs have thoroughly solidified.
3. Plugs are pushed into a 50-ml tube (example: Corning or Falcon polypropylene) containing NDS supplemented with 1 mg/ml Proteinase K (Boehringer-Mannheim) and incubated at 50°C for 24 h. The volume of buffer used is at least twice that of the plugs. Replace the buffer with fresh NDS and enzyme, and continue the incubation for an additional 24 h. Cell lysis can be monitored by the change from an opaque to a transparent appearance of the plugs.
4. Wash the plugs in 50 ml of NDS without Proteinase K, twice for 2 h at 50°C, to remove most of the Proteinase K. Plugs can be stored in NDS at 4°C for at least 1 year.

2. DNA Extraction from Dissected Tissue

The procedure for embedding cells recovered from dissected tissues and organs can be divided into two parts: (1) disrupting the tissue and (2) embedding the dispersed cells, as in the protocol for tissue culture cells. In our hands, the cold-trypsin dispersal method works well.

1. Wash the tissue in ice-cold PBS and chop into 3- to 4-mm pieces using the crossed-knives technique. In a 50-ml tube, wash the pieces two to three times in Hank's balanced salts solution, allowing the pieces to settle by gravity and pouring off the supernatant. Determine the wet weight of the tissue.
2. Place the tube containing the tissue pieces on ice and add 1 ml/100 mg tissue of 0.25% trypsin (Gibco) in CMF (4°C). Incubate the tissue 8 to 12 h or overnight at 4°C. This allows the trypsin to penetrate the tissue without enzymatic activity.
3. Pour off the excess trypsin and incubate the tissue for 20 to 30 min at 37°C. Add 1 ml/100 mg of tissue of prewarmed tissue-culture medium containing fetal calf serum (37°C) and pipette gently until the cells are dispersed. Allow any large pieces of noncellular debris to settle out, approximately 2 to 3 min, and pipette off the suspended cells.
4. Wash cells once in tissue-culture medium supplemented with fetal calf serum, and proceed with the embedding procedure for tissue-culture cells given above.

3. DNA Extraction from Yeast Cells

This procedure is suitable for yeast markers prepared from *Saccharomyces cerevisiae* and *Schizosaccharomyces pombe*. Variations in the procedure for *S. cerevisiae* and *S. pombe* are noted. When using chromosomes from a particular strain of yeast, one should be aware that differences in the size and number of chromosomes exist. Commercially prepared yeast chromosome markers from strains with clearly established chromosome size and number are available from several sources (BRL, Clontech, FMC, Bio-Rad, Immunetics, and others).

1. A single yeast colony is selected from a culture streaked on a YPD agar plate grown at 30°C and used to inoculate 50 ml of YPD medium. The culture is incubated at 30°C overnight.
2. Aliquots (10 ml) of the overnight culture are used to inoculate 2 × 500 ml of YPD in 1-l flasks. Culture incubation is continued at 30°C to an OD_{600} of 0.36 for *Saccharomyces cerevisiae* (approximately 4 × 10^6 cells per milliliter; OD_{600} of 0.3 = 3.3 × 10^6). We use the same OD for *S. pombe*.
3. Cells are harvested by centrifugation at 5000 × g for 10 min and resuspended in SEM buffer at a concentration of 1 × 10^9 cells per milliliter. Add 1 mg/ml Zymolase-100T (Calbiochem) and warm the cell suspension to 37°C.
4. Prepare a solution of 1% low-gelling-temperature agarose in SEM buffer and cool to 37°C. Mix equal volumes of agarose and the yeast-cell suspension and dispense into plug molds. Hold on ice for 5 to 10 min to solidify the plugs.
5. Transfer the plugs to a 50-ml tube containing 10 to 20 ml of STEM buffer supplemented with 1 mg/ml Zymolase-100T (spheroplast formation). Incubate *S. cerevisiae* plugs for 2 to 3 h at 37°C. *Schizosaccharomyces pombe* plugs are incubated overnight as spheroplasts are more difficult to make.
6. Replace the STEM buffer with NDS + 1 mg/ml Proteinase K and incubate for 48 h at 50°, with a change of fresh buffer and enzyme after 24 h.
7. Wash the plugs twice for 2 h with 50 ml of NDS at 50° and store at 4°C in NDS.

4. Preparation of Bacteriophage λ Concatemers

Satisfactory PFE markers, prepared from λ concatemers and embedded in agarose, are also commercially available from the same distributors of yeast chromosome markers. Generally, these are prepared from λ *cI857Sam7,* giving ladders in 48.5-kbp increments. However, we and others have found that when preparing "homemade" markers in the laboratory, commercially prepared λ DNA of the type used to prepare HindIII markers for conventional agarose gels is not suitable for making large λ multimers. Apparently, the cohesive ends are damaged or otherwise unable to form oligomers of sufficient size. For this reason, λ DNA is usually prepared by the large, phage-prep induction method described by Maniatis.[57] Care should be taken with mixing and pipetting to minimize damage. The phage-prep procedure should be followed up to and including the polyethylene glycol precipitation and centrifugation, followed by step 1, given below. Note that ligase is not needed, concatemerization is a passive cohesion process.

1. Resuspend the λ virions in TE, pH 7.5, at 0.88 ml/100 ml of culture. The DNA concentration works out to approximately 20 μg/ml.
2. Mix equal volumes of 1% low-gelling-temperature agarose in PBS and λ DNA at 37°C and dispense into the plug molds. Allow the plugs to solidify on ice for 5 to 10 min.
3. Transfer the plugs to a 50-ml tube, add NDS and 1 mg/ml Proteinase K, and incubate for 48 h at 50°C, as described previously.
4. Equilibrate the λ plugs by washing them three to four times in a tenfold excess of 2X SSC at 50°C to remove the Proteinase K and sarcosine. Decant the buffer and fill the tube with Multimer buffer. Incubate the plugs for 1 to 2 weeks at room temperature (RT) to allow concatemerization. Plugs can then be stored in the same buffer at RT or 4°C, and are stable for at least 1 year. We routinely use one tenth of a plug or 1 μg of DNA in a gel lane and observe multimers of at least 1000 kbp.

Alternatively, λ markers can be prepared by concatemerization in solution, although we have found that markers prepared by this method are less stable than embedded concatemers, presumably because of shearing forces and damage to the ends during processing. Purified λ DNA is diluted to 200 μg/ml in Multimer buffer, and incubated at 37°C for 30 min, followed by incubation at RT overnight. A 5-μl aliquot (1 μg) is sufficient for use as a marker. Markers are stored at 4°C.

B. ENZYMATIC PROCEDURES

Agarose-embedded DNA is a convenient and efficient way to perform restriction digests for many applications. Shearing and DNA losses are minimized, and buffers and enzymes are easily changed without the need for dialysis, phenol extractions, or ethanol precipitations. As indicated in the previous section, the single most important consideration in the enzymatic digestion of DNA plugs is the type of agarose used. Either commercially tested and certified agarose should be purchased or several batches (different lot numbers) of low-gelling-temperature agarose should be screened by preparing DNA plugs and testing for DNA degradation and restriction-digest inhibition. If the agarose batch is compatible with restriction-enzyme digestion, other problems that arise with plug digests will also be seen with solution digests.

Another factor of considerable importance should also be stressed again. DNA plugs are highly sensitive to nuclease contamination. As in the preceding section, wear gloves, sterilize all solutions and aliquot in one-use volumes, autoclave Eppendorf tubes and pipette tips, and handle the plugs with flamed glass coverslips and glass rods or Pasteur pipettes.

With a few modifications, detailed below, we perform restriction digests on plugs according to the enzyme manufacturer's instruction (New England Biolabs, Promega). Solutions and their preparation are described in Table 1. We make and aliquot 10X concentrations of restriction buffers according to the manufacturer's recommendations, but without BSA or reducing agent. These incomplete buffers are used for washing the plugs and in the digestion if buffer is not supplied with the enzyme. Buffers supplied with enzymes are tested before use by incubating a plug slice in the buffer at the appropriate temperature and checking for degradation. We keep these buffers and their enzymes separate from the general population of buffers and enzymes in our lab to avoid possible problems with contamination.

Basically, the process consists of washing the plugs prior to digestion, performing the digestion, and preparing the plugs for the electrophoresis run or storage. The plugs need to be washed prior to digestion to extract detergent, lower the EDTA concentration, and especially to remove and inactivate residual Proteinase K, which will degrade restriction enzymes. A protease inhibitor, phenylmethylsulfonyl fluoride (PMSF), which completely inactivates Proteinase K is included in the washes. Usually, a whole plug is digested with an enzyme, although plug slices can also be processed if care is taken to avoid breaking the slice. Digestion is terminated and the bound enzyme, which could alter the mobility of DNA during electrophoresis, is removed from the DNA by the addition of NDS and Proteinase K. Small slices are cut from the plug with a glass coverslip and loaded on the gel, while the remainder of the digested plug is refrigerated in NDS.

1. Plugs are pooled in a sterile, 10-ml polypropylene tube and washed twice for 1 h in 10 ml of TE, pH 7.5, containing 1 m*M* PMSF, followed by two 30-min washes with 10 ml of TE, pH 7.5, without PMSF. All washes are performed on ice with gentle agitation. Plugs from different DNA sources should, of course, be washed separately.
2. Transfer the plugs to individual Eppendorf tubes and wash for 30 min on ice with 0.5 ml of the appropriate incomplete restriction buffer. We have found that plug breakage is reduced if 2 ml-Eppendorf tubes are used instead of the 1.5-ml type. The plugs, especially whole plugs, are able to lie flat on the bottom of the 2-ml tube rather than bending around the wall of the more conical 1.5-ml tube.

3. Replace the incomplete buffer with complete digestion buffer in the amount indicated and incubate for 20 min on ice to allow the buffer to diffuse through the plug. Treat the agarose plug as if it were a volume of DNA added. For example, a whole plug would account for 100 μl of the final volume. For small slices, adjust the volumes accordingly.

DNA plug	100 μl
10X buffer	20 μl
10 mg/ml BSA	20 μl
Water, other	to 200 μl

The volume of sterile, double-distilled water should be adjusted to account for the enzyme to be added in the next step, as well as the addition of reducing agent and spermidine, if necessary. We add 5 m*M* spermidine to buffers with salt concentrations from 50 to 100 m*M*, and 10 m*M* spermidine to buffers with greater than 100 m*M* salt concentrations. Reducing agent, either 2-mercaptoethanol or dithiothreitol, should be added according to the manufacturer's recommendations. When adding enzyme, keep in mind that the glycerol concentration should not exceed 5%.
4. Add the enzyme and incubate for 20 min on ice to permit diffusion through the plug. Transfer to the appropriate digestion temperature. The amount of enzyme required for complete digestion in a particular time will vary from enzyme to enzyme and should be determined empirically. Complete digestion is usually ensured by the addition of 10 to 20 units of enzyme per μg of DNA for a 6-h incubation. From this set point, amounts and incubation times can be adjusted to economize enzymes. Adding the enzyme in two aliquots, with the second aliquot added halfway through the incubation, works well for less stable enzymes. We also routinely incubate overnight with no detectable degradation.
5. Following digestion, remove the buffer and enzyme and add 0.2 ml of NDS + 1 mg/ml Proteinase K and incubate for 2 h at 50°C. Plugs can be stored at 4°C in NDS. If the plugs are to be loaded on gels directly, remove the NDS and add 0.2 ml of loading buffer. The dye in the loading buffer tinges the transparent plug slices and facilitates loading into the wells.

Partial and double digestions are performed basically as for DNA in solution. While enzymes with the same buffer requirements can be incubated together, we routinely digest separately and remove the first enzyme with Proteinase K, followed by the very important inactivation washes with PMSF, and reequilibration with the new buffer. Partial digestions are best performed by enzyme serial dilution. Controlling partial digests by time is difficult due to the diffusion characteristics of the agarose plugs.

C. ELECTROPHORESIS

The effect of various conditions on pulsed-field electrophoresis has already been discussed (Section IV). However, several practical aspects of preparing, loading, and running PFE gels are important. PFE gels should be poured and run on level surfaces. Since small slices of plugs are pushed to the bottoms of the wells when loaded, a slight deviation from a true level position could cause the DNA to migrate out of the gel. A leveling table should be used for pouring the gel, and if the pulsed-field apparatus does not come equipped with a bubble level, providing a level would be a good investment.

In addition to leveling the gel, one should also be precise in the placement of the comb. Exact placement of the gel in the electrophoresis chamber is required for most PFE instruments because of the electrical field angles involved. A deviation in the positioning of the comb can also result in aberrations of DNA movement through the gel, especially in outer lanes. The comb's position should be marked on the casting plate to ensure reproducibility.

Because our plugs are prepared in a mold with the dimensions of the comb used to make the gel wells, plugs and plug slices fit snugly into the wells with a minimum of air space between the plug and the face of the gel. We fill the wells with loading buffer before loading the plugs. This serves several purposes: (1) the wells are easier to see because of the dye in the buffer, (2) the plugs slide more easily into buffer-filled wells than dry wells, and (3) any air pockets are easily seen and dislodged. After inserting the plugs using coverslips and bent Pasteur pipettes, excess loading buffer is removed, the wells are sealed with 1% agarose, and the gel placed in the electrophoresis unit. As an aid to maintaining a constant buffer temperature, we allow the gel to equilibrate in the circulating buffer for approximately 30 min prior to starting a run.

Following electrophoresis, the gel is stained with 1 μg/ml ethidium bromide in 0.5X TBE for 10 min and destained 20 min in 0.5X TBE. Since the DNA is highly sensitive to nicking in the presence of ethidium bromide, we keep the gel protected from exposure to light by covering with aluminum foil during staining, destaining, denaturation, and neutralization.

D. SOUTHERN BLOTTING OF PULSED-FIELD GELS

The transfer of DNA to nitrocellulose or nylon filters is basically a routine matter similar to procedures for conventional agarose gels. There are, however, two exceptions, nicking the DNA and the length of time for blotting. In our hands, the best transfer results, and therefore the best hybridization signals, are attained with ultraviolet (UV) nicking (either 254 or 300 nm) in the presence of ethidium bromide, rather than acid depurination, and by allowing the buffer transfer to proceed for at least 36 to 40 h. The critical step in the blotting procedure is UV nicking. Large DNA molecules appear to be more highly sensitive to acid nicking than UV nicking, making that method more difficult to monitor in terms of HCl concentration and soaking time, and, hence, reproducibility can vary from gel to gel.

When using the UV nicking method, one must be aware that UV intensity will vary between light sources and over the lifetime of the light bulb. The window for optimum nicking of megabase DNA is fairly narrow, and therefore the light source used for nicking should be titered and routinely monitored by the strength of the hybridization signal. While radiometers can be used to measure the intensity of the light source, readings tend to vary widely. We keep a record of radiometer measurements, but rely on UV titering gels for periodic adjustments in nicking time and by restaining blotted gels to assess transfer. For example, a gel exposed to our 254-nm transilluminator, having an intensity of 2.3 mW/cm^2 (Model J-260 Radiometer, Model J-260-1A 254-nm sensor, Ultra-Violet Products, Inc., San Gabriel, CA) requires approximately 60 s of nicking for maximum signal (Figure 4). When exposing gels to UV light, all exposure times should be calculated, including photographic time. The autorad in Figure 4 was prepared from a UV titering gel in which the lanes, containing equivalent amounts of DNA, were exposed for increasing amounts of time to the UV source. Lane 1 was exposed for 10 s, the time required for photographing the gel. Maximum hybridization with a single-copy probe occurs at approximately 60 s. After 6 min (last lane), no hybridization signal is seen. Presumably, because less DNA is applied to PFE gels (1 to 2 μg per lane, as opposed to conventional genomic gels, 10 μg per lane) and the DNA is of megabase proportion, with too little nicking, the very large DNA fragments are not transferred, while with too much nicking, the probe fails to hybridize well to small fragments. At least one company has advertised an instrument designed to deliver exact-intensity doses suitable for UV cross-linking DNA to filters as well as nicking DNA in agarose gels, eliminating the time and effort required by periodic monitoring and retitering of conventional UV sources (Stratagene).

After nicking the DNA, standard procedures are used for denaturing and neutralizing the gels (Table 2), i.e., 2 × 30 min denaturation at room temperature (RT), with shaking,

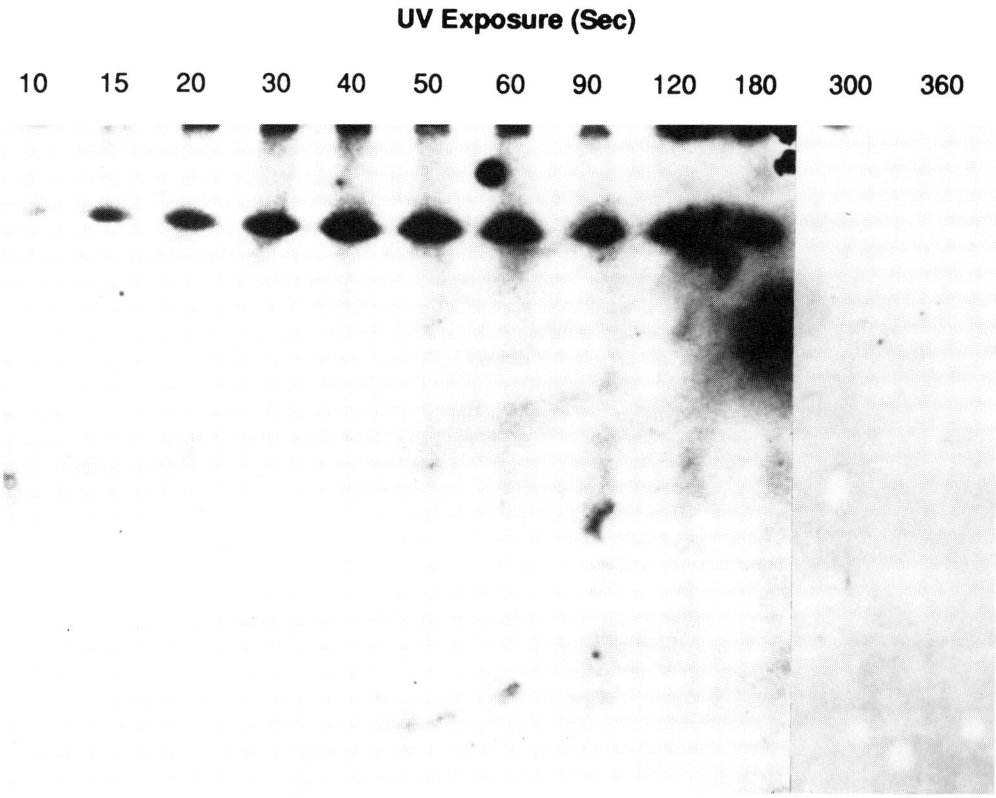

FIGURE 4. Autoradiograph from a blot of a transilluminator titering gel used to determine UV nicking conditions. Human DNA (2 μg per lane) digested with Mlu I was electrophoresed on a 1.5% agarose gel at 10°C in 0.5X TBE. Conditions used were 5 V/cm², 50-s switch time for 36 h. Following a 10-s UV exposure during which a photo was taken, each lane, starting with lane 2, was exposed for an increasing time period to a 254-nm UV source. Total UV exposure is indicated above each lane. Blotting (to Nytran), probe preparation, hybridization, and washes were as described in Tables 2 and 3. The blot was probed with a sequence from the human nerve growth factor receptor gene, which hybridizes to a fragment migrating in the compression zone with the conditions used. The blot was washed to 42°C and placed on film for 6 h (−80°C). The 300- and 360-s lanes are derived from a duplicate gel exposed for longer times to the UV source.

and 2 × 30 min neutralization at RT, with shaking. DNA transfer is performed by capillary action in 15X SSC.[58] Since transfer is dependent on the volume of buffer that passes through the gel and the DNA fragments are large, we transfer at least 36 to 40 h and exchange the upper layers of wet absorbent paper for dry, as needed, to assist buffer movement.

Alternatives to capillary blotting are electrotransfer and vacuum blotting. Both are supposed to cut blotting time to a few hours. The TAFE gel apparatus includes electrotransfer capability in its design.[47,48] Thus far, we have limited experience with electrotransfer of PFE gels, which probably accounts for our limited success with this method. We have not tried vacuum blotting. Despite the time required, capillary blotting has the advantage of not having failed us.

We have used nitrocellulose and nylon filters (Nytran, Schleicher and Schuell), but prefer nitrocellulose for most applications due to the higher signal-to-noise ratio of nitrocellulose (Figures 2B and 3B). Normally, nitrocellulose blots can be reused three to four times, sometimes more, with great care. Nylon filters, with their greater mechanical strength, can be reused more times. The same blotting procedure is normally used for both nitrocellulose and nylon filters. More recently, we have successfully used the alkaline transfer

TABLE 2
Reagents for Southern Blotting and Hybridization of PFE Gels

Denaturation buffer	1.5 *M* NaCl 0.5 *M* NaOH		
Neutralization buffer	1.5 *M* NaCl 0.5 *M* Tris-HCl, pH 7.0		
SSC (1X)	0.15 *M* NaCl 0.015 *M* Na citrate, pH 7.0		
Denhardt's solution (100X)	2 g/100 ml polyvinyl pyrrolidone 2 g/100 ml ficoll 2 g/100 ml BSA, essentially fatty acid free Filter sterilize, store at 4°C		
	Additive	**Stock**	**Final**
Prehybridization mix	Denhardt's	100X	4X
Hybridization mix	SSC	20X	4X
	Formamide	100%	40%
	Dextran sulfate	50%	10%
	Tris-HCl, pH 7.5	1.0 *M*	0.01 *M*
	SDS	20%	0.1%
	SSS DNA[a]	10 mg/ml	0.2 mg/ml

Prepare fresh, enough for prehybridization and hybridization; filter mixture through 0.45-μm filter; boil SSS DNA 10 min, chill on ice, add to prehyb aliquot; make up to final volume with distilled water and probe (2×10^6 cpm/ml)

Stringency washes	2 × 15 min at 42°C, 2X SSC, 0.1% SDS 10 min at 55°C, 0.5X SSC, 0.1% SDS 5 min at 65°C, 0.1X SSC, 0.1% SDS

Check with Geiger counter after 42°C washes; filters can be placed under X-ray film overnight (approximately 12 h) after 42°C wash and rewashed at 55°C and/or 65°C later, if necessary (XAR, Eastman Kodak)

Filter stripping	5 min at 90°C in 0.1X SSC, 0.1% SDS, or 30 min at 75°C in 0.1X SSC, 0.1% SDS Rinse 5 min at RT in 2X SSC

Expose to film 1 to 2 d; darkest specific signal should be absent

[a] Omit from hybridization aliquot.

method to Nytran filters, as described in the Schleicher and Schuell procedure.[59] However, deciding on the best filter type to use is a matter of trade-offs: reusability vs. background vs. ease and time saved in blotting.

E. HYBRIDIZATION

From a review of pulsed-field literature, almost any hybridization protocol can be used, although success with a particular recipe seems to vary from laboratory to laboratory. One should probably start with the same hybridization mixture that works well for conventional genomic blots and modify the recipe as needed for optimal pulsed-field results. We use the

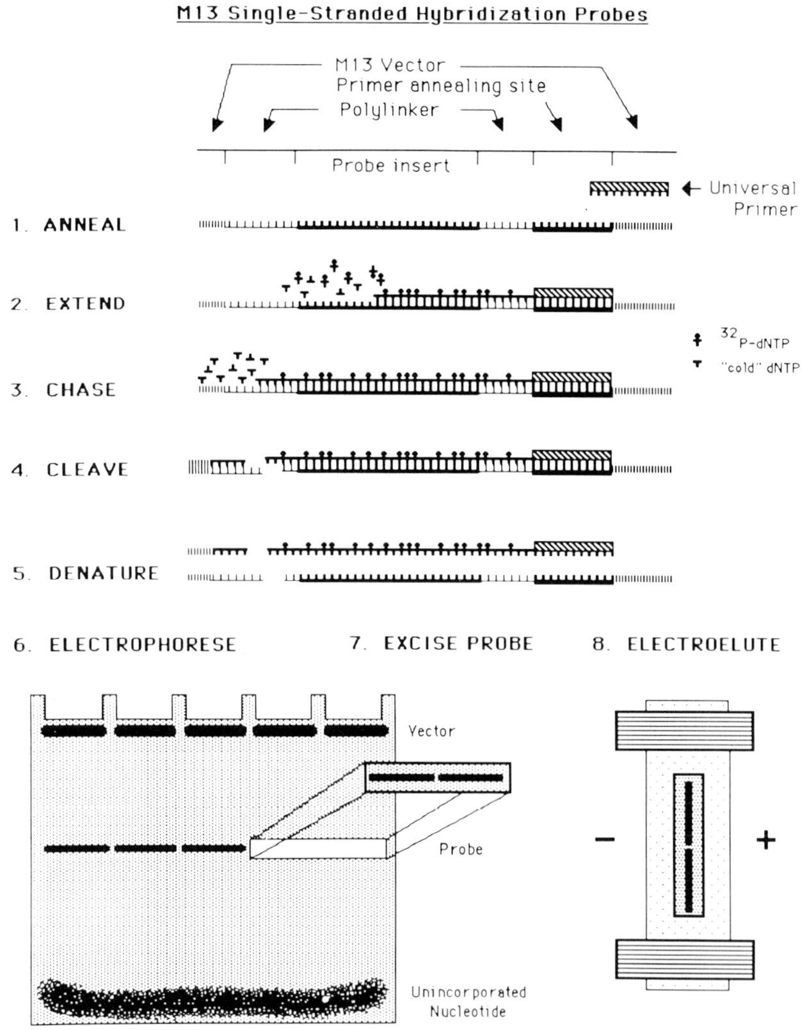

FIGURE 5. Schematic of the protocol for labeling single-stranded M13 probes. For an explanation, see the text and Table 3.

same hybridization mixture for all DNA blots (clone to clone, genomic, and PFE), prehybridization and hybridization, modified only in the amount of carrier DNA and probe added.

For PFE blots, the hybridization solution given in Table 2 is used for a 1- to 3-h prehybridization at 42°C and includes 0.2 mg/ml sonicated salmon sperm DNA. Prehybridization solution is removed and replaced with fresh solution containing 2×10^6 cpm/ml of probe (approximately 2 ng/ml for a 150-bp probe), but omitting the carrier DNA. For a 20×20-cm blot, we like to use 20 to 25 ml of solution for both prehybridization and hybridization, to compensate for the very high specific activity but low mass of the probe. An added benefit is that air bubbles accidently left in the hybridization bag are not as problematic as with low volumes of solution. Blots are hybridized for 18 to 20 h at 42°C.

As with hybridization recipes, different probe-labeling techniques have also been successfully used. The most important aspect of labeling is to achieve a high specific activity, approximately 10^9 cpm/μg or better. After trying standard nick translation[57] and random oligomer labeling,[60] we have developed a technique for preparing single-stranded probes of very high specific activity[61] based on the M13 vector (BRL)[62] (Figure 5). While M13 probes do require more time and effort to make, we believe that this method possesses a number of advantages far outweighing the few hours saved by nick translation or random priming.

Briefly, a 100- to 350-bp sequence from the gene of interest is inserted into the multiple cloning region of the M13mp 18 or 19 vector.[62] Using the 17-mer M13 universal sequencing primer and the modified T7 DNA polymerase, Sequenase (US Biochemicals), a complementary strand is synthesized from the template with a mixture of ^{32}P-labeled and unlabeled nucleotides (Figure 5, Table 3). Following a brief chase with unlabeled dNTPs, the now double-stranded DNA is cut with a restriction enzyme and denatured to release the single-stranded, labeled probe. The reaction mixture is electrophoresed on 6% denaturing acrylamide gels to separate the probe from the vector template and unincorporated nucleotides. Using an autoradiograph of the gel as a guide, the probe is excised from the gel and electroeluted. Purified probe is ready for hybridization without further treatment.

Probes produced by this method can be labeled to very high specific activity through the incorporation of the appropriate radionucleotides. Normally, using two ^{32}P-dNTPs, we achieve a minimum probe activity of 5×10^9 to 10^{10} cpm/μg, compared to 1 to 2×10^9 cpm/μg for random primer labeling, or 1 to 5×10^8 cpm/μg for nick translation. Because of the M13 vector, the probe sequence may already be known, or can be quickly determined, allowing one to tailor the particular combinations of "hot" and "cold" nucleotides to the probe. We have used both high-specific-activity (3000 Ci/mmol) and low-specific-activity (800 Ci/mmol) radionucleotides, in combinations of 1:3, 2:2, and 3:1, labeled:unlabeled nucleotides.

Because the probe is single stranded, there is no competition from the probe's complementary strand during hybridization, and virtually all of the probe is available for hybridization to DNA on the filter. Recovery from the gel yields a probe that is structurally intact and complete with regard to the probe's sequence and length. All molecules are uniformly labeled throughout and are highly purified. The result is a probe that is much more homogeneous than that produced by other methods, producing a very clean, specific signal with little nonspecific interference. Virtually all nonhybridizing probe and nonspecific signal is removed by two 15-min washes with 2X SSC, 0.1% SDS at 42°C (Figure 4). After final stringency washes (Table 2), the signal from most single-copy probes and some partial digestion products can usually be seen on autoradiographs in 12 h or less (Figures 2B and 3B).

In addition to the advantages detailed above, M13 probes have the following benefits:

1. Single-stranded M13 probes are more stable than random primer probes. In the past, we have had little success in using random primer probes prepared and stored longer than 1 to 3 d at −20°C, an effect of radiochemical degradation.[63] However, single-stranded probes have been successfully used when stored at −20°C for more than 1 week.
2. Blots hybridized with M13 probes can be stripped under gentler conditions than we have experienced with random primer or nick-translated probes (Table 2).
3. Although cloning probes into M13 requires a certain amount of effort, often probes requested from other sources are already available in M13 as a result of sequence analysis. Even so, we have routinely been able to process probes from the original, non-M13, double-stranded vector to labeled, single-stranded M13 probe in 4 d.[62]
4. Finally, a distinct advantage conferred by the use of acrylamide gel purification is that the autoradiograph developed to identify the probe band for excision provides a diagnostic guide to the quality of probe and the reagents used. Well-labeled, usable probes can be visualized before use and marginal or defective probes can be discarded before wasting time, effort, and blots on useless hybridizations. With practice, one can also determine the particular reagent that contributed to a defective probe. The

TABLE 3
Preparation of Single-Stranded M13 Probes

Step	Component	Amount	Notes
1. Anneal 2 min at 65°C Cool slowly to 37°C	Template DNA	1 μg	Sequenase and Universal primer (U.S. Biochemicals), 5X annealing buffer, 200 m*M* Tris-HCl, pH 7.5, 100 m*M* $MgCl_2$, 250 m*M* NaCl
	5X buffer	2 μl	
	Universal primer	1 μl	
	Distilled water	to 10 μl	
2. Extension: 20 min at 37°C	100 m*M* DTT	1 μl	DTT, 100 m*M* stock in distilled water
	dGTP + dTTP	2 μl	1 m*M* each dNTP in distilled water (Pharmacia)
	^{32}P-dCTP (25 μCi)	2.5 μl	800 Ci/mmol (Amersham)
	^{32}P-dATP (25 μCi)	2.5 μl	
	Sequenase (1.5 U/μl)	2.0 μl	Sequenase diluted in ice-cold TE, pH 7.5
3. Chase: 20 min at 37°C	dGTP, dTTP, dCTP, dATP	5.0 μl	1 m*M* each dNTP in distilled water (Pharmacia)
4. Cleavage: 2 h at 37°C	10X restriction buffer	5.0 μl	
	Spermidine (optional)	2.0 μl	Spermidine, 100 m*M* stock in distilled water
	Enzyme	2.0 μl	Enzyme concentration, 10 to 20 U/μl
	Distilled water	to 50 μl	
	After 1 h, add additional enzyme	1.0 μl	
5. Add 1:1 (v:v) STOP buffer; boil 5 min			95% formamide, 20 m*M* Na_2EDTA, 0.05% bromophenol blue, 0.05% xylene cyanol FF

6. 6% denaturing acrylamide gel in 1X TBE; electrophorese until bromophenol blue marker is approximately 1 inch from bottom of gel; expose X-ray film to gel for 1 min; develop 1 min (XAR, Kodak, Rochester, NY)

	Component	Amount	Notes
6% acrylamide gel (50 ml)	Acrylamide	2.89 g	Add TBE to 49 ml; mix until in solution (warm slightly); add 1 ml of 10% ammonium persulfate, 0.01 ml TEMED
	Bis	0.1 g	
	Urea	23.3 g	

TBE (1X), 89 m*M* Tris-HCl, 89 m*M* Boric Acid, 1.25 m*M* Na_2EDTA, pH 8.3.

7. Using the autoradiograph as a template, excise probe band, place in dialysis bag, and electroelute in 1X TBE; collect eluant, count

8. Store all stocks except acrylamide mix at −20°C

autorads also serve as a record for comparison of probes labeled at different times and under different conditions.

VI. SUMMARY

A little more than 3 years ago, only a handful of papers were published in all journals describing experiments using PFE methods. In a recent issue of *Genomics,* the number of articles including PFE data nearly equalled that published in all of 1986. Pulsed-field electrophoresis has provided a major impetus to physical mapping of complex genomes, and new applications are being developed on a regular basis. While this review has not concentrated on theory and instrument design as thoroughly as procedural details, those aspects of PFE technology have been well treated in the literature. We have attempted to produce a fairly comprehensive and practical guide to pulsed-field electrophoresis that will be useful to all researchers, from the uninitiated to the veteran "pulsed-fielder" searching for alternatives.

ACKNOWLEDGMENTS

We wish to thank G. Byrne (Yale University) for help and discussion in developing the M13 probe-labeling technique, R. Anand (Oxford University) for helpful advice on crossed-field electrophoresis, and M. Snyder (Yale University) for instruction on CHEF electrophoresis. Special thanks to S. Pafka for photographic services. This work was supported in part by National Institutes of Health Grant GM40883 to FHR and an American Cancer Society grant to CJB.

REFERENCES

1. **Schwartz, D. C., Saffran, W., Welsh, J., Haas, R., Goldenberg, M., and Cantor, C. R.,** New techniques for purifying large DNAs and studying their properties and packaging, *Cold Spring Harbor Symp. Quant. Biol.,* 47, 189, 1982.
2. **Schwartz, D. C. and Cantor, C. R.,** Separation of yeast chromosome-sized DNAs by pulsed field gradient gel electrophoresis, *Cell,* 37, 67, 1984.
3. **Burmeister, M., Monaco, A. P., Gillard, E. F., Van Ommen, G. J. B., Affara, N. A., Ferguson-Smith, M. A., Kunkel, L. M., and Lehrach, H.,** A 10-megabase physical map of human Xp21, including the Duchenne muscular dystrophy gene, *Genomics,* 2, 189, 1988.
4. **Meitinger, T., Boyd, Y., Anand, R., and Craig, I. W.,** Mapping of Xp21 translocation breakpoints in and around the DMD gene by pulsed field gel electrophoresis, *Genomics,* 3, 315, 1988.
5. **Anand, R., Honeycombe, J., Whittaker, P. A., Elder, J. K., and Southern, E. M.,** Clones from an 840-kb fragment containing the 5′ region of the DMD locus enriched by pulsed field gel electrophoresis, *Genomics,* 3, 177, 1988.
6. **Pritchard, C. A., Casher, D., Uglum, E., Cox, D. R., and Myers, R. M.,** Isolation and field-inversion gel electrophoresis analysis of DNA markers located close to the Huntington disease gene, *Genomics,* 4, 408, 1989.
7. **Poustka, A., Lehrach, H., Williamson, R., and Bates, G.,** A long-range restriction map encompassing the cystic fibrosis locus and its closely linked genetic markers, *Genomics,* 2, 337, 1988.
8. **Drumm, M. L., Smith, C. L., Dean, M., Cole, J. L., Iannuzzi, M. D., and Collins, F. S.,** Physical mapping of the cystic fibrosis region by pulsed-field gel electrophoresis, *Genomics,* 2, 346, 1988.
9. **Lawrance, S. K., Smith, C. L., Srivastava, R. Cantor, C. R., and Weissman, S. M.,** Megabase-scale mapping of the HLA gene complex by pulsed field gel electrophoresis, *Science,* 235,, 1387, 1987.
10. **Dunham, I., Sargent, C. A., Trowsdale, J., and Campbell, R. D.,** Molecular mapping of the human major histocompatibility complex by pulsed-field gel electrophoresis, *Proc. Natl. Acad. Sci. U.S.A.,* 84, 7237, 1987.

11. **Bottaro, A., De Marchi, M., Migone, N., and Carbonara, A. O.,** Pulsed-field gel analysis of human immunoglobulin heavy-chain constant region gene deletions reveals the extent of unmapped regions within the locus, *Genomics,* 4, 505, 1989.
12. **Herrmann, B. G., Barlow, D. P., and Lehrach, H.,** A large inverted duplication allows homologous recombination between chromosomes heterozygous for the proximal t complex inversion, *Cell,* 48, 813, 1987.
13. **Smith, C. L., Warburton, P. E., Gaal, A., and Cantor, C. R.,** Analysis of genome organization and rearrangements by pulsed field gradient gel electrophoresis, in *Genetic Engineering,* Vol. 8, Setlow, J. K. and Hollaender, A., Eds., Plenum Press, New York, 1986, 45.
14. **Fountain, J. W., Wallace, M. R., Bruce, M. A., Seizinger, B. R., Menon, A. G., Gusella, J. F., Michels, V. V., Schmidt, M. A., Dewald, G. W., and Collins, F. S.,** Physical mapping of a translocation breakpoint in neurofibromatosis, *Science,* 244, 1085, 1989.
15. **O'Connell, P., Leach, R., Cawthon, R. M., Culver, M., Stevens, J., Viskochil, D., Fournier, R. E. K., Rich, D. C., Ledbetter, D. H., and White, R.,** Two NFl translocations map within a 600-kilobase segment of 17q11.2, *Science,* 244, 1087, 1989.
16. **Drabkin, H., Sage, M., Helms, C., Green, P., Gemmill, R., Smith, D., Erickson, P., Hart, I., Ferguson-Smith, A., Ruddle, F., and Tommerup, N.,** Regional and physical mapping studies characterizing the Greig polysyndactyly 3;7 chromosome translocation, t(3;7) (p21.1; p13), *Genomics,* 4, 518, 1989.
17. **Patterson, M., Kenwick, S., Thibodeau, S., Faulk, K., Mattei, M. G., Mattei, J. F., and Davies, K. E.,** Mapping of DNA markers close to the fragile site on the human X chromosome at Xq27.3, *Nucleic Acids Res.,* 15, 2639, 1987.
18. **Chen, T. L. and Manuelidis, L.,** Neuroblastoma double minutes isolated by pulsed-field gel electrophoresis without prior strand-cleaving treatments, *Genomics,* 4, 430, 1989.
19. **Carle, G. F. and Olson, M. V.,** An electrophoretic karyotype of yeast, *Proc. Natl. Acad. Sci. U.S.A.,* 82, 3756, 1985.
20. **Smith, C. L., Matsumoto, T., Niwa, O., Klco S., Fan, J. B., Yanagida, M., and Cantor, C. R.,** An electrophoretic karyotype for *Schizosaccharomyces pombe* by pulsed field gel electrophoresis, *Nucleic Acids Res.,* 15, 4481, 1987.
21. **Coulson, A., Sulston, J., Brenner, S., and Karn, J.,** Toward a physical map of the genome of the nematode, *Caenorhabditis elegans, Proc. Natl. Acad. Sci. U.S.A.,* 83, 7821, 1986.
22. **Burke D. T., Carle, G. F., and Olson, M. V.,** Cloning of large segments of exogenous DNA into yeast by means of artificial chromosome vectors, *Science,* 236, 806, 1987.
23. **Lai, E., Birren, B. W., Clark S. M., Simon, M. I., and Hood, L.,** Pulsed field gel electrophoresis, *BioTechniques,* 7, 34, 1989.
24. **Anand, R.,** Pulsed field gel electrophoresis: a technique for fractionating large DNA molecules, *Trends Genet.,* 2, 278, 1986.
25. **Smith, C. L., Lawrance, S. K., Gillespie, G. A., Cantor, C. R., Weisman, S. M., and Collins, F. S.,** Strategies for mapping and cloning macroregions of mammalian genomes, *Methods Enzymol.,* 151, 461, 1987.
26. **Carle, G. F. and Olson, M. V.,** Orthogonal-field-alternation gel electrophoresis, *Methods Enzymol.,* 155, 468, 1987.
27. **Van Ommen, G. J. B. and Verkerk, J. M. H.,** Restriction analysis of chromosomal DNA in a size range up to two million base pairs by pulsed field gradient electrophoresis, in *Human Genetic Diseases,* Davies, K. E., Ed., IRL Press, Oxford, 1986, 1131.
28. **Barlow, D. P. and Lehrach, H.,** Genetics by gel electrophoresis: the impact of pulsed field gel electrophoresis on mammalian genetics, *Trends Genet.,* 3, 159, 1987.
29. **Gemmill, R. M., Coyle-Morris, J. F., McPeek, F. D., Ware-Uribe, L. F., and Hecht, F.,** Construction of long-range restriction maps in human DNA using pulsed field gel electrophoresis, *Gene Anal. Tech.,* 4, 119, 1987.
30. **Bentley, K. L., Ferguson-Smith, A. C., and Ruddle, F. H.,** A review of genomic physical mapping, *Cancer Surv.,* 7, 267, 1988.
31. **Lumpkin, O. J., Dejardin, P., and Zimm, B. H.,** Theory of gel electrophoresis of DNA, *Biopolymers,* 24, 1573, 1985.
32. **Slater, G. W. and Noolandi, J.,** On the reptation theory of gel electrophoresis, *Biopolymers,* 25, 431, 1986.
33. **Lalande, M., Noolandi, J., Turmel, C., Rousseau, J., and Slater, G. W.,** Pulsed-field electrophoresis: application of a computer model to the separation of large DNA molecules, *Proc. Natl. Acad. Sci. U.S.A.,* 84, 8011, 1987.
34. **Southern, E. M., Anand, R., Brown, W. R. A., and Fletcher, D. S.,** A model for the separation of large DNA molecules by crossed field electrophoresis, *Nucleic Acids Res.,* 15, 5925, 1987.

35. **Deutsch, J. M.,** Theoretical studies of DNA during gel electrophoresis, *Science,* 240, 922, 1988.
36. **Smith, S. B., Aldridge, P. K., and Callis, J. B.,** Observation of individual DNA molecules undergoing gel electrophoresis, *Science,* 243, 203, 1989.
37. **Lai, E., Birren, B. W., Clark, S. M., and Hood, L.,** Relaxation intervals alter the mobility of large DNA molecules in pulsed field gel electrophoresis, *Nucleic Acids Res.,* 16, 10,376, 1988.
38. **Schwartz, D. C. and Korval, M.,** Conformational dynamics of individual DNA molecules during gel electrophoresis, *Nature,* 338, 520, 1989.
39. Guide to biotechnology products and instruments, *Science,* 243(II), G7, 1989.
40. **Carle, G. F. and Olson, M. V.,** Separation of chromosomal DNA molecules from yeast by orthogonal-field-alternation gel electrophoresis, *Nucleic Acids Res.,* 12, 5647, 1984.
41. **Chu, G., Vollrath, D., and Davis, R. W.,** Separation of large DNA molecules by contour-clamped homogeneous electric fields, *Science,* 234, 1582, 1986.
42. **McPeek, F. D., Coyle-Morris, J. F., and Gemmill, R. M.,** Separation of large DNA molecules by modified pulsed field gradient gel electrophoresis, *Anal. Biochem.,* 156, 274, 1986.
43. **Cantor, C. R., Warburton, P., Smith, C. L., and Gaal, A.,** Voltage ramp pulsed field gel electrophoresis separation of large DNA molecules, in *Electrophoresis,* Dunn, M. J., Ed., VCH Verlagsgesellschaft, Weinheim, 1986, 161.
44. **Carle, G. F., Frank, M., and Olson, M. V.,** Electrophoretic separations of large DNA molecules by periodic inversion of the electric field, *Science,* 232, 65, 1986.
45. **Ellis, T. H. N., Cleary, W. G., Burcham, K. W. G., and Bowen, B. A.,** Ramped field inversion gel electrophoresis: a cautionary note, *Nucleic Acids Res.,* 15, 5489, 1987.
46. **Larson, J. J., Nicholson, A. W., and Siegel, A.,** Field inversion of large DNA fragments using an inexpensive unit, *BioTechniques,* 5, 228, 1987.
47. **Gardiner, K., Laas, W., and Patterson, D.,** Fractionation of large mammalian DNA restriction fragments using vertical pulsed-field gradient gel electrophoresis, *Somatic Cell Mol. Genet.,* 12, 185, 1986.
48. **Stewart, G., Furst, A., and Avdalovic, N.,** Transverse alternating field electrophoresis, *BioTechniques,*, 6, 68, 1988.
49. **Vollrath, D. and Davis, R.,** Resolution of DNA molecules greater than 5 megabases by contour-clamped homogeneous electric fields, *Nucleic Acids Res.,* 15, 7865, 1987.
50. **Ferris, S., Freeby, S., Zoller, P., Ragsdale, C., and Stevens, A.,** A megabase DNA electrophoresis system, *Biotechnol. Lab.,* 7, 36, 1989.
51. **Serwer, P.,** Gel electrophoresis with discontinuous rotation of the gel: an alternative to gel electrophoresis with changing direction of the electric field, *Electrophoresis,* 8, 301, 1987.
52. **Clarke, S. M., Lai, E., Birren, B. W., and Hood, L.,** A novel instrument for separating large DNA molecules with pulsed homogeneous electric fields, *Science,* 241, 1203, 1988.
53. *FMC Bioproducts Source Book,* FMC Corporation, Rockland, ME, 1988.
54. **Upcroft, J. A., Boreham, P. F. L., and Upcroft, P.,** Different grades of agarose affect electrophoretic migration of large DNA molecules, *Nucleic Acids Res.,* 17, 3315, 1989.
55. **Zernik, J. and Lichtler, A.,** Borate-buffer-related effects on the electrophoretic mobility of linear DNA fragments in agarose, *BioTechniques,* 5, 411, 1987.
56. **Cook, P. R.,** A general method for preparing intact nuclear DNA, *EMBO J.,* 3, 1837, 1984.
57. **Maniatis, T., Fritsch, E. F., and Sambrook, J.,** *Molecular Cloning: A Laboratory Manual,* Cold Spring Harbor Laboratory, Cold Spring Harbor, NY, 1982.
58. **Southern, E. M.,** Detection of specific sequences among DNA fragments separated by gel electrophoresis, *J. Mol. Biol .,* 98, 503, 1975.
59. Transfer and Immobilization of Nucleic Acids to S&S Solid Supports, Schleicher and Schuell, Keene, NH.
60. **Feinberg, A. P. and Vogelstein, B.,** A technique for radiolabeling DNA restriction endonuclease fragments to high specific activity, *Anal. Biochem.,* 132, 6, 1983.
61. **Bentley, K. L., Bieberich, C. J., Byrne, G., and Ruddle, F. H.,** A method for radiolabeling single-stranded probes in M13 to high specific activity, in preparation.
62. *M13 Cloning/Dideoxy Sequencing Instruction Manual,* Bethesda Research Laboratories Life Technologies, Inc., Gaithersburg, MD, 1986.
63. **Hodgson, C. P., Fisk, R. Z., and Willet, L. B.,** Radiochemical degradation of nucleic acids by 32-phosphorus, *BioTechniques,* 6, 208, 1988.

Chapter 8

MAPPING THE HUMAN GENOME

G. Mark Lathrop and Yusuke Nakamura

TABLE OF CONTENTS

I. INTRODUCTION

Discovery of the chromosomal location of a gene that harbors a mutation responsible for a disease of unknown biochemical origin is the first step toward identification and characterization of that gene's biological function. Genes associated with several human disorders, including Huntington's disease,[1] cystic fibrosis,[2-4] polycystic kidney disease,[5] von Recklinghausen neurofibromatosis,[6] multiple endocrine neoplasia 2a,[7,8] and familial adenomatous polyposis[9,10] have been localized to specific chromosomal regions through linkage to polymorphic DNA markers. The localization of a gene can provide essential information for its characterization; e.g., the gene harboring mutations causing Duchenne muscular dystrophy was isolated on the basis of knowledge of its specific location on the X chromosome, first obtained from linkage studies, and its function can now be investigated.[11] Identification of the genes responsible for chronic granulomatous disease[12] and retinoblastoma[13] has followed a similar path, assisted by knowledge derived from linkage analysis.

Systematic efforts to develop and map polymorphic DNA markers have contributed greatly to many recent successes in the application of linkage methods to studies of genetic disease. Primarily linkage maps with an average genetic distance of about 15 centimorgans (cM) between markers have now been constructed for nearly all human chromosomes. Subsets of mapped markers can be selected, to allow systematic searches for linkage to a disease gene. High-resolution linkage maps with 1- to 2-cM spacing between adjacent markers will soon be available for many chromosomal regions, to allow rapid identification of a small chromosomal segment containing a mutant gene and to provide anchor points for the subsequent construction of physical maps of regions defined by linked markers.

This chapter describes techniques for linkage studies in humans. It includes a protocol for isolation of probes and detection of RFLPs; in addition, it describes the construction of linkage maps on the basis of genotypic data derived from reference families and discusses the application of maps to the study of human genetic diseases.

II. PROBES FOR LINKAGE STUDIES

Restriction fragment-length polymorphisms (RFLPs) are the basis for most linkage studies in human genetics. The number of mapped polymorphic markers reported at the Human Gene Mapping Workshops has increased from 24 in 1980 to 1391 in 1987.[14,15] Most reported RFLPs can be divided into two classes. Conventional RFLPs, shown in several examples of disease mapping to follow, are generated by a base substitution within the recognition site of a restriction enzyme. RFLPs of a second type are called VNTR (for "variable number of tandem repeats") markers.[16] VNTRs are typically more informative than site polymorphisms for linkage studies because they can exist in multiple allelic forms (3 to $>$100) and consequently have very high heterozygosity (defined as the probability that two alleles obtained randomly from a population sample will be distinct from one another). VNTR loci reflect differences in the number of repeat elements in a tandemly repeated, 2- to $\sim$100-base-long sequence.

A third type of polymorphism of potential importance in linkage studies reflects variation at loci that contain "GT polymers".[17] Within DNA sequences determined by the polymerase chain reaction (PCR), polymorphisms are often detectable in a high-resolution acrylamide gel, similar to a DNA-sequencing gel; the difference in size between alleles can be very small (at minimum, only two base pairs). Although this method is laborious, it is likely to be very useful for developing polymorphisms within or near a cloned gene, where RFLPs are often very difficult to detect.

III. PROTOCOLS FOR SCREENING THE GENOME FOR POLYMORPHISMS

A. SITE POLYMORPHISMS

1. Digest DNA prepared from six or more unrelated individuals with restriction enzymes. We use a selection of nine enzymes; one set consists of enzymes (*Msp*I, *Taq*I, *Pst*I, *Bgl*II) that detect RFLPs very frequently; enzymes in the second set (*Eco*RI, *Bam*HI, *Hind*III and *Pvu*II) detect RFLPs less frequently, but are economical.

10X restriction-enzyme buffer	10 μl
DNA	(5.5 μg)
Enzyme	50 U (tenfold excess over DNA)
0.1 *M* spermidine	5 μl
H_2O	______
	(100 ul)

Transfer 10 μl of the reaction mixture and add 0.5 μg (in 1 μl) of λ DNA as a control for digestion. When several blots are planned, multiply the volume of the reaction mixture.

2. Store overnight at a temperature most effective for each enzyme.
3. Verify complete digestion. This is a critical step in the isolation procedure because partially digested DNA produces patterns that resemble RFLPs. Run a control of λ DNA in an agarose gel and examine it for complete digestion; if the digestion is not complete, add more enzyme to the DNA or purify the sample more stringently.
4. Prepare the Southern blot(s); the protocol is presented elsewhere in this volume.
5. Hybridize the blots with DNA probes that contain repetitive elements (e.g., *Alu*I sequences), to eliminate heavy "background", according to the following protocol (phage or cosmid clones can then function as single-copy probes, although about 5% of the cosmid clones will still produce some background).

Prehybridization

50%	formamide
5X	SSC
50 m*M*	Na_2PO_4 (pH 6.5)
10X	Denhardt's solution
250 μg/ml	human placental DNA (sonicated; average size, ~500 bp)
2%	dextran sulfate
5%	SDS

Keep at 42°C overnight (minimum). Blots not used immediately can be stored at 4°C.

Hybridization

50%	formamide
5X	SSC
50 m*M*	Na_2PO_4 (pH 6.5)
1X	Denhardt's solution
250 μg/ml	human placental DNA
5%	dextran sulfate
5%	SDS

Note: The probe should be prehybridized in the hybridization solution at 42°C for 1 h after denaturation, and membranes should be added to the hybridization solution. Maintain the blots at 42°C for 12 to 36 h.

Washing

2X SSC, 0.1% SDS at room temperature for 10 min
0.1X SSC, 0.1% SDS at 65°C for 15 min, twice

Nearly 50% of cosmid clones (mean insert size, 40 kb), or almost 15% of phage clones (mean insert size, 15 kb), will show RFLPs with one or more of the enzymes listed above.

B. VNTR MARKERS

Nearly 3% of randomly chosen cosmid clones show the VNTR type of polymorphism that results in a multiallelic polymorphic system and higher heterozygosity. For isolating VNTR markers more efficiently, we designed a method of screening with synthetic oligonucleotides contructed on the basis of known VNTR sequences, for selecting cosmids that contain tandemly repeated elements with partial homology to the synthetic probes.

This approach has proven to be highly efficient for isolating VNTR markers: 30% of the cosmids selected with oligonucleotides, especially those containing GGNNGTGGG (the core sequence common to known VNTR markers), exhibit VNTR polymorphism; this efficiency is greater by a power of ten than that for random screening. Steps and conditions for hybridization with oligonucleotides are as follows:

1. Calculation of melting temperature (T_m) for each oligonucleotide:

   T_m (°C) = 4 × (the number of G-C pairs) + 2 × (the number of A-T pairs)

   For example, the T_m for GGCTAGTAGCGTAGG is 4 × 9 + 2 × 6 = 50°C.
2. Prehybridization

   - 5X SSC
   - 50 m*M* Tris-HCl (pH 7.4)
   - 50 μg/ml yeast tRNA
   - 10X Denhardt's

   at a temperature 10°C below T_m for 2 to 16 h.
3. Hybridization
   Add 1 × 10^5 to 10^6 cpm of end-labeled oligonucleotide (~1 to 3 × 10^6 cpm/pmol) to the prehybridization mixture and maintain the prehybridization temperature for 6 to 16 h.
4. Washing

   - Wash in 5X SSC at room temperature for 5 min.
   - Wash in 5X SSC at 3 to 8°C below T_m for 5 min (several times).

Check the radioactivity left on the membrane after each washing. If you are uncertain whether washing is complete, expose a film. For VNTR screening, we recommend *Sau*3AI, *Hae*III, *Hin*fI, or *Alu*I in addition to the nine-enzyme panel because they cut DNA frequently; the smaller segments allow better resolution of alleles.

Figure 1 shows a typical VNTR polymorphism, detected with a whole-cosmid probe, cTHH33. Polymorphisms are detected with several enzymes, and the patterns with all enzymes are similar among six individuals. The average heterozygosity of VNTR markers is almost 70%, roughly twice that of site-polymorphic markers.

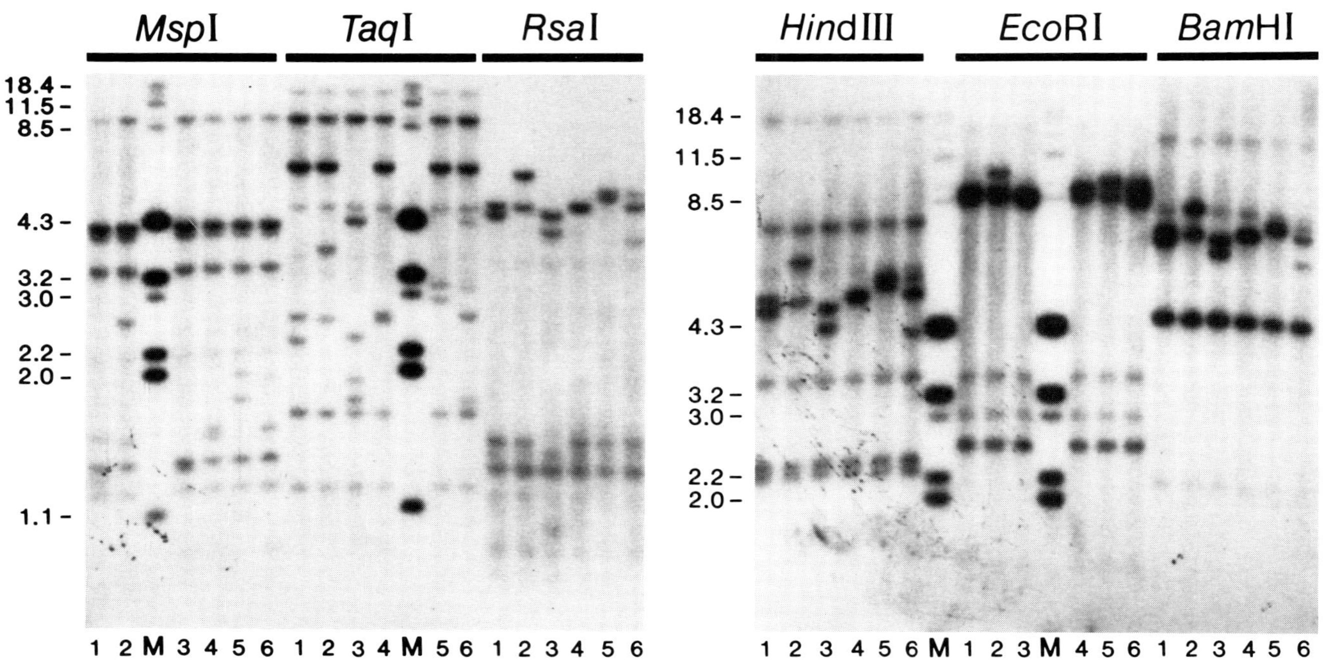

FIGURE 1. Autoradiograms of Southern transfers of DNA from six unrelated individuals (numbered at bottom of each lane; M = molecular size marker lane, with size in kilobases identified at left of each film), digested with six different enzymes and probed with radiolabeled whole-cosmid cTHH33. Multiple alleles can be distinguished for each enzyme. (From Nakamura, Y., Leppert, M., O'Connell, P., Wolff, R., Holm, T., Culuer, M., Martin, C., Fujimoto, E., Hoff, M., Kumlin, E., and White, R., *Science*, 235, 1616, 1987. With permission.)

IV. STATISTICAL METHODS FOR LINKAGE ANALYSIS

Investigations of linkage in humans are hampered by the absence of controlled crosses, and by the difficulty of obtaining large numbers of informative meioses for study. Consequently, efficient statistical methods are essential for analysis of linkage data. Standard techniques for the construction of genetic maps and the localization of disease genes now include lod-score calculations,[18] multilocus linkage analysis,[19] and the location score method.[20]

A. LOD SCORES

Linkage is usually assessed initially for pairs of loci by the method of lod scores. For completely informative data, e.g., when the parental alleles are distinct at each locus, and the grandparents' genotypes can be used to infer phase, application of the lod-score method requires only a handheld calculator. Chromosomes transmitted to the children can be scored for the presence or absence of recombination between two loci. Suppose that recombination between the two loci occurs with probability theta (θ), where θ is the unknown recombination fraction. If r recombinants are observed in n meioses, then the likelihood of the data, L (the probability of r/n recombinants as a function of θ), follows a binomial density function. The value of θ is estimated by the principle of maximum likelihood, i.e., by obtaining the value of θ that maximizes the likelihood function. In this case, the maximum likelihood estimate is $\theta = r/n$, the observed recombination rate.

The lod score, z, is defined for a given value as $z = \log_{10} [L(\theta)/L(0.5)]$, i.e., the logarithm of the ratio of the likelihood (the relative probabilities of the observations, given linkage between the two loci) at a recombination rate of θ, to the alternative of free recombination ($\theta = 0.5$). A large value of the lod score for $\theta < 0.5$ indicates that the number of observed recombinants is incompatible with the hypothesis that the loci are unlinked. The maximum lod score, z_{max}, is found at θ. For the example above:

$$\begin{aligned} z_{max} &= r\log_{10}(r) - (n - r)\log_{10}(n - r) - n\log_{10}n - n\log_{10}(0.5) \\ &= r\log_{10}(r) - (n - r)\log_{10}(n - r) - n\log_{10}n + 0.3n \end{aligned}$$

Most linkage studies in humans include families in which the parental phase is not uniquely determined or in which the data are not fully informative because the genotypes of the parents are the same or are unknown at some of the loci being tested. As shown in Figure 2, chromosomes cannot generally be scored as recombinant or nonrecombinant with such limited information. Nevertheless, one can obtain the likelihood function for a single family by summing the probabilities of the observations for each of the possible parental genotypes and phase possibilities, and obtain the overall lod score as the product of the individual likelihoods for each family (see Figure 2). Further complications arise as a consequence of incomplete penetrance for disease phenotypes, the recessive or dominant nature of disease-causing alleles (DNA markers at arbitrary loci are, of course, usually codominant), and the difficulty of quantifying some phenotypic data.[21] However, genotypic data can also be incorporated into a likelihood function by summing the probabilities of each potential combination of genotype and phase within a family.

The statistical procedures for estimation of recombination fractions and calculation of the lod scores from incomplete data are similar to those for the binomial situation, but computer programs are usually required for evaluation of the likelihood equations. Linkage analysis for general pedigrees and arbitrary phenotypes has been implemented in computer programs such as LIPED[22] and LINKAGE,[21] which are freely distributed in order to make these techniques widely available to the scientific community.

B. TESTS OF LINKAGE AND CONFIDENCE INTERVALS

The likelihood-ratio statistic, defined as $-2 \ln [L(\theta_{max})/L(0.5)]$, is distributed as a χ^2 with 1 degree of freedom when the two loci are unlinked. Therefore, a statistical test of

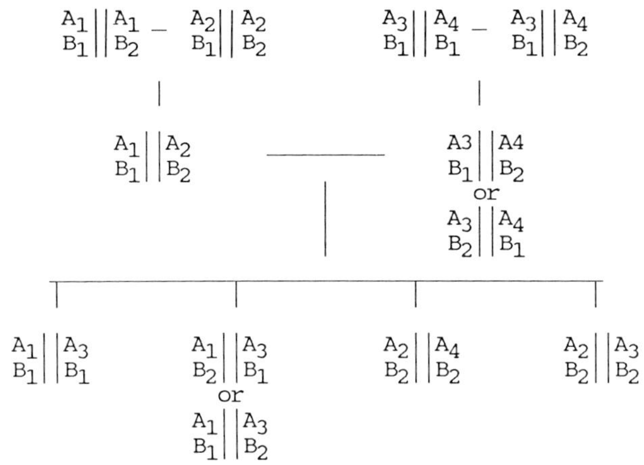

Likelihood (ignoring constants):

$$\text{Phase 1} \quad (1-\theta)^2 \times [(1-\theta)\theta + \theta(1-\theta)] \times (1-\theta)^2 \times (1-\theta)\theta$$

$$\text{Phase 2} \quad (1-\theta)\theta \times [\theta^2 + (1-\theta)^2] \times (1-\theta)\theta \times (1-\theta)^2$$

$$\text{Total likelihood} = \text{Phase 1 likelihood} + \text{Phase 2 likelihood}$$

$$= 2\,(1-\theta)^6\,\theta^2 + (1-\theta)^4\,\theta^2[\theta^2+(1-\theta)^2]$$

$$= 2\,(1-\theta)^4\,\theta^2\,[\theta^2 + 3(1-\theta)^2]$$

FIGURE 2. Example of a likelihood calculation for two codominant marker loci and a simple family structure. Alleles are labeled A_1, A_2, etc. at the first locus and B_1, B_2, etc. at the second locus. The grandparental genotypes determine the allelic phase in the first parent. Two possibilities of phase are shown for the second parent; each has equal probability, given the grandparental genotypes. The parental origin of chromosomes transmitted to the first, third, and fourth children can be identified (the chromosome transmitted from the first parent is given before that transmitted from the second parent). For each phase, the chromosomes can be scored into recombinant (probability 0) and nonrecombinant classes $(1 - \theta)$. The second child has received two possible chromosome combinations; this is taken into account by summing the probabilities for each. The likelihood is proportional to the product of probabilities over children for each phase, summed over phases. The likelihood equation, shown in the figure, has a maximum at $\theta = 0.26$ (by numeric calculation). The maximum lod score for the family is calculated as the logarithm of the ratio of the likelihoods at $\theta = 0.26$ and 0.5, given $z_{max} = 0.35$ for this family. For a sample composed of several families, the total likelihood is the product of the likelihoods of individual families, and the combined lod score is the sum of the individual lod scores.

linkage could be based on the comparison of this statistic with the critical values of the χ^2 distribution. Since lod scores are calculated using $\log_{10}$ for historical reasons, the critical values need to be translated to this scale for use with most published data; this can be achieved by dividing the likelihood ratio statistic by 2 ln(10), approximately 4.6. Thus, at a significance level of 0.01, the hypothesis of no linkage can theoretically be rejected when the likelihood ratio statistic is >3.9, equivalent to $z_{max} > 0.85$.

In practice, the more stringent condition of $z_{max} > 3$ has been adopted to exclude false positive results. The criterion of $z_{max} > 3$ was introduced by Morton[18] to adjust type 1 statistical error for the prior probability that two loci are on the same chromosome in humans. Although the usefulness of the $z_{max} > 3$ limit has been demonstrated empirically in many studies, adoption of less stringent significance levels could be entertained in some special circumstances, e.g., to investigate linkage to a candidate gene or to test for linkage of a disease gene to X-linked markers.

In large samples, the maximum likelihood estimate of θ is approximately normal. Although the large sample distribution could be used to obtain an approximate confidence interval, a different technique has been recommended by the Human Gene Mapping Workshop committee on linkage analysis.[23] The recommended technique consists of defining the upper and lower limits of the confidence region at points 1 lod unit below the peak lod score. This approach has the advantage that it reflects the nonsymmetrical distribution of the estimate for infrequent recombination in small samples.

C. MULTILOCUS LINKAGE ANALYSIS

Construction of a genetic map, or placement on the map of a gene that harbors mutation(s) responsible for a disease, requires that genetic distance be estimated between several loci that can be arranged in linear order. Statistical evidence for gene order can be obtained from the distribution of recombination events, taking account of several loci simultaneously. The assumption of different gene orders changes the number of postulated recombination events, given the observed meioses, thus providing statistical evidence in favor of one order against another (Figure 3). Two orders are compared statistically by their likelihood ratio, or relative odds, defined as the ratio of their likelihoods evaluated at the maximum likelihood recombination estimates.[19] Odds of 1000:1 or greater are considered to be strong evidence against one order vs. another; odds of 100:1, compared to the maximum order, are often used as a limit for defining rejection of alternative orders.

Likelihood evaluations and recombination estimates are obtained by multilocus analysis, incorporating information on the segregation of many markers in a chromosomal region.[19,20] Statistical techniques for multilocus likelihood calculations, with codominant markers in three-generation reference families under a single gene order, were introduced by Lathrop et al.;[24,25] subsequently, related methods were described for X-linked disease[26] and for nuclear pedigrees that consist of children, parents, and grandparents.[27]

D. LOCATION SCORES

Once a map of a given chromosomal region has been constructed, it can be used to localize new markers or disease genes by the location score method.[20] In this approach, the previously constructed map is assumed to be fixed, and likelihoods for the placement of the new locus are calculated by multilocus methods. The location score for a position, d, on the map is defined as $-2 \ln [L(d)/L(\text{unkinked})]$, where L(unlinked) is the likelihood of the data if the new locus is unlinked to the map. For two loci, the location score is equivalent to the likelihood-ratio statistic; it is directly related to the lod score, as discussed above. Therefore, 13.8 (equivalent to a lod score of 3) should be an appropriate threshold for declaring linkage with the location score method. One-lod-unit confidence intervals for the location are constructed from two-locus recombination estimates.

Location scores can be presented graphically, as shown in Figures 4 through 6 for genes associated with diseases that will be discussed later in this chapter — familial adenomatous polyposis, cystic fibrosis, and X-linked spastic paraplegia. The maximum likelihood location of an unknown gene, assuming the fixed map of markers in the region, is the position that yields the highest peak on the likelihood curve. Typically, different maxima are observed for different orders; the tested orders are compared by reporting the likelihood ratios, or the odds, at the maxima under the alternative orders compared to the maximum likelihood position.

The location score method is particularly useful for the localization of a disease gene when the genotypic data contributed by an affected family are limited. Data from different studies can be combined, because the location score curves are additive if the same genetic map is assumed for the markers.

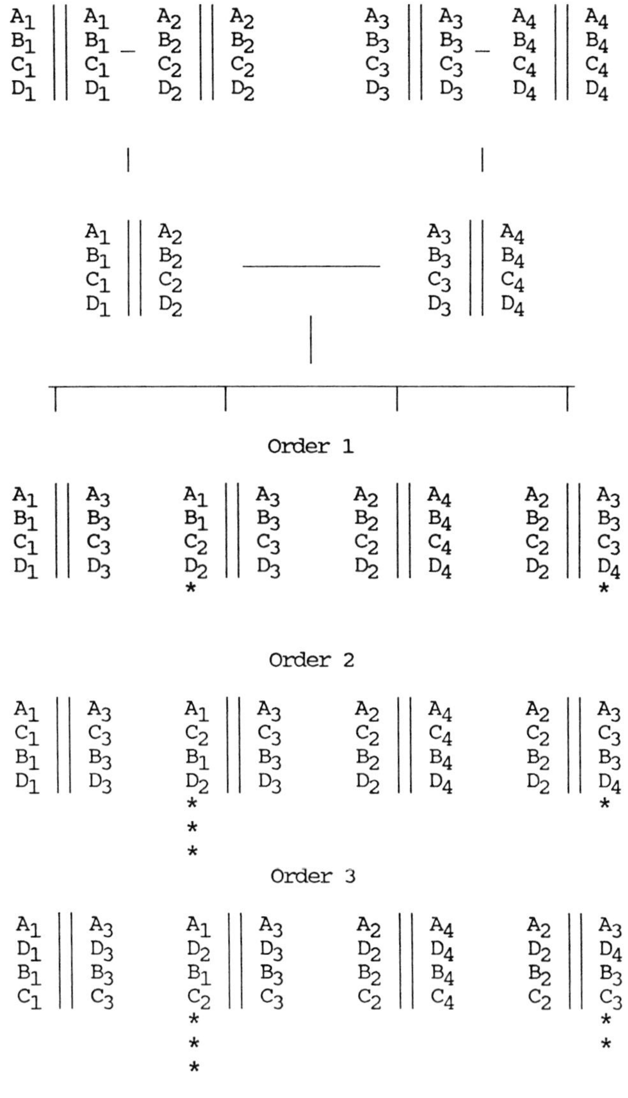

FIGURE 3. Example of a single family showing how different gene orders change the recombination events required to explain the observed segregation patterns. Offspring chromosomes from a single family are shown under three different possible gene orders. Recombinant chromosomes are indicated by the presence of an asterisk (one for each postulated recombination event). The order ABCD gives the minimum number of recombination events (2) in the family. Changing the orientation of BC to obtain ACBD requires that two new recombination events be postulated, for a total of four. The alternative of ADBC requires five recombinants. As the recombination fraction for linked markers is <0.5, a change in gene order that introduces a new postulated recombination event typically decreases the likelihood. The actual likelihoods of the different orders depend upon the estimates of the recombination rates in each case.

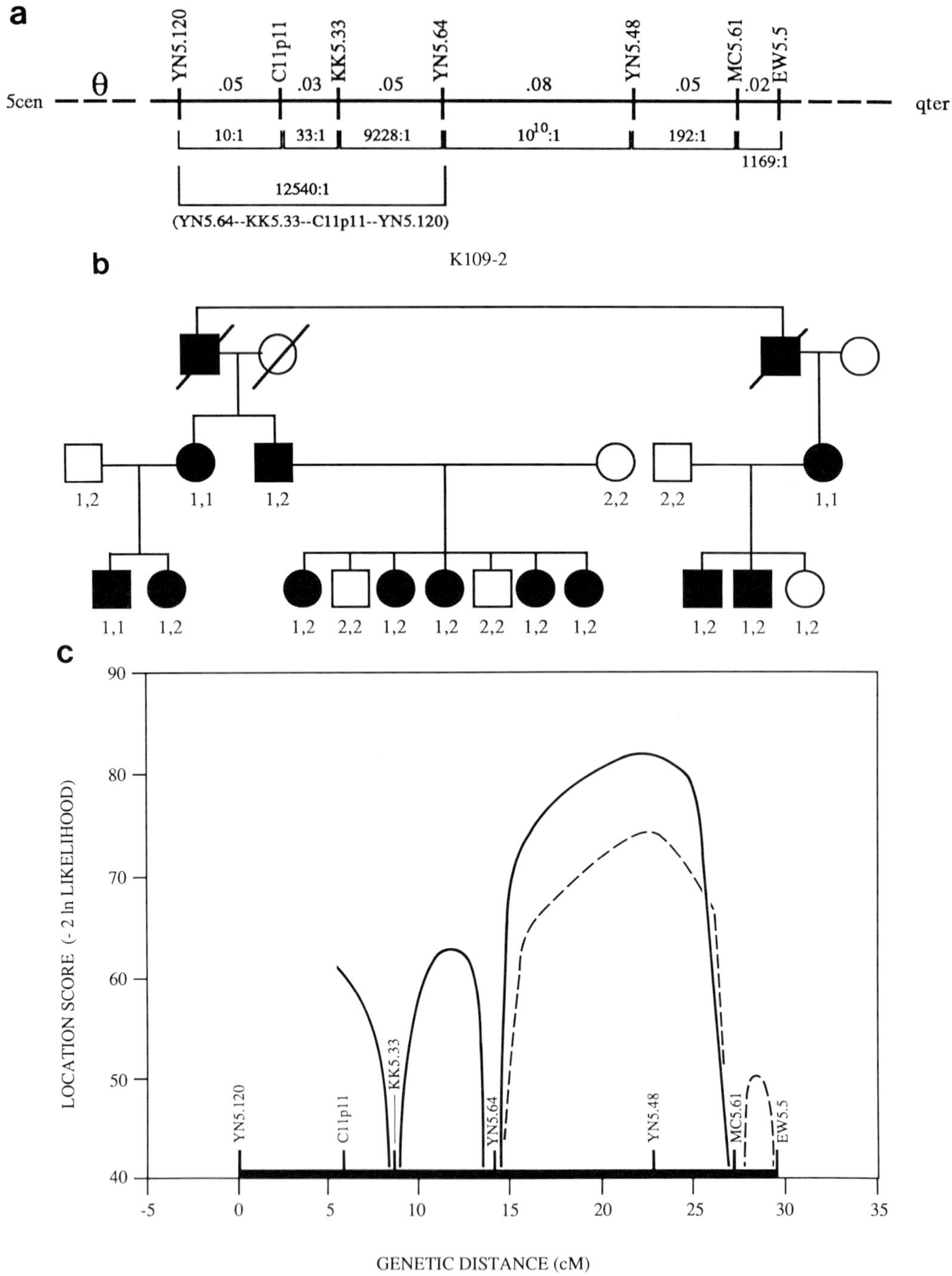

FIGURE 4. Linkage of the gene causing familial adenomatous polyposis coli (APC) to markers on chromosome 5. (a) High-resolution map of seven markers in the APC region, showing odds against inversion of adjacent loci. Odds are 12,540:1 against reversing the orientation of the subgroup containing the proximal four loci. (b) Typical cosegregation pattern of a linked marker (YN5.48) and disease in an extended pedigree; (c) Location score curves showing the maximum likelihood placement of the APC locus within the map of seven closely linked markers. The solid curve indicates the location score with respect to the subset of loci Cllpll, KK5.33, YN5.64, and YN5.48; the dashed line indicates the location score calculated with respect to the subset YN5.64, YN5.48, MC5.61, and EW5.5. Both curves indicate a maximum likelihood location very close to the YN5.48 locus. (From Nakamura, R., Lathrop, G. M., Leppert, M., Dobbs, M., Wasmuth, J., Wolff, E., Carlson, M., Fujimoto, E., Krapcho, K., Woodward, S., Hughes, J., Burt, R., Gardner, E., Lalovel, J.-M., and White, R., *Am. J. Hum. Genet.*, 43, 638, 1988. With permission.)

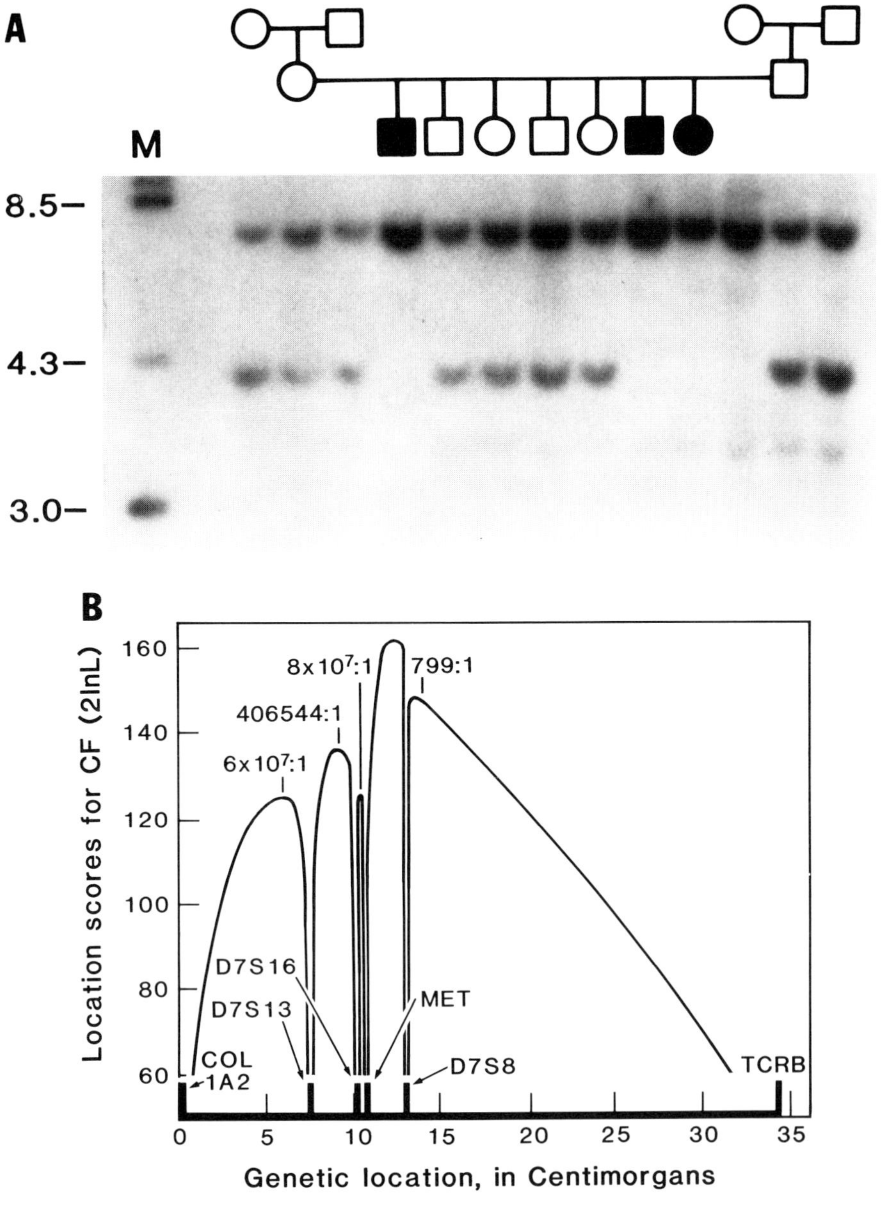

FIGURE 5. Linkage of the gene causing cystic fibrosis (CF) to markers on chromosome 7. (A) Typical cosegregation pattern of a linked marker (metH) and disease in an extended pedigree. CF-affected children,1,1 homozygotes for the probe, are designated by filled symbols; all of their unaffected siblings are heterozygous carriers of the CF allele. (B) Map of six markers in the region of CF (bold line on horizontal axis) and location scores showing the maximum likelihood placement of the CF locus. (From Lathrop, G. M., Farrall, M., O'Connell, P., Wainwright, B., Leppert, M., Nakamura, Y., Lench, N., Kruyer, H., Dean, M., Park, M., Vande Woude, G., Lalouel, G. M., Williamson, R., and White, R., *Am. J. Hum. Genet.*, 42, 38, 1988. With permission.)

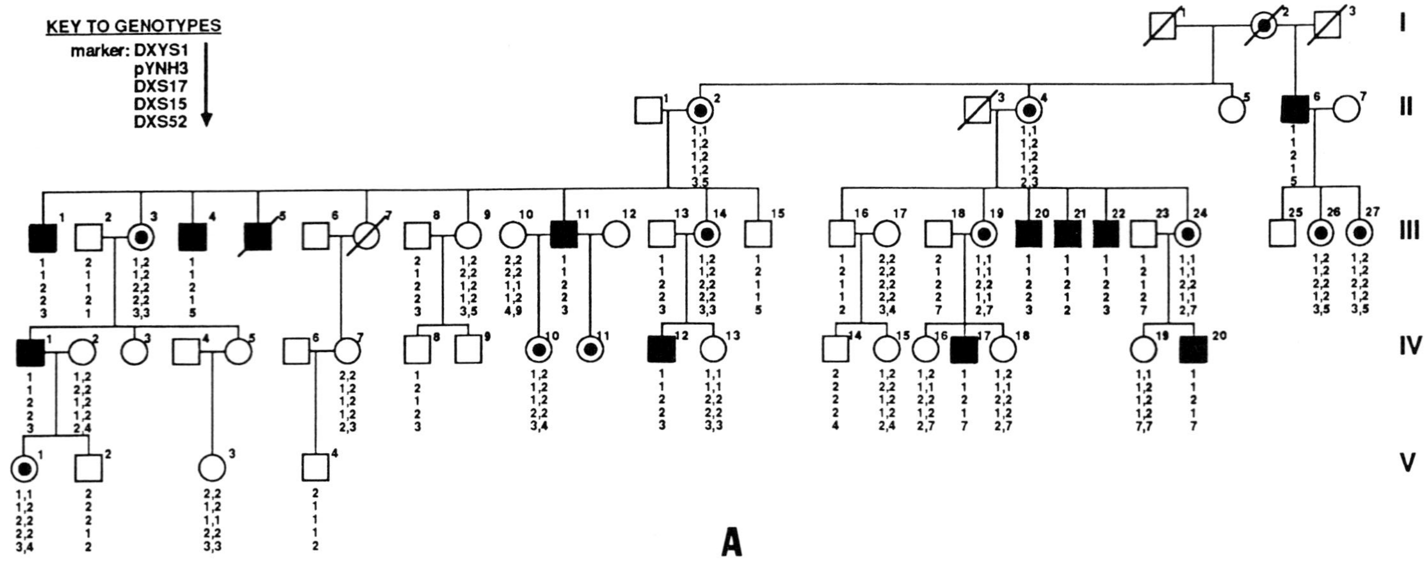

FIGURE 6. Linkage of genes causing X-linked spastic paraplegia to DNA markers. (A) Typical cosegregation pattern of linked markers with the disease gene in an extended pedigree; see text for discussion. (B) Location score curves showing different maximum likelihood locations for the genes responsible for SP in two extended pedigrees. One gene lies close to markers in Xq27-28 (solid line); the other (dashed line) lies among markers in Xq13-22. (From Keppen, L. D., Leppert, M., O'Connell, P., Nakamura, Y., Stauffer, D., Lathrop, M., Lalouel, J.-M., and White, R., *Am. J. Hum. Genet.*, 41, 933, 1987. With permission.)

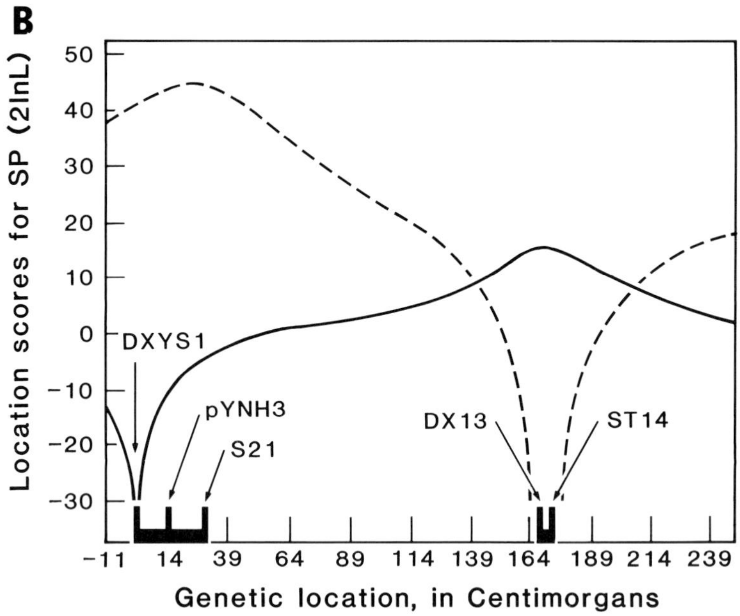

FIGURE 6B.

E. MISCELLANEOUS STATISTICAL TOPICS

Good family material is essential for the success of linkage studies. Therefore, one of the first steps in evaluating the feasibility of a study is to determine the probability of detecting linkage with the families available for study. Although exact-power calculations are difficult for general pedigrees, simple family structures can be investigated by a combination of analytic and simulation methods.[64] For a recessive trait, approximately 35 families containing two affected sibs would be required to assure a 95% probability of detecting linkage ($z_{max} > 3$), at 10% recombination distance, for a marker locus with 100% heterozygosity; for a 2-allele polymorphism with equal gene frequencies, approximately 45 such families are required.

Achieving the same power when studying a dominant disease gene requires, respectively, 35 and 95 phase-known meioses that are potentially informative for the disease locus. With phase-unknown data, detection of linkage to a dominant disease gene requires larger samples. For example, approximately 60 phase-unknown families with 4 children each (240 potentially informative meioses) are required to assure a 95% probability of detecting linkage between a two-allele, equal-frequency polymorphism and a fully penetrant disease locus. Reduced pentrance decreases the power to detect linkage for a dominant disease, as information is lost due to the probability that unaffected children carry the disease mutation. In the phase-unknown example, 60 families each with 2 affected and 2 unaffected offspring offer an approximately 90% probability of detecting linkage with a penetrance parameter of 90% and a 30% probability with penetrance of 50%.

One approach to the problem of inadequate sample size is to use positive lod scores (>2 or >1) as indications of possible linkage and type other markers in the region in order to obtain confirmation at less distance. Also, one can investigate extended pedigrees, which usually are more informative than nuclear families for linkage studies. Simulation methods are available for evaluating the information available from complete pedigrees.[28]

Lod scores in extended pedigrees are often very sensitive to estimates of gene frequency at the maker locus when genotypes are not available for one or more ascending generations. This is because when an allele is rare and therefore unlikely to have been introduced twice into the pedigree, the probability increases that a similar allele at the marker locus is identical by descent in two distantly related, affected individuals. Extended pedigrees studied for recessive disease usually contain consanguinity loops and are affected to an even greater degree by assumptions with respect to gene frequencies. As frequencies are often poorly estimated for populations representative of the disease group, we recommend that analyses be repeated with several different gene-frequency assumptions, to verify linkage.

The possibility of genetic heterogeneity should also be considered in linkage studies, Ott[29] has discussed statistical tests of heterogeneity in detail. If genetic heterogeneity is present, the number of families needed to detect linkage may be much larger than the number needed for a single-gene disorder.

V. LINKAGE MAPS OF CHROMOSOMES

Primary genetic linkage maps with markers spaced approximately 15 cM apart are now available for most human chromosomes. These maps can be used to select markers for establishing the regional localization of any disease gene for which sufficient family material is available. Primary maps will provide anchor points for construction of high-resolution linkage maps in which markers are spaced, on average, 1 to 2 cM apart. Most such maps that are available today have been established in areas adjacent to a disease gene, after an initial localization. However, systematic efforts are now underway to obtain high-resolution maps of entire chromosomes, for linkage studies and as an aid for construction of a physical map of the genome.

A. REFERENCE FAMILIES

The rapid progress in obtaining primary linkage maps has been greatly aided by efforts to establish a panel of reference families. Marker loci are most efficiently studied in nuclear families with large sibships (including grandparents, when possible, to verify parental genotypes and to provide phase information). A number of such families were initially ascertained in the Utah population, from whom cell lines were established to provide a permanent source of DNA.[30] So that other laboratories could have free access to the same reference material, the CEPH (Centre d'Etude du Polymorphism Humain) in Paris has undertaken to distribute DNA from families selected largely from the Utah panel.[31] The CEPH reference panel consists of 40 families (27 from Utah, 10 from France, 3 from Venezuela, and 1 Amish) containing 656 potentially informative meioses.

More than 50 laboratories worldwide are collaborating to map the human genome by genotyping new markers in the CEPH panel, and a genotypic database of over 1500 marker loci characterized in the 40 families has been constituted. CEPH collaborators receive DNA samples from the panel free of charge. In return, they undertake to use the panel to characterize DNA probes in all parents, and to type the complete family whenever the marker is informative for linkage (i.e., when one or both parents are heterozygotes). The decision to use the same family material allows markers studied in different laboratories to be combined in a single linkage map. Collaborations of this type are presently underway for most chromosomes.

In our studies, we have supplemented the CEPH panel with an additional 19 families to obtain a total of 1212 potentially informative meioses. The extended reference panel of 59 families has proved very useful for precise ordering of loci and for estimation of recombination distances. The larger number of meioses available in the extended family panel will make resolution of ambiguous gene orders in a 1-cM linkage map more feasible.

B. CONSTRUCTION OF PRIMARY LINKAGE MAPS

Primary maps start with anchor loci that have been assigned to a particular chromosome, or to a specific chromosomal region, by physical methods. Placements for other loci are obtained by linkage to one of the physical anchor points (or to other markers as the number of linked loci increases). Since physical data to determine gene order are usually limited, techniques of multilocus linkage analysis are used both to infer gene order and to estimate recombination distances in the map. Linkage results often serve to confirm physical localizations of anchor loci.

When a map contains a small number of loci, the maximum likelihood order can be inferred by comparison of the likelihoods of all possibilities; with seven or fewer loci, the total number of gene orders to compare is not greater than 2520. However, as the number of loci increases, comparison of all likelihoods is no longer feasible; for example, the total number of possible gene orders for eight loci is 20,160, and exceeds 1.8 million for ten loci. Therefore, strategies for limiting the number of gene orders to be tested are needed for efficient construction of a genetic linkage map.

A heuristic approach to this problem has been developed and implemented in the computer program GMS, described by Lathrop et al.[32] Under the GMS algorithm, a linkage group that has been localized to a particular chromosome or chromosomal region is given an initial trial order using one of several algorithms, most of them based on two-locus recombination estimates; several of these algorithms have been reviewed by Kammerer and MacCluer.[33] The trial order is used to obtain multilocus recombination estimates between adjacent loci. These determine a division of loci into subgroups of closely linked markers, for tests of gene order within and between subgroups. Different permutations and orientations of the subgroups are tested using the maximum likelihood order within each subgroup; alternative orders within each subgroup are tested for those permutations and orientations with high relative likelihood. The best-supported order is chosen for a new trial order, and the above steps are repeated until convergence.

Examples of the application of the GMS algorithm can be found in primary maps published from our laboratory and cited below. Other groups have adopted similar approaches to map construction.[34-36]

C. WHAT THE MAPS REPRESENT

The inference of gene order, and recombination estimates for linkage data, are subjected to statistical variation. Several techniques are available for representing the analytical results and for illustrating the degree of confidence that can be assigned to the conclusions.

An alternative order is usually rejected statistically when odds of >100:1 are found against it, compared to the best-supported order. With the extended family panel, the placement of the majority of our markers is well determined under this criterion. However, some orders remain unresolved and so are kept as possible alternatives until further data become available. Generally, we report the odds against the inversions of all adjacent loci, since the large majority of unresolved orders involve a change in the orientation of closely linked, adjacent markers. This approach also allows us to indicate areas in which the inferred gene order is well supported, as high odds against the inversion of adjacent loci are characteristic of such regions.

In addition, we use approximate confidence intervals for the placement of loci in the map, to indicate the precision of the recombination and gene-order estimates simultaneously. Confidence intervals are calculated by fixing the recombination distances between loci at the maximum likelihood estimates under the best-supported gene order. Each locus is removed in turn from the map, and location scores are calculated for it with respect to the remaining loci. All locations falling within 1 lod unit of the maximum are included in the confidence region for the locus. Alternative locations both within and outside the best-supported order may be contained in the confidence region. Application of these techniques to markers on human chromosome 17 yielded the map illustrated in Figure 7.

Primary linkage maps have now been constructed for most human chromosomes, by

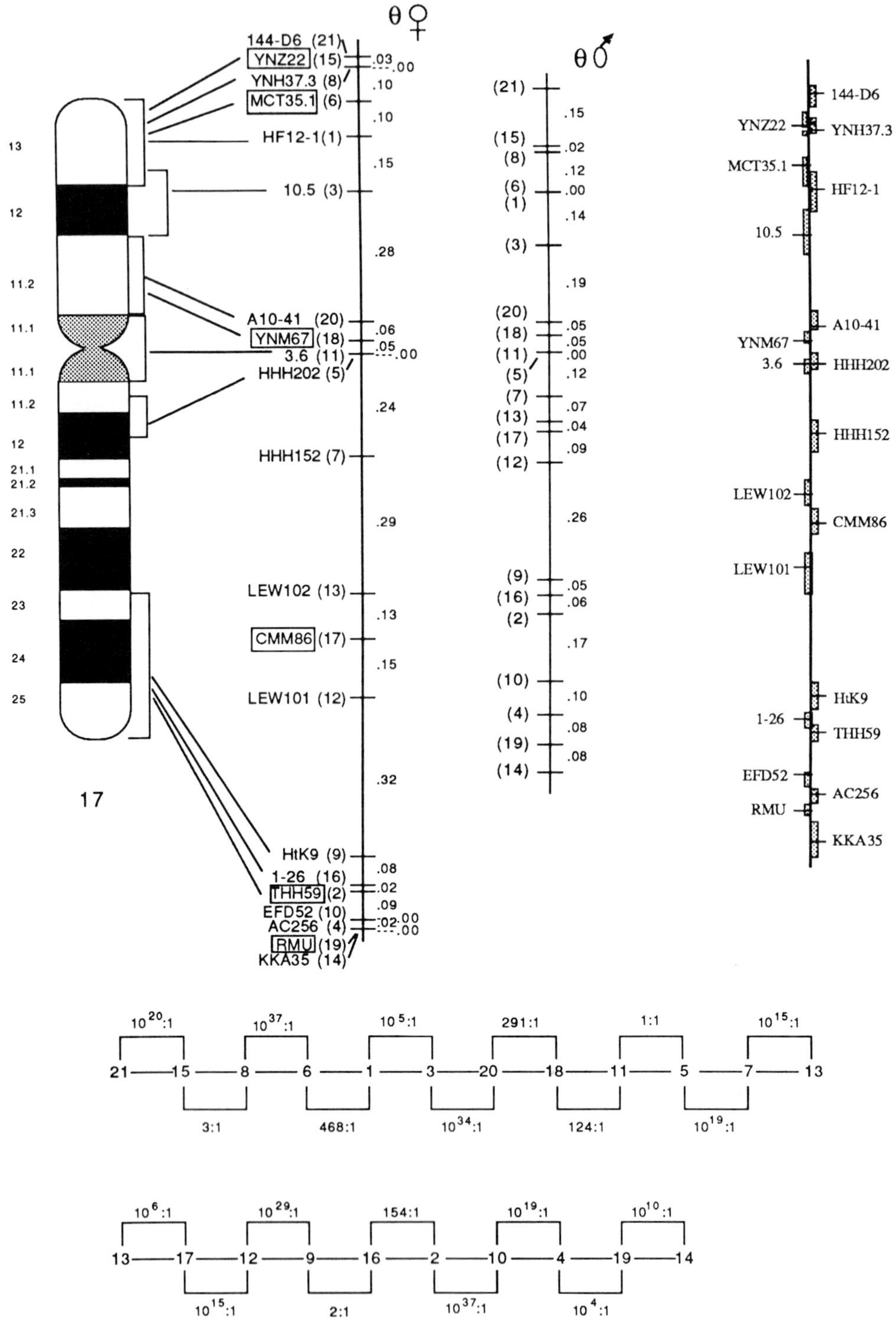

FIGURE 7. Male and female genetic maps of 21 markers on chromosome 17, with physical locations of selected loci indicated on the karyogram. Boxed loci on the female map constitute the subset of markers selected for linkage studies in disease families. Odds against inversions of adjacent loci are shown below, and confidence intervals at right. Maps are scaled in centimorgans; θ = recombination fraction. (Adapted from Nakamura, Y., Lathrop, G. M., O'Connell, P., Leppert, M., Barker, D., Wright, E., Skolnick, M., Kondoleon, S., Litt, M., Lalouel, J.-M., and White, R., *Genomics,* 2, 302, 1988. With permission.)

several laboratories working with the CEPH panel or with other families. The extended family panel used in Salt Lake City has been used to produce maps of chromosomes 1,[37] 2,[38] 5,[10] 7,[39] 9,[40] 10,[32,41] 11,[30] 12,[42] 13,[43] 15,[44] 17,[45] 18,[46] 19,[47] 20,[48] and X.[49] Donis-Keller et al.[35] typed a subset of 23 families from the CEPH panel for their linkage map covering parts of all human chromosomes.

The many mapped markers that have now been genotyped on the CEPH families have increased the value of this panel as a resource for localization of polymorphic genes or new marker loci. Donis-Keller et al.[35] estimated that over 90% of RFLPs could be linked to markers included in their map. Once linkage is detected, the genetic map can be used to provide a rapid regional localization of the new locus by the method of location scores. An example of this approach is given in Bowcock et al.[50]

D. SEX-SPECIFIC DIFFERENCES IN GENETIC DISTANCE

Primary genetic maps have produced new insights into the patterns of sex-specific differences in the occurrence of crossing over during meiosis. Although most chromosomal regions show an excess of observed recombination in female meioses, not all do. White et al.[30] first described a region of three- to fourfold excess recombination in males, on chromosome 11p. Subsequently, our studies have shown areas of statistically significant male excess in regions of other chromosomes, results that have been confirmed by other studies on the CEPH panel.[35]

To investigate these patterns on each chromosome, we routinely use three different statistical hypotheses to allow for sex-specific variation in a linkage analysis.[42] These are (1) the assumption of equal male and female recombination, (2) the assumption of a constant ratio of female/male genetic distance throughout a chromosome or chromosomal region, and (3) a variable ratio of female/male genetic distance in each interval defined by adjacent markers. The likelihood ratio statistic comparing models 1 and 2 is considered to have a χ^2 distribution with 1 degree of freedom for statistical tests. For comparison of models 2 and 3, the likelihood ratio statistic is considered to have a χ^2 distribution with $n - 2$ degrees of freedom, where n is the number of loci included in the test. Further statistical investigations are underway to examine the properties of these tests and to evaluate the influence of different patterns of sex-difference on the construction of a genetic map.

E. HIGH-RESOLUTION LINKAGE MAPS

Primary maps are the basis for constructing high-resolution linkage maps in which adjacent markers may be spaced every 1 to 2 cM. This spacing should lead to rapid identification of markers that flank a small chromosomal segment that contains a disease gene, once a regional localization has been obtained. Moreover, a high-resolution linkage map can be a valuable tool for ordering cosmids into a physical map.

Although a complete 1-cM map of the human genome could contain as few as 3000 to 4000 loci, it is likely that a larger number of markers will be mapped initially in order to ensure coverage. Analysis of a simple statistical model that assumes isolation of markers, taken from a uniform distribution on a chromosome-specific basis, indicates that 7000 to 8000 loci would be needed to ensure that 90% of the 1-cM intervals in the genome would have at least one marker.[64]

As illustrated in Figure 3, the ordering of adjacent loci is the most difficult aspect of constructing a high-resolution map from linkage data. Although cytogenetic techniques and other methods of physical mapping could be helpful for ordering some closely linked loci, linkage data on the extended reference panel should provide enough information to order most adjacent pairs of markers if they exhibit sufficient heterozygosity in the population sample.

Calculations that assume a placement of 3500 markers uniformly 1 cM apart show that,

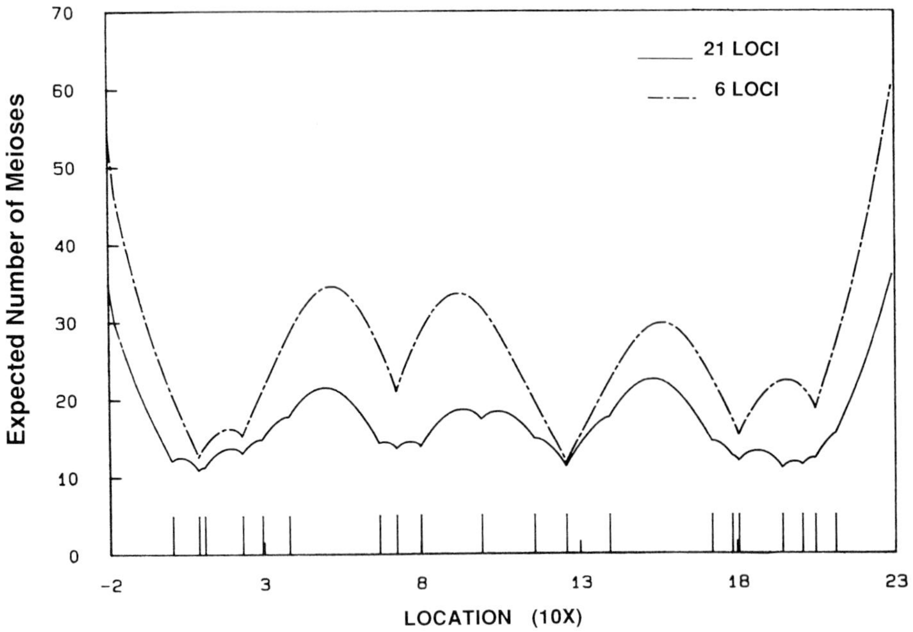

FIGURE 8. An upper bound on the expected number of meioses required to achieve a location score of 13.8 for a dominant disease, with all 21 loci in the chromosome 17 map (solid curve), and with a subset of 6 loci (dashed curve) (YNZ22, MCT35.1, YNM67, CMM86, THH59, RMU). The expected number of meioses depends on the placement of the disease gene. Phase-known data are assumed. (From White, R., Leppert, M., Nakamura, R., O'Connell, P., Lathrop, M., and Lalouel, J.-M., in *Cellular and Molecular Biology of Human Carcinogenesis*, Bautwell and Riegel, Eds., Academic Press, San Diego, in preparation. With permission.)

with 95% probability, the orders of 99% of adjacent loci will be resolved by at least two informative recombination events for markers with 75% heterozygosity, and 79% of orders will be resolved for markers with 50% heterozygosity.[64] Two recombination events supporting the same order are generally considered necessary to reduce the possibility of errors due to laboratory mistyping. Highly informative marker loci are clearly important, not only for determining order from linkage data but also for applying the map to localization of disease genes.

An example of a well-resolved, high-resolution linkage map, constructed from data on the extended reference panel for a small region of a chromosome (14q32), is given in Reference 51.

F. SELECTION OF APPROPRIATE MARKERS FOR LINKAGE STUDIES

Primary genetic maps provide information to assist one's choice of markers for efficient linkage studies in disease pedigrees. The goal is to assure coverage of each chromosome with a minimum number of markers, taking account of the family information available for study. Other important criteria for selection of markers are the ease of use and the number of different enzymes required to detect a polymorphism.

Given a primary linkage map, an initial subset of markers having a high degree of efficiency for detection of linkage with a disease locus can be selected by considering an idealized study of phase-known data for a rare dominant disease. An example for chromosome 17 is given in Figure 8. Here, we used an approximation to calculate an upper bound on the number of meioses needed to achieve an expected location score of 13.8 (similar to an expected lod score of 3) for a disease gene at different positions on the map, taking account of the gene frequencies and the placement of 21 markers on a complete primary map of this chromosome.[45] No more than 23 meioses would be needed to achieve an expected location score >13.8 for a fully penetrant mutant gene lying within the region spanned by the two most distal markers in the map. Estimates for the number of meioses needed to detect linkage outside this region cannot be obtained because, in the absence of specific telomeric markers, the distal intervals not covered by the map are unknown.

To select a set of markers on chromosome 17 for linkage studies, the expected numbers of meioses needed to detect linkage with various subsets of markers was compared to the number of meioses needed for a linkage analysis with all 21 loci in the region interior to the two most distal markers. One subset of six loci, with average heterozygosity of 77%, gave a maximum expected number of 35 phase-known meioses (see Figure 8). Although the chromosome 17 map contains eight highly informative VNTR loci, they are not uniformly distributed over the chromosome, so only four of them are used; the two other markers in the subset are site polymorphisms. Maps for other chromosomes also exhibit clustering of VNTRs, often in telomeric or centromeric regions.

Markers selected for linkage studies on other chromosomes are reported in the publications cited above.

VI. LINKAGE STUDIES FOR HUMAN GENETIC DISEASE

Genetic linkage maps obtained from reference families are important not only for the initial selection of markers for linkage studies; they also assist in refining the localization of the gene carrying a mutation responsible for the disease. As well as narrowing the search for candidate genes, the refined localization can be an aid for genetic counselling, and for investigations of genetic heterogeneity. Examples follow of the application of genetic maps to linkage studies of dominant, recessive, and X-linked diseases.

A. AUTOSOMAL-DOMINANT DISORDERS: ADENOMATOUS POLYPOSIS COLI (APC)

The recent localization of APC to chromosome 5q21-22 illustrates the use of mapping data from reference families to obtain a refined localization, after initial detection of linkage in extended pedigrees. APC is characterized by hundreds to thousands of polyps in the colon, in a precancerous stage that can develop to a cancer unless surgical treatment intervenes. Clear evidence of multiple polyps is usually seen before age 40, and dominant Mendelian inheritance of polyposis is observed in families. APC has been divided on clinical grounds into familial polyposis coli (FPC) and Gardner's syndrome (GS). Extracolonic manifestations are seen with GS in addition to polyposis; these include multiple osteomas, fibrous desmoid or epidermoid cysts, and congenital hypertrophy of retinal pigment epithelium. The differences between FPC and GS originally raised the possibility that they represented distinct disease entities, with potentially different genetic causes.

Following the report of a partial deletion of the long arm of chromosome 5 (q13-15 or q15-22) in a patient with GS and mental retardation,[52] two groups[9,10] mapped the gene responsible for APC to chromosome 5 by linkage to a DNA marker, C11p11 (D5S71), that had been localized to chromosome 5q21-22 by physical methods. The combined maximal lod score from the two studies was 6.7 at $\theta = 0.0$.

To provide more precise information on the localization of the APC gene, Nakamura et al.[53] screened a cosmid library constructed from a somatic cell hybrid line whose human component consisted of 40% of the long arm of chromosome 5. Six cosmids characterized in the reference panel of families showed close linkage to C11p11. These data were used to construct a map of the region with the gene order and distance shown in Figure 4a. Except for alternative placements of the two markers closest to C11p11, the order of the loci was well determined on the basis of the likelihood-ratio criterion.

The seven marker loci (including C11p11) were characterized in six extended APC pedigrees. Members of three of the pedigrees exhibited symptoms typical of GS, while affected individuals in the other three pedigrees showed classical familial polyposis without extracolonal disorders. Figure 4b illustrates the typical cosegregation pattern between the genotype at one of the marker loci and the phenotype (affected or unaffected) of each sampled member of family K109. All members of the pedigree who inherited allele 1 of marker YN5.48 from an affected mother or father had polyps in their colons. This suggests that one of the chromosome 5s carrying allele 1 of the marker in this pedigree also carries a mutation in a nearby gene that causes adenomatous polyposis.

Statistical analysis of all six families showed that five of the markers had combined lod scores >3 with APC; the other two markers (YN5.120 and EW5.5) had a maximum combined lod score >2. Two markers, C11p11 and YN5.48, exhibited no recombination with APC (lod scores of 3.32 and 8.25, respectively). Location scores for the placement of the APC locus with respect to the map of markers on chromosome 5 were calculated in order to obtain a more precise localization of the disease gene. For computational efficiency, the location scores were calculated twice: first according to a map containing the subset of markers C11p11-KK5.33-YN5.64-YN5.48, and second with the subset YN5.64-YN5.48-MC5.61-EW5.5 (Figure 4c). Despite our continued failure to observe recombination between APC and C11p11, these results strongly supported the localization of APC near YN5.48 (odds >40000:1 against a placement in the region of C11p11), which could be as much as 17 cM distal to C11p11.

In addition to giving a much-refined estimate for the location of APC, the study by Nakamura et al.[53] provided important evidence regarding the question of genetic heterogeneity between FPC and GS, an issue that had not been resolved by the earlier studies that detected linkage to C11p11: the probe YN5.48 showed a signifiant lod score (>3.0 at θ = 0) in either FPC or GS families. Although the possibility that two different defective genes lie very close together cannot yet be excluded, it is more likely that separate mutations within the same gene are responsible for the different phenotypes.

B. RECESSIVE DISORDERS: CYSTIC FIBROSIS

Cystic fibrosis (CF) is a recessive disorder that has been localized through linkage studies in nuclear families with multiple affected offspring. The discovery of the genetic location of the CF gene was of major importance due to the high prevalence (1/2000 to 1/3000) and the lethal nature of the disease.

The first marker reported to have a significant linkage to CF was the enzyme paraoxonase.[54] Although the chromosomal location of paraoxonase was unknown, the discovery of linkage between CF and a DNA marker (D7S15) that was subsequently localized to chromosome 7 placed CF on this chromosome.[2] As both markers showed approximately 15% recombination with CF, this information did not provide an accurate localization of the gene on the chromosome; however, it led to the rapid discovery of a very close linkage between CF and two other chromosome 7 markers, J3.11 (D7S8)[3] and MET.[4] The pattern of cosegregation of CF with the MET gene is shown in Figure 5A.

A collaborative study then suggested that the CF gene lies between MET and J3.11.[55] The combined data led to construction of a map of six markers within the CF region, genotyped in the extended reference panel, and the refined placement of the CF gene by the method of location scores[32] as shown in Figure 5B. Pulsed-field gel electrophoresis later confirmed the placement of the marker loci that were closely linked to CF.[56,57] The placement of additional markers in the region between MET and J3.11, by linkage studies in recombinant CF families, has demonstrated the usefulness of genetic mapping approaches for further refining the placement of the gene and for narrowing the search for potential candidate sequences.[58]

C. X-LINKED DISORDERS: SPASTIC PARAPLEGIAS

The early development of a linkage map of the X chromosome,[49] and the small target this chromosome represents for linkage studies (approximately 5% of the total genome), have greatly aided in the localization of X-linked disease.

Figure 6A shows a pedigree in which specific alleles or three genetic markers (DXYS1, pYNH3, and DXS17) cosegregate with an X-linked form of hereditary spastic paraplegia (HSP), as described by Keppen et al.[59] In this kindred, the disease phenotype includes hyperreflexia and spastic gait, but the intelligence of affected individuals is normal. Affected male offspring all carry alleles 1,1, and 2 of markers DXYS1, YNH3, and DXS17, respectively, an observation which suggests that the disease gene was introduced into this pedigree in association with that haplotype. Females are not likely to be affected because they each have also inherited a normal gene at the HSP locus from their fathers. However, a male child whose mother carries the defective allele has approximately a 50% chance to be affected. HSP gives a lod score of 4.48 at $\theta = 0$ with pYNH3 in this family and 4.00 with DXS17 at $\theta = 0$.

Kenwrick et al.[60] reported linkage data on another extended pedigree, in which X-linked spastic paraplegia is associated with mental retardation, congenital malformations, and optic nerve atrophy. They reported complete linkage to two different chromosome X markers, DX13 and ST14. To obtain further information on the localization of the genes involved in HSP in each of the two pedigrees, data on five markers — genotyped in the extended reference panel described earlier in this chapter — were used to construct a map of Xq. Three of the same markers had been studied by Kenwrick et al.,[60] and all five had been characterized in family K313. The location scores shown in Figure 6B, calculated separately for the two data sets, gave significant evidence for genetic heterogeneity ($\chi^2 = 21.08$; $p < 0.00001$).

VII. CONCLUSIONS

Primary genetic linkage maps for most human chromosomes provide powerful tools for studies of human molecular genetics. Mapped markers have led to preliminary localizations of many genes harboring deleterious mutations. Moreover, the strategy that consists of studying both disease families and reference panels, in order to obtain refined linkage maps of chromosome regions that contain genes of interest, is proving to be an effective approach to the twin problems of precisely localizing the gene and characterizing the mutations responsible for the disease. This technique is likely to be effective also for the study of multifactorial diseases. In addition to the applications cited here, mapped sets of genetic markers have also proven to be a way to investigate loss of heterozygosity in several types of cancer cells.[61]

REFERENCES

1. **Gusella, J., Wexler, N., Conneally, P., Naylor, S., Anderson, M., Tanzi, R., Watkins, P., Ottina, K., Wallace, M., Sakaguchi, A., Young, A., Shoulson, I., Bonilla, E., and Martin, J.,** A polymorphic DNA marker genetically linked to Huntington's disease, *Nature,* 306, 234, 1983.
2. **Tsui, L.-C., Buchwald, M., Barker, D., Braman, J. C., Knowlton, R., Schumm, J. W., Eiberg, H., Mohr, J., Kennedy, D., Plavsic, N., Zsiga, M., Markiewicz, D., Akots, G., Brown, V., Helms, C., Gravius, T., Parker, C., Rediker, K., and Donis-Keller, H.,** Cystic fibrosis locus defined by a genetically linked polymorphic DNA marker, *Science,* 230, 1054, 1985.
3. **Wainwright, B., Scambler, P., Schmidtke, J., Watson, E. A., Law, H. Y., Farrall, M., Cooke, H. J., Eiberg, H., and Williamson, R.,** Localization of cystic fibrosis to human chromosome 7cen-q22, *Nature,* 318, 384, 1985.

4. **White, R., Woodward, S., Leppert, M., O'Connell, P., Nakamura, Y., Hoff, M., Herbst, J., Lalouel, J.-M., Dean, M., and Vande Woude, G.,** A closely linked genetic marker for cystic fibrosis, *Nature,* 318, 382, 1985.
5. **Reeders, S. T., Breuning, K. E., Davies, K. D. Nicholls, R., Jarman, A., Higgs D., Pearson, P., and Weatherall, D.,** A highly polymorphic DNA marker linked to adult polycystic kidney disease on chromosome 16, *Nature,* 317, 542, 1985.
6. **Barker, D., Wright, E., Nguyen, K., Cannon, I., Fain, P., Goldgar, D., Bishop, T., Carey, J., Baty, B., Kivlin, J., Willard, H., Waye, J., Greig, G., Leinwand, L., Nakamura, Y., O'Connell, P., Leppert, M., Lalouel, J.-M., White, R., and Skolnick, M.,** Gene for von Recklinghausen neurofibromatosis is in the percicentromeric region of chromosome 17, *Science,* 236, 1100, 1987.
7. **Mathew, C., Chin, K., Easton, D., Thorpe, K., Carter, C., Liou, G., Fong, S.-L., Bridges, C., Haak, H., Kruseman, A., Schifter, S., Hansen, H., Telenius, H., Telenius-Berg, M., and Ponder, B.,** A linked genetic marker for multiple endocrine neoplasia type 2A on chromosome 10, *Nature,* 328, 527, 1987.
8. **Simpson, N., Kidd, K., Goodfellow, P., McDermid, H., Myers, S., Kidd, J., Jackson, C., Duncan, A., Farrer, L., Brasch, K., Castiglione, C., Genel, M., Gerner, J., Greenberg, C., Gusella, J., Holden, J., and White, B.,** Assignment of multiple endocrine neoplasia type 2A to chromosome 10 by linkage, *Nature,* 328, 428, 1987.
9. **Bodmer, W., Bailey, C., Bodmer, J., Bussey, H., Ellis, A., Gorman, P., Lucibello, F., Murday, V., Rider, S., Scambler, P., Sheer, D., Solomon, E., and Spurr, N.,** Localization of the gene for familial adenomatous polyposis on chromosome 5, *Nature,* 328, 614, 1987.
10. **Leppert, M., Dobbs, M., Scambler, P., O'Connell, P., Nakamura, Y., Stauffer, D., Woodward, S., Burt, R., Hughes, J., Gardner, E., Lathrop, M., Wasmuth, J., Lalouel, J.-M., and White, R.,** The gene for familial polyposis coli maps to the long arm of chromosome 5, *Science,* 238, 1411, 1987.
11. **Monaco, A., Neve, R., Colletti-Feener, C., Bertelson, C., Kurnit, D., and Kunkel, L.,** Isolation of candidate cDNAs for portions of the Duchenne muscular dystrophy gene, *Nature,* 323, 646, 1986.
12. **Royer-Pokora, B., Kunkel, L. M., Monaco, A. P., Goff, S., Newburger, P., Baehner, R., Cole, F., Curnutte, J., and Orkin, S.,** Cloning the gene for an inherited human disorder — chronic granulomatous disease — on the basis of its chromosomal location, *Nature,* 322, 32, 1986.
13. **Friend, S., Bernards, R., Rogelj, S., Weinberg, R., Rapaport, J., Albert, D., and Dryja, T.,** A human DNA segment with properties of the gene that predisposes to retinoblastoma and osteosarcoma, *Nature,* 323, 643, 1986.
14. **Pearson, P. L., Kidd, K. K., and Willard, H. F.,** Report of the committee on human gene mapping by recombinant DNA techniques, Human Gene Mapping 9, *Cytogenet. Cell Genet.,* 46, 390, 1987.
15. **Kidd, K. K., Bowcock, A. M., Pearson, P. L., Schmidtke, J., Willard, H. F., Track, R. K., and Ricciuti, F.,** Report of the committee on human gene mapping by recombinant DNA techniques, Human Gene Mapping 9.5, *Cytogenet. Cell Genet.,* 47, 132, 1988.
16. **Nakmura, Y., Leppert, M., O'Connell, P., Wolff, R., Holm, T., Culver, M., Martin, C., Fujimoto, E., Hoff, M., Kumlin, E., and White, R.,** Variable number of tandem repeat (VNTR) markers for human gene mapping, *Science,* 235, 1616, 1987.
17. **Weber, J. and May, P.,** Abundant class of human DNA polymorphisms which can be typed using the polymerase chain reaction, *Am. J. Hum. Genet.,* 44, 388, 1989.
18. **Morton, N. E.,** Sequential tests for detection of linkage, *Am. J. Hum. Genet.,* 7, 277, 1955.
19. **Lathrop, G. M., Lalouel, J. M., Julier, C., and Ott, J.,** Multilocus linkage analysis in humans: detection of linkage and estimation of recombination, *Am. J. Hum. Genet.,* 37, 482, 1985.
20. **Lathrop, G. M., Lalouel, J.-M., Julier, C., and Ott, J.,** Strategies for multilocus linkage analysis in humans, *Proc. Natl. Acad. Sci. U.S.A.,* 81, 3443, 1984.
21. **Lathrop, G. M. and Lalouel, J.-M.,** Simple calculation of lod-scores on small computers, *Am. J. Hum. Genet.,* 36, 460, 1984.
22. **Ott, J.,** A computer program for linkage analysis of general human pedigrees, *Am. J. Hum. Genet.,* 28, 528, 1976.
23. **Conneally, P. M., Edwards, J. H., Kidd, K. K., Lalouel, J.-M., Morton, N. E., Ott, J., and White, R.,** Report of the committee on methods of linkage analysis and reporting, Human Gene Mapping 8, *Cytogenet. Cell Genet.,* 40, 356, 1985.
24. **Lathrop, G. M., Lalouel, J.-M., and White, R.,** Calculation of human linkage maps: likelihood calculations for multilocus linkage analysis, *Genet. Epidemiol.,* 3, 39, 1986.
25. **Lathrop, G. M. and Lalouel, J.-M.,** Efficient computations in multilocus linkage analysis, *Am. J. Hum. Genet.,* 42, 498, 1988.
26. **Clayton, J.,** A multipoint linkage analysis program for X-linked disorders, with the example of Duchenne muscular dystrophy and seven DNA probes, *Hum. Genet.,* 73, 68, 1986.
27. **Lander, E. S. and Green, P.,** Construction of multilocus genetic linkage maps in humans, *Proc. Natl. Acad. Sci. U.S.A.,* 84, 2363, 1987.

28. **Boehnke, M.,** Estimating the power of a proposed linkage study: a practical computer simulation approach, *Am. J. Hum. Genet.,* 39, 513, 1986.
29. **Ott, J.,** *Analysis of Human Genetic Linkage,* Johns Hopkins University Press, Baltimore, 1985.
30. **White, R., Leppert, M., Bishop, T., Barker, D., Berkowitz, J., Brown, C., Callahan, P., Holm, T., and Jerominski, J.,** Construction of linkage maps with DNA markers for human chromosomes, *Nature,* 313, 101, 1985b.
31. **Dausset, J.,** Le centre d'étude du polymorphisme humain, *La Presse Med.,* 15, 1801, 1986.
32. **Lathrop, G. M., Nakamura, Y., Cartwright, P., O'Connell, P., Leppert, M., Jones, C., Tateishi, H., Bragg, T., Lalouel, J.-M., and White, R.,** A primary genetic map of markers for human chromosome 10, *Genomics,* 2, 157, 1988.
33. **Kammerer, C. M. and MacCluer, J. W.,** Empirical power of three preliminary methods of ordering loci, *Am. J. Hum. Genet.,* 43, 964, 1988.
34. **Lander, E. S., Green, P., Abramson, J., Barlow, A., Daly, M. J., Lincoln, S., and Newburg, L.,** MAPMAKER: an interactive computer package for constructing primary genetic linkage maps of experimental and natural populations, *Genomics,* 1, 174, 1987.
35. **Donis-Keller, H., Green, P., Helms, C., Cartinhour, S., Weiffenbach, B., Stephens, K., Keith, T., Bowden, D., Smith, D., Lander, E., Botstein, D., Akots, G., Rediker, K., Gravius, T., Brown, V., Rising, M., Parker, C., Powers, J., Watt, D., Kauffman, E., Bricker, A., Phipps, P., Muller-Kahle, H., Fulton, T., Ng, S., Schumm, J., Braman, J., Knowlton, R., Barker, D., Crooks, S., Lincoln, S., Daly, M., and Abrahamson, J.,** A genetic linkage map of the human genome, *Cell,* 51, 319, 1987.
36. **Dracopoli, N. C., Stanger, B. Z., Ito, C. Y., Call, K. M., Lincoln, S. E., Lander, E. S., and Houseman, D. E.,** A genetic linkage map of 27 loci from PND to FY on the short arm of chromosome 1, *Am. J. Hum. Genet.,* 43, 462, 1988.
37. **O'Connell, P., Lathrop, M., Nakamura, Y., Leppert, M., Ardinger, R., Murray, J., Lalouel, J.-M., and White, R.,** Twenty-eight loci form a continuous linkage map of markers for human chromosome 1, *Genomics,* 4. 12, 1989.
38. **O'Connell, P., Lathrop, G. M., Nakamura, Y., Leppert, M., Lalouel, J.-M., and White, R.,** Twenty loci form a continuous linkage map of markers for human chromosome 2, submitted.
39. **Lathrop, G. M., O'Connell, P., Leppert, M., Nakamura, Y., Farrall, M., Tsui, L.-C., Lalouel, J.-M., and White, R.,** Twenty-five loci form a continuous linkage map of markers for human chromosome 7, submitted.
40. **Lathrop, G. M., Nakamura, Y., O'Connell, P., Leppert, M., Woodward, S., Lalouel, J.-M., and White, R.,** A mapped set of genetic markers for human chromosome 9, *Genomics,* 3, 361, 1988.
41. **Nakamura, Y., Lathrop, M., Bragg, T., Leppert, M., O'Connell, P., Jones, C., Lalouel, J.-M., and White, R.,** An extended genetic linkage map of markers for human chromosome 10, *Genomics,* 3, 389, 1988.
42. **O'Connell, P., Lathrop, G. M., Law, M., Leppert, M., Nakamura, Y., Hoff, M., Kumlin, E., Thomas, W., Elsner, T., Ballard, L., Goodman, P., Azen, E., Sadler, J. E., Cai, G. Y., Lalouel, J.-M., and White, R.,** A primary genetic linkage map for human chromosome 12, *Genomics,* 1, 93, 1987.
43. **Leppert, M., Cavenee, W., Callahan, P., Holm, T., O'Connell, P., Thompson, K., Lathrop, M., Lalouel, J.-M., and White, R.,** A primary genetic map of chromosome 13q, *Am. J. Hum. Genet.,* 39, 425, 1986.
44. **Nakamura, Y., Lathrop, G. M., O'Connell, P., Leppert, M., Lalouel, J.-M., and White, R.,** A mapped set of DNA markers for human chromosome 15, *Genomics,* 3, 342, 1988.
45. **Nakamura, Y., Lathrop, G. M., O'Connell, P., Leppert, M., Barker, D., Wright, E., Skolnick, M., Kondoleon, S., Litt, M., Lalouel, J.-M., and White, R.,** A mapped set of DNA markers for human chromosome 17, *Genomics,* 2, 302, 1988c.
46. **O'Connell, P., Lathrop, M., Leppert, M., Nakamura, Y., Muller, U., Lalouel, J.-M., and White, R.,** Thirteen loci form a primary genetic linkage map of markers for human chromosome 18, *Genomics,* 3, 367, 1988.
47. **Nakamura, Y., Lathrop, G. M., O'Connell, P., Leppert, M., Lalouel, J.-M., and White, R.,** A primary map of 10 DNA markers and 2 serological markers for human chromosome 19, *Genomics,* 3, 67, 1988.
48. **Nakamura, Y., Leppert, M., O'Connell, P., Lathrop, M., Lalouel, J.-M., and White, R.,** A genetic linkage map of markers for human chromosome 20, submitted.
49. **Drayna, D. and White, R.,** The genetic linkage map of the human X chromosome, *Science,* 230, 753, 1985.
50. **Bowcock, A. M., Kidd, J. A., Lathrop, G. M., Daneshvar, L., Sehgal, P., May, L. T., Kidd, K. K., and Cavalli-Sforza, L. L.,** The human "interferon-β_2/hepatocyte stimulating factor/interleukin-6" gene: DNA polymorphism studies and its localization to chromosome 7p21, *Genomics,* 3, 8, 1988.
51. **Nakamura, Y., Lathrop, G. M., O'Connell, P., Leppert, M., Lalouel, J.-M., and White, R.,** A small region of human chromosome 14 shows a high frequency of recombination, *Genomics,* 4, 76, 1989.

52. **Herrera, L., Kakati, S., Gibas, E., Pietrzak, E., and Sandberg, A.,** Brief clinical report. Gardner syndrome in a man with an interstitial deletion of 5q, *Am. J. Med. Genet.*, 25, 473, 1986.
53. **Nakamura, Y., Lathrop, G. M., Leppert, M., Dobbs, M., Wasmuth, J., Wolff, E., Carlson, M., Fujimoto, E., Krapcho, K., Woodward, S., Hughes, J., Burt, R., Gardner, E., Lalouel, J.-M., and White, R.,** Localization of the genetic defect in familial adenomatous polyposis within a small region of chromosome 5, *Am. J. Hum. Genet.*, 43, 638, 1988.
54. **Eiberg, H. J., Mohr, K., Schmiegelow, K., Nielsen, L., and Williamson, R.,** Linkage relationships of paraoxonase (PON) with other markers: indication of PON-cystic fibrosis synteny, *Clin. Genet.*, 28, 265, 1985.
55. **Beaudet, A., Bowcock, A., Buchwald, M., Cavalli-Sforza, L., Farrall, M., King, M.-C., Klinger, K., Lalouel, J.-M., Lathrop, G. M., Naylor, S., Ott, J., Tsui, L.-C., Wainwright, B., Watkins, P., White, R., and Williamson, R.,** Linkage of cystic fibrosis to two tightly linked DNA markers: joint report from a collaborative study, *Am. J. Hum. Genet.*, 39, 681, 1986.
56. **Poustka, A. M., Lehrach, H., Williamson, R., and Bates, G.,** A long range restriction map encompassing the cystic fibrosis locus and its closely linked genetic markers, *Genomics*, 2, 346, 1988.
57. **Drumm, M. L., Smith, C. L., Dean, M., Cole, J. L., Iannuzzi, M. C., and Collins, F.,** Physical mapping of the cystic fibrosis region by pulsed-field gel electrophoresis, *Genomics*, 2, 346, 1988.
58. **Farrall, M., Wainwright, B. J., Feldman, G. L., Beaudet, A., Sretenovic, Z., Halley, D., Simon, M., Dickerman, L., Devoto, M., Romeo, G., Kaplan, J.-C., Kitzis, A., and Williamson, R.,** Recombination between IRP and cystic fibrosis, *Am. J. Hum. Genet.*, 43, 471, 1988.
59. **Kappen, L. D., Leppert, M., O'Connell, P., Nakamura, Y., Stauffer, D., Lathrop, M., Lalouel, J.-M., and White, R.,** Etiological heterogeneity in X-linked spastic paraplegia, *Am. J. Hum. Genet.*, 41, 933, 1987.
60. **Kenwrick, S., Ionasescu, V., Ionasescu, G., Searby, C., King, A., Dubowitz, M., and Davies, K. E.,** Linkage studies of X-linked recessive spastic paraplegia using DNA probes, *Hum. Genet.*, 73, 264, 1986.
61. **Ponder, B.,** Cancer gene losses in human tumours, *Nature*, 335, 400, 1988.
62. **Lathrop, G. M., Farrall, M., O'Connell, P., Wainwright, B., Leppert, M., Nakamura, Y., Lench, N., Kruyer, H., Dean, M., Park, M., Vande Woude, G., Lalouel, J. M., Williamson, R., and White, R.,** Refined linkage map of chromosome 7 in the region of the cystic fibrosis gene, *Am. J. Hum. Genet.*, 42, 38, 1988.
63. **White, R., Leppert, M., Nakamura, Y., O'Connell, P., Lathrop, M., and Lalouel, J.-M.,** Mapping genes that cause disease, in *Cellular and Molecular Biology of Human Carcinogenesis*, Bautwell and Riegel, Eds., Academic Press, San Diego, in preparation.
64. Unpublished data.

Chapter 9

FRAGMENT-LENGTH POLYMORPHISMS FOR FORENSIC SCIENCE APPLICATIONS*

Bruce Budowle, Dwight E. Adams, and Robert C. Allen

TABLE OF CONTENTS

* This is publication number 89-04 of the Laboratory Division of the Federal Bureau of Investigation. Names of commercial manufacturers are provided for identification only and inclusion does not imply endorsement by the Federal Bureau of Investigation.

I. INTRODUCTION

In 1983, Lynda Mann's body was found in Leicestershire, England. She had been raped and then murdered. Three years later, the body of Dawn Ashworth, who also had been raped and murdered in a similar fashion, was discovered within 1 mile of the location of Lynda Mann's body. Within a week of the discovery of Ashworth's body, the local authorities arrested Richard Buckland. The authorities were convinced that Buckland, who had confessed to the crimes committed on the girls, was the perpetrator. However, Dr. Alec Jeffreys (University of Leicester, England) demonstrated that the DNA "fingerprint"[1,2] derived from Buckland's blood did not match the DNA fingerprint obtained from the semen found at both crime scenes.[3] However, the semen derived from both crime scenes yielded identical DNA profiles. Subsequently, the police began searching for the individual who had the same DNA profile obtained from the semen. They instituted a voluntary program whereby blood samples were donated by the 5500 males residing within the locality. Most contributors (88%) were eliminated as suspects using the traditional forensic genetic marker tests — ABO[4] and phosphoglucomutase-1.[5,6] The remaining 12% of the samples were subjected to DNA fingerprinting by the Home Office Central Research Establishment in Aldermaston, England. None of the DNA profiles obtained from the samples matched the one derived from the semen stains. The police working on information gathered by traditional investigative practices arrested Colin Pitchfork. He had convinced a friend to substitute for him as a source for the voluntary blood donor program. It was determined later that Pitchfork's DNA fingerprint was identical to the profile obtained from the crime scene semen samples.

During the same time period (1987), the Lifecodes Corporation (Valhalla, NY) used its DNA "Print Test"[7,8] to associate a suspect, Tommy Lee Andrews, with semen evidence derived from a rape victim in Florida. Hence, DNA typing in forensics was born.

The DNA typing technology used in the above cases was the technique that detects restriction fragment-length polymorphisms (RFLPs). This approach affords the forensic scientist the potential to identify a body fluid/tissue donor to the exclusion of all other individuals, except for the case of monozygotic twins. RFLPs are variations in the sizes of DNA fragments generated by restriction endonuclease digestion. Although of different sizes, the fragments of a particular polymorphism share similar or homologous nucleotide sequences.

The term "fingerprinting" has been applied inappropriately to describe DNA profiles or patterns obtained for characterization of evidentiary materials. It unfortunately implies that any time any DNA typing technique is used, unique identification is conferred. For example, sex typing provides an exclusion value of only 50%, and due to the limited quantities of DNA yielded from evidentiary materials, even the RFLP method described by Jeffreys[1] may not achieve unique identification.[9] Therefore, the authors prefer the more appropriate terms DNA types, profiles, or patterns.

It is not possible to address explicitly all aspects of RFLP methodology in this chapter. Therefore, the first section of this chapter will deal with the more interesting and informative aspects of the methodology that have improved the utility of RFLP analysis, particularly for forensic science. The second portion of the chapter will describe research in progress by the authors to improve the utility of fragment-length polymorphism (FLP) technology.

II. RESTRICTION FRAGMENT-LENGTH POLYMORPHISMS

A. TECHNOLOGY

The RFLP technique is conceptually a very simple technique based on the strong and specific interaction of a single-stranded piece of DNA with its complementary sequence.[10] Once purified from other cellular materials, the DNA is subjected to phosphoester cleavage

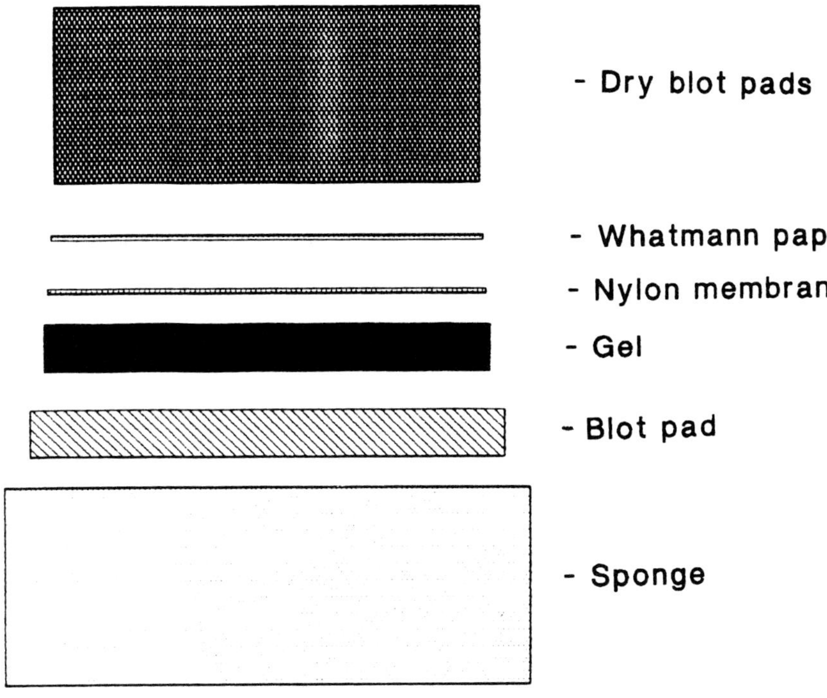

FIGURE 1. There are many configurations on the basic theme of Southern blotting.[11] This is a diagram of the transfer apparatus used in our laboratory. The sponge contains 0.4 *M* NaOH. The blot pad overlaid on the sponge is soaked in 0.4 *M* NaOH for 30 min. The gel has been washed in 0.4 *M* NaOH for 30 min. Transfer proceeds for 6 h.

at specific sites (restriction sites) by the action of a type II restriction endonuclease. After digestion with a restriction endonuclease, the resulting DNA fragments are fractionated, based upon size, by electrophoresis. During electrophoresis, the gel matrix (normally agarose) exerts a sieving effect on DNA molecules such that smaller fragments migrate faster than larger fragments. Then, the fractionated array of DNA fragments is transferred out of the gel onto a nitrocellulose or nylon membrane by Southern blotting.[11] Southern blotting relies on the flow, by capillary action, of a transfer solution from a reservoir through the agarose gel to a membrane overlaid by a stack of dry paper towels or blot pads. The DNA fragments are carried along with the flow of transfer solution out of the gel and onto the membrane, where they readily bind in the same array as they occurred in the gel (Figure 1). At some point, the DNA fragments must be denatured to yield single-stranded DNA; otherwise, target sites will not be available for interaction with a probe (single-stranded complementary piece of labeled DNA or RNA). Most procedures effect denaturation of the fractionated DNA *in situ* prior to Southern blotting. Thus, DNA is single stranded when it is fixed to the membrane. Following a prehybridization wash to block unoccupied binding sites on the membrane, the membrane is exposed to a radioactively labeled probe. Under appropriate conditions, the probe can hybridize with the membrane-bound, single-stranded DNA fragments that contain a complementary sequence. Subsequently, the membrane is washed to remove any unhybridized probe. Washing the hybridized membrane under conditions of low stringency (high salt, low temperature) will permit a greater degree of mismatched hybrids to remain, while washing at high stringency conditions (low salt, high temperature) will eliminate to a greater degree the mismatching. After the posthybridization washing steps have been completed, the membrane is placed in direct contact with a photographic film that is sensitive to the decay products of radioactive molecules. After an

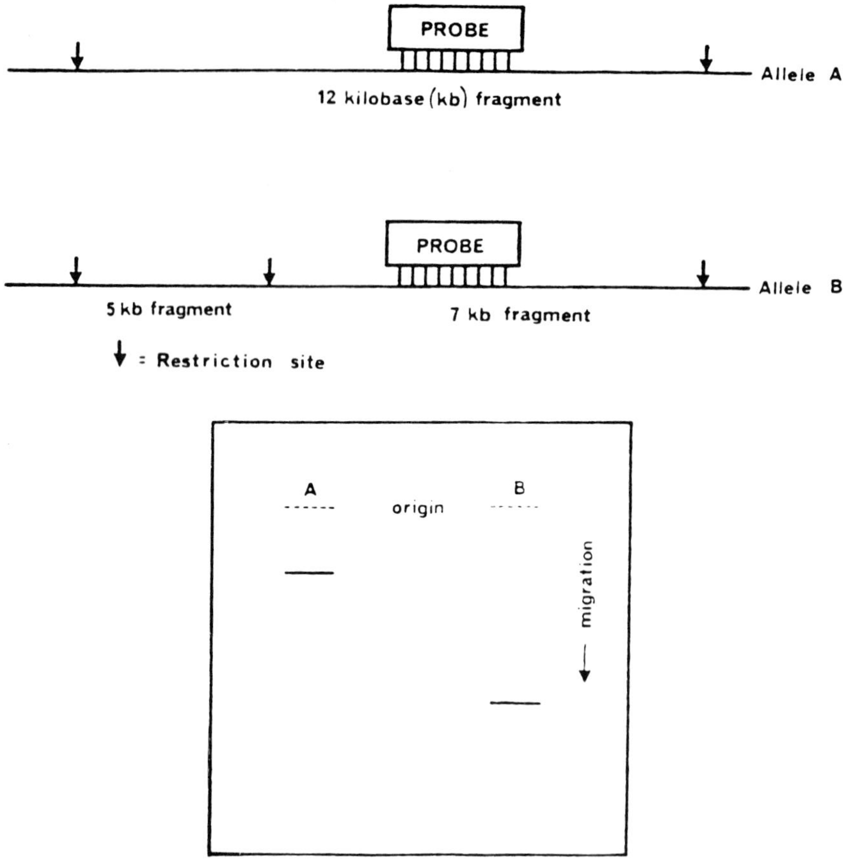

FIGURE 2. Generic example of RFLP caused by a point mutation. The arrows indicate restriction sites (top) and migration of the restriction fragments after gel electrophoresis (bottom).

appropriate exposure time, the film will bear a latent image of the positions of the radioactive probe — target DNA duplexes on the membrane.

B. TYPES OF RFLPs

The major types of DNA polymorphisms that can be detected by RFLP analysis for forensic purposes (i.e., point mutations, mini-satellites, and variable number of tandem repeats, VNTRs, at a single locus) are more adequately addressed in other chapters in this book and therefore will only be discussed briefly here.

1. Point Mutations

The majority of RFLPs used for clinical research purposes are due to point mutational events that lead to the creation or elimination of restriction sites. A generic example of this type of RFLP is shown in Figure 2. The restriction sites are defined by the arrows. For allele A, there are 12,000 base pairs (bp) between the restriction sites, but for allele B, a point mutation has occurred such that an additional cleavage site has been created. Since the probe hybridizes only to a portion of the DNA fragment, a smaller fragment will be observed for allele B. A specific example in man is the β-globin gene.[12] Within the nucleotide sequence that encodes the production of the normal hemoglobin molecule is the sequence C-C-T-G-A̲-G-G-A-G. Individuals carrying the sickle-cell anemia allele have the sequence C-C-T-G-T̲-G-G-A-G, in which a single base substitution has taken place. The restriction endonuclease Mst II recognizes and cleaves DNA at sites containing the sequence C-C-T-N-A-G-G-. Thus, Mst II can cleave the sequence in the normal allele, but not in the sickle-cell allele. The RFLP analysis reveals a 1150-bp fragment for the normal hemoglobin sequence and a 1350-bp fragment for the sickle-cell allele.

2. VNTRs

Although RFLPs generated by the elimination or creation of restriction sites are simple and straightforward approaches for screening potential variation among individuals, they do not provide high levels of discrimination (certainly not at the level of unique identification). Observations by Wyman and White[13] revealed that RFLPs could be generated due to varying numbers of repeat core sequences confined to a specific locus. The length of each fragment is a function of the number of tandem repeats contained within it. Because these VNTRs are inherited in a Mendelian fashion, the number of repeats can be considered as an allele based on size. Some VNTR loci, such as D2S44 (YNH24)[14,15] and D1S7 (MS1),[16] have been reported to exhibit a high degree of polymorphism such that almost the entire population is heterozygous.

3. Mini-Satellites

In contrast, mini-satellites are variable numbers of tandem arrays of short repeat units dispersed throughout the genome.[1] For the mini-satellite probe system discovered by Jeffreys et al.[1] (i.e., 33.15 and 33.6), the membrane-bound DNA is reacted with the probe under low stringency conditions. This permits probing of many loci simultaneously. Using this technique, the variability is so great that all individuals, except monozygotic twins, carry a unique DNA profile. The patterns initially were so impressive that the government forensic laboratories in England pursued this approach for typing evidentiary material.[2,9] However, for the forensic laboratories within the Federal Bureau of Investigation and the Royal Canadian Mounted Police, RFLP typing for VNTRs confined to a single locus seemed more appropriate.

4. Single Locus vs. Low Stringency Mini-Satellite Analysis

Although the information provided by the use of Jeffreys' multilocus probes makes them appear quite attractive for use in criminal cases, they possess limitations that must be considered. First, the sensitivity of detection of these probes is considerably less than that of VNTR probes: 1 μg of high molecular weight DNA or semen stains and bloodstains that are 2.5 and 3.0 cm in diameter, respectively, are required to obtain a sufficient quantity of DNA for a multilocus profile.[17] Since the quantity of DNA derived from evidentiary materials can be limiting, the lower sensitivity of detection of these mini-satellite probes will reduce their utility. This is particularly true for bloodstains, saliva stains, and forcibly removed hairs. In contrast, VNTR probes, such as YNH24 (D2S44), CMM101 (D14S13)[18] (R. White, Howard Hughes Medical Institute, Salt Lake City, UT), MS1 (D1S7) (Cellmark Diagnostics, Gaithersburg, MD), V1(D17S79) (Lifecodes Corporation, Valhalla, NY), 3′HVR, L355, and PR365 (Collaborative Research, Inc., Bedford, MA), can detect RFLPs analyzing as little as 20 to 100 ng of genomic DNA.[16,19] In fact, we have been able to generate RFLP patterns using these VNTR probes from as little as 10-μl saliva stains, one forcibly removed hair, 1-μl semen stains, and 2- to 5-μl bloodstains. Second, the large number of bands revealed by multilocus probes, while advantageous for individualizing specimens, can make interspecimen pattern comparison difficult. This problem is evident especially when different quantities of genomic DNA per sample are evaluated and/or several individuals have contributed to the sample. The single-locus probe systems that produce one or two band patterns per individual provide a much simpler pattern for interpretation of the complex situations frequently encountered when more than one individual may have contributed to the sample.

Third, these multilocus probes recognize sequences common to a variety of unrelated species, including birds, dogs, cats, and rice,[20-24] thus necessitating alternative strategies for species determination within the forensic laboratory. On the other hand, the single-locus VNTRs have been determined by us to cross-react only with higher primates. Fourth, the mutation rate observed for these mini-satellites is extremely high,[25] thereby making interpretation of forensic paternity cases complicated. Only genetic marker systems with low mutation rates should be considered in a paternity context. Fifth, the statistics are not as straightforward (in a Mendelian genetic context) regarding multilocus analysis as for VNTR single-locus systems, and, consequently, have been questioned.[26] A final consideration concerns DNA degradation. The mini-satellite probes of Jeffreys generate profiles with fragments ranging in size between a few hundred and 23,000 bp. As forensic samples are exposed to environmental insults, the recoverable DNA can be degraded to such an extent that the high molecular weight DNA is less than 20,000 bp in size.[27] Selection of probe systems that recognize small fragments will increase the number of conclusive calls when forensic samples are analyzed.

C. PRACTICAL IMPROVEMENTS OF THE RFLP METHODOLOGY

Most protocols for RFLP analysis are based on historical approaches[10,11,28-31] and therefore may not necessarily exploit more efficacious conditions. Because of this, the methodology can be inordinately time consuming, labor intensive, costly, less than reliable at times, and can provide low sensitivity of detection. To address these concerns, our laboratory has investigated a number of steps commonly employed in the RFLP methodology. These include extraction of DNA, choice of restriction endonuclease, capillary transfer solution, prehybridization and hybridization conditions, choice of nylon membrane, and, already discussed, the selection of single-locus VNTR probes.

1. Sample Requirements

The DNA derived from body-fluid stains obviously must be free of contaminants that can inhibit restriction endonuclease activity, be of sufficient molecular weight, and be of sufficient quantity to be adequate for the analytical demands of the RFLP methodology. At times, these qualities will be beyond the control of the analyst. The age of the materials, the original quantity of body fluid/tissue deposited, the environmental insult(s), and the substrate on which the biological material has been deposited all impact on the efficiency of the final RFLP typing. For example, attempts to isolate high molecular weight DNA from extremely aged material such as guagga[32] and mummies[33] have been fruitless. An additional consideration is the source of the biological specimen. It can be anticipated that, generally, semen would provide more DNA than a similar quantity of blood, which contains mostly anucleated cells.

2. Extraction

Another factor to consider regarding DNA yield is the rehydration of the body-fluid stain sample to be extracted. Fortunately, extraction of DNA from body-fluid stains is quite similar to standard procedures.[2,9,27,33] An improved yield of DNA can be demonstrated by lengthening the rehydration (or solubilization) time period.[66]

Swabs used to collect semen from the vaginal cavity often will contain cellular material representative of both the female rape victim and the male contributor. Gill et al.[2] reported on a procedure to effectively isolate intact sperm from other contaminating cells by exploiting differences in membrane integrity between cell types. Female vaginal epithelial cells may be preferentially lysed by incubation in a SDS/protease K solution. In contrast, the membrane of a sperm cell contains a high degree of cross-linked, thiol-rich proteins;[34] thus, the sperm nuclei are resistant to lysis in the absence of a reducing agent. After initial lysis of the female epithelial cells, the unlysed sperm are subjected to centrifugation and washed. The purified sperm nuclei pellet then is lysed by exposure to a SDS/protease K solution in the presence of a strong reducing agent, such as dithiothreitol. Presumably, this differential extraction procedure initially was employed to eliminate the obscuring effects of more than one contributor to the mini-satellite profile. DNA loss during the initial extraction can reduce the efficiency of the differential extraction procedure. However, it may not be necessary to perform this type of extraction in all cases where single-locus VNTR analysis is applied.

3. Restriction Endonuclease

Restriction endonucleases are bacterial enzymes that cleave double-stranded DNA. The class II restriction endonucleases will cleave DNA only if a specific sequence of bases is present and will cut the DNA at the particular site. Several factors must be considered when opting for a restriction enzyme for RFLP analysis of evidentiary material. HaeIII appears to be an appropriate enzyme. It recognizes the four-nucleotide sequence G-G-C-C and forms a blunt-end cut between the G and C. Since the samples encountered at crime scenes are not of a pristine quality, it was deemed necessary to select a hardy restriction endonuclease. Blakesley et al.[35] demonstrated, and we have confirmed, that HaeIII is extremely hardy. Under less than optimum conditions, such as temperatures as high as 70°C, ionic strengths varying from 25 to 150 m*M*, and a pH range of 6 to 9, HaeIII cuts at its restriction site with no observable adverse effects (Figure 3). Thus, HaeIII will be more able to digest properly genomic DNA containing cellular and environmental contaminants, compared to many other restriction endonucleases (Figure 4). A further consideration is that HaeIII, recognizing a four-base sequence, generally will cleave DNA into smaller fragments than restriction endonucleases, such as PstI and EcoRI, that recognize larger sequence sites. Due to the inherent nature of the electrophoretic separation, the smaller the DNA fragments, the more likely subtle size differences can be resolved. This has the effect of increasing the discrimination power of a particular VNTR system. Also, the DNA derived from evidentiary materials, at times, will be somewhat degraded. Thus, the smaller the DNA fragments containing the target site of interest, the greater the potential of providing genetic information for characterization of forensic samples. The probe L1077 (Collaborative Research, Inc., Bedford, MA) is an example of such a VNTR. After HaeIII digestion, the fragments containing the probe's complementary sequence are all less than 1000 bp in length (Figure 5). Finally, the methylation scenario of mammals (5-methyl CpG)[36,37] has no impact on the ability of HaeIII to recognize and cleave its particular restriction site.[38] Therefore, DNA derived from different body tissues of an individual will not likely produce different DNA profiles.

4. Blotting Procedures

Reed and Mann[39] demonstrated that Southern blotting in a low ionic strength, alkali environment was more effective for binding DNA to positively charged nylon membranes (Zeta probe, Bio-Rad, Richmond, CA). Our data support these observations when human genomic DNA is analyzed. The transfer solution routinely used is 0.4 *M* sodium hydroxide. Sodium hydroxide denatures DNA for subsequent probing. By transferring DNA out of the gel in an alkali environment, a single-stranded nature for DNA is maintained at the time of fixation to the membrane. The alkali transfer approach eliminates the need for a gel neutralization step prior to Southern blotting, provides an environment that readily binds DNA to positively charged nylon membranes, and uses an easily prepared and stable transfer solution. These observations do not preclude the use of alternative Southern blotting procedures in other laboratories. In our hands, we found this procedure to have a positive impact on the signal intensity of the RFLP pattern for human genomic DNA, but not for plasmid or viral DNA analyses. In contrast, high-salt-concentration transfers provided better

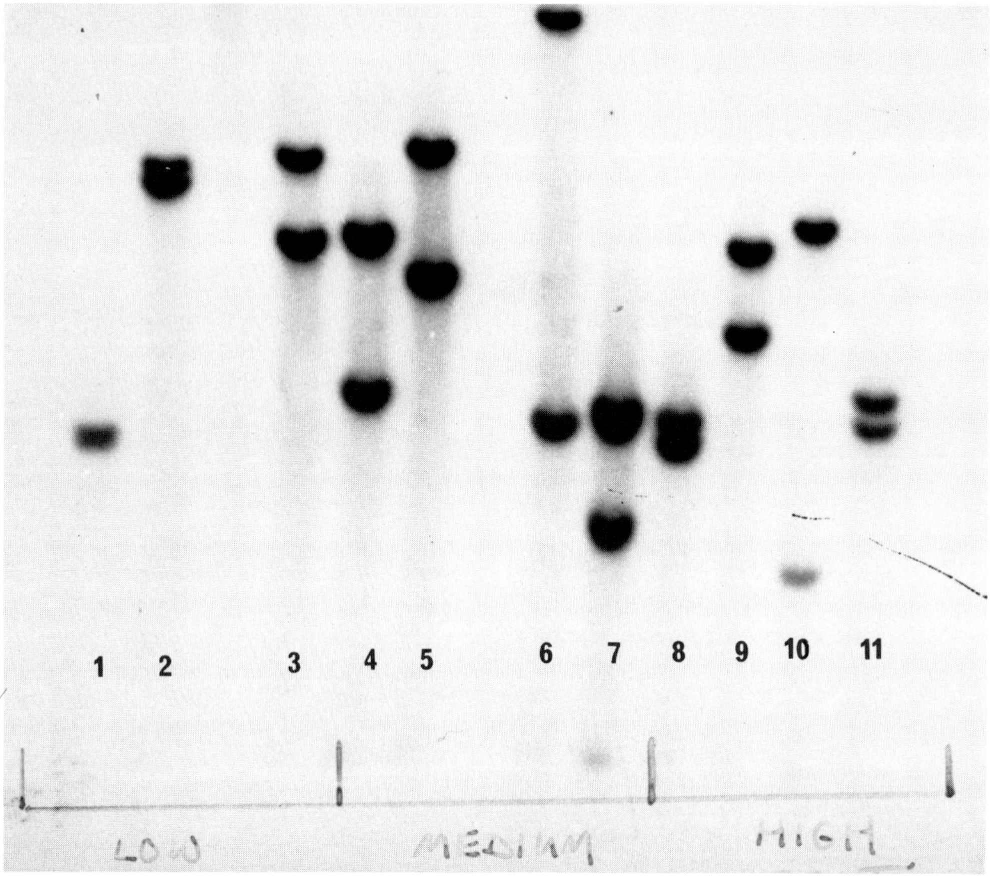

FIGURE 3. RFLP of HaeIII digests of human genomic DNA. The digest was carried out at 56°C. Samples 1 to 3 were digested in a low salt environment; samples 4 to 7 were digested in a medium salt environment; samples 8 to 11 were digested in a high salt environment. The locus was D2S44.

sensitivity of detection for plasmid and viral DNA analyses. In a similar fashion, Figure 6 shows that neutral-charged nylon membranes such as Hybond N (Amersham, Arlington Heights, IL) have a negative impact on the intensity of the RFLP signal (probably due to DNA binding efficiency). However, when DNA was transferred to the neutral-charged nylon membrane in a high-salt-concentration solution, the overall signal was increased (although not as high as that produced by alkali transfer and positively changed membranes). Observations such as these demonstrate that there is more than one approach to achieving a desirable RFLP pattern. In regard to standardization, particularly for comparing genetic profiles between forensic laboratories, only the restriction enzyme would have to be standardized. Also, the above observations may explain the varying effectiveness of the procedure that different laboratories have experienced.

5. Prehybridization

The prehybridization washing of a membrane appears to be more an historical step and has been proven to be unnecessary. Figure 7 clearly demonstrates that the membrane simply can be placed in the hybridization solution and the probe immediately added. There was no loss of sensitivity of detection and no increase in background staining when the prehybridization wash was eliminated. No prehybridization was most effectively achieved by

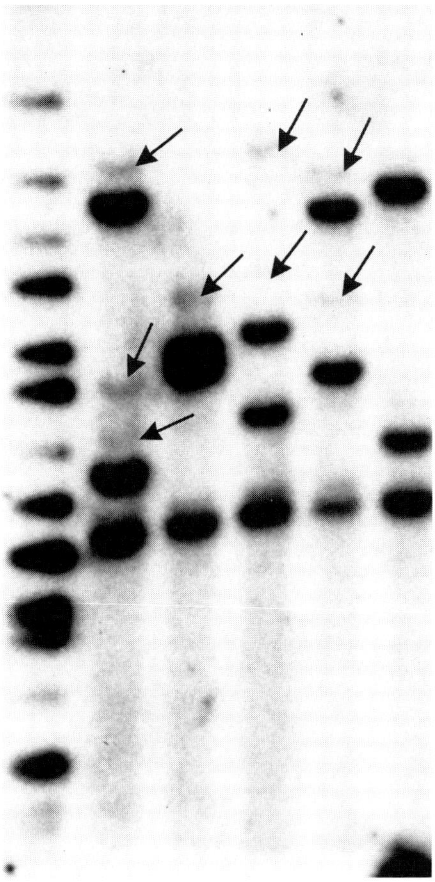

FIGURE 4. RFLP of HinfI digests of human genomic DNA. The arrows indicate partial digestion problems. Longer autoradiographic exposures of HaeIII digests of similar quantities of DNA do not yield these bands. This digest was performed at 37°C in a high salt environment. The locus was D2S44.

following the suggestion of Church and Gilbert,[40] and using 7% SDS in the hybridization solution. The high concentration of SDS immediately blocks the unoccupied sites of the membrane and also appears to increase the sensitivity of detection of the system (Figures 8 and 9) (possibly by probe networking[29] or improved conditions for hybridization kinetics). An additional advantage was that the prehybridization step which consumed an extra day was eliminated, reducing overall assay time.

6. Hybridization

Many components used for hybridization appear to be superfluous. Components such as Denhardt's solution,[10] milk,[42] and dextran sulfate[29] may have had some import when first employed, but can be replaced with less complicated, less costly, and more effective reagents. We have developed a three-component hybridization solution by borrowing portions of the procedures reported by Church and Gilbert,[40] Amasino,[42] and Reed and Mann.[39] The hybridization solution contains only 7% SDS, 10% polyethylene glycol (mol wt = 8000), and phosphate buffer. The effects of 7% SDS have already been discussed. It appears that polyethylene glycol has the same effect as dextran sulfate (Figure 10), but for less cost. There have been four major advantages obtained using this simplified hybridization solution: (1) increased sensitivity of detection; (2) reduced membrane background; (3) reduced costs; and (4) easier trouble-shooting of defective reagents.

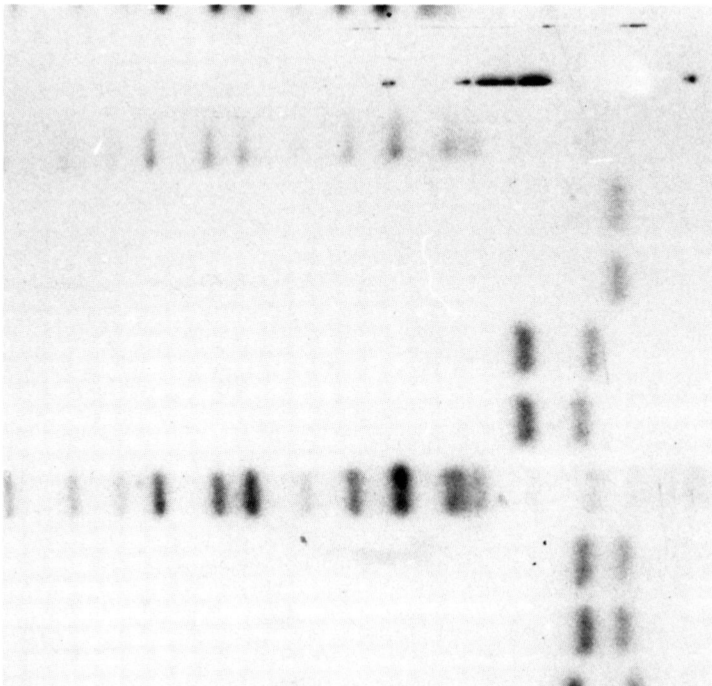

FIGURE 5. RFLP of L1077 (Collaborative Research, Inc., Bedford, MA) HaeIII digests yield fragments less than 1 kb in length.

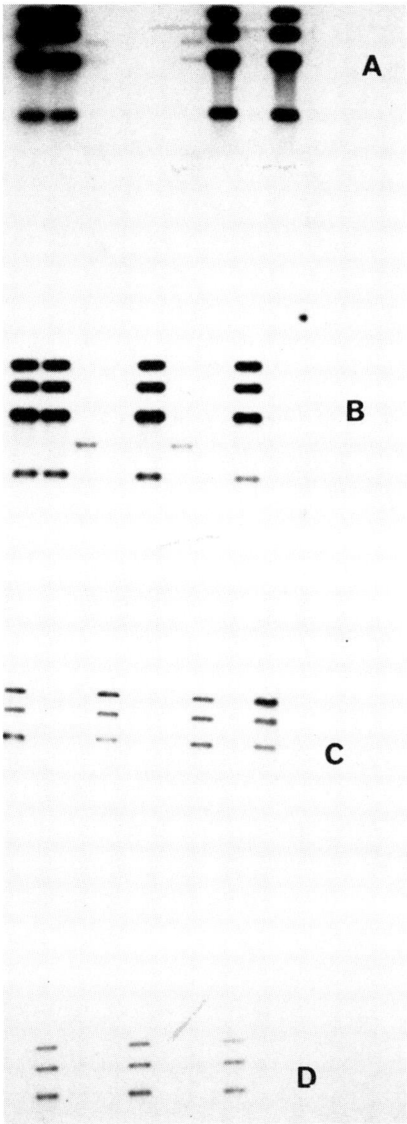

FIGURE 6. Comparison of sensitivity of detection (or signal intensity) provided by use of different nylon membranes. (A) Zeta Probe from Bio-Rad, Richmond, CA; (B) Biotrace from Gelman; (C) an amphoteric membrane supplied by MSI; (D) a neutral-charged membrane (Hybond N) from Amersham, Arlington Heights, IL.

7. Radioactivity

Finally, it is important to stress that the high sensitivity of detection we have experienced is partially due to the use of radioactively labeled probes. From a handling, safety, and particularly for technology transfer considerations, it would be desirable to use nonisotopic detection systems. However, our experience and that of others[43] has been that the commercially available nonisotopic labeling systems provide sensitivity of detection levels that are usually one to two orders of magnitude less sensitive than autoradiography.

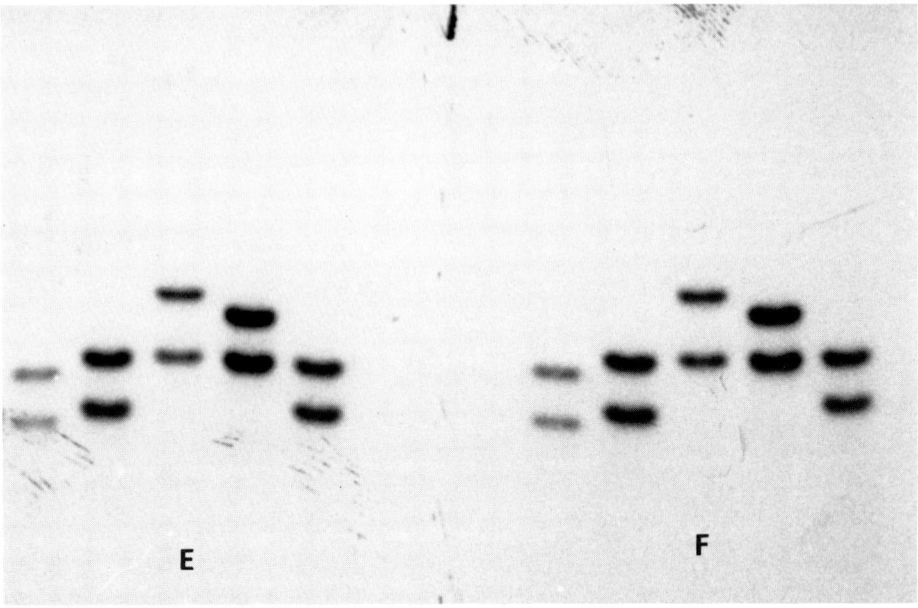

FIGURE 7. Membrane E was subjected to prehybridization for 2 h. Membrane F was not subjected to a prehybridization wash. There was no increase of background staining when the prehybridization step was eliminated. The locus was D2S44.

III. RESEARCH

The first section of this chapter addressed the RFLP methodology, with particular concern for forensic science. While RFLP technology is an effective technique and will figure prominently in forensic (as well as clinical) analyses, in its present form, this methodology will not likely be adopted by a majority of forensic laboratories, particularly small laboratories. The technique entails tedious sample preparation, numerous steps, and long incubation times. The technology cannot be transferred to laboratories that are incapable of providing adequate manpower and economic resources. Further, for laboratories (especially clinical diagnostic laboratories) that require results in a relatively short time period, the RFLP technique, again in its present form, will not likely replace the protein-electrophoretic genetic marker, immunological, and/or culture-based assays. Finally, although the sensitivity of detection of the RFLP technique is good, there are still some samples that provide DNA of a quantity and quality that is inadequate for RFLP analysis. Therefore, the second half of this chapter will address two research areas that may obviate some of the above shortcomings of the RFLP methodology.

A. NONFILM-BASED RADIATION DETECTORS

Radioactivity-based assay systems are the most sensitive, and may be the only approach that will provide results when assaying small quantities of DNA. However, film-based autoradiography suffers from long exposure times — from a few hours to several days. A radioactive detection approach that could be exploited to decrease the assay time would have obvious advantages.

Film is an inefficient detector for β-particles. Over 95% of the β-particles pass through the film without being detected.[44,45] An intensifying screen is necessary to enhance the overall signal, but with the negative result of wider bands and having to deal with low-intensity reciprocity failure.[44,46] Film has a limited dynamic range and, as mentioned before, film detectors suffer from long exposure times.

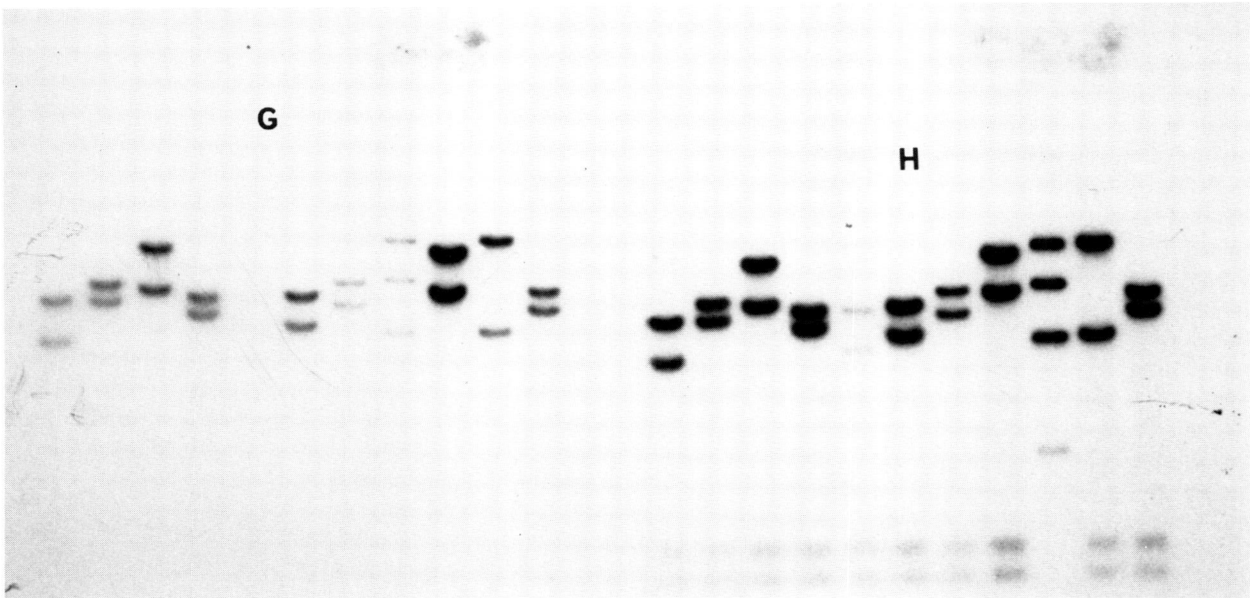

FIGURE 8. Membrane G was subjected to a hybridization solution containing 1% SDS. Membrane H was subjected to a hybridization solution containing 7% SDS. The locus was D2S44.

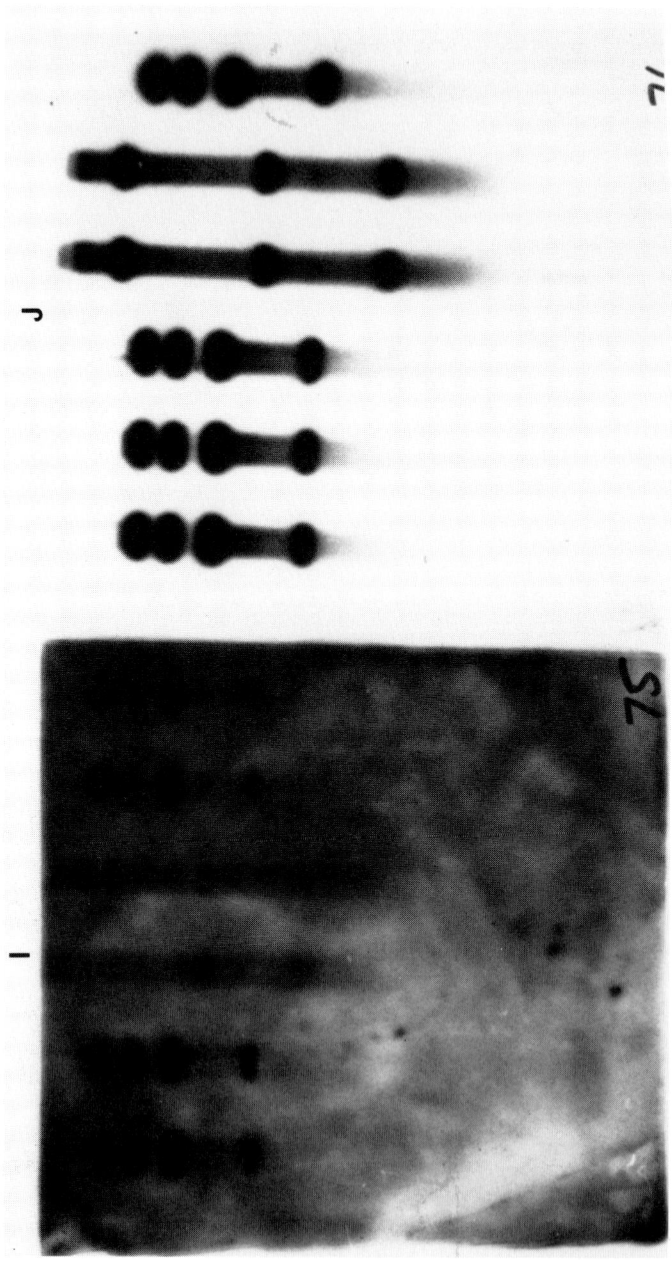

FIGURE 9. Long autoradiographic exposures can detect low-level background staining. Membrane I was subjected to a hybridization solution containing 1% SDS. Membrane J was subjected to a hybridization solution containing 7% SDS. The autoradiographic exposure was for 4 d. It appears that 7% SDS effectively removes background staining.

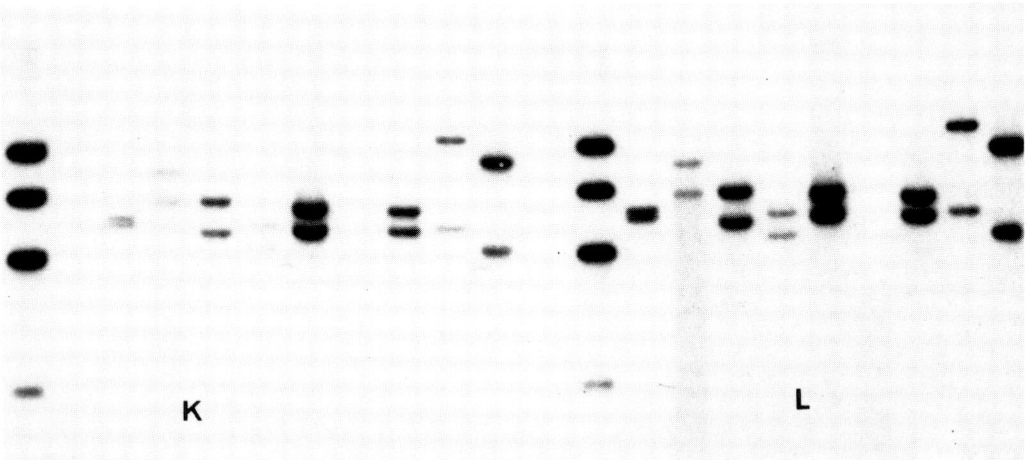

FIGURE 10. Membrane K was subjected to a hybridization solution containing 10% dextran sulfate. Membrane L was subjected to a hybridization solution containing 10% polyethylene glycol (mol wt = 8000).

There are two basic alternative approaches to the film-based radiation detection system. One is an electronic radiation sensor based on the ability of β-particles to ionize gas molecules.[45] The Ambis (San Diego, CA) and Betagen (Waltham, MA) systems exploit the electronic detector approach. The other detection system is based on phosphor particle sensors.[47-51] β-Particles emitted from ^{32}P interact with storage phosphors and are converted into a latent image in the form of trapped electrons. Phosphors are capable of storing absorbed energy in a quasistable state. When the storage phosphor is stimulated with a laser, the trapped sites luminesce, and the emission can be detected by photomultiplier tubes.

Both detection systems are superior to X-ray film detection of β-particles. They have a higher detective quantum efficiency, which increases their sensitivity of detection over film. With a greater sensitivity of detection, the assay time is greatly reduced. The dynamic range of both approaches is approximately three to four orders of magnitude wider than that of film. This, in effect, allows multiple optimum exposures to be observed on one image. Since the digital signals can be normalized for a wide range of exposure conditions, the output signals can occupy the same range. Thus, the overexposed bands observed on film can be eliminated. Finally, computer processing of the digital signal permits contrast enhancement, spatial frequency enhancement, and superior digital manipulation, compared to film.

While both approaches are superior to film, the electronic detector suffers from one shortcoming. The electronic detector has limited throughput. Only one membrane can be exposed and digitized at a time. This would be unacceptable for high-volume laboratories that may have to analyze a multitude of membranes at any given time. The phosphor detector has the same throughput as film. Thus, the analytical flow of the laboratory would not have to be altered.

1. Photostimulable Phosphors

The FBI and Eastman Kodak have, for the past year, collaborated to investigate the utility of photostimulable storage phosphor screens for autoradiography. Membranes were placed into cassettes containing phosphor screens for various exposure times. The process is similar to that for film exposure, but since low-intensity reciprocity failure[44,46] is not a consideration, exposure is performed at ambient temperature. After exposure, the latent image held within the phosphor screen is read in a laser-stimulated photo-detector device[50]

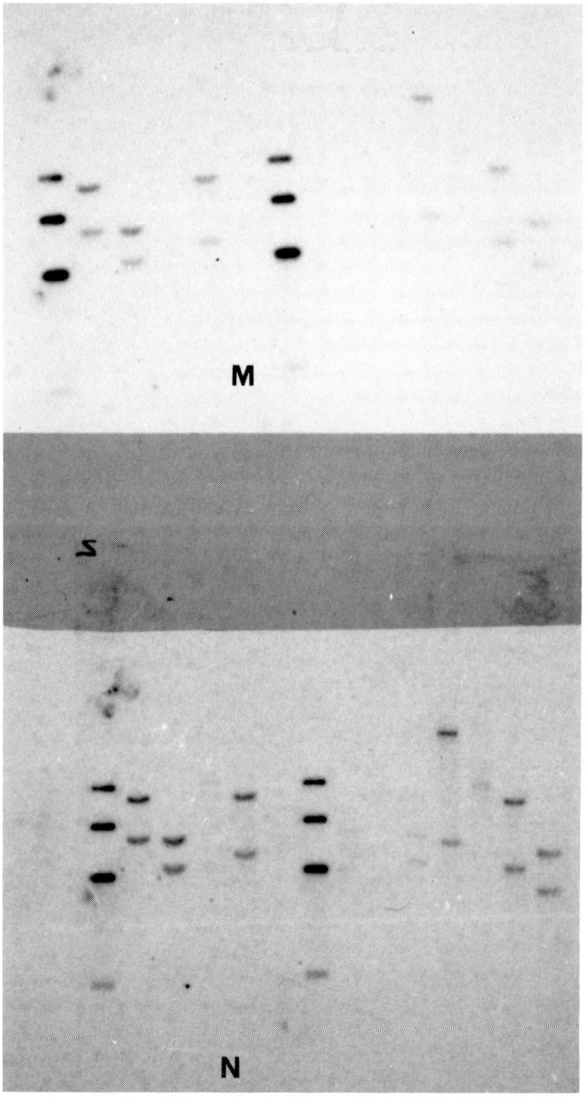

FIGURE 11. Image M was produced by standard film-based detection using Kodak XAR film and Dupont Cronex Lightning Plus intensifying screens. The exposure time was 24 h. Image N was generated using photostimulable phosphors. The exposure time was 4 h.

in a 90-s to 5-min time period. Figure 11 shows an example of phosphor detection of an RFLP. In one eighth the time period, the signal is comparable to X-ray film (with intensifying screens), but with reduced band width. With reduced assay time, an improved (resolved) image, and no overexposures, phosphor-generated images will have a positive impact on the molecular biology field, particularly clinical diagnoses. The extent of the utility of phosphors for autoradiography is currently the subject of intense investigation.

B. POLYMERASE CHAIN REACTION

Regardless of the advantages of radioactive-labeled probes, especially in light of the above autoradiographic detection system, DNA tests will have to be configured with non-isotopic assays to gain widespread acceptance in diagnostic laboratories. Further, when evidentiary samples are exposed to environmental adversities, the DNA may degrade to a level that would be inadequate for RFLP analysis. A method by which short target sequences of DNA could be quantitatively amplified to provide an analytical level sufficient for analysis would be extremely useful for the characterization of specimens of limited quantitative and qualitative DNA yield, as well as eliminate the need for isotopic detection. The polymerase chain reaction (PCR) may be such a technique.[52-56]

The PCR methodology is characterized by a three-step process: (1) target DNA is denatured by exposure to an elevated temperature (94°C); (2) short, single-stranded oligonucleotide primers of known DNA sequence are annealed to regions that flank the target DNA sequence, and (3) a DNA polymerase, utilizing the four deoxyribonucleoside triphosphates, will synthesize a new strand of DNA whose sequence will be directed by the template strand to which the primer was attached. Each cycle of denaturation, priming, and elongation potentially doubles the quantity of target-site DNA. Repetitive cycles, normally 15 to 30 cycles, of the PCR reaction are performed until the desired quantity of DNA is generated.

It becomes obvious that for PCR to be applied to the amplification of a DNA polymorphic site, the sequence of the target site, or at least the flanking regions of the target site, must be known in advance. Otherwise, proper oligonucleotide primers cannot be synthesized. Therefore, a search for well-defined polymorphic sites is a necessity. Examples of such sites are the β-globin and HLA-DQ α-loci.[52-54]

1. Fragment-Length Polymorphisms

Fragment-length polymorphisms (FLPs) can provide a good deal of genetic information and, with the appropriate analytical electrophoretic system, can be implemented readily in most laboratories. Higuchi et al.[57] have demonstrated that PCR-amplified mitochondrial DNA products containing deletions can be typed by electrophoretic separation in agarose gels in the presence of the fluorescent dye ethidium bromide. The East Asian nine-base deletion of Region V was utilized to demonstrate this ability.[58] They amplified this region and produced two different-sized fragments — a 120-bp fragment and a 111-bp fragment (corresponding to the 9-bp detection). After separating these fragments by electrophoresis, the bands of different sizes were visualized under UV light.

White,[67] Jeffreys,[59] Waye,[68] and Boerwinkle and Chan[60] are attempting to amplify several variable-number tandem-repeat loci and subsequently subject the fragments to electrophoresis for rapid typing. This approach would combine the increased sensitivity of the PCR technique with the information content of VNTR typing — so-called AMP-FLPs.

2. High-Resolution Electrophoresis of PCR Products

A rehydratable polyacrylamide gel bonded to a mylar backing has been developed to separate AMP-FLPS. A sulfate/borate (leading/trailing ion) multizonal electrophoresis system using 450-μm-thick rehydratable polyacrylamide gels[62,63] (5 to 8% T, 3% C) is used, and the separation carried out on a horizontal flatbed electrophoretic system. Samples, diluted one tenth the ionic strength of the resolving gel, are placed directly onto the gel surface in 0.5 to 10.0-μl volumes. DNA in excess of 10,000 bp in length will enter the gel ahead of the moving boundary. After the samples migrate into the gel, they are zone sharpened by the passage of the moving boundary and then unstacked in the order of their size. This produces much sharper bands than is possible with commonly used continuous-zone electrophoresis when separation is carried out over 12 cm in length. A mylar sheet is placed on the gel surface to prevent evaporation of the gels. The edges of the gel are maintained under a reduced ionic strength to increase mobility along the edge, which effectively eliminates any smile effect. The trailing and counter ions are contained in polyacrylamide plugs, which are placed under the electrodes and directly on the separating gel surface at the anodal and cathodal ends of the gel. Such gel-to-gel contact prevents the production of a junction potential, which can occur between dissimilar materials, and allows a precision of electro-

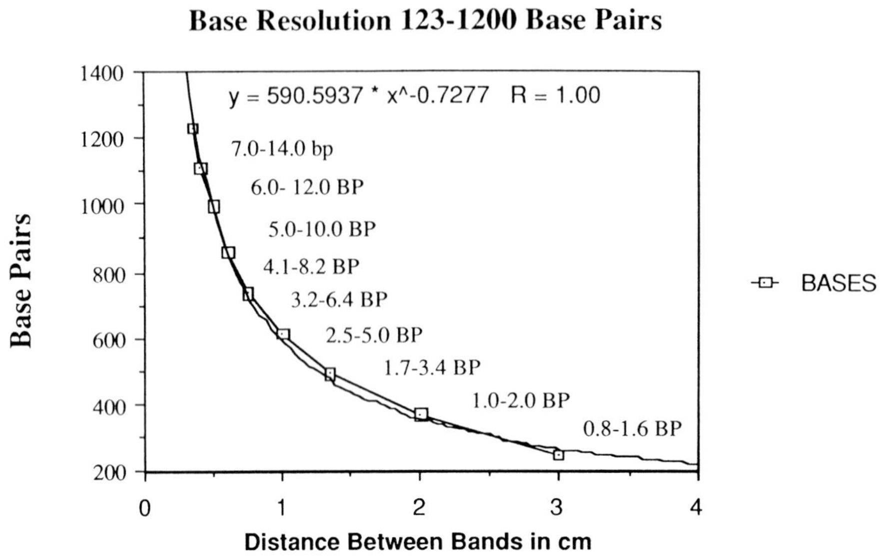

FIGURE 12. Base-pair resolution of a 6% polyacrylamide gel with a 15-cm interelectrode wick distance.

chemical conditions not available when paper or other materials serve as a bridge to a buffer reservoir. Separations on gels with interelectrode wick distances of 10 to 20 cm can be carried out in 2 to 4.5 h, respectively. A 40-min silver stain assay[64,65] provides a sensitivity of detection level of 1 to 4 pg/mm band width.

Figures 12 and 13 demonstrate the results that can be obtained with this gel system. A 20-cm gel can provide resolution of 1 to 10 bp for DNA fragments of 1 to 1000 bp in length. Thus, an analytical procedure is available that can provide a high degree of genetic characterization within 1 d. The major advantages of this electrophoretic system, compared with submarine gel electrophoresis, is that DNA samples with different concentrations do not migrate at different rates. Samples diluted over 100-fold migrated evenly. Additionally, characteristics such as a more sensitive assay than ethidium bromide staining, a nontoxic assay procedure, reduced band width, and permanent records make this procedure desirable. It is anticipated that a procedure such as this can be implemented readily in many laboratories. An intense investigation is underway for suitable fragment-length polymorphisms amenable to PCR amplification.

3. Intralaboratory Contamination

Prior to the use of any analytical assay to genetically characterize PCR products, intralaboratory contamination will have to be controlled. With the RFLP method, contamination of a sample with background DNA cannot be detected and, thus, is not a problem; however, PCR can amplify background DNA to a level where the problem of contamination can become acute. An example of sample-to-sample contamination would be the following. Samples A and B contain genomic DNA at concentrations of 10 μg/ml and 1 ng/ml, respectively. If 0.1 μl of sample A were transferred inadvertently to B (e.g., via Eppendorf-tip contamination), then 1 ng of A would now be in B. Thus, both A and B would be amplified equally. An example of an amplified product (C) contaminating an unamplified product (D) also should be considered. If, after PCR, sample C contained 10^{13} copies per milliliter of a particular product and 0.1 μl were transferred to the 10 μg/ml sample D, then C could contaminate D with 1000 more copies of a particular target site than contained in all of sample D (10^9 copies of C vs. 10^6 copies of D). Approaches are being investigated

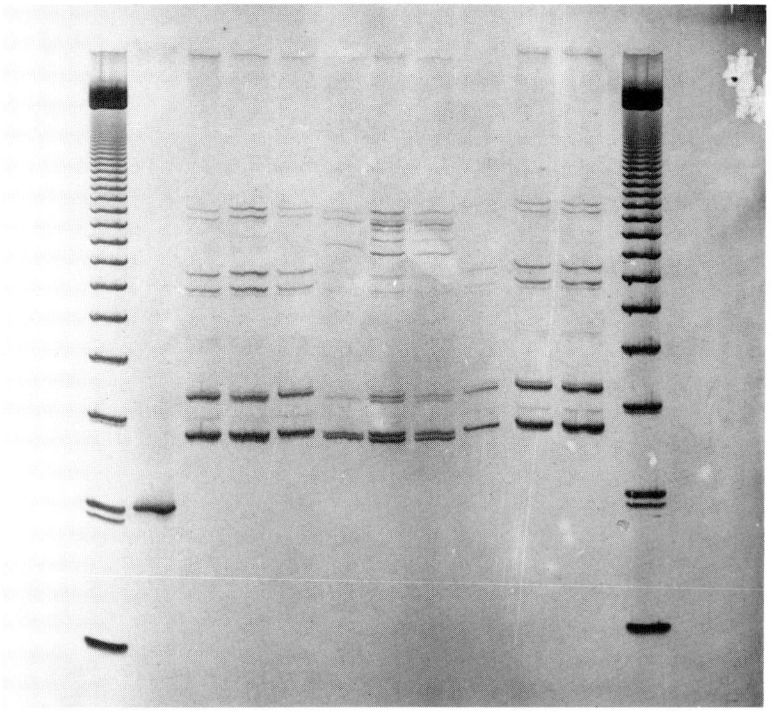

FIGURE 13. An AMP-FLP silver-stained gel demonstrating PCR-amplified microsatellite VNTRs.

presently to address this contamination issue. These include screw-cap tubes, centrifugation prior to opening the tubes, performing sample preparation and PCR in different laboratory areas, dedicated pipets, positive-displacement pipets, and robotics.

IV. CONCLUSIONS

Both RFLPs and AMP-FLPs can be useful for the genetic characterization of biological specimens. Extensive research is providing data to make these procedures robust, less tedious, less costly, and more sensitive — all with reduced assay times. These methods already are providing discrimination powers that far exceed protein-genetic marker typing for such body/fluid tissues as blood, semen, saliva, and hair. It is anticipated that in the near future the basic research methods in molecular biology will make their way into the application-oriented laboratories and play a major role not only in the forensic genetic analyses discussed here, but also in many other levels of biology and medicine.

REFERENCES

1. **Jeffreys, A. J., Wilson, V., and Thein, S. L.,** Individual-specific fingerprints of human DNA, *Nature,* 316, 76, 1985.
2. **Gill, P., Jeffreys, A. J., and Werrett, D. J.,** Forensic applications of DNA fingerprints, *Nature,* 318, 577, 1985.
3. **Gill, P. and Werrett, D. J.,** Exclusion of a man charged with murder by DNA fingerprinting, *Forensic Sci. Int.,* 35, 145, 1987.

4. **Fiori, A. and Benciolini, P.,** The ABO grouping of stains from body fluids, *Z. Rechtsmed.*, 70, 214, 1972.
5. **Spencer, N., Hopkinson, D. A., and Harris, H.,** Phosphoglucomutase polymorphism in man, *Nature,* 204, 742, 1964.
6. **Bark, J. E., Harris, M. J., and Firth, M.,** Typing the common phosphoglucomutase variants using isoelectric focusing — a new interpretation of the phosphoglucomutase system, *J. Forensic Sci. Soc.*, 16, 115, 1976.
7. **Kanter, E., Baird, M., Shaler, R., and Balazs, I.,** Analysis of restriction fragment length polymorphisms in deoxyribonucleic acid (DNA) recovered from dried bloodstains, *J. Forensic Sci.*, 31, 403, 1986.
8. **Giusti, A., Baird, M., Pasquale, S., Balazs, I., and Glassberg, J.,** Application of deoxyribonucleic acid (DNA) polymorphisms to the analysis of DNA recovered from sperm, *J. Forensic Sci.*, 31, 409, 1986.
9. **Gill, P., Lygo, J. E., Fowler, S. J., and Werrett, D. J.,** An evaluation of DNA fingerprinting for forensic purposes, *Electrophoresis,* 8, 38, 1987.
10. **Denhardt, D. T.,** A membrane-filter technique for the detection of complementary DNA, *Biochem. Biophys. Res. Commun.*, 23, 641, 1966.
11. **Southern, E. M.,** Detection of specific sequences among DNA fragments separated by gel electrophoresis, *J. Mol. Biol.*, 98, 503, 1975.
12. **Studencki, A. G., Conner, B. J., Impraim, C. C., Teplitz, R. L., and Walker, R. B.,** Discrimination among the human beta^A, beta^S, and beta^C-globin genes using allele-specific oligonucleotide hybridization probes, *Am. J. Hum. Genet.*, 37, 42, 1985.
13. **Wyman, A. R. and White, R.,** A highly polymorphic locus in human DNA, *Proc. Natl. Acad. Sci. U.S.A.*, 77, 67, 1980.
14. **Nakamura, Y., Leppert, M., O'Connell, P., Wolff, R., Holm, T., Culver, M., Martin, C., Fujimoto, E., Hoff, M., Kumlin, E., and White, R.,** Variable number of tandem repeat (VNTR) markers for human gene mapping, *Science,* 235, 1616, 1987.
15. **Nakamura, Y., Gillilan, S., O'Connell, P., Leppert, M., Lathrop, G. M., Lalouel, J.-M., and White, R.,** Isolation and mapping of a polymorphic DNA sequence pYNH24 on chromosome 2 (D2S44), *Nucleic Acids Res.*, 15, 10073, 1987.
16. **Wong, Z., Wilson, V., Patel, I., Povey, S., and Jeffreys, A. J.,** Characterization of a panel of highly variable minisatellites cloned from human DNA, *Ann. Hum. Genet.*, 51, 269, 1987.
17. **Werrett, D. J.,** DNA fingerprinting, *Int. Crim. Police Rev.*, September/October, 21, 1987.
18. **Nakamura, Y., Culver, M., Gill, J., O'Connell, P., Leppert, M., Lathrop, G. M., Lalouel, J.-M., and White, R.,** Isolation and mapping of a polymorphic DNA sequence pMLJ14 on chromosome 14 (D14S13), *Nucleic Acids Res.*, 16, 381, 1988.
19. **Budowle, B., Deadman, H. A., Murch, R. S., and Baechtel, F. S.,** An introduction to the methods of DNA analysis under investigation in the FBI Laboratory, *Crime Lab. Dig.*, 15, 8, 1988.
20. **Hill, W. G.,** DNA fingerprinting applied to animal and bird populations, *Nature,* 327, 98, 1987.
21. **Wetton, J. H., Carter, R. E., Parkin, D. T., and Walters, D.,** Demographic study of a wild house sparrow population by DNA fingerprinting, *Nature,* 327, 147, 1987.
22. **Dallas, J. F.,** Detection of DNA fingerprints of cultivated rice by hybridization with a human minisatellite DNA probe, *Proc. Natl. Acad. Sci. U.S.A.*, 85, 6831, 1988.
23. **Hoelzel, A. R. and Amos, W.,** DNA fingerprinting and scientific whaling, *Nature,* 333, 305, 1988.
24. **Rogstad, S. H., Patton, J. C., and Scheal, B. A.,** A human minisatellite probe reveals RFLPs among individuals of two angiosperms, *Nucleic Acids Res.*, 16, 11378, 1988.
25. **Wong, Z., Wilson, V., Jeffreys, A. J., and Thein, S. L.,** Cloning a selected fragment from a human DNA fingerprint: isolation of an extremely polymorphic minisatellite, *Nucleic Acids Res.*, 14, 4605, 1986.
26. **Lynch, M.,** Estimation of relatedness by DNA fingerprint, *Mol. Biol. Evol.*, 5, 584, 1988.
27. **Gill, P.,** A new method for sex determination of the donor of forensic samples using a recombinant DNA probe, *Electrophoresis,* 8, 35, 1987.
28. **Gillespie, D. and Spiegelman, S.,** A quantitative assay for DNA-RNA hybrids with DNA immobilized on a membrane, *J. Mol. Biol.*, 12, 829, 1965.
29. **Wahl, G. M., Stein, M., and Stark, G. R.,** Efficient transfer of large DNA fragments from agarose gels to diazobenzyloxymethyl — paper and rapid hybridization by using dextran sulfate, *Proc. Natl. Acad. Sci. U.S.A.*, 76, 3683, 1979.
30. **Gatti, R. A., Concannon, P., and Salser, W.,** Multiple use of southern blots, *BioTechniques,* 2, 148, 1984.
31. **Maniatis, T., Fritsch, E. F., and Sambrook, J.,** *Molecular Cloning. A Laboratory Manual,* Cold Spring Harbor Laboratory, Cold Spring Harbor, NY, 1982, 1.
32. **Higuchi, R., Bowman, B., Freiberger, M., Ryder, O. A., and Wilson, A. C.,** DNA sequences from the quagga, an extinct member of the horse family, *Nature,* 312, 282, 1984.
33. **Paabo, S.,** Molecular cloning of ancient Egyptian mummy DNA, *Nature,* 314, 644, 1985.

34. **Calvin, H. I. and Bedford, J. M.,** Formation of disulphide bonds in the nucleus and accessory structures of mammalian spermatozoa during maturation in the epididymis, *J. Reprod. Fertil.*, Suppl. 13, 65, 1971.
35. **Blakesley, R. W., Dodgson, J. B., Nes, I. F., and Wells, R. D.,** Duplex regions in single-stranded phiX174 DNA are cleaved by a restriction endonuclease from *Haemophilus aeqyptius, J. Biol. Chem.*, 252, 7300, 1977.
36. **Adams, R. L. P. and Burdon, P. H.,** The methylation machinery: DNA methyl transferases, in *Molecular Biology of DNA Methylation*, Rich, A., Ed., Springer-Verlag, New York, 1985, chap. 5.
37. **Adams, R. L. P. and Burdon, P. H.,** Methylation in higher eukaryotes, in *Molecular Biology of DNA Methylation*, Rich, A., Ed., Springer-Verlag, New York, 1985, chap. 7.
38. **Roberts, R. J.,** Restriction enzymes and their isoschizomers, *Nucleic Acids Res.*, 16 (Suppl.), r271, 1988.
39. **Reed, K. C. and Mann, D. A.,** Rapid transfer from agarose gels to nylon membranes, *Nucleic Acids Res.*, 13, 7207, 1985.
40. **Church, G. M. and Gilbert, W.,** Genomic sequencing, *Proc. Natl. Acad. Sci. U.S.A.*, 81, 1991, 1984.
41. **Johnson, D. A., Gautsch, J. W., Sportsman, J. R., and Elder, J. H.,** Improved techniques utilizing non-fat dry milk for analysis of proteins and nucleic acids transferred to nitrocellulose, *Gene Anal. Technol.*, 1, 3, 1984.
42. **Amasino, R. M.,** Acceleration of nucleic acid hybridization rate by polethylene glycol, *Anal. Biochem.*, 152, 304, 1986.
43. **Zwadyk, P., Cooksey, R. C., and Thornsberry, C.,** Commercial detection methods for biotinylated gene probes: comparison with 32 P-labeled DNA probes, *Curr. Microbiol.*, 14, 95, 1986.
44. **Hahn, E. J.,** Autoradiography: a review of basic principles, *Am. Lab.*, July 1, 1, 1983.
45. **Sullivan, D. E., Auron, P. E., Quigley, G. J., Watkins, P., Stanchfield, J. E., and Bolon, C.,** The nucleic acid blot analyzer. I. High speed imaging and quantification of ^{32}p-labeled blots, *BioTechniques*, 5, 672, 1987.
46. **Babcock, T. A.,** A review of methods and mechanisms of hypersensitization, *AAS Photo-Bull.*, 13, 1, 1976.
47. **Sonoda, M., Takano, M., Miyahara, J., and Kato, H.,** Computed radiography utilizing scanner laser stimulated luminescence, *Radiology*, 148, 833, 1983.
48. **Merritt, C. R., Tutton, R. H., Bell, K. A., Bluth, E. I., Kalmar, J. A., Matthews, C. C., Miller, K. D., Kogutt, M. S., and Balter, S.,** Clinical application of digital radiography: computed radiographic imaging, *Radiographics*, 5, 397, 1985.
49. **Bilderback, D., Moffat, K., Owen, J., Rubin, B., Schildkamp, W., Szebenyi, D., Smith Temple, B., Volz, K., and Whiting, B.,** Protein crystallographic data acquisition and preliminary analysis using Kodak storage phosphor plates, *Nucl. Instrum. Methods Phys. Res.*, A266, 636, 1988.
50. **Whiting, B. R., Owen, J. F., and Rubin, B. H.,** Storage phosphor x-ray diffraction detectors, *Nucl. Instrum. Methods Phys. Res.*, A266, 628, 1988.
51. **Amemiya, Y. and Miyahara, J.,** Imaging plate illuminates many fields, *Nature*, 336, 89, 1988.
52. **Saiki, R. K., Scharf, S., Faloona, F., Mullis, K. B., Horn, G. T., Erlich, H. A., and Arnheim, N.,** Enzymatic amplification of beta-globin genomic sequences and restriction site analysis for diagnosis of sickle cell anemia, *Science*, 230, 1350, 1985.
53. **Erlich, H. A., Sheldon, E. L., and Horn, G. T.,** HLA typing using DNA probes, *BioTechnology*, 4, 975, 1986.
54. **Saiki, R. K., Bugawan, T. L., Horn, G. T., Mullis, K. B., and Erlich, H. A.,** Analysis of enzymatically amplified beta-globin and HLA-DQalpha DNA with allele-specific oligonucleotide probes, *Nature*, 324, 163, 1986.
55. **Li, H., Gyllensten, U. B., Cui, X., Saiki, R. S., Erlich, H. A., and Arnheim, N.,** Amplification and analysis of DNA sequences in single human sperm and diploid cells, *Nature*, 335, 414, 1988.
56. **Oste, C.,** Polymerase chain reaction, *BioTechniques*, 6, 162, 1988.
57. **Higuchi, R., Von Beroldingen, C. H., Sensabaugh, G. F., and Erlich, H. A.,** DNA typing from single hairs, *Nature*, 332, 543, 1988.
58. **Wrischnik, L. A., Higuchi, R. G., Stoneking, M., Erlich, H. A., Arnheim, N., and Wilson, A. C.,** Length mutations in human mitochondrial DNA: direct sequencing of enzymatically amplified DNA, *Nucleic Acids Res.*, 15, 529, 1987.
59. **Jeffreys, A. J., Wilson, V., Neumann, R., and Keyte, J.,** Amplification of human minisatellites by the polymerase chain reaction: towards DNA fingerprinting of cells, *Nucleic Acids Res.*, 16, 10953, 1988.
60. **Boerwinkle, E. and Chan, L.,** Rapid typing of tandemly repeated hypervariable loci by the polymerase chain reaction, *Am. J. Hum. Genet.*, 43, A138, 1988.
61. **Allen, R. C.,** Polyacrylamide gel electrophoresis with discontinuous buffers at a constant pH, in *PAGE and PAGIF*, Allen, R. C. and Maurer, H. R., Eds., Walter de Gruyter, Berlin, 1974, chap. 2.4
62. **Allen, R. C. and Lack, P. M.,** Standardization in isoelectric focusing on ultrathin-layer rehydratable polyacrylamide gels, in *New Directions in Electrophoretic Methods*, ACS Symp. Ser. 335, Jorgenson, J. W. and Phillips, M., Eds., American Chemical Society, Washington, D.C., 1987, chap. 8.

63. **Allen, R. C., Lack, P. M., Graves, G., and Budowle, B.,** Rehydrated polyacrylamide gels: a comparison with conventionally cast gels, in *Electrophoresis '86,* Dunn, M., Ed., VCH Publishers, Weinheim, 1986, 462.
64. **Merril, C., Goldmann, D., Sedman, S., and Ebert, M.,** Ultrasensitive stain for proteins in polyacrylamide gels shows regional variation in cerebrospinal fluid proteins, *Science,* 211, 1437, 1981.
65. **Budowle, B. and Scott, E.,** Transferrin subtyping of human bloodstains, *Forensic Sci. Int.,* 28, 269, 1985.
66. **Hoar, D.,** personal communication.
67. **White, R.,** personal communication.
68. **Waye, J.,** personal communication.

III. PROTEIN-NUCLEIC ACID INTERACTIONS

2. Electrophoresis Buffers

The buffers typically used in DNA electrophoresis are "TBE" (90 m*M* Tris, 90 m*M* boric acid, and 2.5 m*M* EDTA, pH 8.3 at 22°C) or "TEAC" (40 m*M* Tris base, 20 m*M* Na acetate, 18 m*M* NaCl, and 2 m*M* EDTA, plus HCl to give a pH of 8.1 at 22°C). TBE has given suitable results in gel retardation assays of DNA-protein interactions as well. A potential disadvantage of borate-containing buffers is that they can, in some cases, lead to artifacts due to interactions between borate ions and the protein of interest.[13] Such an effect has been seen in studies of pure albumin,[14] where two bands were observed upon electrophoresis, and appears to be the cause of spurious double bands found in my own lab in experiments with *E. coli* RNA polymerase and DNA fragments containing the galactose promoter. To remedy this problem, other buffers such as TEAC, Tris-acetate-EDTA (40 m*M* Tris base, 20 m*M* Na acetate, and 2 m*M* EDTA, pH 8.1 at 22°C), or Tris-EDTA (10 m*M* Tris-HCl and 1 m*M* EDTA, pH 8.0 at 22°C) have been used.

In most cases, a buffer having a pH of about 8 is appropriate. However, Carey[15] recently found that the *E. coli* tryptophan repressor (TrpR), a dimer composed of identical 12,500-Da subunits, itself migrates in a TBE gel with a mobility similar to that of the 90-bp DNA fragment of choice; therefore, there was little retardation of the complex with respect to free DNA. To remedy this problem, a buffer of 10 m*M* NaH_2PO_4, pH 6.0, was used. This pH is very near the isoelectric point of TrpR, and the complexes and free DNA could be easily resolved.

It may be necessary to add various factors to the electrophoresis buffer in order to maintain the activity of the protein. Thus, for TrpR, 0.1 m*M* L-tryptophan is needed; for *E. coli* CAP, cAMP at 20 μ*M* is used in the gel and reservoir buffers.

Relatively low ionic strength buffers are the norm in electrophoresis experiments. High salt concentrations in the buffer are generally undesirable; in such circumstances, only a small fraction of the current is carried by the macro ions, and substantial heating may occur during the run. On the other hand, at very low ionic strengths, there may be unacceptably large broadening of the DNA-containing bands.[13] In some situations, however, the choice of electrophoresis buffer is dictated by more specific considerations. For example, in studies with purified or partially purified systems, one goal may be to assess the stoichiometry or other properties of the complexes themselves. Because DNA-protein interactions are very sensitive to salt concentrations, artifacts are minimized if the electrophoresis buffer is the same as or resembles that in the solution of interest. *In vitro* assays of transcription by *E. coli* RNA polymerase are often performed in a buffer containing 20 m*M* Tris, pH 8.0 at 22°C, 100 m*M* KCl (or NaCl), and 10 m*M* $MgCl_2$. However, good transcription can also be obtained in a lower salt buffer, e.g., 14 m*M* Tris, pH 8.0, 13 m*M* NaCl, and 3 m*M* $MgCl_2$. Since DNA and complexes migrate adequately in the latter buffer, it is possible to use gels to study the initiation complexes in a buffer which supports transcription.

The ability to find an electrophoresis buffer which closely approximates the solution buffer of interest may not be critical. In cases where tested (for instance, with proteins involved in regulating transcription at *E. coli* promoters), it appears that complexes found in the gel reflect the situation in the reaction buffer placed atop the gel. The CAP protein shows substantial nonspecific binding to DNA, as well as interaction with specific sites at catabolite-sensitive promoters. If CAP is mixed with DNA in TBE buffer, centrifugation studies show that, at this low ionic strength, the nonspecific binding is so strong that all CAP molecules will be bound to the DNA. Conversely, at 100 m*M* KCl and 3 to 10 m*M* $MgCl_2$, there is specific binding of CAP to the *lac* or *gal* promoters, but essentially no nonspecific binding at the concentrations of protein and DNA normally used. If a solution of CAP plus *lac* DNA fragments in the higher salt buffer is placed atop a gel and run using TBE as the gel and reservoir buffer, only specific complexes are seen. Thus, any effects of mixing as the sample is layered into the well, or changes in buffer concentration as electrophoresis begins, do not appreciably affect the results of the experiment — the concentration and properties of CAP-DNA complexes seen in the gel are those in the original solution.

The low ionic strength buffers used in electrophoresis will tend to enhance the stability of DNA-protein complexes which enter the gel. A cofactor such as cAMP or Mg^{2+}, if needed for binding, can also be included in the electrophoresis buffer. Nevertheless, where conditions in the reaction solution are much different from those in the gel, the possibility exists that the complexes found in the gel might differ in some way from those in the initial solution; thus, it is desirable to have other means for verifying the stoichiometries, etc. derived from electrophoresis data.

3. Effects of the Gel Matrix

Separation of complexes from unbound DNA is based on the sieving effect of the gel matrix, since the average pore size of the gel presumably approximates the size of the DNA and complexes. Under such conditions, it is of interest to ask whether the gel itself affects the DNA-protein interactions in the complex. It is conceivable, for example, that the polyacrylamide or agarose matrix could increase the degree of structure in a significant fraction of the solvent, which, in turn, could alter the thermodynamic characteristics of the nucleic acid-protein interaction. Alternatively, the binding parameters might be affected by a "caging" phenomenon. If the complex is contained in a polymeric cage within the gel and the DNA and protein are not free to move apart when they dissociate, the effective affinity constant in the gel will be higher than the true value in free solution.

Effects of the gel on the complexes have been studied using *E. coli* RNA polymerase plus DNA fragments containing the CAP-independent mutant *lac* UV5 promoter, and with CAP plus the wild-type *lac* promoter.[16] In one experiment, the products of transcription in the gel were compared with those made in solution. To do this, polymerase-*lac* UV5 complexes were made and electrophoresed into a 5% polyacrylamide gel for 30 min. (Heparin was added to the sample before electrophoresis to remove nonspecifically bound protein molecules; transcriptionally competent complexes are stable in the presence of heparin.) The gel buffer was 14 m*M* Tris, 13 m*M* NaCl, and 3 m*M* $MgCl_2$, pH 7.9. The run was interrupted and a solution containing the four ribonucleoside triphosphates, one of which was radioactive (α-^{32}P), was placed atop the gel lane and electrophoresis was resumed. The NTPs move rapidly down the gel and overtake the band of complexes. After further electrophoresis, the gel was autoradiographed and a band of radioactivity was observed, well separated from the free nucleotides, near the band of DNA-protein complexes. This radioactivity was eluted from the gel and analyzed under denaturing conditions on a 12% polyacrylamide gel in TBE buffer plus 7 *M* urea. Full-length runoff transcripts were seen, and the transcription pattern was quite similar to that found when the reaction was done in free solution, in the same buffer. Thus, the properties of the complexes in terms of their ability to catalyze the synthesis of RNA molecules are not changed when they are in a polyacrylamide gel.

The issue of caging was addressed by comparing the dissociation rate of CAP-*lac* DNA complexes in the gel with that in free solution. To do this, CAP promoter complexes were made in a buffer containing 20 m*M* Tris, 100 m*M* KCl, 3 m*M* $MgCl_2$, pH 7.9, and 20 μM cAMP and were electrophoresed into a gel made in TBE buffer plus 5 μM cAMP. In the higher salt buffer, nonspecific DNA-CAP interactions are minimal, and a clean band of specific complexes is seen. To follow dissociation in the gel (in TBE buffer), aliquots were placed on the gel at different times. As electrophoresis proceeds, the complex band becomes less intense, since some dissociation occurs during the run. By measuring the amounts of complexes remaining after various times in the gel, it is straightforward to evaluate the dissociation rate constant. The off-rate for complexes in solution was determined by means described briefly here and in more detail below. Radioactively labeled promoter fragments and CAP were incubated in TBE buffer plus 5 μM cAMP (at DNA excess, to eliminate nonspecific binding). At time zero, a tenfold excess of unlabeled *lac* DNA was added and the amounts of complex remaining were assayed by gel retardation over a period of 60 min. The results show the same dissociation constant for the complexes, well within experimental error, in the gel and in solution. Furthermore, the dissociation rate was the same for DNA fragments of different lengths (64, 203, and 789 bp). Therefore, when a complex dissociates in the gel, the electric field causes the DNA to move rapidly away from the protein, so that reassociation does not occur. We conclude that there is no caging effect in 5% polyacrylamide gels, at least for CAP-DNA interactions, though such an effect is not ruled out at higher polyacrylamide concentrations.

4. Detecting and Analyzing Bands in the Gel

It is easiest to locate bands by assays for DNA. Two techniques are commonly used to detect DNA in the gel. The DNA may be made radioactive, usually by labeling at the 5′ end using T4 polynucleotide kinase and γ-^{32}P-ATP.[17] In this case, following electrophoresis, the gel is removed from the apparatus and may be dried onto a filter paper under vacuum and heat (for instance, in a Bio-Rad gel dryer) and then is placed flat against X-ray film and exposed for an appropriate length of time. A band containing 1000 cpm can easily be seen following overnight exposure. To speed up the process about fivefold a Kronex enhancing screen may be used, though at a cost of somewhat reduced resolution. The effects of the screen are obtained only if exposure is done at −70°C. It is also possible to do autoradiography on a wet gel; the gel is left on one glass plate and is encased in plastic wrap before being placed against the film. This is convenient if one wishes to extract bands from the gel after locating them by autoradiography. Pieces of paper containing radioactive ink spots are taped to the plate next to the gel; this produces spots on the exposed film which can be used to align the film and gel so that the desired bands can be accurately excised.

A second approach for identifying DNA-containing bands is staining with ethidium bromide (EtBr). Here, the gel is submerged in a tray containing a solution of EtBr at ~1 μg/ml. After about 10 min, bands of DNA can be seen under UV light, since EtBr intercalated into DNA has a characteristic fluorescence. By this means, one can easily detect as little as 30 ng of DNA. The gel can be photographed after being placed on a transilluminator box, which sends light at ~300 nm through the gel from below. Several companies (e.g., Fotodyne) sell photographic apparatus specifically built for this purpose. Care must be exercised if the DNA is to be removed from the gel and used further, because cleavage of phosphodiester bonds is enhanced in the presence of EtBr and UV light. To remove the dye from the DNA, the solution is shaken with multiple batches of buffer-saturated *n*-butanol.

III. APPLICATIONS

In this section, a number of areas in which the gel mobility assay is being used are discussed. While many of the experiments are described in specific detail, it should be kept in mind that variations on these approaches are quite common, depending on the precise needs of the investigator. A fairly broad survey is presented here, but no attempt has been made to be all-inclusive. The assay is a versatile one, which readily lends itself to modification for new purposes.

The gel retardation technique can yield either quantitative or qualitative information about nucleic acid-protein interactions. Quantitative characterization of equilibrium and kinetics properties is best done with purified components (though crude extracts can be used in some instances). Only a few systems are at the stage of understanding where extensive thermodynamic measurements are essential. However, as more systems are studied in detail *in vitro*, the gel shift approach will find increasing use for determination of binding constants and of kinetic parameters. Thus, this section begins with a description of how the assay has been applied to the *E. coli lac* regulatory system, as an example of what can be done with other systems in the future.

A. QUANTITATIVE APPLICATIONS

1. Stoichiometry

To assess the stoichiometry of the DNA-protein complexes, it is obviously necessary to quantify the protein in the band. This is easily done if the protein can be radioactively labeled. For example, a bacterial protein might be isolated from cells grown in the presence of ^{35}S or ^{14}C. Labeling with tritium via the sodium borohydride reaction works well in some cases, but in others the protein seems to be inactivated during the labeling process. A concern in using radioactive protein is possible quenching in the presence of the gel. (This is not usually an issue with labeled DNA, for two reasons. First, DNA is often labeled with the powerful isotope ^{32}P. Second, a reference lane containing a known amount of DNA can be run for calibration; this may be more difficult to do for the protein, which, in fact, may barely enter the gel by itself if its net charge is small.) An estimate of the degree of quenching can be obtained by dissolving an appropriately sized piece of the gel with tissue solubilizer (such as NEN Protosol) and comparing the number of counts per minute found when radioactive protein is added to this solution with the number of counts per minute in a control sample which contains identical components except for the solubilized gel.

In studies of CAP-DNA interactions,[18] ^{3}H-CAP was prepared and complexes of active CAP with ^{32}P-*lac* DNA fragments were isolated from inactive protein by chromatography on Sephadex G-100, after which a gel shift assay was done. Lanes were sliced, solubilized, and counted for both ^{3}H and ^{32}P, from which it was concluded that the complex has a 1:1 stoichiometry.

Another approach to quantifying proteins in the complex involves cutting out the band, soaking the excised piece of gel in buffer containing sodium dodecyl sulfate (SDS), casting this gel slice atop another polyacrylamide gel, and electrophoresing the denatured proteins in a second dimension. Calibration lanes containing known amounts of the protein can be run on the same gel. Following Coomassie blue or silver staining, the gels can be scanned and the level of the protein of interest determined. The stoichiometry of its binding to DNA can then be assessed, provided the corresponding amount of DNA is known from measurements of radioactivity or of the intensity of EtBr fluorescence. (In the latter case, one must be certain that the presence of the bound protein does not affect the EtBr signal.) The intensity of an EtBr band can be derived by using a densitometer to scan the negative of a photograph of the gel, provided the exposure lies within the linear range of the film. Note also that electrophoresis in SDS in a second dimension can be used qualitatively, simply to identify those polypeptide chains which are present in the complexes.

It has also been possible to evaluate the degree of ligand binding in complexes. This was done in the *lac* promoter-RNA polymerase-CAP-cAMP system using ^{3}H-cAMP.[19] Following electrophoresis, a peak of radioactivity was found at the band of complexes. This was cut from the gel and assayed in a scintillation counter, with appropriate corrections made for potential quenching effects. This experiment was initially done without cyclic nucleotide in the electrophoresis buffer, which yields a minimum value for the amount bound in the complex (since any cAMP which dissociated during the run would be lost). In some cases, the binding of the ligand may be sufficiently tight that no significant losses occur during electrophoresis; in others, the ligand will need to be included in the gel buffer to ensure equilibrium binding to the complexes, in spite of the fact that unlabeled ligand will reduce the number of counts found in the complexes. In this situation, the ligand-protein interaction must still be rather strong, so that the ligand concentration in the buffer can be low enough that a significant number of counts above background are detectable in the complex band.

2. Equilibrium Constants

For a simple binding reaction, R + P ↔ RP (where R, for example, could be RNA polymerase, P could be promoter DNA, and RP represents the complex), the association constant is

$$K = \frac{RP}{[R][P]}$$

If R_T and P_T are known input (total) concentrations, then

$$K = \frac{RP}{[R_T - RP][P_T - RP]}$$

Often one component, say R, is present in excess, so that $[R_T - RP] \sim R_T$. Then,

$$K = \frac{RP}{R_T[P_T - RP]} = \frac{RP/P_T}{R_T[1 - RP/P_T]} = \frac{F}{R_T[1 - F]}$$

where F is the fraction of DNA in complex. The value of K then can be calculated from the measured value of F. Note, however, that values of F very near to zero or unity are not conducive to an accurate determination of K. The gel retardation assay yields directly the concentration of free DNA in a DNA-protein mixture, provided no change in the level of free DNA occurs in the short time period while the DNA is entering the gel. Knowing the input level of DNA, one can calculate the amount in complex (hence, F). Thus, the gel method resembles the nitrocellulose filter assay,[20] which has been a mainstay in studies of this type. In the filter assay, the concentration of complexes is inferred from radioactivity retained on the filter. Under favorable circumstances, negligible dissociation of complexes takes place during the electrophoresis process, the sum of the DNA in the free and complex bands equals the known total amount in the solution, and [RP] is directly measured.

In a typical experiment for determination of the equilibrium constant, solution conditions are chosen so that variations in the degree of binding (F) occur as the levels of protein and DNA are changed. To establish this situation may entail adjusting the ionic strength. The gel-binding assay permits study over a wide range of component concentrations. Using radioactively labeled DNA allows detection of as little as a few picomoles of DNA. At such low concentrations, loss of protein by adherence to tubes and pipette tips may become significant. The upper limit on concentration is usually dictated by solubility problems which may arise or by the availability of DNA fragments or protein.

For the binding of CAP to the *lac* promoter, Fried and Crothers[5] used threefold serial dilutions of protein-DNA mixtures. The starting solutions contained 10 n*M* CAP, 11 n*M* *lac* DNA, and 1 μ*M* cAMP. (In this case, there was less radioactive DNA in each diluted solution, hence, they used different exposure times to be certain that band intensities did not exceed the linear response range of the X-ray film.) In principle, K can be determined from a single experiment in which F is measured for a given R_T and P_T, but multiple data points obviously are required to ensure an accurate result.

Once a binding constant has been determined, competition experiments can be performed to evaluate K for related systems.[21] In such cases, it may be convenient to mix DNA fragments of different lengths and containing different promoters, then to determine the degree of protein binding to each. If K_u represents the unknown association constant for the reaction with DNA 1, and K_k, the known constant for the reaction involving DNA 2, then

$$K_u = K_k \frac{[\text{complex with DNA 1}]}{[\text{free DNA 1}]} \times \frac{[\text{free DNA 2}]}{[\text{complex with DNA 2}]}$$

If the complexes are stable during electrophoresis, and either both DNAs are radioactive or EtBr staining is used, then the band intensities yield all the concentrations on the right side of this equation; hence, K_u can be calculated. If only one DNA fragment is labeled, the unknown binding constant is derived by measuring the change in the level of binding to the "known" DNA in the presence of the competitor. Equations for evaluating K_u are somewhat complex, depending on the relative amounts of protein and of each DNA fragment.

Of course, an alternative approach to determining the equilibrium constant is via measurement of association and dissociation rates for the reaction, as discussed below; the ratio of these kinetics parameters yields the thermodynamic binding constant.

3. Kinetics

Gel mobility shift assays are quite useful for ascertaining kinetic characteristics of nucleic acid-protein interactions. Solution conditions must be adjusted so that the reaction rates are in a range for convenient measurement. This requires that at least 15 s (and preferably 60 s or more) separates each time point at which a measurement is to be made. There is, in principle, no limit on longer time points except for possible protein denaturation and the like, which would be common to all such lengthy experiments. To slow association reactions to a manageable rate, the temperature may be lowered, the salt concentration increased, or the reactant concentrations decreased. At elevated salt, dissociation of DNA-protein complexes is often faster; dissociation reactions are usually first-order processes and, hence, are independent of the concentration of complexes.

a. Association Reactions

To study association processes, DNA and protein are mixed at time zero, at convenient concentrations in the chosen buffer. At various times, the reaction is stopped by the addition of a suitable agent to sequester unreacted protein. (Note: in all applications to date, the extent of reaction has been determined by detection of DNA. In the unlikely event that detection of protein, and not of DNA, is necessary, then the reaction should be quenched with excess unlabeled protein.) If the DNA is radioactive, excess unlabeled DNA is always appropriate for quenching the reaction. If the DNA is to be detected by ethidium bromide staining, one may use an excess of DNA of different length (but still containing the specific protein binding site under study). For bacterial RNA polymerases, heparin is often used as a quenching agent because it binds strongly to RNA polymerase and removes it from further reaction. Nonspecific polymerase-DNA complexes are not stable in the presence of heparin, but the transcriptionally competent complexes of interest are usually not affected. Any effects of heparin on transcription complexes can be assayed and corrected for by comparing data from experiments at different concentrations of heparin. It is important to be sure that no significant dissociation of complexes formed occurs after quenching, during loading of the sample. Therefore, at the beginning of a series of experiments, the dissociation rate of the complexes should be estimated as described below.

When the reaction has been terminated, the mixture is placed in the well of the gel and electrophoresis is begun. As indicated above, provided no changes in the solution occur during the startup period of the electrophoresis, the levels of free DNA and complex in the sample are measured by the gel retardation assay. Thus, either directly or by difference (input DNA − free DNA), one derives data for the concentration of complexes as a function of time, precisely the information needed to evaluate the rate constant for the association process. As in all such experiments, analysis of the results is simplified if one reactant is in excess (tenfold or more). For example, for a reaction $R + P \rightarrow RP$ with forward rate constant k_a, when R is in excess and $[R] \sim R_T$, a plot of $\ln(1 - F)$ vs. time, where F is the fraction of DNA in complex, has a slope of $-k_aR_T$. For the formation of heparin-stable transcription complexes involving bacterial RNA polymerases and promoter-containing DNA fragments, a more realistic reaction mechanism is[22]

$$R + P \overset{K_B}{\longleftrightarrow} (RP)_c \overset{k_2}{\longrightarrow} (RP)_o$$

Here, $(RP)_c$ represents a "closed" complex intermediate in which the enzyme has recognized the promoter region in a double-helical configuration, while $(RP)_o$ denotes "open" complexes in which RNA polymerase is tightly bound, poised to initiate transcription in the presence of nucleoside triphosphates. In many instances, the closed complex reaction can be characterized by an equilibrium constant K_B, while the isomerization to form the open complex is essentially an irreversible process governed by the forward rate constant, k_2.

The gel mobility assay provides data on the rate of $(RP)_o$ formation. McClure[23] has shown that for the above mechanism, where RNA polymerase is in excess, $\tau = 1/K_B R_T k_2 + 1/k_2$. Here, a plot of $\ln(1 - F)$ vs. t, where F is the fraction of complexes formed at time t, yields $1/\tau$. Thus, τ is measured at several input concentrations of RNA polymerase and a plot of τ vs. $1/R_T$ is made; the slope of this plot is $1/K_B k_2$, the intercept is $1/k_2$, and both parameters can be determined.

b. Dissociation Reactions

Measurement of dissociation rates is done by allowing complexes to form, then adding an excess of a sequestering agent at time zero and assaying for the level of complexes remaining as the reaction progresses. Thus, after the reaction has proceeded for a given amount of time, one tenth volume of "loading buffer" containing dyes plus glycerol or Ficoll is added as described above and the sample is immediately placed in the gel well and electrophoresed. Reaction buffer conditions and temperature may be adjusted to achieve a convenient off-rate, though a wide range of times is acceptable. For the most accurate results, only a minimum amount of dissociation should occur in the 60 s or so needed for free DNA to enter the gel at the start of electrophoresis. The sequestering agent can be DNA containing the protein binding site of interest. This DNA may be unlabeled if radioactive DNA is used to form the complexes; if ethidium bromide is used to detect bands, it can be of a different and distinguishable length, compared to DNA in the complexes. Reactions involving bacterial RNA polymerases may be quenched with heparin.

Results from a typical dissociation experiment are shown in Figure 2. Here, complexes of RNA polymerase and ^{32}P-labeled DNA containing the *lac* UV5 promoter were formed after incubation for 10 min at 37°C in a buffer containing 20 m*M* Tris, pH 8.0 at 22°C, 100 m*M* KCl, 10 m*M* $MgCl_2$, 0.1 m*M* EDTA, and 5% glycerol. The protein is in slight excess, to be sure that virtually all the DNA is initially in complex. Samples 25 μl in volume were placed in separate plastic reaction tubes and maintained at 37°C. Dissociation was initiated by the addition of heparin to 100 μg/ml. After a given time, 2.5 μl of 10X loading buffer was added with gentle mixing and electrophoresis begun in a 5% polyacrylamide gel in a vertical Watson apparatus at 15 V/cm. The gel was run at room temperature with a cooling fan to prevent undue heating. Figure 2A follows the dissociation reaction over a 90-min period, showing the decrease in the level of complexes concomitant with an increase in the amount of free DNA. The dots above and to the left of the lanes were produced from radioactive India ink spots and allow the gel to be aligned with the autoradiogram so that the bands can be cut out of the wet gel and assayed for radioactivity. The fraction of complex (F) at each time (t) is then calculated and a graph is made of $\ln F$ vs. t (Figure 2B). The slope of this plot is the negative of the first-order rate constant which characterizes the dissociation process.

It is evident from Figure 2 that electrophoresis of the various reaction mixtures was started at different times. This was done to increase the accuracy of the data. It is not desirable that all samples be available for loading at precisely the same time, since several minutes could elapse between loading of the first and last samples, during which time

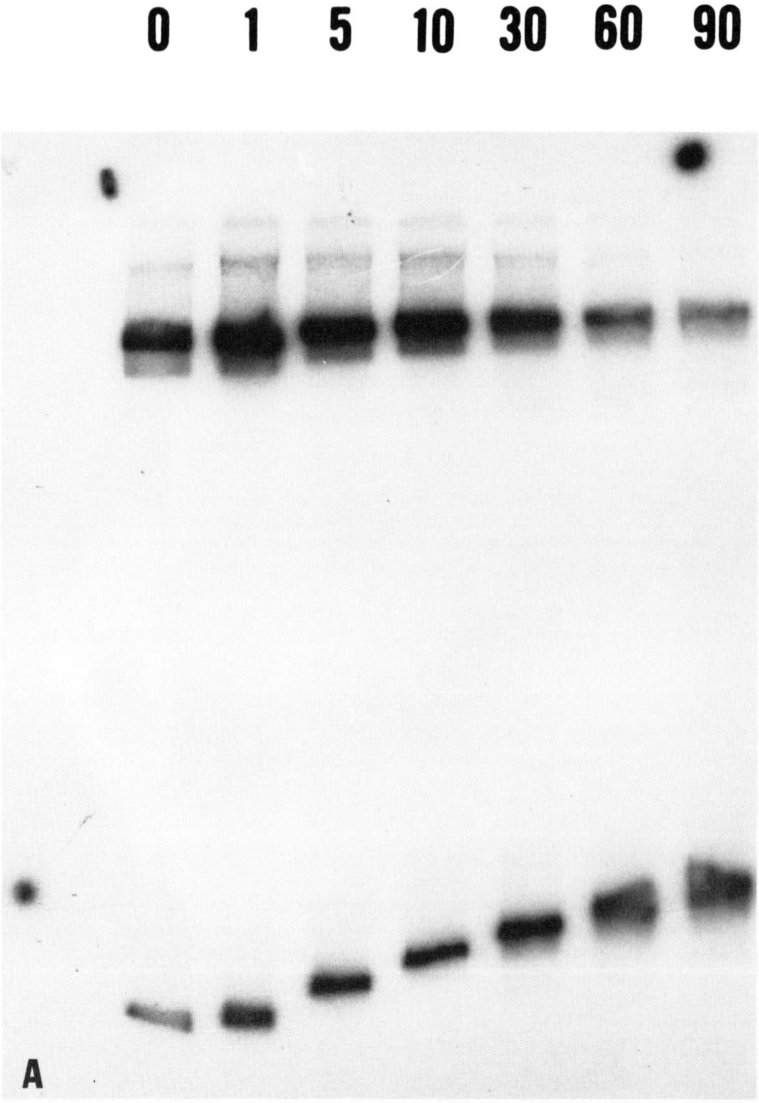

FIGURE 2. Data from a dissociation rate experiment, *E. coli* RNA polymerase-*lac* UV5 promoter complexes. (A) Autoradiogram of gel showing diminution of complexes with time (in minutes) after heparin was added to initiate the dissociation process; (B) graph of data as log (percent complexes remaining) vs. time. The half-life for the dissociation under the conditions of this experiment (see text) is approximately 60 min. (Figure courtesy of Donald Lorimer, University of Illinois — Chicago.)

significant additional dissociation might occur. Thus, the experimental protocol was such that the samples were ready to be loaded at staggered time intervals — here, every 2 min. In this experiment, dissociation was initiated by adding heparin first to the solution which would provide the "90-min" time point, then 28 min later the "60-min" reaction was started, 28 min later the "30-min" reaction was begun, 18 min later the "10-min" reaction started, etc. An alternative approach would be to add heparin to a single solution containing preformed complexes and to remove aliquots for electrophoresis at various times. In this case, solutions corresponding to the longest time points will be run on the gel for the least amount of time. As long as adequate separation of free DNA and complexes is obtained by electrophoresis, this protocol works well also.

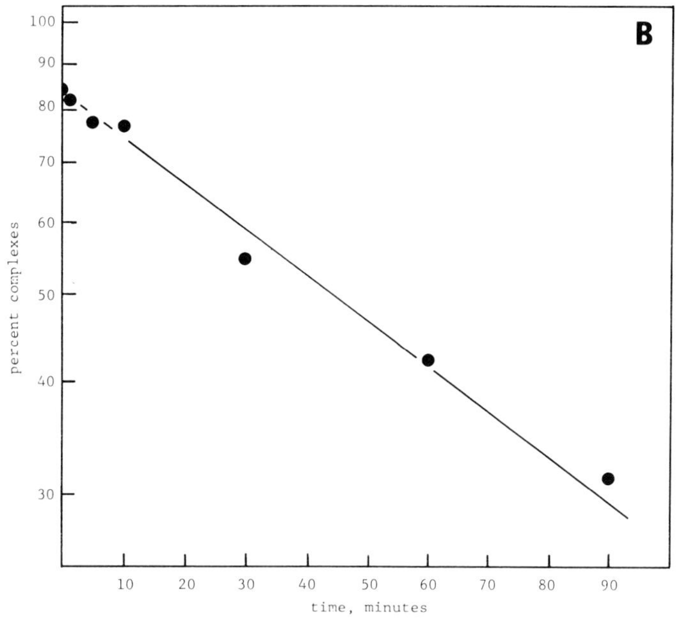

FIGURE 2B.

The same considerations apply to measurements of association rates. The reaction can be initiated by mixing components in a single tube; then portions are removed at appropriate times, quenched, and electrophoresed. Alternatively, a series of tubes can be used in which the reaction is started at various times such that each reaction will have proceeded for its desired period at about the same time; all can then be quenched, loaded onto the gel, and electrophoresed.

B. QUALITATIVE APPLICATIONS

1. Searching for DNA Sites which Bind to a Specific Protein

It may be of interest to ascertain at which sites a protein binds in a segment of DNA containing on the order of a few thousand base pairs. For example, one might wish to locate promoter regions in a plasmid. To do this, the DNA is digested with restriction enzyme(s) into smaller fragments ranging in length from, say, 50 to 800 base pairs (bp), incubated with the protein, and assayed by the gel retardation technique. If RNA polymerase is involved, one looks for heparin-insensitive complexes. Figure 3 shows an experiment using the CAP protein and a *Hae*III digest of a 789-bp sequence containing the wild-type *lac* promoter. In this ethidium bromide-stained gel, lane 1 shows the digested DNA alone, lane 2 shows the results of electrophoresing a mixture of CAP plus this DNA digest, and lane 3 contains molecular weight standards. Comparing lanes 1 and 2 reveals that under these conditions CAP interacts with only one fragment in the digest, which happens to be the 203-bp sequence known to contain the natural CAP binding site. The band of complexes is the uppermost band in lane 2, and is not present in lane 1.

These types of experiments were previously done using filter assays;[24] complexes retained on the filter were washed off, and the DNA was recovered and assayed by gel electrophoresis. The gel mobility method is preferred to filter assays here, since it is more direct and easier to use.

2. Searching for DNA-Binding Proteins in Crude Extracts

This is currently one of the most widespread applications of the gel shift assay. The

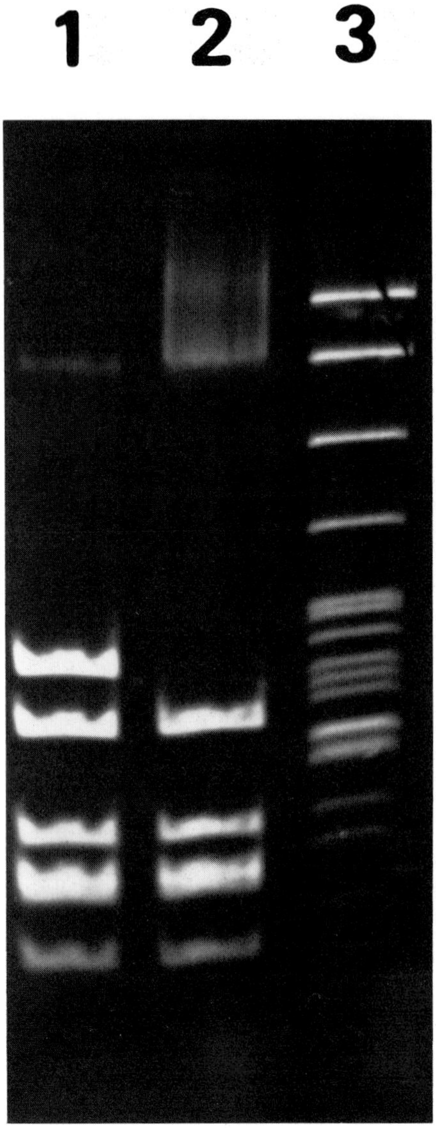

FIGURE 3. Detecting DNA fragments having a specific binding site. Lane 1: *Hae*III digest of 789-bp *E. coli* sequence containing the wild-type lactose promoter region; lane 2: same solution as in lane 1, except that CAP protein was added; lane 3: molecular weight standards. Reaction conditions: [CAP] = 7.6×10^{-7} *M*, [DNA] = 0.9×10^{-7} *M* promoters; buffer was 10 m*M* Tris, 100 m*M* KCl, 8 m*M* $MgCl_2$, 0.1 m*M* EDTA, and 20 μ*M* cAMP, pH 8.0, at 22°C. The electrophoresis buffer was TBE plus 20 μ*M* cAMP.

experiment is, in principle, quite simple. A radioactive fragment containing the DNA sequence of interest is added to a cell extract and the mixture is electrophoresed; a protein specifically bound to the labeled fragment will alter its mobility, which can be easily detected by autoradiography.[8,25,26]

Precautions must be taken to avoid problems from other proteins in the extract which bind nonspecifically to DNA sequences. In fact, if the labeled fragment alone is added to the extract, one finds that a large fraction of the radioactive material will barely enter the

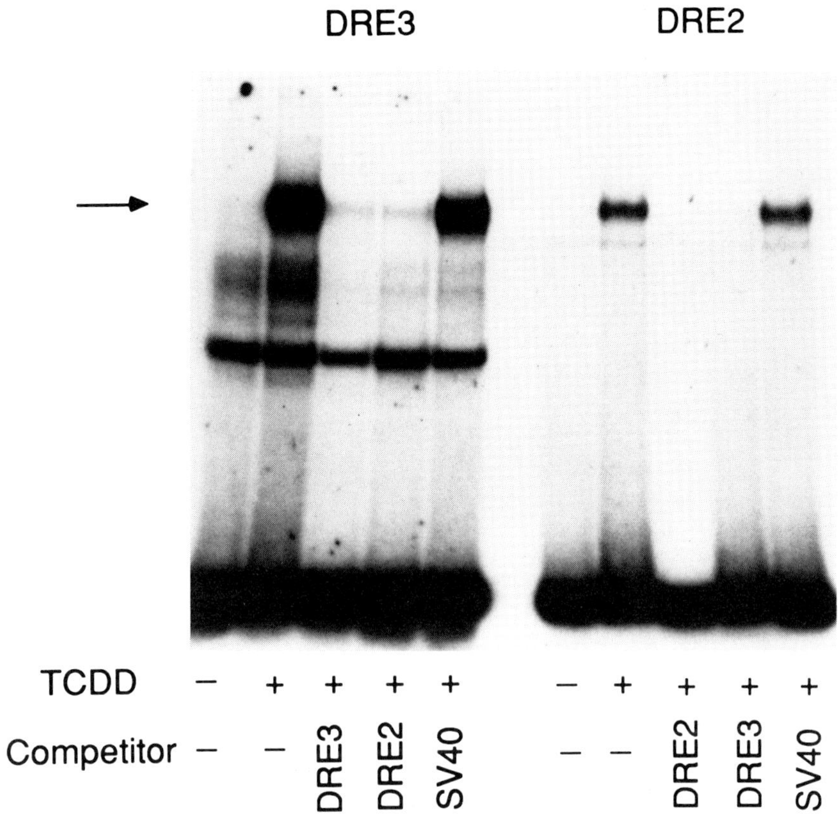

FIGURE 4. Gel retardation assay to identify DNA-binding protein(s) in crude extracts. Experimental details are given in the text. Arrow indicates band of inducer-dependent complexes of DNA fragments and the receptor protein. This band is seen for DNA fragments containing the enhancer-like sequences DRE3 or DRE2, and can be competed away with an excess of either unlabeled fragment. Complexes are found only in the presence of the inducer, TCDD. A heterologous enhancer sequence from SV40 does not interact with this protein. The additional band in the left lanes is likely due to the binding of another protein, present at relatively high levels in the extract, to the DRE3 fragment. (Figure courtesy of Dr. Michael Denison, Michigan State University, East Lansing.)

gel, since many protein molecules are bound to each DNA fragment. This situation can be remedied by adding an unlabeled competitor DNA to ''soak up'' the nonspecifically binding protein. The competitor should not contain the DNA sequence of interest. Calf thymus DNA, *E. coli* DNA, or poly d(I-C) are often used. Titration experiments show that as more nonspecific DNA is added to the crude extract, the amount of radioactivity remaining at the top of the gel diminishes and bands of free DNA and of complex appear. The optimum level of competitor is reached when the radioactivity in the free DNA and complex bands is maximized; further addition of nonspecific DNA can lead to competition by this DNA for the protein under study and, hence, would be undesirable.

Data from a gel mobility assay using crude extracts are presented in Figure 4. The particular experiment depicted involves DNA fragments containing ''dioxin-responsive elements'' (DRE2, DRE3) from the flanking region of the cytochrome P-450 gene in mouse hepatoma cells.[27] These sequences, which have enhancer-like properties, stimulate transcription from the cytochrome P-450 promoter and from other promoters as well. The gel retardation assay was used to show that nuclear extracts contain a protein which interacts with the DREs in an inducer-dependent manner, i.e., a receptor protein in complex with

TCDD (2,3,7,8-tetrachlorodibenzy-*p*-dioxin) binds specifically to the DRE sequences, while the receptor alone shows little, if any, DNA binding.

Thus, in Figure 4, we see that there is no receptor binding to the 33-bp DRE3 fragment in the absence of TCDD, but that complexes are formed when the inducer is added. The radioactive complexes are not seen if a 50-fold excess of cold DRE3 is added or if the same amount of an unlabeled DRE2-containing fragment is present. Similarly, the DRE2 fragment (30 bp long) forms specific complexes with the receptor protein in the extract if TCDD is present; these radioactive complexes are, of course, competed out when excess cold DRE2 or DRE3 is added. Interestingly, DNA fragments containing the SV40 enhancer sequence do not compete with the DREs for receptor-TCDD complexes.

Some technical points regarding Figure 4 are of interest. This experiment was done using a 4% polyacrylamide gel; the electrophoresis buffer was 6.7 m*M* Tris-HCl, pH 7.5, 3.3 m*M* Na acetate, and 1 m*M* EDTA. The binding reaction contained, in 25 μl, 5 μl of nuclear cell extract (3 μg total protein), 0.1 to 0.3 ng (10,000 cpm) of ^{32}P-labeled fragment, in 25 m*M* HEPES, pH 7.5, 0.7 m*M* EDTA, 1 m*M* dithiothreitol, 70 m*M* KCl, and 10% glycerol. Poly d(I-C) was at 1.8 μg/25 μl, this having been determined to be optimal. The bands at the bottom of the gel are the free DNA fragments; specific complexes are indicated by the arrow. The extra band in the left lanes apparently is due to another protein in the extract which binds to the DRE3 fragment. Other data (not shown) indicate that the mobility of the DNA-protein complex is about the same for fragments ranging in length from 20 to 80 bp, so it seems that the degree of electrophoretic retardation in this system reflects primarily the size of the protein in the complex. In general, one should use as short a fragment as possible, to reduce potential problems from nonspecific binding to sequences on the fragment which flank the binding site under study.

A modification of this approach might find use in situations where mutant DNA regions have been fully or partially characterized. For example, certain strains of *E. coli* are known to have mutations near a *glg* promoter sequence; these strains are constitutive for expression of the operon, perhaps because a repressor protein will no longer bind there. However, no such repressor has yet been found. One could search for the putative repressor in cell extracts by probing with a radioactive fragment containing the wild-type (presumed) operator region and using an excess of unlabeled *mutant* DNA to remove interference from all other proteins; the mutant DNA would not compete at all for the repressor protein being sought.

When the gel mobility assay is used with crude extracts, additional analysis of the DNA-protein complexes is possible. For instance, the use of gel retardation in conjunction with DNAse I footprinting is described below. In principle, one might try to characterize the bound proteins by cutting out the complex band and doing SDS-polyacrylamide gel electrophoresis in a second dimension. This will be complicated by the likely presence of other proteins near the radioactive band in the gel, which have migrated to that position either on their own or bound to, say, poly d(I-C). The degree of protein contamination can be reduced if partial fractionation of the cell extract is feasible prior to the gel shift assay or if the nonspecific DNA used is of large size so that it does not move as far into the gel as do the complexes of interest. In such situations, it may be possible to assess the molecular weight(s) of the polypeptide chain(s) in the complex. This would involve comparing an SDS gel run with the band of complexes with a control SDS gel in which the corresponding section of a lane run without added specific DNA fragments is electrophoresed. The latter would contain peptides which are present in the retardation gel near the complexes, but which are not part of the complexes. By difference, then, those peptides involved in specific binding to the DNA of interest would be revealed.

3. Use of Gel Shift Assays in "Footprinting" Experiments

A footprinting experiment reveals which DNA phosphodiester linkages are protected from attack by enzymes or chemicals when the DNA sequence is bound to a protein molecule. Complexes containing DNA which has been radiolabeled at one end are exposed to reagents such as DNAse I,[28] hydroxyl radicals,[29] or 1,10-phenanthroline-copper ion,[30] each of which has a ''nuclease'' activity that results in cleavage of the DNA chain. Conditions are selected so that only limited digestion occurs (one nick or less per DNA molecule). Proteins are removed by phenol extraction, and the DNA is denatured and run on a ''sequencing gel'', a denaturing gel containing 7 *M* urea and TBE buffer. Comparing the pattern obtained from a solution containing DNA only with that from a mixture of complexes reveals a clear (band-free) region at the position where protein was bound, and the DNA was not accessible to the reagent. In this way, the location of the protein binding site is ascertained.

Footprinting techniques were developed for components in free solution; however, their sensitivity is considerably enhanced when coupled with gel retardation.[31] For example, if a significant fraction of DNA is not in complex, cleavage of the unbound DNA leads to a high background level, and the protected region will become obscured. Likewise, if more than one type of complex exists in the solution, the protected sequences may not be discernible from the denaturing gel. These problems can be avoided if, following reaction of the complexes and a free DNA control with the ''nuclease'', the solutions are electrophoresed in the usual gel mobility assay. It has been found that single nicks generally do not affect the mobility of the DNA-protein complexes; hence, complexes will be separated from any unbound DNA. The bands are located by autoradiography and DNA eluted from the gel. Elution can be effected by the crush-and-soak method (in which the section of gel containing the band is physically mashed and the nucleic acid allowed to diffuse out into surrounding buffer) or by use of commercially available electroelution devices. The purified DNA is then run on a denaturing gel, to compare bands derived from the isolated complexes and from the free DNA. If more than one complex band is present, each may be analyzed in this way and the binding sites compared.

With certain reagents, it may be desirable to perform the reactions on DNA in complexes *in the gel*. For example, footprinting with 1,10-phenanthroline-copper ion has been done by allowing the reagent to diffuse into the polyacrylamide matrix following electrophoresis.[32] The bands may first be located by autoradiography of the wet gel, then cut out and immersed in a solution containing 1,10-phenanthroline and cupric sulfate. The reaction is quenched with 2,9-dimethyl-1,10-phenanthroline/ammonium acetate/EDTA, and the DNA is eluted, denatured, and analyzed on a sequencing gel. Alternatively, because these reagents are inexpensive, the entire gel can be immersed in the 1,10-phenanthroline-copper solution for the desired time and the reaction quenched, after which the bands are identified by autoradiography, eluted, etc. The latter procedure avoids possible problems from protein denaturation during autoradiography, which would be a concern if a long exposure were needed for weak bands. In their recent studies using this approach, Kuwabara and Sigman[32] found two interesting results. First, the reactivity of DNA with the reagent shows the same sequence-dependent pattern in solution and in the gel, thus confirming the notion that the gel matrix does not alter the DNA conformation. Second, the patterns obtained in solution and in the gel were different for RNA polymerase-*lac* UV5 promoter complexes. The differences were in the region -3 to -6 (relative to the transcriptional start point, $+1$), which is known to be in a single-stranded configuration when RNA polymerase is bound in an open complex. The discrepancy has been attributed to the lack of Mg^{2+} in the gel buffer. Footprints derived from bands in gel slices which had been incubated with added $MgCl_2$ corresponded precisely to those seen in transcription mixtures in solution. Thus, the data obtained in the absence of Mg^{2+} may reflect the properties of a closed complex. The authors note that isolation of DNA-protein complexes in a solid matrix may provide an important new way to study structural changes caused by diffusible ligands.

4. Methylation Protection Experiments

Proteins bound tightly to specific DNA regions can also protect the DNA from reaction with other chemical reagents, such as dimethylsulfate (DMS).[33] When G or A residues are methylated, they become sensitive to phosphodiester bond cleavage in the reactions of Maxam-Gilbert chemistry.[34] Thus, a methylation protection experiment would involve radiolabeling DNA at one end, exposing DNA-protein complexes (and free DNA as a control) to DMS to a limited extent, purifying the DNA and subjecting it to Maxam-Gilbert cleavage, and then comparing the banding patterns on a denaturing gel. The gel shift assay can be used here just as it is in nuclease protection experiments, to isolate bands of modified complexes, thus reducing background levels and permitting analysis of solutions in which more than one type of complex may be present. Likewise, the methylation reaction can be done in the gel slice by diffusing in DMS, quenching at the desired time, eluting the DNA, etc. Straney and Crothers[35] have done this with RNA polymerase-*lac* UV5 complexes. They report that two complex bands, each apparently containing one molecule of protein and one molecule of DNA, are seen under certain conditions. These complexes, called O_u and O_l, differ in their rates of methylation. Comparing data obtained when the protection experiment was done in the gel with that from free solution indicates that the different rates of methylation for the complexes do not result from their interconversion in the solution.

5. Exonuclease III Digestion Experiments

Location of protein binding sites using this technique involves incubating the complexes with ExoIII until the single-stranded nuclease is inhibited in its digestion of one strand of the DNA when it encounters the protein of interest tightly bound at its specific site.[36] Since the ExoIII digestion will substantially change the mobility and character of the complexes, it is not done prior to electrophoresis. It has proved possible to perform ExoIII digestions by diffusing the enzyme into gel slices. Control of reaction time is achieved by soaking the gel in ExoIII-containing solution, adding $MgCl_2$ to start the digestion process, and quenching at the desired time with EDTA.[35] This approach allowed evaluation of alternative hypotheses regarding the properties of the O_u and O_l complexes mentioned above.

6. Transcription in Gel Slices

In a previous section of this chapter, it was indicated that transcription reactions will occur in the gel if nucleotide are layered atop the gel lane and allowed to catch up with RNA polymerase-promoter complexes during electrophoresis.[16] It is also possible to diffuse nucleotides into gel slices and to elute and analyze the RNA made therein. This approach was quite useful in characterizing short initiation products made from the O_u and O_l RNA polymerase-*lac* UV5 "open" complexes.[37] Significant differences in the relative amounts of specific oligonucleotides made from these complexes were reported.

7. Interference Experiments

In interference experiments, the DNA is modified before the protein of interest is added; determination of those modifications that inhibit protein binding indicates points of close approach (contacts) between the protein and the DNA.[38] The reagents commonly used are dimethyl sulfate, which methylates G (and A) and ethylnitrosylurea, which ethylates the phosphodiester backbone. In this approach, separation of complexes from free DNA is mandatory. Nitrocellulose filters have been used in the past, but gel retardation is now the method of choice. After limited modification of DNA which has been radiolabeled at one end, and incubation with the protein under study, the solution is electrophoresed and the complex and DNA bands are located. The bands are excised, the DNA is eluted and any protein is removed, and the DNA is subjected to Maxam-Gilbert chemistry to cleave the backbone at the points of modification, followed by electrophoresis on a sequencing gel.

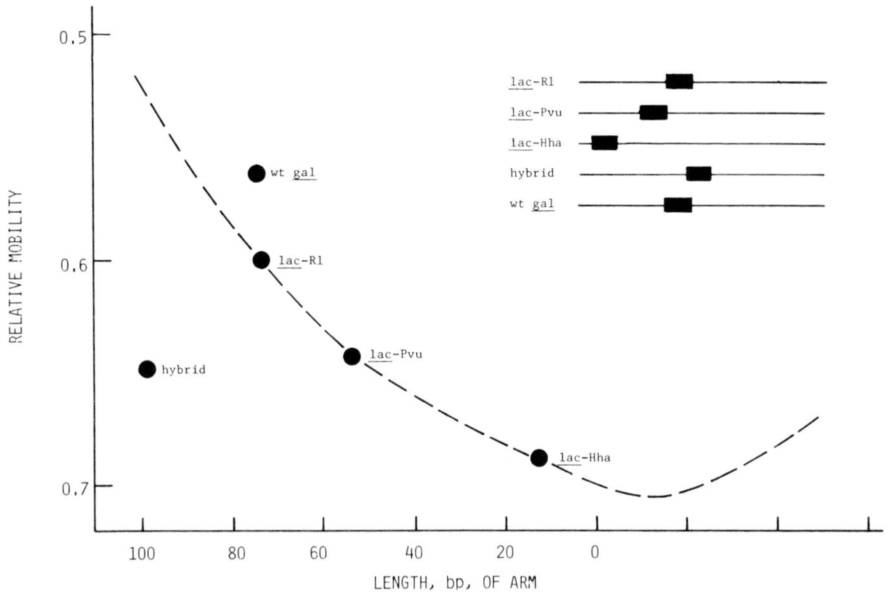

FIGURE 5. Relative mobilities of CAP-DNA complexes. All DNA molecules are very nearly the same length; the location of the protein binding site within a fragment is indicated by the darkened box. The three *lac* fragments were generated by digestion with the restriction enzymes *Eco*RI, *Pvu*II, or *Hha*I of cloned multiple tandem sequences containing the wild-type *lac* promoter. The "hybrid" fragment is described in the text. The length of the "arm" is the distance from the left end of the fragment to the center of the CAP binding site. Note that lower mobilities correspond to higher points on the vertical axis.

Comparison of the patterns for the complexes and unbound DNA reveals which modifications prevent the protein from binding and, hence, are contact points.

8. DNA Bending

DNA fragments which are curved or bent move anomalously in gels relative to fragments of the same length which are straight.[39] The curvature can be intrinsic to the DNA itself or can be induced by binding of a protein. The CAP-cAMP complex causes substantial bending when bound to the lactose promoter.[9] This bending can be observed by electrophoresis of "circularly permuted" DNA fragments generated by digestion with different restriction enzymes of a cloned DNA sequence containing multiple tandem copies of the *lac* promoter region. When the CAP binding site is in the center of the fragment, the average end-to-end distance of the bent DNA is at a minimum and the gel mobility is lowest. The mobility is higher when the protein is bound near either end of the DNA molecule. The types of information which can be gleaned from gel retardation assays of bent DNA molecules are indicated in Figure 5, which shows the relative mobilities of CAP complexes with several DNA fragments, each 200 to 208 bp in length, containing protein binding sites as shown by the dark boxes. All the DNA molecules have the same electrophoretic mobility when free, indicating that they have little, if any, intrinsic curvature. Results for CAP-*lac* fragment complexes are consistent with those of Wu and Crothers;[9] the mobility is least when the bend is in the center of the fragment. The CAP binding site at −30 to −50 of the wild-type galactose promoter is at virtually the same position in the "wt *gal*" fragment as is the CAP site in the "*lac* R1" fragment, yet the CAP-wt *gal* complex moves with a significantly reduced relative mobility. We conclude that *gal* DNA in complex with CAP is even more bent than is *lac* DNA in such complexes. The "hybrid" fragment is composed of *lac* sequences to −49 and *gal* sequences downstream of −43, connected by a 6-bp *Eco*RI linker. This fragment is known to bind CAP at the *lac* site (−50 to −70), but we see from its mobility that the DNA is not bent as much as it would be in CAP-wild-type *lac* DNA complexes. Thus, although the CAP protein protects its binding site from DNAse I, etc. only in the −50 to −70 region of the hybrid DNA, additional sequences outside this region are needed to achieve the full protein-induced bending effect, a finding also reported by others.[40] Clearly, a remarkable amount of information can be deduced from gel shift assays.

9. Other Applications

A few recent uses to which the gel assay has been put are presented here, to indicate its versatility.

Hope and Struhl[41] made use of the fact that the DNA binding of the eukaryotic transcriptional activator protein GCN4, as well as of certain truncated GCN4 derivatives, can be analyzed by gel retardation. Complexes of DNA fragments containing promoters of yeast amino acid biosynthetic genes with the smaller polypeptides have higher electrophoretic mobilities than does the complex of DNA with native GCN4. These authors mixed wild-type GCN4 and a smaller derivative protein with a promoter DNA fragment and observed an additional band of intermediate mobility which they interpret to contain a mixed protein molecule comprised of one full-length and one shortened polypeptide. This suggests that functional GCN4 protein binds to promoter regions as a dimer.

Heumann and colleagues[42] have studied DNA fragments containing the early T7 promoter A1 (at which *E. coli* RNA polymerase initiates). They report several types of complexes — holoenzyme-DNA, core enzyme-DNA, etc. — which migrate differently in gel electrophoresis. The gel retardation assay should prove useful in further analysis of the interactions of RNA polymerase with T7 A1 promoter constructs designed to produce arrested elongation complexes. It will be of interest to compare the electrophoretic mobilities of open preinitiation complexes, core-polymerase complexes, and elongating structures from which the sigma factor of RNA polymerase has dissociated.

Recently, a correlation has been presented that relates the mobility of a protein-DNA complex (relative to that of free DNA) and the molecular weight of the bound protein.[43] So as long as the migration of the complex is not anomalous because of DNA bending, it may be possible to estimate the molecular weight of the protein from the degree of retardation. This may permit one to deduce whether a particular protein is binding as a monomer or a dimer, etc.

The gel shift assay has been used as the final step in purification of a specific DNA-binding protein from chicken liver.[44] Here, the protein was prepurified using a variety of columns, then incubated with a 100-bp DNA fragment containing two consensus recognition sequences, and electrophoresed in a polyacrylamide gel. The preparative gel was run using about 350-fold more protein and DNA, compared to what is used in analytical gel assays. This step produced a further threefold purification, although it was not reported that the protein was recovered from the gel. Elution of complexes and subsequent isolation of the protein may be feasible; agarose gels may be of use in such purification procedures.

IV. CONCLUSIONS

The gel mobility shift assay has been put to a wide range of uses in the past several years, and new applications are appearing with regularity. For the future, a number of trends seems likely. The value of the gel assay for studying eukaryotic transcription factors will continue to grow, heightened by the fact that, in many cases, significant numbers of amino acids can be removed from these proteins without seriously impacting on DNA binding properties. One expects that even more unique ways will be developed to use the assay to characterize DNA-binding proteins from eukaryotic cells. As more proteins are purified, from bacterial systems as well, their interactions with DNA will be studied in detail at the thermodynamic and kinetics level. The movement of nucleic acid-protein complexes (and of DNA alone) through gels is not well understood in theoretical terms, and work in this area will proceed. In some instances, we can make use of unexpected shifts in gel mobilities without knowing the precise origin of the shifts, but additional insight into the complexes will come if we know why they migrate as they do. Finally, there is every reason to believe that the gel shift approach will be used for studies of RNA-protein binding and will reveal as much about such systems as it has about DNA-protein interactions.

REFERENCES

1. **Garner, M. M. and Revzin, A.,** A gel electrophoresis method for quantifying the binding of proteins to specific DNA regions: application to components of the *Escherichia coli* lactose operon regulatory system, *Nucleic Acids Res.*, 9, 3047, 1981.
2. **Fried, M. and Crothers, D. M.,** Equilibria and kinetics of *lac* repressor-operator interactions by polyacrylamide gel electrophoresis, *Nucleic Acids Res.*, 9, 6505, 1981.
3. **Garner, M. M. and Revzin, A.,** The use of gel electrophoresis to detect and study nucleic acid-protein interactions, *Trends Biochem. Sci.*, 11, 395, 1986.
4. **Crothers, D. M.,** Gel electrophoresis of protein-DNA complexes, *Nature*, 325, 464, 1987.
5. **Fried, M. G. and Crothers, D. M.,** Equilibrium studies of the cyclic AMP receptor protein-DNA interaction, *J. Mol. Biol.*, 172, 241, 1984.
6. **Shanblatt, S. H. and Revzin, A.,** Kinetics of RNA polymerase-promoter complex formation: effects of nonspecific DNA-protein interactions, *Nucleic Acids Res.*, 12, 5287, 1984.
7. **Fried, M. G. and Crothers, D. M.,** Kinetics and mechanism in the reaction of gene regulatory proteins with DNA, *J. Mol. Biol.*, 172, 263, 1984.
8. **Singh, H., Sen, R., Baltimore, D., and Sharp, P. A.,** A nuclear factor that binds to a conserved sequence motif in transcriptional control elements of immunoglobulin genes, *Nature*, 319, 154, 1986.
9. **Wu, H.-M. and Crothers, D. M.,** The locus of sequence-directed and protein-induced DNA bending, *Nature*, 308, 509, 1984.
10. **McEntee, K., Weinstock, G. M., and Lehman, I. R.,** Binding of the *rec*A protein of *Escherichia coli* to single- and double-stranded DNA, *J. Biol. Chem.*, 256, 8835, 1981.
11. **Green, P. J., Kay, S. A., and Chua, N.-H.,** Sequence-specific interactions of a pea nuclear factor with light-responsive elements of the rbcS-3A gene, *EMBO J.*, 6, 2543, 1987.
12. **Hendrickson, W. and Schleif, R. F.,** Regulation of the *Escherichia coli* L-arabinose operon studied by gel electrophoresis DNA binding assay, *J. Mol. Biol.*, 178, 611, 1984.
13. **Cooper, T. G.,** *The Tools of Biochemistry*, John Wiley & Sons, New York, 1977, 198.
14. **Cann, J. R.,** Multiple electrophoretic zones arising from protein-buffer interaction, *Biochemistry*, 5, 1108, 1966.
15. **Carey, J.,** Gel retardation at low pH resolves *trp* repressor-DNA complexes for quantitative study, *Proc. Natl. Acad. Sci. U.S.A.*, 85, 975, 1988.
16. **Revzin, A., Ceglarek, J. A., and Garner, M. M.,** Comparison of nucleic acid-protein interactions in solution and in polyacrylamide gels, *Anal. Biochem.*, 153, 172, 1986.
17. **Maniatis, T., Fritsch, E., and Sambrook, J.,** *Molecular Cloning — A Laboratory Manual*, Cold Spring Harbor Laboratory Press, Cold Spring Harbor, NY, 1982, 122.
18. **Fried, M. G. and Crothers, D. M.,** CAP and RNA polymerase interactions with the *lac* promoter: binding stoichiometry and long range effects, *Nucleic Acids Res.*, 11, 141, 1983.
19. **Garner, M. M. and Revzin, A.,** Stoichiometry of catabolite activator protein/adenosine cyclic 3′,5′-monophosphate interactions at the *lac* promoter of *Escherichia coli*, *Biochemistry*, 21, 6032, 1982.
20. **Jones, O. W. and Berg, P.,** Studies on the binding of RNA polymerase to polynucleotides, *J. Mol. Biol.*, 22, 199, 1966.
21. **Kolb, A., Spassky, A., Chapon, C., Blazy, B., and Buc, H.,** On the different binding affinities of CRP at the *lac, gal,* and *mal*T promoter regions, *Nucleic Acids Res.*, 11, 7833, 1983.
22. **Chamberlin, M. J.,** The selectivity of transcription, *Annu. Rev. Biochem.*, 43, 721, 1974.
23. **McClure, W. R.,** Rate limiting steps in RNA chain initiation, *Proc. Natl. Acad. Sci. U.S.A.*, 77, 5634, 1980.

24. **Reznikoff, W. S.,** Formation of RNA polymerase-*lac* promoter open complex, in *RNA Polymerase,* Losick, R. and Chamberlin, M., Eds., Cold Spring Harbor Laboratory Press, Cold Spring Harbor, NY, 1976, 441.
25. **Strauss, F. and Varshafsky, A.,** A protein binds to a satellite DNA repeat at three specific sites that would be brought into mutual proximity by DNA folding in the nucleosome, *Cell,* 37, 889, 1984.
26. **Schneider, R., Gander, I., Müller, U., Mertz, R., and Winnacker, E. L.,** A sensitive and rapid gel retention assay for nuclear factor I and other DNA-binding proteins in crude nuclear extracts, *Nucleic Acids Res.,* 14, 1303, 1986.
27. **Denison, M. S., Fisher, J. M., and Whitlock, J. P., Jr.,** Inducible, receptor-dependent protein-DNA interactions at a dioxin-responsive transcriptional enhancer, *Proc. Natl. Acad. Sci. U.S.A.,* 85, 2528, 1988.
28. **Schmitz, A. and Galas, D. G.,** The interaction of RNA polymerase and *lac* repressor with the *lac* control region, *Nucleic Acids Res.,* 6, 111, 1979.
29. **Tullius, T. D. and Dombroski, B. A.,** Hydroxyl radical "footprinting": high-resolution information about DNA-protein contacts and application to lambda repressor and *cro* protein, *Proc. Natl. Acad. Sci. U.S.A.,* 83, 5469, 1986.
30. **Spassky, A., Rimsky, S., Buc, H., and Busby, S.,** Correlation between the conformation of *Escherichia coli* −10 hexamer sequences and promoter strength: use of orthophenanthroline cuprous complex as a structural index, *EMBO J.,* 7, 1871, 1988.
31. **Hendrickson, W.,** Protein-DNA interactions studied by the gel electrophoresis DNA binding assay, *BioTechniques,* 3, 198, 1985.
32. **Kuwabara, M. D. and Sigman, D. S.,** Footprinting DNA-protein complexes in situ following gel retardation assays using 1,10-phenanthroline-copper ion: *Escherichia coli* RNA polymerase-*lac* promoter complexes, *Biochemistry,* 26, 7234, 1987.
33. **Humayun, Z., Kleid, D., and Ptashne, M.,** Sites of contact between λ operators and λ repressor, *Nucleic Acids Res.,* 4, 1595, 1977.
34. **Maxam, A. M. and Gilbert, W.,** Sequencing end-labeled DNA with base-specific cleavages, *Methods Enzymol.,* 65, 499, 1980.
35. **Straney, D. C. and Crothers, D. M.,** Comparison of the open complexes formed by RNA polymerase at the *Escherichia coli lac* UV5 promoter, *J. Mol. Biol.,* 193, 279, 1987.
36. **Shalloway, D., Kleinberger, T., and Livingston, D. M.,** Mapping of SV40 replication origin region binding sites for the SV40 T antigen by protection against exonuclease III digestion, *Cell,* 20, 411, 1980.
37. **Straney, D. C. and Crothers, D. M.,** Intermediates in transcription initiation from the *E. coli lac* UV5 promoter, *Cell,* 43, 449, 1985.
38. **Sienbenlist, U., Simpson, R. B., and Gilbert, W.,** *E. coli* RNA polymerase interacts homologously with two different promoters, *Cell,* 20, 269, 1980.
39. **Hagerman, P. J.,** Sequence-directed curvature of DNA, *Nature,* 321, 449, 1986.
40. **Liu-Johnson, H.-N., Gartenberg, M. R., and Crothers, D. M.,** The DNA binding domain and bending angle of *E. coli* CAP protein, *Cell,* 47, 995, 1986.
41. **Hope, I. A. and Struhl, K.,** GCN4, a eukaryotic transcriptional activator protein, binds as a dimer to target DNA, *EMBO J.,* 6, 2781, 1987.
42. **Heumann, H., Metzger, W., and Niehörster, M.,** Visualization of intermediary transcription states in the complex between *Escherichia coli* DNA-dependent RNA polymerases and a promoter-carrying DNA fragment using the gel retardation method, *Eur. J. Biochem.,* 158, 575, 1986.
43. **Bading, H.,** Determination of the molecular weight of DNA-bound protein(s) responsible for gel electrophoretic mobility shift of linear DNA fragments exemplified with purified viral *myb* protein, *Nucleic Acids Res.,* 16, 5241, 1988.
44. **Rupp, R. A. W. and Sippel, A. E.,** Chicken liver TGGCA protein purified by preparative mobility shift electrophoresis (PMSE) shows a 36.8 to 29.8 kd microheterogeniety, *Nucleic Acids Res.,* 15, 9707, 1987.

Chapter 11

DNA FOOTPRINTING TECHNIQUES: APPLICATIONS TO EUKARYOTIC NUCLEAR PROTEINS

Urs Pauli, Kenneth Wright, André van Wijnen, Gary Stein, and Janet Stein

TABLE OF CONTENTS

I. INTRODUCTION

Over the past several years, significant progress has been made in our understanding of eukaryotic gene transcription and mechanisms operative in transcriptional control. To a large extent, such advances have been a direct result of approaches that have been developed for identification and characterization of protein-DNA interactions in regulatory regions of genes associated with a broad spectrum of biological processes.

Particularly important in establishing the functional significance of observed protein-DNA binding in promoter regulatory elements is the balanced application of both *in vitro* and *in vivo* approaches. *In vitro* methods such as footprinting,[1] electrophoretic mobility shift analysis[2,3] (see Chapter 10, this volume), and DNA affinity chromatography[4,5] are necessary for fractionation of nuclear proteins and assessment of binding with sequence specificity. However, identification of protein-DNA interactions in intact cells is necessary to establish the biological relevance of protein binding observed in cell-free systems. Thus, in this chapter we will describe approaches to examining the interactions of nuclear proteins with gene regulatory sequences both *in vivo* and under cell-free conditions; we will use our work with the cell cycle-regulated human histone genes as examples to illustrate methods for establishing protein-DNA relationships in promoter sequences.

The expression of the cell cycle-regulated histone genes is temporally as well as functionally coupled with DNA replication, with control mediated at both the transcriptional and posttranscriptional levels (for a review, see Reference 6). The histone genes we have examined are transcribed throughout the cell cycle, exhibiting a three- to fivefold enhancement in transcription during the initial part of the S phase.[7,8] This increase in transcription is accompanied by modifications in chromatin structure,[9,10] including changes in DNAse I and S1 nuclease sensitivity; therefore, these genes have been well characterized with respect to structure, organization, and regulation of expression.

II. *IN VIVO* ASSAYS FOR PROTEIN-DNA INTERACTIONS

A. NATIVE GENOMIC BLOTTING

1. Background and General Considerations

For more than a decade, nucleases (DNAse I and II, micrococcal nuclease, S1 nuclease, exonuclease III, etc.) have been used to analyze the accessibility of DNA sequences in chromatin and thereby to provide valuable insight into the basic elements of chromatin structure. Early characterization of chromatin structure led to the discovery of several aspects of nucleoprotein organization, among them the basic nucleosomal repeat,[11] nuclease-sensitive and hypersensitive sites,[12] and regions of otherwise altered structure, e.g., regions where phased nucleosomes can be detected.[13] Comparisons of the nuclease sensitivity of different active and inactive genes have provided the basis for correlation of the transcriptional state of a gene and its structure. Accessibility of chromatin to nucleases can be functionally interpreted; for example, it has been well established that active genes, potentially active genes, and genes that were active at some point display DNAse I sensitive sites (for reviews, see References 14 and 15).

The resolution of such nuclease-accessible sites by gel electrophoresis has played a crucial role in the identification of different types of DNAse I hypersensitive sequences and the relationship of these to regulatory function. With the native genomic blotting technique described here, resolution is increased by at least a factor of three over the traditional Southern blot assay using agarose gels, and a double-stranded DNA region of 400 to 700 bp can be screened in one autoradiograph. In addition, high-resolution analysis of different DNAse I hypersensitive sites indicates that this method is equally suited to the identification of protein-DNA contacts as regions protected from nuclease digestion (see below). Therefore, the native genomic blotting procedure allows a high-resolution analysis of DNAse I hypersensitive sites as well as the identification of contacts between proteins and DNA in functional regions of a gene on the same autoradiograph. This procedure also provides data that reflect structural requirements for the generation of hypersensitive sites and, hence, may provide insight into the functional significance of these sites in active and potentially active genes.[16] It should also be noted that various nucleases generate chromatin digestion patterns that can be distinguished by the native genomic blotting technique and thereby reflect the heterogeneity of elements that contribute to chromatin structure. However, native genomic blotting does not provide the single nucleotide resolution analysis of protein-DNA interactions that can be obtained with genomic sequencing (see Section II.B, this chapter).

The native genomic blotting method is schematically outlined in Figure 1 and an autoradiograph of a "blot" in which the H4 histone gene promoter was examined by the procedure is shown in Figure 2 (for a complete analysis, see References 17 and 18).

2. Method

a. DNAse I Digestion of Isolated Nuclei

Isolated nuclei are resuspended in reticulocyte sample buffer (RSB) (10 m*M* Tris-HCl, pH 7.4, 10 m*M* NaCl, and 3 m*M* $MgCl_2$) at 6×10^8 nuclei per milliliter for digestion with DNAse I. Nuclei can be washed at this point with increasing concentrations of NaCl to determine the dissociation point of DNA-binding proteins (see Figure 2C), as described by Pauli et al.[19] Digestion of nuclei is carried out at concentrations ranging from 0.1 to 45 μg/ml (as DNA) for 10 min at 37°C. The reaction is terminated by the addition of Na_2EDTA, pH 8 (25 m*M* final concentration), SDS (0.5%), and Proteinase K (Boehringer-Mannheim) (0.5 mg/ml), and the sample is incubated at 37°C overnight. The DNA is extracted with phenol/chloroform/isoamyl alcohol (25:24:1), precipitated with ethanol, and stored at 4°C.

b. Digestion of the DNA and Separation on Native Polyacrylamide Gels

DNA (12 to 15 μg) is digested with the appropriate restriction endonuclease to carry out indirect labeling. The genomic DNA is then ethanol precipitated and lyophilized in a SpeedVac (Savant) to dryness without further purification. When there is a high salt concentration in the digestion buffer, only a limited amount of salt should be added prior to ethanol precipitation, since excess salt may prohibit DNA from entering the gel. DNA pellets are dissolved in a small amount of Tris-EDTA buffer (TE) (10 m*M* Tris-HCl, pH 7.5, and

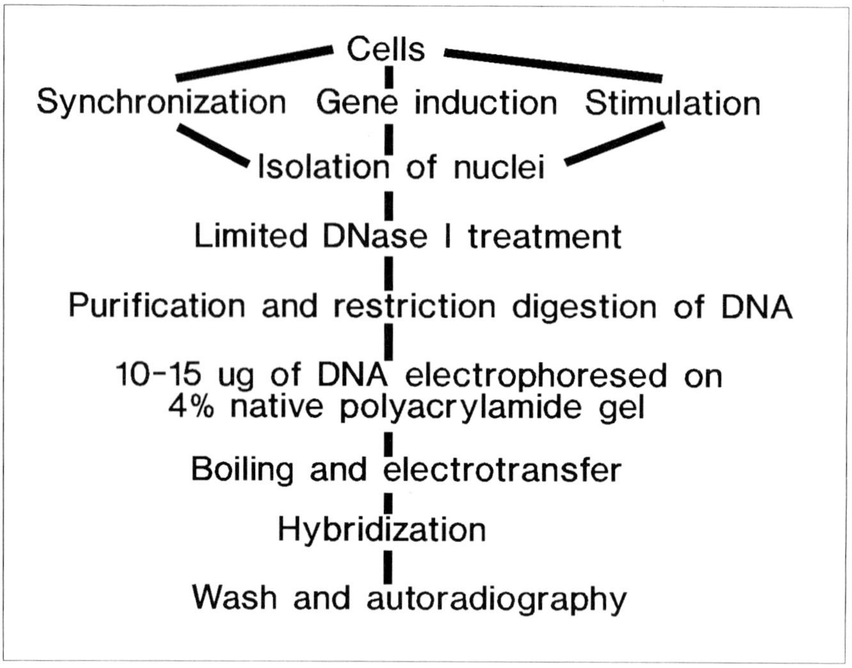

FIGURE 1. Flow chart diagram of the native genomic blotting technique.

1 m*M* Na_2EDTA, pH 7.5) and 0.1 volume of dye mix is added (30% Ficoll, 10 m*M* Tris-HCl, 20 m*M* EDTA, 0.5% bromophenol blue, and 0.5% xylene cyanol, pH 7.5).

A native 4% polyacrylamide gel (acrylamide to bis-acrylamide = 40:1) in 50 m*M* Tris-borate-EDTA (TBE) (50 m*M* Tris-HCl, 50 m*M* boric acid, and 2 m*M* Na_2EDTA, pH 8.3) is prepared for running the DNA samples. We usually use a gel of 14 cm × 16 cm × 1.5 mm. The gel should be polymerized for at least 4 h or overnight because the ratio of acrylamide to bis-acrylamide is high. Preelectrophoresis should be carried out for 1 to 2 h at 100 V.

After the DNA is loaded, electrophoresis is continued for 3 to 4 h at 100 V or until the bromophenol blue dye is located 2 to 3 cm from the bottom of the gel. The glass plates are then separated carefully and the gel is sealed in an air-tight plastic bag (Chiswick Trading, Sudbury, MA) capable of withstanding temperatures of more than 90°C. It is important that buffer is excluded from the bag because this will interfere with the DNA denaturing step and, in addition, result in loss of DNA from the gel. The plastic bag is incubated in a covered, boiling water bath for 10 min and then quenched immediately in ice water to prevent renaturation. The gel is removed from the bag and the DNA is electroblotted onto a nylon membrane, essentially as in the genomic sequencing procedure (see below). Transfer time is 90 min because of the thickness of the gel.

c. *Preparation of DNA Probes*

The DNA probes used for hybridization of the native genomic blot can be prepared essentially according to the method of Church and Gilbert[20] (and described in the next section of this chapter). The primary requirement for the probe is that it should be of high specific activity (>5 × 10^8 cpm/μg). In our hands, a scaled-down version of the Church and Gilbert reaction is effective (40 μCi instead of 250 μCi of radiolabeled dNTP; phage DNA and primer scaled down proportionally; other components are the same as in the full reaction described below), and it is not necessary to purify the labeled fragment from a gel — it is simply chromatographed on a BioGel A15m column and heated prior to hybridization. We have also successfully used a random primer extended probe.[21] All the probes gave identical results; however, with the scaled-down M13 probe, the exposure time of the autoradiogram had to be lengthened by a factor of two, and with the random oligo primer extended probe, by a factor of four.

d. Hybridization, Washing, and Autoradiography of the Filters

Hybridization of the "blots" is performed essentially according to Church and Gilbert.[20] The hybridization buffer contains 1% crystalline-grade bovine serum albumin, 1 m*M* Na_2EDTA, 0.5 *M* $NaHPO_4$, pH 7.2, and 7% SDS [1 *M* $NaHPO_4$ (pH 7.2) stock is composed of 134 g of $Na_2HPO_4 \cdot 7H_2O$ and 4 ml of 85% H_3PO_4 per liter]. Carrier DNA, RNA, pyrophosphate, dextran sulfate, Ficoll, and polyvinylpyrrolidone are not required in these hybridizations. Church and Gilbert report that the bovine serum albumin has only a twofold effect on background. The filter is prehybridized for 5 min at 65°C, and the buffer is removed and the probe added. After hybridization at 65°C for 8 to 24 h, the filter is washed two times in 0.5% fraction V-grade bovine serum albumin, 1 m*M* Na_2EDTA, 40 m*M* $NaHPO_4$, pH 7.2, and 5% SDS for 5 min, and then eight times in 1 m*M* Na_2EDTA, 40 m*M* $NaHPO_4$, pH 7.2, and 1% SDS for 5 min each. The wash solutions are at 65°C prior to use, but the agitations are done at room temperature. Hybridizations can be done in plastic bags or in plastic rotating cylinders in an oven (e.g., Bachofer, Germany). For washing, Church and Gilbert[20] recommend 1 l of volume per 30- × 40-cm filter. After washing, filters are placed between two sheets of Whatman 541 paper to remove excessive liquid and then placed, still damp, between two layers of Saran Wrap™ (see below). Autoradiography is done at −70°C with an intensifying screen for 1 to 4 d.

B. GENOMIC SEQUENCING

1. General Considerations and Background

In contrast to the native genomic blotting method, the genomic sequencing technique reveals protein-DNA interactions at single nucleotide resolution. However, the region that can be analyzed in one gel is limited to about 100 to 200 bases, compared with the 400 to 700 bp that are resolvable in one gel by the native genomic blotting method.

With genomic sequencing, the contacts of proteins with guanine residues of the DNA are reflected by an altered reactivity of the guanine residue to chemical cleavage (i.e., by dimethylsulfate [DMS]). Either cleavage is decreased (protection) or increased (enhancement) when the result is compared with deproteinized DNA that has been chemically cleaved in the same way. Interestingly, interactions of nucleosomes are not detected with this method, suggesting that a strong protein-G residue interaction, indicative of a sequence-specific binding site, is necessary to generate DMS protection or DMS enhancement.

One of the principal advantages of the genomic sequencing method is that intact cells are reacted with DMS for only a brief interval. Thus, the protein-DNA interactions revealed are an accurate reflection of the *in vivo* situation and meaningful from a biological standpoint. Figure 3 schematically illustrates the genomic sequencing procedure. Shown in Figure 4 are the results of a genomic sequencing analysis of the promoter region of a cell cycle-regulated human H4 histone gene;[19] indicated are (1) the probes used, (2) results of the upper-strand analysis of guanine interactions with protein(s), and (3) the interpretation of these results. We have used the same human H4 histone gene to illustrate identification and characterization of protein-DNA interactions by both the native genomic blotting and genomic sequencing methods; thus, the results shown in Figures 2 and 4 are directly comparable.

In the following protocol for genomic sequencing, we have described only DMS treatment of cells. When genomic sequencing is carried out with DNA from DNAse I-digested nuclei, the chemical cleavage step is omitted and the same amount of DNA is fractionated electrophoretically in sequencing gels.

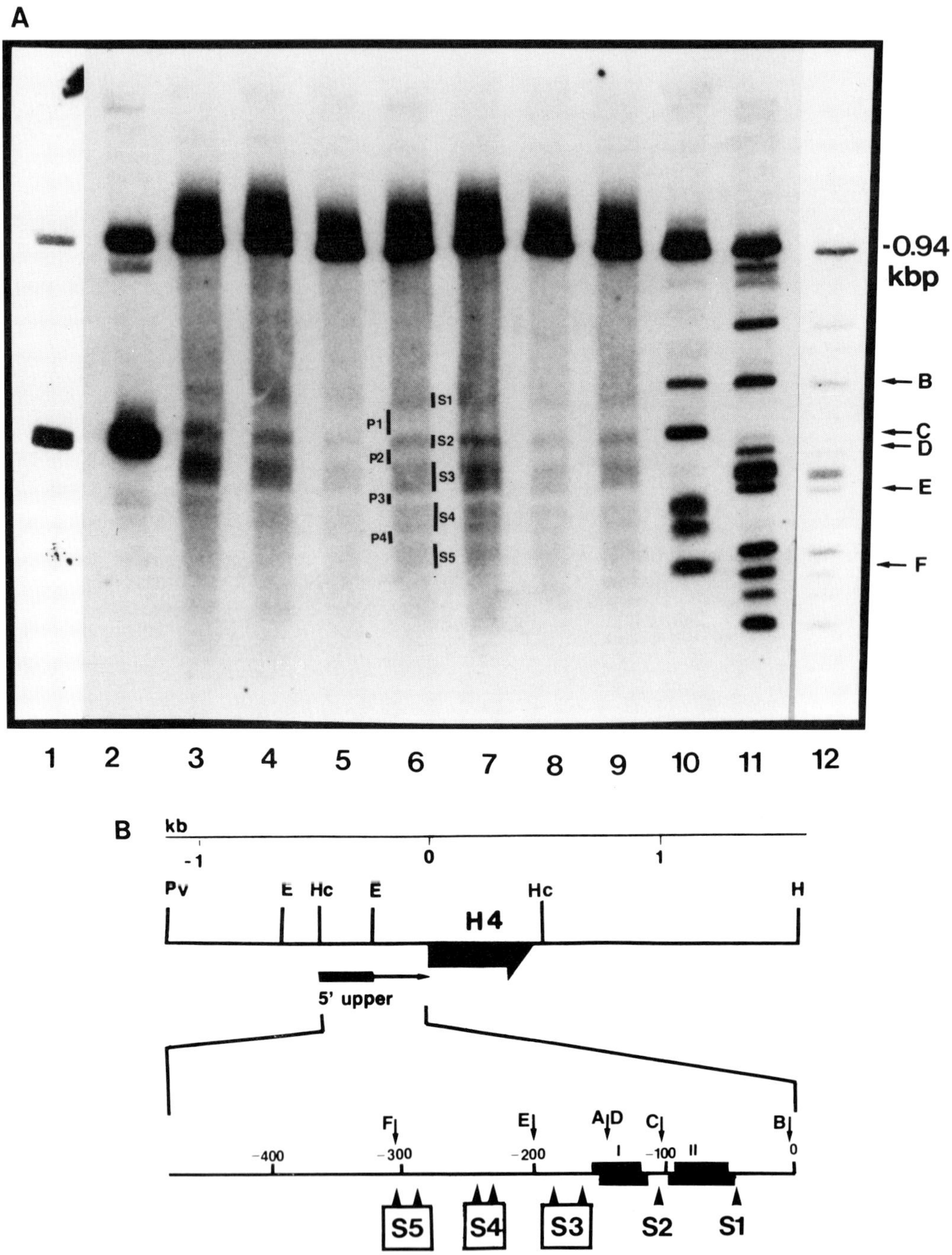

FIGURE 2. High-resolution analysis of a histone H4 gene with the native genomic blotting method. (A) Lanes 2 and 10: HincII-digested HeLa DNA from nuclei treated with TaqI and MspI, respectively. Lane 1: lighter exposure of lane 2; Lanes 3 to 9: HincII-digested DNA from nuclei of thymidine-blocked cells, early S phase (1 and 3 h after release of the thymidine block), late S phase (6 and 8 h after release), and mitosis/G1 phase (10 and 12 h after release), which were treated with DNAse I concentrations of 5 to 10 μg/ml; Lane 11: deproteinized HeLa DNA digested with HincII and partially digested with MnlI; Lane 12: lighter exposure of lane 11. The number on the right indicates the size of the HincII fragment of this histone H4 gene. Letters indicate cleavage of the DNA by restriction enzymes at known sequences: A = TaqI site at −132 bp, B = MnlI site +3 bp, C = MspI site at −97 bp, D = MnlI site at −136 bp, E = MnlI site at −198 bp, and F = MspI site at −307 bp, with respect to the ATG codon. The areas of hypersensitivity are labeled S1 to S5 and the areas of protection are labeled P1 to P4. (Figure 2 is continued on next page.)

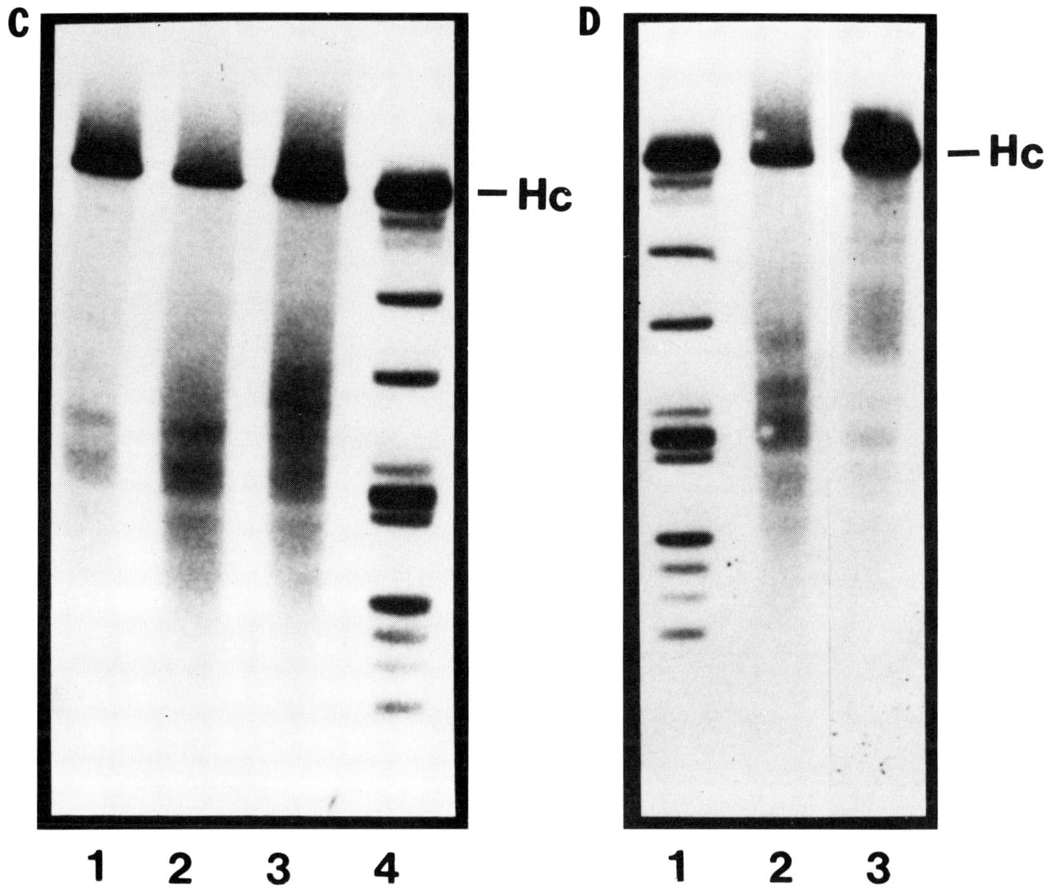

FIGURE 2C and D.

FIGURE 2B, C, and D. (B) Top: restriction map of the human H4 histone gene. Black arrow: direction and extent of transcription. Restriction sites: Pv = PvuII; E = EcoRI; Hc = HincII; H = HindIII. Bottom: expansion of the promoter region from the HincII site at −472 bp to the ATG codon. Open boxes: regions of DNAse I hypersensitivity; arrows point to darker bands seen in these regions. Letters A to F are the same as in (A) and are displayed at their position in the sequence. Solid bars show the protection from DNAse I digestions on the upper strand (above line) and lower strand (below line), as determined by the genomic sequencing technique. (A and B from Pauli, U., Chrysogelos, S., Stein, J., and Stein, G., *Proc. Natl. Acad. Sci. U.S.A.*, 85, 16, 1988.) (C) The DNAse I pattern of salt-washed nuclei. Lanes 1 to 3: DNA from nuclei which were washed with 0.05, 0.18, and 0.44 *M* NaCl, respectively, prior to DNAse I digestion with 2 μg/ml DNAse I; Lane 4: partial MnlI digest; Hc = HincII. (D) Comparison of S1 nuclease and DNAse I pattern. Lane 1: partial MnlI digest; Lane 2: DNA from nuclei digested with 7.5 μg/ml of DNAse I; Lane 3: DNA from nuclei digested with 5000 U/ml of S1 nuclease; Hc = HincII. (C and D from Pauli, U., Chrysogelos, S., Stein, J., and Stein, G., *BioTechniques*, 6(2), 142, 1988. With permission.)

2. Method

a. DMS Treatment of Cells and Isolation of High Molecular Weight DNA

The starting material should be 10^7 to 10^8 cells that are grown either as monolayers or in suspension. As few as 10^7 monolayer cells will yield sufficient DNA to run one or two genomic sequencing gels.

Suspension cells are harvested by centrifugation at the normal incubation temperature and at the lowest possible speed to prevent formation of a compact pellet that could interfere with the uniform accessibility of DMS to all cells. It is advisable to perform preliminary experiments to establish optimal centrifugation and DMS reaction conditions. Cells are suspended in 3 to 5 ml of fresh medium or PBS + 60 m*M* Tris (per liter: 8 g of NaCl,

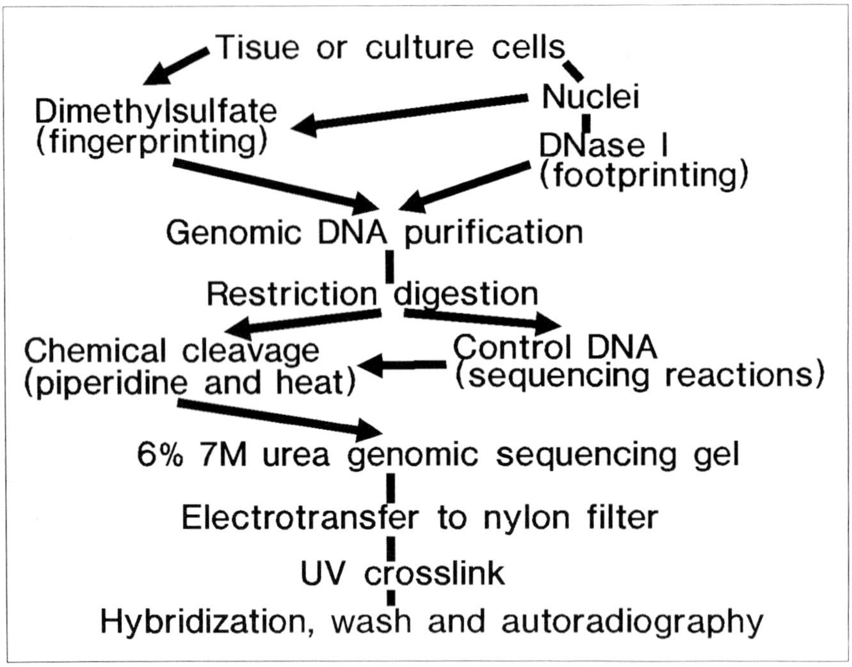

FIGURE 3. Flow chart diagram of the genomic sequencing method.

0.2 g of KH_2PO_4, 0.2 g of KCl, 1.15 g of Na_2HPO_4, and 60 m*M* Tris-HCl, pH 7.5). Tris is needed to buffer the sulfuric acid that is generated when dimethylsulfate (DMS) is degraded in an aqueous solution.

Because DMS is anhydrous, it will pick up moisture from the air and therefore should be stored under nitrogen. *Caution:* DMS is very toxic and mutagenic! Working in a fume hood and wearing gloves are always necessary. Pipette tips and solutions that were in contact with DMS should be discarded into a container filled with water. (DMS degrades in water after a few days.) All steps should be carried out using newly made solutions or reagents that have been used only for genomic sequencing. Contamination with plasmid DNA can ruin experiments, since 1 ng of plasmid is 1000 times more DNA than the amount of a single copy gene in 40 μg of DNA!

DMS (Eastman Kodak) is added to the cell suspension to a final concentration of 0.5% (50-ml plastic conical tubes are well suited for this reaction) and the suspension is gently shaken for 60 s at the same temperature at which the cells were grown. It may be necessary to vary the reaction time for different cell types within the range of 30 to 90 s. Conditions for DMS treatment may also need to be modified, based on the age of the reagent.

The DMS reaction is quenched by the addition of at least a tenfold excess of ice-cold PBS + Tris and the cells are collected by gentle centrifugation. Cells are resuspended in a small volume of ice-cold PBS + Tris and, after resuspension, the tube is filled with ice-cold PBS + Tris. It is important not to vortex the cells, to avoid damage that might affect the yield of DNA. When suspending the pellet, it is important to remove all the remaining DMS to avoid any further reaction with DNA. The washing step with PBS + Tris is repeated twice, and the final cell pellet is dissolved in 3 to 5 ml of 100 m*M* EDTA and 60 m*M* Tris-HCl, pH 7.5. SDS is added to a final concentration of 0.1% to lyse the cells and Proteinase K (Boehringer-Mannheim) is added to a concentration of 300 to 500 μg/ml. The viscous solution is left overnight at room temperature. (Viscosity is an indication that sufficient cells were present to yield the required amount of DNA.) Incubation for several hours at 37°C will result in degradation of large DNA.

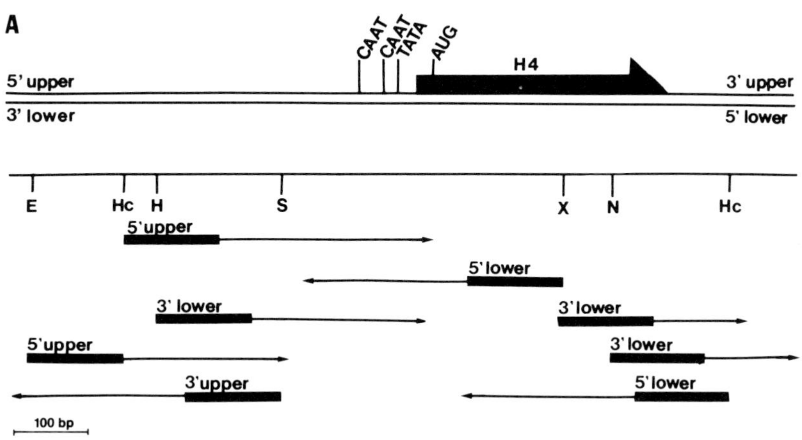

FIGURE 4. Complete analysis of a histone H4 gene for protein-DNA contacts. (A) Top: H4 histone gene, including flanking regions. Arrow: direction and extent of transcription. Important elements in the promoter and the start of the coding region are shown. Restriction sites: EcoRI; Hc = HincII; H = HindIII; S = SmaI; X = XmnI; and N = NcoI. Bottom: M13 probes used. One end of the probe always corresponds to an indicated restriction site. (B) Genomic DNA was restricted with HincII; 220 bases were run off the bottom of the gel. The hybridization probe was the 5′ Hinc upper probe. The guanine pattern: Lane 1 = control plasmid; Lanes 2 to 4 = DNA from cells of early S phase, mid S phase, and mitotic/G1 phase, respectively. Arrows on the left indicate protected guanine residues. (C) DNAse I digestion pattern from nuclei. Lanes 1 to 4: DNA from nuclei of early S phase, mid S phase, mitotic/G1 phase, and subsequent S-phase cells, respectively; Lane 5: DNA from nuclei of exponentially growing cells; Lanes 6 to 8: deproteinized HeLa DNA digested with 100 to 400 pg of DNAse I per μg of DNA. Arrowheads on the right indicate protection from DNAse I digestion. (D) Summary of protein-DNA interactions for the H4 histone gene. The coding sequence is enclosed in solid lines. Protected guanine residues are marked with open circles. Sites I and II are marked by solid horizontal lines. The start of transcription is marked by a dotted line. (From Pauli, U., Chrysogelos, S., Stein, G., Stein, J., and Nick, H., *Science,* 236, 1308, 1987. Copyright 1987 by the AAAS. With permission.)

For monolayer cultures grown on 100-mm cell culture plates, the procedure is modified as follows. Growth medium is carefully aspirated with a pipette. Cells are washed with PBS + Tris, which is carefully pipetted onto the plate to prevent dislodging the cells. DMS (0.5% final concentration in 10 ml of PBS + 60 m*M* Tris) is poured onto the cell layer and the plate is gently shaken for 1 min. Then the reaction solution is aspirated with a pipette and ice-cold PBS is added to quench the reaction. (Careful pouring of the cold solution is necessary to avoid dislodging cells from the plate). This washing solution is left on the plate for about 5 to 10 s, the washing step is repeated twice, and then 3 to 5 ml of the EDTA, Tris, SDS, Proteinase K solution is added to lyse the cells, as described above. The lysate is placed on a rocking table until the next plate is ready for lysis. At that time, the lysate is scraped from the plate with a rubber policeman and poured onto the newly prepared plate for further lysis. This procedure prevents an extensive increase of the volume.

A phenol/chloroform extraction (1:1) is performed at room temperature by gently rocking the tube back and forth. Phenol is redistilled and equilibrated with TE (10 m*M* Tris-HCl, pH 7.5, and 0.1 m*M* EDTA). Chloroform contains isoamylalcohol at a 24:1 ratio. If the DNA is too viscous to be removed from the phenol/chloroform layer, additional TE is added to the aqueous phase and the extraction is repeated. A wide-bore pipette tip (cut off) should be used to aspirate the DNA to minimize shearing. Traces of phenol are extracted from the aqueous phase with a large volume of chloroform.

Sodium acetate, pH 7.5, is added to the DNA solution to a final concentration of

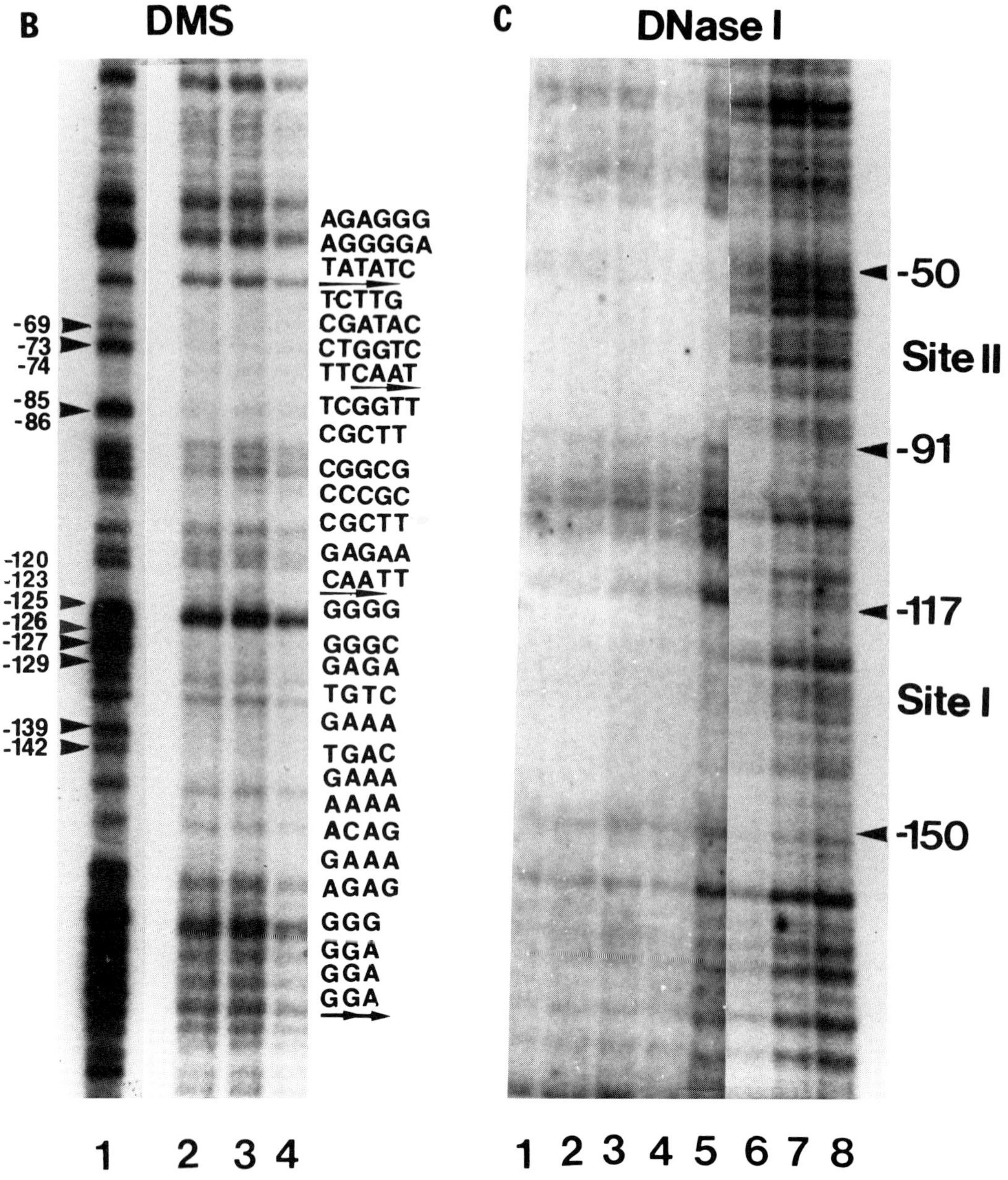

FIGURE 4B and C.

300 m*M* and the aqueous phase is overlayered with cold ethanol (−20°C). A siliconized glass rod is used to spool the DNA. DNA should be visible and gelatinous. If the DNA is white, it is too dehydrated and might not readily solubilize in TE. The DNA is redissolved by pipetting 3 to 5 ml of TE over the spooled DNA on the rod. Contaminating RNA which is also spooled is digested with RNAse A and RNAse T1 (DNAses in the RNAse solutions are heat inactivated by boiling for 10 min) at 37°C for 2 to 4 h. The DNA is again subjected to phenol/chloroform extraction, chloroform extraction, and spooling (see above). RNA must be removed for accurate OD_{260} readings for quantitation of DNA to be loaded on the genomic sequencing gel. After this second spooling, the DNA is dialyzed overnight at 4°C against a large volume of TE (4 l of TE per two samples).

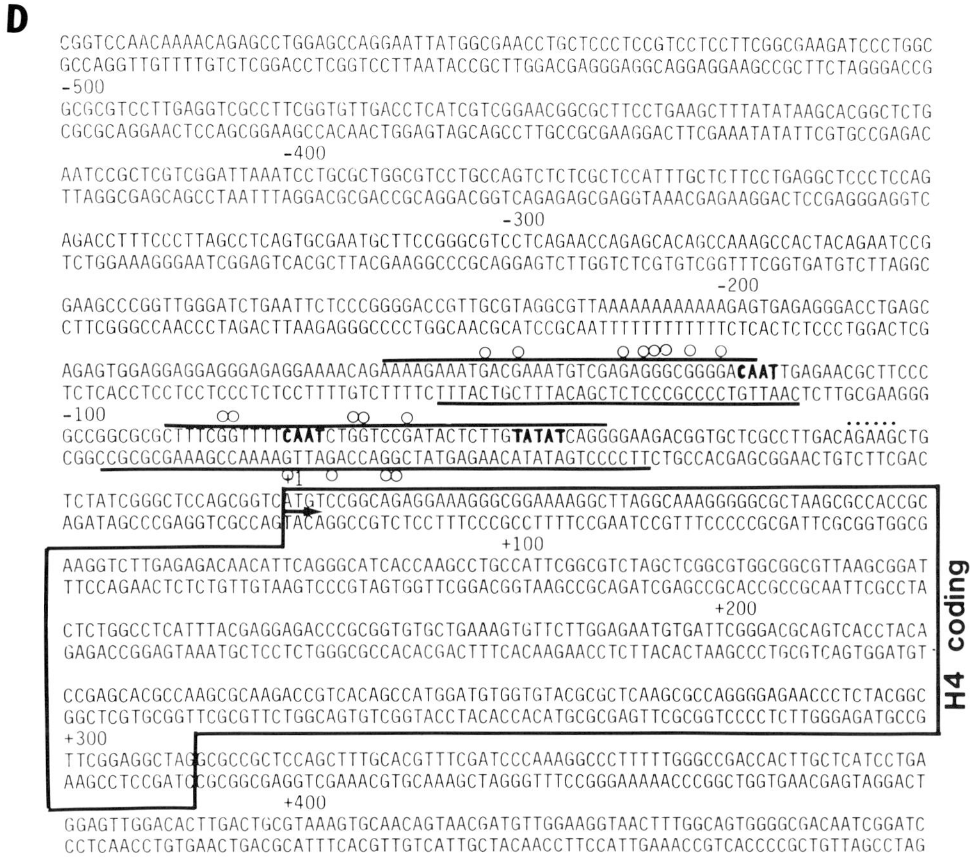

FIGURE 4D.

b. DNA Concentration and Restriction Digestion

Isolated DNA is concentrated by 1-butanol extraction to a volume of approximately 1 ml and is subsequently extracted with ether to remove all traces of butanol. The ether is then removed by placing the sample in a 50°C water bath or by passing a stream of nitrogen over the solution. The DNA can be stored for more than 1 year at 4°C in a microcentrifuge tube without significant degradation.

For analysis of the DMS reaction, 40 to 60 μg of genomic DNA are aliquoted and purged with nitrogen for 1 min to remove any traces of ether before digestion. Normally, restriction enzyme digestions are performed in a 100-μl volume using the buffers recommended by the supplier. If the DNA is too dilute, it should be further concentrated by butanol extraction. Ethanol precipitation should be avoided because large DNA fragments will not readily redissolve. It is our experience that this procedure always permits complete digestion of the genomic DNA. After digestion, the DNA can be ethanol precipitated after the addition of ammonium acetate to a final concentration of 2.5 *M*.

As a control, 60 μg of deproteinized DNA (i.e., DNA from cells that have not been treated with DMS, but which has otherwise been isolated identically) are cleaved with DMS according to the procedure of Maxam and Gilbert.[22] The DNA is ethanol precipitated with ammonium acetate (see above) and analyzed in the same way as the samples isolated from treated cells.

c. Chemical Cleavage of the Genomic DNA

DNA is evaporated to dryness in a SpeedVac (Savant) for 5 min and suspended in freshly diluted 1 *M* piperidine (Aldrich; the dilution is made in a glass vial). The microcentrifuge tubes containing the DNA are placed in a rack where the caps are tightly pressed down by a covering plate and submersed in a 90°C water bath for 30 min. Subsequently, the tubes are transferred to dry ice to freeze the samples, which are then lyophilized to dryness in the SpeedVac. Water (100 μl) is added and the samples are frozen and dried again. The same procedure is repeated with 50 μl of H_2O. After this final lyophilization, the pellet is resuspended in 5 μl of formamide/dye solution (1 ml of recrystallized formamide + 25 μl of 0.5 *M* EDTA + 0.5 mg of bromophenol blue [added as powder] and 0.5 mg of xylene cyanol [added as powder]). The samples are briefly vortexed and centrifuged in a microcentrifuge for a few seconds. This procedure is repeated five to six times to ensure complete solubilization of the pellet.

d. Genomic Sequencing Gel and Transfer to Nylon Membrane

The genomic sequencing gel (60 cm long, 45 cm wide, and 1.2 mm thick) is cast according to the protocol of Church and Gilbert.[20] Acrylamide (Bio-Rad) should be >99.9% pure, and the ratio of acrylamide to bis-acrylamide should be 40:1. A 6% gel stock solution is made and used for about ten gels. The gel is always poured the day before it is used for electrophoresis to ensure complete polymerization. Preelectrophoresis is performed for 2 h at 40 W, and the DNA separation run is continued for 20 h at 24 W.

DNA samples for genomic sequencing are submersed in boiling water for 3 min and frozen on dry ice. DNA is stable while frozen. All the wells of the gel are flushed just before the first sample is loaded, to remove excess urea. Each sample is thawed immediately prior to being loaded onto the gel with a 10-μl gas-tight Hamilton syringe (1701SNWG, 31 gauge, 5 cm, point 3 needle). Such a syringe is necessary because the DNA solution is viscous. *Note:* Samples do not appear to enter the gel for about 1 h, but after that they look normal. The length of the electrophoresis is determined by the xylene cyanol dye marker, which migrates equivalent to a 120-nucleotide DNA segment in a 6% sequencing gel. Normally, 100 to 200 bases are run off the gel and, therefore, a region 100 to 350 bases away from the restriction site can be analyzed.

After completion of electrophoresis, the gel plates are carefully separated and a precut sheet of Whatman 541 paper is placed on top of the gel. The gel is cut along the edges of the paper and placed in the electrotransfer device (Polytech Products, Somerville, MA). TBE buffer (40 m*M* Tris-borate/EDTA, pH 8.3) is applied to the top of the gel from a squeeze bottle and a prewetted (TBE buffer), precut Genescreen nylon filter (New England Nuclear) is laid on top of the gel. The filter is wetted with TBE buffer and then the excess TBE is squeezed out between the filter and the gel; this is a necessary step for complete and uniform DNA transfer. Electrotransfer is carried out as described by Church and Gilbert[20] at 108-V DC for 60 min, either in a cold room or with a cooling system driven by a circulating pump. The buffer in the transfer box is 40 m*M* TBE.

e. UV Irradiation for Binding of the DNA to the Filter

The nylon filter is removed from the gel, placed with the DNA side down on a tightly stretched sheet of Saran Wrap™, and irradiated with the transfer side facing the UV light. The UV box is lined with a reflective metal surface and contains six UV bulbs (General Electric, GE G15T8); the distance from the filter to the bulbs is 30 cm. The required time for optimal DNA binding to the filter must be established experimentally with a test sequencing gel (in our hands, UV irradiation time is approximately 2 to 5 min). Insufficient irradiation may result in poor binding, whereas excessive irradiation may degrade the DNA into small pieces which no longer hybridize. The filter can be stored at 4°C in the dark for several months.

f. Cloning of the Probe and Large-Scale Isolation of Single-Stranded Recombinant Phage

To probe the nylon filter, a fragment of about 100 to 200 bp, located at or very near the restriction site corresponding to the restriction enzyme used to digest the genomic DNA, is cloned into the SmaI site of the M13mp19 phage (indirect end labeling). The nomenclature[20] of the clones is based on the sequence obtained in the sequencing gel (see Figure 2B), e.g., the 5′ HincII upper probe displays the 5′ upper sequence in the autoradiogram.

The SmaI site is chosen for cloning because any blunt-ended fragment can be easily inserted and the site is close to the position where the primer anneals. The cloned templates should be sequenced prior to large-scale production.

For preparation of phage stock, a single plaque is picked and grown (in JM101 cells) for 6 h in 2xTY medium (per liter: 16 g of Bacto tryptone, 10 g of yeast extract, and 10 g of NaCl). The cells are pelleted by centrifugation in a microcentrifuge for 5 min, and the supernatant containing the recombinant phage particles can be stored at 4°C for several weeks.

For preparation of template DNA, JM101 cells are grown in 50 ml of 2xTY medium to an OD_{550} of 0.4; 10 ml of these cells are transferred to a 500-ml prewarmed culture flask containing 2xTY, and the culture is grown to $OD_{550} = 0.4$. Recombinant phage solution (1 ml) is added and incubation is continued overnight at 37°C. Cells are pelleted by centrifugation at 5000 rpm for 20 min, and phage in the supernatant is precipitated with one fifth volume of 20% polyethylene glycol (PEG 8000, Eastman) and 2.5 *M* NaCl at 4°C for at least 40 min. The precipitate is pelleted by centrifugation (5000 rpm, 20 min, 4°C) and the supernatant is discarded. The centrifuge bottles are turned upside down to remove all traces of liquid, and then the pellet is dissolved in 15 ml of TE by vortexing. This solution is centrifuged (10,000 rpm, 20 min) to remove residual cell particles, and the supernatant is saved for further isolation of the template. The phage is precipitated again using PEG/NaCl as above, and the pellet is dissolved in 1.5 ml of TE and aliquoted to three 1.5-ml microcentrifuge tubes. SDS is added to each tube to a final concentration of 1% and, after vortexing, the DNA is extracted with phenol/chloroform (one time) and then chloroform (two to three times). It is important not to take any material from the interphase when removing the aqueous phase during the extractions. The DNA is ethanol precipitated in the presence of ammonium acetate, dissolved in TE, and precipitated again with ethanol. The yield should be a few milligrams of single-stranded phage DNA from a 500-ml culture. The DNA is resuspended at approximately 2 mg/ml in 10 m*M* Tris HCl and 1 m*M* EDTA, pH 7.5. The OD_{260}/OD_{280} ratio is normally between 1.4 and 1.7.

g. Probe Synthesis

Phage DNA (5 μg, 2 to 3 μl) and primer (80 ng, 15 to 20 μl) are mixed and one fourth volume of Tris/NaCl solution (0.16 *M* Tris-HCl, pH 7.5, and 1 *M* NaCl) is added to bring the NaCl concentration to 0.2 *M* for annealing. The tubes are submersed for 2 min in a 90°C water bath and then submersed immediately in a 50°C water bath. Again, tight sealing of the microcentrifuge tube caps is ensured by using a special rack. Annealing is carried out for 45 min, the tubes are briefly centrifuged, and then each of the following is added: 1 μl of 0.5 *M* β-mercaptoethanol, 1 μl of 10 mg/ml BSA (Fraction V), 1 μl of 3 m*M* dTTP/dGTP/dCTP, 1 μl of 0.2 *M* $MgCl_2$, and 250 μCi of ^{32}P-α-dATP (Amersham, 3000 Ci/mmol). The reaction is started with the addition of 5 U of *Escherichia coli* DNA polymerase I large (Klenow) fragment (Boehringer-Mannheim). Incubation is continued for 45 min at 37°C and the reaction is then quenched by the addition of 2 μl of 0.5 *M* EDTA and at least twice the reaction volume of formamide/dye solution (Section II.B.2.c). The tube is boiled for 3 min and the sample is then loaded directly onto a hot (~50°C surface temperature) 6% polyacrylamide gel (6% acrylamide, 0.15% bis-acrylamide, 7 *M* urea,

50 m*M* TBE, pH 8.3, 0.1% ammonium persulfate, and 0.1% TEMED; gel dimensions: length, 10 cm; thickness, 3 mm). Preelectrophoresis is carried out for 10 to 15 min (100 to 150 mA) and the electrophoretic fractionation is carried out for approximately another 10 min (bromophenol dye marker has migrated 3 to 4 cm into the gel). The current for the run is 100 to 150 mA to generate the required heat.

The probe is loaded just after boiling, while the gel is running, to avoid reannealing of the synthesized fragment with the M13 template. After electrophoresis is complete, one plate is removed and the gel is covered with Saran Wrap™. The gel is then exposed for 3 min to a sheet of Polaroid film (Polaroid, Type 55), and after the film is developed, the radioactive probe band is identified and excised from the gel. The probe fragments should be ~90 to 150 nucleotides in length and will migrate near the xylene cyanol marker. If the probe is too long, the reaction conditions should be adjusted by increasing the primer and/or template DNA concentrations.

h. Isolation of Single-Stranded Probe and Hybridization

The gel slice is placed into a 50-ml plastic tube and ground into small pieces with a siliconized glass rod, and 5 to 8 ml of hybridization buffer is then added. Hybridization buffer is prepared by mixing 20.8 ml of 1 *M* sodium phosphate buffer, pH 7.2, 14 ml of 20% SDS, 5.2 ml of 8 m*M* EDTA, and 0.4 g of BSA (Sigma, A-7511). The 1 *M* $NaHPO_4$ (pH 7.2) stock consists of 134 g of $Na_2HPO_4{\cdot}7H_2O$ and 4 ml of H_3PO_4 per liter. The tube is incubated at 50°C and shaken for 2 h to elute the DNA quantitatively and then centrifuged (2500 rpm, 5 min) to sediment the polyacrylamide. The probe mixture is poured immediately into a 20-ml syringe fitted with a 0.45-μm Millex-GV or Gelman Acrodisc filter and filtered to remove the residual polyacrylamide. By this procedure, a hybridization solution of 10 to 50 μCi/ml is obtained, and the probe has a specific activity of up to 4×10^9 cpm/μg DNA. Routinely, 80% of the radioactive label is incorporated into the probe. Hybridization is performed at 65°C in a plexiglass tube attached to a rotator inside a hybridization oven (Bachofer). This procedure promotes equal distribution of the probe DNA and reduces the volume of the hybridization solution required. The filters are prehybridized for 1 to 2 h and then hybridized overnight with the radioactive single-stranded probe.

i. Washing of the Filters and Autoradiography

After completion of hybridization, the filters are washed six times at room temperature with 1-l aliquots of a solution of 40 m*M* sodium phosphate, pH 7.2, 20 m*M* EDTA, and 1% SDS. The washing buffer is at 65°C only initially when it is poured into the tray. Filters are dried between two sheets of Whatman 3MM paper, covered with Saran Wrap™, and exposed to X-ray film (Kodak, X-AR) with an enhancing screen for 3 to 5 d at −70°C.

III. *IN VITRO* ASSAYS FOR PROTEIN-DNA COMPLEXES

Detection of specific protein-DNA interactions *in vitro* is a prerequisite for the identification, characterization, and purification of transcription factors. The gel retardation assay (electrophoretic mobility shift assay),[2,3,23,24] described elsewhere in this book, is an invaluable tool in the identification of such DNA-protein complexes using unfractionated protein preparations. In addition, by combining this assay with reciprocal deletion analysis (stairway assay),[25] whereby the binding of proteins to progressively shortened DNA fragments is studied, it is possible to determine the boundaries of DNA-protein interaction sites at the 20- to 50-nucleotide level of resolution.[25-27]

Establishing the exact recognition sequence of a DNA binding protein at the single nucleotide level can be accomplished by a combination of enzymatic and chemical protection techniques ("footprinting") that are based on the steric shielding of the chemical bonds in a DNA recognition sequence from enzymatic or chemical cleavage due to the binding of a sequence-specific DNA binding protein. These techniques include DNAse I footprinting,[28] exonuclease III protection analysis,[29] methidium-propyl-EDTA·iron(II),[30-32] 1,10-phenanthroline-copper ion,[33,34] and hydroxyl radical footprinting.[32,35]

DNAse I footprinting (described in detail below) and exonuclease III protection involve enzymatic digestions of the DNA. Because the enzymes and the DNA binding proteins are large in relation to the DNA, steric hindrance reveals a footprint, usually 15 to 30 nucleotides, extending to both sides of the actual protein-DNA contacts. A disadvantage of these methods is the preference for digesting some bases or sequences and therefore generating an uneven sequence ladder, even in the absence of any DNA binding proteins. This problem can obscure some protein-DNA contacts and make boundaries difficult to determine.

Chemical protection techniques are being employed to alleviate some of these problems. Methidium-propyl-EDTA·iron(II) [MPE-Fe(II)] and 1,10-phenanthroline-copper ion (OP-Cu) cleave the DNA backbone by the metal-dependent generation of highly reactive, reduced oxygen moieties. The metal is presented close to the DNA by intercalation of the methidium compound in the case of MPE-Fe(II); for the reaction with OP-Cu, the 1,10-phenanthroline binds in the minor groove. The footprint revealed by both methods is due to steric hindrance between the DNA binding protein and the rather large chemical compound and is similar in size to that generated by DNAse I. MPE-Fe(II) has no base or sequence preference and therefore provides a uniform control pattern for easy determination of the protein boundary. OP-Cu does have minor fluctuations in cutting efficiency, depending on the conformation of the DNA, but in general gives a uniform control pattern. Hydroxyl radical footprinting is similar to the other chemical footprinting methods; however, the hydroxyl radical-generating compound, $Fe(EDTA)^{2-}$, is kept away from the DNA by electrostatic repulsion. The highly reactive and very small hydroxyl radical is the only moiety to approach the DNA. Inhibition of cutting occurs only at points of very close protein-DNA contact. The footprint generated is much smaller than those from the other methods and can yield information about the side and groove of the DNA with which the protein is interacting. As with MPE-Fe(II), hydroxyl radical footprinting has no base or sequence preferences. One complication with hydroxyl radical footprinting is the presence of glycerol.[36] Glycerol is present at concentrations as high as 20% in most high-salt nuclear extracts and will act as a hydroxyl radical scavenger. The concentration of glycerol must be lowered below 0.5% in order to get an adequate level of cutting. This may cause problems with the stability of some DNA binding proteins.

Points of intimate contact between the protein and the individual nucleotides of its recognition sequence can be determined by a number of chemical protection or interference assays ("fingerprinting"). Methylation protection analysis using dimethylsulfate (DMS fingerprinting)[37] is the most commonly used technique and is based on the shielding by a DNA binding protein from DMS-mediated methylation of the N7 position of guanine residues (in the major groove) and the N3 position of adenine residues (in the minor groove).

Alkylation interference techniques,[37,38] such as methylation interference (DNA methylated by DMS) and ethylation interference (modification of the DNA sugar-phosphate backbone by 1-ethyl-1-nitrosourea), are based on the principle that the addition of alkyl groups to the bases or the sugar-phosphate backbone prior to the binding reaction can sterically (or otherwise) prevent the interaction of a protein with its cognate sequence. However, unlike DMS fingerprinting experiments, interference techniques require that unbound DNA (not complexed with proteins during a binding reaction, and thus enriched in molecules containing modified nucleotides in the DNA binding domain) is separated from DNA-protein complexes that contain modified nucleotides not interfering with the interaction. Apart from the above methods, contact points with thymines can be investigated using a bromouracil (BrdU) substitution technique involving UV irradiation.[39]

In vitro DNA-protein interaction assays are usually performed with protein preparations derived from a high-salt nuclear extract,[40-42] or a whole-cell extract,[43] and a DNA fragment spanning the region of interest carrying a terminal ^{32}P label. The most important aspect of *in vitro* DNA binding assays is the amount of binding site saturation that can be obtained with a given protein preparation. For instance, incompletely saturated binding sites may not be detectable in footprinting assays due to the background signal from the unbound DNA. Full saturation of a binding site can be achieved by carefully adjusting binding conditions, such as the concentration of DNA binding sites, nonspecific competitor DNA, and salt ions. However, this condition is seldom obtained with unfractionated protein preparations due to the low abundance of most transcription factors and the presence of nonspecific DNA binding activities.

The complete occupancy of a binding site by a specific protein is facilitated by using protein preparations that have been partially purified using liquid chromatography procedures.[4,5,44] Alternatively, it is possible to perform footprinting experiments with unfractionated nuclear proteins, provided a method is used that separates DNA-protein complexes from unbound DNA molecules. Two methods can be used for separating DNA-protein complexes from free DNA: (1) nitrocellulose filter binding[5,45,46] and (2) gel electrophoresis.[2,3,23,24] The filter-binding technique functions on the basis that most proteins adhere to nitrocellulose membranes, while double-stranded DNA does not. Hence, when a protein-DNA binding reaction is passed through nitrocellulose, the free DNA passes through, while the proteins and protein-DNA complexes remain associated with the filter. The electrophoretic method for separating free DNA and protein-bound DNA is a scaled-up version of the gel retardation assay (see Chapter 10).

In the sections below, we will discuss DNAse I footprinting, DMS fingerprinting, and DMS interference, the most frequently used methods for the initial determination of the recognition sequence of a DNA binding protein. Examples are taken from our studies of the binding of transcription factors to the proximal promoters of human histone genes.

A. DNAse I FOOTPRINTING

1. General Background

DNAse I footprinting[28] delineates the outer limits of the protein-DNA binding site. Subsequent to the formation of the protein-DNA complex, the enzyme DNAse I is introduced to the binding mixture and allowed to react with the DNA to produce a single-strand break by cleavage of the phosphodiester bond under conditions that produce, on the average, 1 randomly placed nick per 200 nucleotides. The bound DNA is separated from the free DNA if necessary, purified, denatured, and separated on a sequencing gel alongside a chemical sequencing ladder.[22] The denaturation generates single-stranded radiolabeled fragments with a length corresponding to the distance from the labeled end of the fragment to the first DNAse I cleavage site. It should be noted that the DNAse I-generated fragments migrate one half base pair faster than fragments generated during sequencing reactions because chemical cleavage produces a 3′ hydroxyl group, while DNAse I leaves a 3′ phosphate group. DNAse I digestion conditions are optimized so that a ladder of DNAse I cleavage products of approximately equal intensity over the region of interest is generated on the autoradiogram. When the lane representing the DNAse I-digested protein-DNA mixture is compared to a control lane representing DNAse I digestion of DNA in the absence of protein, a loss in intensity of the bands corresponding to the area of protein-DNA contact can be observed. This area of DNAse I protection is generated by the steric hindrance of the DNAse I enzyme by the bound protein. The size of the bound protein usually protects a region extending beyond both sides of the actual recognition sequence.[47]

2. Method

a. Preparation of Radiolabeled DNA Probe

Restriction endonuclease sites are chosen so as to generate from the cloned DNA a fragment of unique size (80 to 200 nucleotides) containing the region of interest. Restriction enzymes that yield 5′ overhanging ends are preferred because they are labeled more efficiently by the phosphorylating enzyme. Plasmid DNA (50 to 100 μg) is digested to completion with a first enzyme, extracted with phenol/chloroform and ethanol precipitated. The precipitated DNA is resuspended in 50 μl of 50 m*M* Tris-HCl and 0.1 m*M* EDTA, pH 8 and dephosphorylated with 1 U of calf intestinal alkaline phosphatase (CIP, Boehringer-Mannheim) by incubation at 37°C for 30 to 60 min. The CIP enzyme is inactivated by adding 50 μl of 0.1 *M* EGTA and heating at 65°C for 30 min. The sample is extracted with phenol/chloroform and precipitated with ethanol. The DNA precipitate is resuspended in 10 m*M* Tris-HCl and 1 m*M* EDTA, pH 7.6 (TE buffer), to a final concentration of 2 μg/μl, and the DNA solution is stored at −20°C. Aliquots (10 μg) of this stock are used for subsequent labeling reactions.

The labeling reaction is performed with 5 μl of dephosphorylated DNA (10 μg), 4 μl of 10X kinase buffer (50 m*M* Tris-HCl, pH 7.6, 10 m*M* $MgCl_2$, 10 m*M* β-mercaptoethanol, and 10 μg/ml bovine serum albumin) and 11 μl (80 to 120 μCi) of γ-^{32}P-ATP (Amersham, 3000 Ci/mmol). T4 polynucleotide kinase (U.S. Biochemical) is added (5 U) to the mixture and the sample is incubated for 30 to 60 min at 37°C. The volume of the solution is increased to 100 μl by the addition of TE buffer to facilitate handling and extracted with phenol/chloroform. The DNA is precipitated in the presence of 0.3 *M* sodium acetate (NaAc, 10 μl of a 3 *M* NaAc stock solution) with 2.5 volumes of 95% ethanol (room temperature) and the sample is centrifuged for 15 to 30 min in an Eppendorf microcentrifuge. The DNA is resuspended, digested with the secondary restriction enzyme, and prepared for gel electrophoresis by the addition of 0.5% SDS, 0.5% glycerol, and bromophenol blue tracking dye. Fragments are separated in a 4% native polyacrylamide gel, the gel is covered with plastic, and the bands are visualized by autoradiography of the wet gel for 5 min. The fragment of interest is excised from the gel, stained by the addition of 100 to 200 μl of tracking dye on top of the gel segment, and transferred to a dialysis bag. The dialysis bag is filled with 1 ml of 0.25X TBE buffer (1X TBE buffer = 89 m*M* Tris, 89 m*M* borate, and 2 m*M* EDTA, pH 8.3), and the DNA is recovered by electrophoresis at 25 mA until the tracking dye is no longer visible in the gel segment (10 to 30 min). The electroeluate is precipitated with ethanol and resuspended in 100 μl of TE buffer. The concentration of the DNA is estimated as follows: the size of the original plasmid DNA is typically 3 to 6 kilobases, the molecular weight of the plasmid DNA (using the approximation 1 bp = 660 g/mol) is approximately 2 to 4 $\times$ 10^6 g/mol, and the molar amount of any DNA fragment derived from 10 μg of plasmid DNA is therefore (e.g., 10×10^{-6} g $\div$ 2×10^6 g/mol) between 2.5 and 5×10^{-12} mol (2.5 to 5 pmol). The yield of the DNA fragment can be estimated by comparing DNA samples before and after electroelution using autoradiography or Cerenkov counting. Yields are typically between 40 and 80%. If 50% recovery is assumed, then the molarity of the DNA solution is 12 to 25 fmol/μl (i.e., 2.5 pmol $\times$ 50% per 100 μl). Typically, 10 to 100 fmol of DNA are used in footprinting and fingerprinting experiments.

b. Direct DNAse I Footprinting

Protein extract (0, 2, 4, and 6 μl) is preincubated on ice for 15 min, with 0.1 μg of sonicated *E. coli* DNA or 1 μg of poly (dI-dC)·poly (dI-dC) (Pharmacia) as nonspecific competitor DNA, in a final volume of 10 μl containing 12 m*M* HEPES, pH 7.9, 60 m*M* KCl, 5 m*M* $MgCl_2$, 0.5 m*M* dithiothreitol (DTT), and 12% glycerol. The binding reaction is initiated by the addition of 1 μl (5 to 25 fmol) of the radiolabeled DNA fragment and incubated at 20°C for 15 min. DNAse I (1 μl of a 25- to 75-μg/ml dilution, 5 U/μg, Boehringer-Mannheim or Worthington) is added. After 90 s the reaction is terminated by adding 50 μl of a solution containing 0.65% SDS and 20 m*M* EDTA and immediately extracted with phenol/chloroform. A stock of 1 mg/ml DNAse I may be stored at −20°C and diluted just prior to use in 20 m*M* Tris-HCl, pH 7.6, 50 m*M* NaCl, 1 m*M* DTT, 100 μg/ml bovine serum albumin (BSA), and 50% glycerol. The amount of DNAse I to be used is determined empirically for each binding reaction. The length of the DNAse I digestion time is usually kept constant and minimized (30 to 120 s), as DNA-protein complexes may dissociate and associate during prolonged incubation times, allowing DNAse I to digest DNA sequences at the recognition site. Usually, the control lane (i.e., no protein extract) requires severalfold less DNAse I (1 μl of 5 μg/ml dilution) to give the same digestion pattern as the DNA-protein mixture because the DNA is more accessible to DNAse I.

c. DNAse I Footprinting Using Electrophoretically Separated DNA-Protein Complexes

The DNAse I footprinting protocol described above is effective only when the DNA binding protein of interest is fairly abundant and has a relatively high specific DNA binding activity, as is the case for a number of ubiquitous transcription factors.[32,38,44] In other cases, it is necessary to separate the bound DNA from the unbound DNA (representing, respectively, the "footprint DNA" and the "naked DNA" fragments). This separation is done by performing a preparative-scale gel retardation assay (see Chapter 10). Although many different electrophoresis and DNA binding conditions can be used, the protocol described below is based on the conditions described by Strauss and Varshavsky[23] and Singh et al.[24]

Prior to preparative gel retardation electrophoresis, it is necessary to optimize the binding conditions of a given protein preparation and probe DNA with respect to protein concentration, nonspecific competitor DNA concentration, salt concentration, and the divalent cation requirement of the protein of interest. The conditions described here were optimized for the binding of a factor designated HiNF-B that binds to the proximal promoter of a human H1 histone gene.[26]

Binding reactions are essentially a 25-fold scale-up of an optimized gel retardation electrophoresis experiment, performed in a volume of 500 μl containing 50 μg of nuclear extract protein, 1 ng of a 130-bp ^{32}P-labeled probe spanning the HiNF-B binding site, and 2 μg of poly (dI-dC)·poly (dI-dC) in a binding buffer containing 12% glycerol, 60 m*M* KCl, and 12 m*M* HEPES/NaOH, pH 7.9. In addition, 5 m*M* $MgCl_2$ is added, as DNAse I has a divalent cation requirement, although magnesium ions are not usually required for the formation of DNA-protein complexes during binding reactions. After binding (incubation for 30 min at 24°C), the mixture is pulse digested with 5 U DNAse I for 1 min at 24°C and the digestion is quenched by the addition of EDTA to a final concentration of 10 m*M*. The sample is immediately loaded into ten 8-mm-wide wells of a 4% polyacrylamide gel (30:1 acrylamide to bis-acrylamide) that has been preelectrophoresed for 2 h at 200 V with a low-ionic-strength electrophoresis buffer (1 m*M* EDTA, 3.3 m*M* NaAc, and 6.6 m*M* Tris, pH 7.9). Electrophoresis is at 25 mA (200 V) for 1 to 2 h, depending on the separation between complexed DNA and free DNA. After electrophoresis, the wet gel is exposed overnight at 4°C.

Radioactive bands corresponding to free DNA and complexed DNA are excised from the gel, stained with bromophenol blue by pipetting 100 to 200 μl of tracking dye onto the gel segment, and electroeluted in dialysis bags using 0.25X TBE buffer as the electrophoresis buffer. The electroeluate is precipitated by the addition of 0.1 volume of 3 *M* NaAc and 2.5 volumes of 95% ethanol at room temperature in microcentrifuge tubes, followed by centrifugation at 4°C for 30 min. No carrier nucleic acid (tRNA or salmon sperm DNA) needs to be added, as the acrylamide contaminating the electroeluate aides in the precipitation of the DNA. The DNA pellets are resuspended in TE buffer, pooled, and reprecipitated until one pellet is obtained per sample.

Excess acrylamide present in the ethanol precipitates is removed by cetyltrimethylammonium bromide (CTAB) precipitation. The pellet from each tube is dissolved in 200 μl of 0.3 *M* NaAc by mechanical agitation. Subsequently, 50 μl of 1% CTAB (Sigma) and 1 μl (10 mg/ml) of yeast tRNA are added and the mixture is incubated on ice for 10 min. The addition of yeast tRNA is essential, as most of the acrylamide will remain in solution and no longer aid in the precipitation of the DNA. The mixture is then centrifuged for 15 min at 4°C and the supernatant containing most of the polyacrylamide is removed. The pellet containing CTAB-precipitated DNA and tRNA is dissolved in 100 μl of TE buffer and precipitated again by adding 10 μl of 3 *M* NaAc and 250 μl of 95% ethanol, followed by centrifugation for 30 min at 4°C. The final precipitate is rinsed with 70% ethanol and dried by exposure to air or vacuum.

An alternative method for the isolation of the radiolabeled DNA and DNA-protein complexes after preparative gel retardation electrophoresis is by electrotransfer of the entire gel to a DEAE membrane.[48] The preparative gel retardation is performed in TGE electrophoresis buffer (1X TGE = 50 m*M* Tris-HCl, pH 8.5, 380 m*M* glycine, and 2 m*M* EDTA) either at high salt conditions (1X TGE) or low salt conditions (0.5X TGE). After electrophoresis, the gel is electrophoretically transferred in 0.5X TGE buffer to the DEAE membrane (NA45, Schleicher and Schuell) in a TE Series Transphor (Hoefer Scientific) at 200 mA for 2 h at 4°C. After transfer, the membrane is exposed to X-ray film for 3 to 16 h at room temperature. Radioactive bands on the membrane corresponding to each species of DNA-protein complex or free DNA are excised and pooled into one microcentrifuge tube for each complex. The DNA is eluted from the membrane by the addition of 400 μl of elution buffer (10 m*M* Tris-HCl, pH 8.0, 1 m*M* EDTA, and 1 *M* NaCl) and heating to 65°C for 30 min. The eluate, containing 60 to 80% of the DNA, is removed and the elution is repeated with another 400 μl of elution buffer. The eluates are pooled and the DNA is precipitated with 2.5 volumes of 95% ethanol and 10 μg of yeast tRNA for >4 h at −20°C. This method alleviates the need for time-consuming CTAB precipitations. It also allows a more precise excision of the radioactive bands, thus increasing the footprint signal over the background noise.

d. Gel Electrophoresis

The final ethanol precipitates of the DNA samples obtained during either the direct or indirect DNAse I footprinting experiments are resuspended in 4 to 12 μl of recrystallized formamide (Sigma) containing a small amount of xylene cyanol and bromophenol blue dye. The samples (3 to 6 μl) are boiled for 3 min and quickly cooled on dry ice, then thawed and loaded onto a 40-cm-long denaturing 8% polyacrylamide, 8 *M* urea gel next to Maxam and Gilbert sequencing reactions[22] of the same fragment. Gel electrophoresis is performed at 50 W constant power until the bromophenol blue has migrated to within 3 cm of the bottom. The gel is then dried on Whatman 3MM paper and autoradiographed at −70°C using a Cronex Lightning Plus enhancing screen (Du Pont).

B. DMS FINGERPRINTING

1. General Background

DMS fingerprinting[37,38] takes advantage of the base-specific chemical cleavage reactions for guanine and adenine residues developed by Maxam and Gilbert.[22] Dimethylsulfate (DMS) methylates the N7 position of the guanine base and, less efficiently, the N3 position of the adenine base. The modified base is treated with piperidine at high temperature (90°C), resulting in a break of the adjacent glycosidic bond, which ultimately leads to the cleavage of the DNA. Proteins that bind in very close contact with the reactive position of the guanine or adenine residue can shield these bases from reaction with the DMS, resulting in a DMS protection. In rare cases, the protein may form a hydrophobic pocket around the nucleotide residue, which generates a more stable environment for the DMS and leads to a higher frequency of reaction at that position, resulting in a DMS enhancement. Protections or enhancements of guanine residues indicate that the protein is located within the major groove of the DNA where the N7 position is protruding. However, interactions with adenine residues indicate the protein is within the minor groove where the N3 position is exposed. These contact points should lie within the DNAse I footprint of the protein, but as these interactions require a very close contact with either a guanine or adenine residue, they may not be present in all DNAse I footprints.

2. Method

a. Preparation of Radiolabeled DNA Probe

The same probes generated for DNAse I footprinting are used for DMS fingerprinting.

b. Direct DMS Fingerprinting

The preincubation and binding reaction are exactly the same as for the DNAse I footprinting. DMS (Eastman Kodak) is diluted 25-fold in water immediately prior to use and 1.65 μl of the dilution is added to the reaction vial (0.5% final concentration). The chemical modification is allowed to proceed for 45 s at 20°C and then stopped by the addition of 3.2 μl of DMS stop solution (1.5 *M* NaAc, pH 7, 1 *M* β-mercaptoethanol, and 100 μg/ml tRNA), 48 μl of cold (−20°C) ethanol, and freezing in a dry ice/ethanol bath for 10 min. The samples are pelleted by centrifugation, resuspended in 100 μl of 0.3 *M* NaAc, and precipitated again with 300 μl of ethanol. The DMS-modified DNA is then cleaved with piperidine.[22]

c. DMS Fingerprinting Using Electrophoretically Separated DNA-Protein Complexes

As for DNAse I footprinting, DMS fingerprinting requires a protein preparation that contains a high specific binding activity of the protein of interest. Because unbound DNA molecules in the binding reaction cause background bands in the protection pattern, these molecules should be separated from the complexed DNA fragments by gel retardation electrophoresis. Binding conditions are the same as in the DNAse I footprinting section (III.A.2.c.), with the exception that magnesium ions usually can be omitted. DMS reactions are performed by adding 2.5 μl of undiluted DMS to a 500-μl, preparative-scale gel retardation binding reaction. The reaction is not quenched, but immediately loaded on a low-ionic-strength gel (see Section III.A.2.c.). After isolation of the bands corresponding to unbound and bound DNA, and CTAB precipitation, the DMS-modified DNA is cleaved by piperidine. The alternative method for isolating the radioactive bands by electrotransfer, described in the DNAse I footprinting section, can also be employed here.

d. Piperidine Cleavage and Gel Electrophoresis

The precipitated, modified DNA is resuspended in 100 μl of 20 m*M* ammonium acetate, pH 7.0, and 0.1 m*M* EDTA and heated at 90°C for 15 min. Piperidine (10 μl of 10 *M* stock, Fisher) is added to each tube, mixed, and heated at 90°C for 30 min. The piperidine reaction is stopped by quickly (5 s) spinning down the sample, freezing on dry ice, and lyophilizing to dryness in a rotary vacuum dryer (Savant, SpeedVac). The samples are subsequently resuspended in 50 μl of water, lyophilized, resuspended in 25 μl of water, lyophilized, and resuspended in 4 μl of recrystallized formamide with tracking dye. Gel electrophoresis on a denaturing polyacrylamide gel and autoradiography are the same as above (Section III.A.2.d.).

C. METHYLATION INTERFERENCE

1. General Background

The difference between DMS protection and methylation interference has been discussed earlier (Section III, introduction). The practical difference between these techniques is that in the case of DMS protection analysis, the DNA is modified during the binding reaction, while methylation interference is performed by modifying the DNA with DMS prior to the binding reaction. Second, this method requires that the complexed DNA is separated from free DNA, which is optional in DMS protection analysis.

2. *Method*

The ^{32}P-labeled probe (sufficient amounts for a 25-fold scaled-up gel retardation binding reaction) is treated with 1 μl of undiluted DMS in 200 μl of 50 m*M* sodium cacodylate, pH 8.0, and 1 m*M* EDTA[22] for 5 to 10 min at 24°C. After ethanol precipitation, the modified probe is used in a 25-fold scaled-up preparative gel retardation binding reaction and processed as in "indirect" DMS fingerprinting (Section III.B.2.c.).

NOTES ADDED IN PROOF

1. Ward et al. (A1) have described the use of RNA probes for genomic sequencing.
2. Additional helpful information about genomic sequencing can be found in *A Laboratory Guide to Genomic Sequencing* (A2).
3. Recently, procedures that take advantage of polymerase chain reaction (PCR) methodologies have been developed for examining *in vivo* protein-DNA interactions (A3,A4). These approaches may circumvent some of the technical complexities of the original method and have the decided advantage of requiring significantly smaller amounts of DNA (and therefore fewer cells).

REFERENCES

1. **Topol, J., Ruden, D. M., and Parker, C. S.,** Sequences required for *in vitro* transcriptional activation of a *Drosophila* hsp 70 gene, *Cell,* 42, 527, 1985.
2. **Garner, M. M. and Revzin, A.,** A gel electrophoresis method for quantifying the binding of proteins to specific DNA regions: application to components of the *Escherichia coli* lactose operon regulatory system, *Nucleic Acids Res.,* 9, 3047, 1981.
3. **Fried, M. and Crothers, D. M.,** Equilibria and kinetics of lac repressor-operator interactions by polyacrylamide gel electrophoresis, *Nucleic Acids Res.,* 9, 6505, 1981.
4. **Kadonaga, J. T. and Tjian, R.,** Affinity purification of sequence-specific DNA binding proteins, *Proc. Natl. Acad. Sci. U.S.A.,* 83, 5889, 1986.
5. **Rosenfeld, P. and Kelly, T. J.,** Purification of nuclear factor I by DNA recognition site affinity chromatography, *J. Biol. Chem.,* 261, 1398, 1986.
6. **Stein, G. S., Plumb, M. A., Stein, J. L., Marashi, F. F., Sierra, L. F., and Baumbach, L. L.,** Expression of histone genes during the cell cycle in human cells, in *Recombinant DNA and Cell Proliferation,* Stein, G. S. and Stein, J. L., Eds., Academic Press, Orlando, FL, 1984, 107.
7. **Plumb, M., Stein, J., and Stein, G.,** Coordinate regulation of multiple histone mRNAs during the cell cycle in HeLa cells, *Nucleic Acids Res.,* 11, 2391, 1983.
8. **Baumbach, L. L., Stein, G. S., and Stein, J. L.,** Regulation of human histone gene expression: transcriptional and posttranscriptional control in the coupling of histone messenger RNA stability with DNA replication, *Biochemistry,* 26, 6178, 1987.
9. **Chrysogelos, S., Riley, D. E., Stein, G., and Stein, J.,** A human histone H4 gene exhibits cell cycle-dependent changes in chromatin structure that correlate with its expression, *Proc. Natl. Acad. Sci. U.S.A.,* 82, 7535, 1985.
10. **Moreno, M. L., Chrysogelos, S. A., Stein, G., and Stein, J.,** Reversible changes in the nucleosomal organization of a human H4 histone gene during the cell cycle, *Biochemistry,* 25, 5364, 1986.
11. **Kornberg, R. D.,** Chromatin structure: a repeating unit of eight histone molecules and about 200 DNA base pairs, *Science,* 184, 868, 1974.
12. **Wu, C.,** The 5′ ends of *Drosophila* heat shock genes in chromatin are hypersensitive to DNase I, *Nature,* 286, 854, 1980.

13. **Bryan, P. N., Hoffstetter, H., and Birnstiel, M.,** Nucleosome arrangement on tRNA genes of *Xenopus laevis, Cell,* 27, 459, 1981.
14. **Eissenberg, J. C., Cartwright, I. L., Thomas, G. H., and Elgin, S. C. R.,** Selected topics in chromatin structure, *Annu. Rev. Genet.,* 19, 485, 1985.
15. **Elgin, S. C. R.,** The formation and function of DNAse I hypersensitive sites in the process of gene activation, *J. Biol. Chem.,* 263, 19, 259, 1988.
16. **Fritton, H., Igo-Kemenes, T., Nowock, J., Strech-Jurk, U., Theisen, M., and Sippel, A.,** Alternative sets of DNAse I hypersensitive sites characterize the various functional states of the chicken lysozyme gene, *Nature,* 311, 163, 1984.
17. **Pauli, U., Chrysogelos, S., Stein, J., and Stein, G.,** Native genomic blotting: high resolution mapping of DNAse I hypersensitive sites and protein-DNA interactions, *Proc. Natl. Acad. Sci. U.S.A.,* 85, 16, 1988.
18. **Pauli, U., Chrysogelos, S., Stein, J., and Stein, G.,** Native genomic blotting: a novel approach to mapping DNAse I hypersensitive sites and protein-DNA interactions at high resolution, *BioTechniques,* 6(2), 142, 1988.
19. **Pauli, U., Chrysogelos, S., Stein, G., Stein, J., and Nick, H.,** Protein-DNA interactions *in vivo* upstream of a cell cycle-regulated human H4 histone gene, *Science,* 236, 1308, 1987.
20. **Church, G. M. and Gilbert, W.,** Genomic sequencing, *Proc. Natl. Acad. Sci. U.S.A.,* 81, 1991, 1985.
21. **Roberts, C. and Wilson, G. L.,** Simplified random priming technique for the radiolabeling of intermediate-sized DNA probes, *Focus,* 7(3), 16, 1985.
22. **Maxam, A. M. and Gilbert, W.,** Sequencing end-labeled DNA with base-specific chemical cleavages, *Methods Enzymol.,* 65, 499, 1980.
23. **Strauss, F. and Varshavsky, A.,** A protein binds to a satellite DNA repeat at three specific sites that would be brought into mutual proximity by DNA folding in the nucleosome, *Cell,* 37, 889, 1984.
24. **Singh, H., Sen, R., Baltimore, D., and Sharp, P.,** A nuclear factor that binds to the conserved sequence motif in transcriptional elements of immunoglobulin genes, *Nature,* 319, 154, 1986.
25. **van Wijnen, A. J., Stein, J. L., and Stein, G. S.,** A nuclear protein with affinity for the 5′ flanking region of a human H4 histone gene, *Nucleic Acids Res.,* 15, 1679, 1987.
26. **van Wijnen, A. J., Wright, K. L., Massung, R. F., Gerretsen, M., Stein, J. L., and Stein, G. S.,** Two target sites for protein binding in the promoter region of a cell cycle regulated human H1 histone gene, *Nucleic Acids Res.,* 16, 571, 1988.
27. **Cohen, R. B., Sheffery, M., and Kim, C. G.,** Partial purification of a nuclear protein that binds to the CCAAT box of the mouse alpha(1)-globin gene, *Mol. Cell. Biol.,* 6, 821, 1986.
28. **Galas, D. and Schmitz, A.,** DNAse footprinting: a simple method for the detection of protein-DNA binding specificity, *Nucleic Acids Res.,* 5, 3157, 1978.
29. **Wu, C.,** Two protein-binding sites in chromatin implicated in the activation of heat shock genes, *Nature,* 309, 229, 1984.
30. **Hertzberg, R. P. and Dervan, P. B.,** Cleavage of DNA with methidiumpropyl-EDTA-iron(II): reaction conditions and product analyses, *Biochemistry,* 23, 3934, 1984.
31. **Sawadago, M. and Roeder, R. G.,** Interaction of a gene-specific transcription factor with the adenovirus major late promoter upstream of the TATA box region, *Cell,* 43, 165, 1985.
32. **Pruijn, G. J. M., van Miltenburg, R. T., Claessens, J. A. J., and van der Vliet, P. C.,** Interaction between the octamer-binding protein nuclear factor III and the adenovirus origin of DNA replication, *J. Virol.,* 62, 3092, 1988.
33. **Spasski, A. and Sigman, D. S.,** Nuclease activity of 1,10-phenanthroline-copper ion. Conformational analysis and footprinting of the lac operon, *Biochemistry,* 24, 8050, 1985.
34. **Kuwabara, M. D. and Sigman, D. S.,** Footprinting DNA-protein complexes in situ following gel retardation assays using 1,10-phenanthroline-copper ion: *Escherichia coli* RNA polymerase-lac promoter complexes, *Biochemistry,* 26, 7234, 1987.
35. **Tullius, T. D. and Dombroski, B. A.,** Hydroxyl radical "footprinting": high-resolution information about DNA-protein contacts and application to lambda repressor and Cro protein, *Proc. Natl. Acad. Sci. U.S.A.,* 83, 5469, 1986.
36. **Tullius, T., Dombroski, B., Churchill, M., and Kam, L.,** Hydroxyl radical footprinting: a high-resolution method for mapping protein-DNA contacts, *Methods Enzymol.,* 155, 537, 1987.
37. **Sienbenlist, U. and Gilbert, W.,** Contacts between *Escherichia coli* RNA polymerase and an early promoter of phage T7, *Proc. Natl. Acad. Sci. U.S.A.,* 77, 122, 1980.
38. **De Vries, E., van Driel, W., van den Heuvel, S. J. L., and van der Vliet, P. C.,** Contact point analysis of the HeLa nuclear factor I recognition site reveals symmetrical binding at one side of the DNA helix, *EMBO J.,* 6, 161, 1987.
39. **Ogata, R. and Gilbert, W.,** Contacts between the lac repressor and thymines in the lac operator, *Proc. Natl. Acad. Sci. U.S.A.,* 74, 4973, 1977.

40. **Challberg, M. D. and Kelly, T. J., Jr.**, Adenovirus DNA replication in vitro, *Proc. Natl. Acad. Sci. U.S.A.*, 76, 655, 1979.
41. **Dignam, J. D., Lebovitz, R. M., and Roeder, R. G.**, Accurate transcription initiation by RNA polymerase II in a soluble extract from isolated mammalian nuclei, *Nucleic Acids Res.*, 11, 1475, 1983.
42. **Shapiro, D. J., Sharp, P. A., Wahli, W. W., and Keller, M. J.**, A high efficiency HeLa cell nuclear transcription extract, *DNA*, 7, 47, 1988.
43. **Manley, J. L., Fire, A., Cano, A., Sharp, P. A., and Gefter, M. L.**, DNA-dependent transcription of adenovirus genes in a soluble whole-cell extract, *Proc. Natl. Acad. Sci. U.S.A.*, 77, 3855, 1980.
44. **Briggs, M. R., Kadonaga, J. T., Bell, S. P., and Tjian, R.**, Purification and biochemical characterization of the promoter-specific transcription factor, Spl, *Science*, 234, 47, 1987.
45. **Riggs, A. D., Suzuki, H., and Bourgeois, S.**, Lac repressor-operator interaction (I): equilibrium studies, *J. Mol. Biol.*, 48, 67, 1970.
46. **Riggs, A. D., Bourgeois, S., and Cohn, M.**, Lac repressor-operator interaction (II): effects of galactosides and other ligands, *J. Mol. Biol.*, 53, 401, 1970.
47. **Suck, D. and Oefner, C.**, Structure of DNase I at 2.0 Å resolution suggests a mechanism for binding to and cutting DNA, *Nature*, 321, 620, 1986.
48. **Singh, H., Lebowitz, J., Baldwin, A., and Sharp, P.**, Molecular cloning of an enhancer binding protein: isolation by screening of an expression library with a recognition site DNA, *Cell*, 52, 415, 1988.

REFERENCES ADDED IN PROOF

A1. **Ward, C., Bolden, A., and Weissbach, A.**, Genomic sequencing of the 5′-flanking region of the mouse β-globin major gene in expressing and nonexpressing mouse cells, *J. Biol. Chem.*, 265, 3030, 1990.
A2. **Saluz, H. P. and Jost, J.-P.**, *A Laboratory Guide to Genomic Sequencing: The Direct Sequencing of Native Uncloned DNA*, Birkhäuser, Basel, Switzerland, 1987.
A3. **Saluz, H. and Jost, J.P.**, A simple high-resolution procedure to study DNA methylation and *in vivo* DNA-protein interactions on a single-copy gene level in higher eukaryotes, *Proc. Natl. Acad. Sci., U.S.A.*, 86, 2602, 1989.
A4. **Mueller, P. R. and Wold, B.**, *In vivo* footprinting of oa muscle specific enhancer by ligation mediated PCR, *Science*, 246, 780, 1989.

Chapter 12

RNA FOOTPRINTING

David S. McPheeters

TABLE OF CONTENTS

I. INTRODUCTION

The specific recognition of RNA molecules by proteins is a complex and poorly understood process that occurs during a large number of biochemical reactions. An important aspect of understanding this process is the identification of the site(s) on an RNA molecule to which a protein specifically binds. Many different techniques, including photochemical cross-linking,[1,2] mutational analysis,[3-5] isotope exchange,[2] toeprinting,[6,7] phylogenetic comparison,[8,9] and damage selection experiments,[10-13] have been used to identify where specific interactions with proteins occur on RNA molecules. The most widely used technique for this purpose is footprinting (for examples, see Table 1). Footprinting relies on the ability of a bound protein (or other ligand) to protect portions of an RNA molecule from nuclease digestion or chemical modification. In this chapter, I will discuss in detail a technique for RNA footprinting where the sites of enzymatic cleavage or chemical modification are assayed using primer extension.

The specific interaction of proteins with RNA may involve recognition of RNA primary,[3,4,14] secondary,[3,8,12,15] and tertiary[2,3] structure. Therefore, it is essential to have some knowledge of how the unbound RNA which will be used for footprinting is folded. Several excellent reviews on the prediction and determination of RNA secondary structure have recently been published.[16,17] Many of the chemical and enzymatic probes used in the footprinting assay are sensitive to RNA structure, so structural[17] and footprint analyses of an RNA binding site may be done simultaneously. Comparative sequence analysis can also be used to identify conserved features of RNA structure which potentially may be recognized by proteins.[8,9] Comparative sequence analysis is easily accomplished using the primer extension assay discussed below.

The primer extension technique for RNA footprinting has several advantages, as well as disadvantages, compared to other methods currently available for footprinting protein (or other ligand) binding sites on RNA molecules. The various techniques available for RNA footprinting are summarized below, and the advantages and disadvantages of each technique are discussed.

A. FOOTPRINTING WITH INTERNALLY LABELED RNA

The first RNA footprinting experiments were done in the early 1970s[18-20] using the technique outlined in Figure 1. Internally ^{32}P-labeled RNA is first incubated with protein under conditions which allow the formation of specific protein-RNA complexes (step i). These complexes are then treated extensively with RNAse (step ii) and the ^{32}P-labeled RNA fragments still bound to the protein after RNAse digestion are isolated (step iii) and sequenced (step iv) using fingerprint analysis.[21] Separation of the bound ^{32}P-RNA fragments from the unbound fragments is the most crucial step in this procedure and is usually accomplished by differential centrifugation, immunoprecipitation, native gel electrophoresis, or filter binding techniques.[13,18-20,22] This method of footprinting is limited to the identification of "simple" RNA binding sites where continuous regions of RNA are contacted by the bound protein. Extensive RNAse digestion of more complex RNA binding sites, those in which the protein contacts regions of the RNA are separated from one another in the primary structure, may cause dissociation of protein-RNA complexes and preclude identification of the bound fragments.

B. FOOTPRINTING WITH END-LABELED RNA

The folding of an RNA molecule often plays a crucial role in its recognition by proteins, and extensive cleavage or chemical modification of an RNA is likely to disrupt this folding. By limiting the amount of chemical or enzymatic probe added to a footprinting reaction so that each individual RNA molecule is modified or cleaved only once, the chance of large

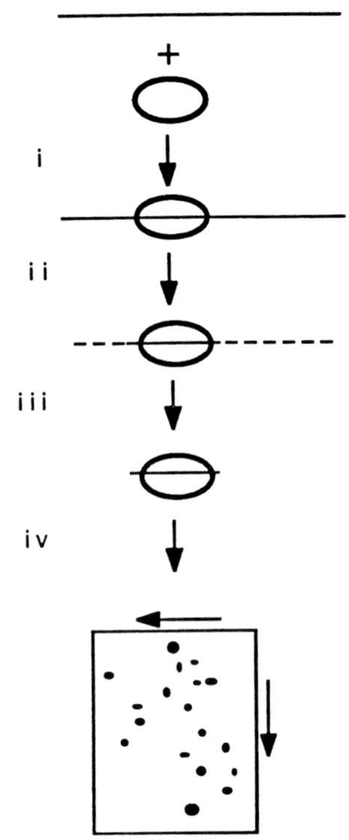

FIGURE 1. Footprinting with internally labeled RNA. See text for details.

alterations in RNA structure and disruption of complex protein binding sites is reduced. When limited enzymatic cleavage, or chemical modification followed by cleavage of the RNA at modified bases, is performed with a 5′ or 3′ ^{32}P end-labeled RNA, a nested set of end-labeled RNA fragments is produced which can be separated by gel electrophoresis. Footprinting using limited enzymatic cleavage or chemical modification of an end-labeled RNA is schematically diagrammed in Figure 2. The techniques for analyzing 5′ or 3′ end-labeled RNA molecules are identical, except that the RNA fragments detected originate from opposite ends of the molecule. Two reactions, one in which purified end-labeled RNA alone (A) and another with both RNA and protein together (B), are incubated under conditions which allow the RNA to adopt a native conformation and permit the formation of specific protein-RNA complexes (step i). The accessibilities of individual bases in the bound and unbound RNAs are then probed by limited chemical modification or limited enzymatic digestion (step ii). After termination of the probing reactions (step iii), chemically modified RNAs must be subsequently treated in order to hydrolyze the RNA specifically at modified residues. The lengths of the end-labeled RNA fragments are then determined by gel electrophoresis. Base-specific cleavage reactions[23-27] done with unbound, end-labeled RNA serve as size markers on the gel.

The binding site of a protein is revealed by comparing changes in the sensitivity of each nucleotide to the probe used in the presence and absence of the protein. The bases in the RNA contacted by the protein will usually exhibit protection from the probing reagent. The extent of protection seen will depend on several factors, including the concentration of the protein in the footprinting mixture, the protein's binding affinity for the RNA, and whether the association of the protein with the RNA is stable or transitory in nature (i.e., the dissociation rate of complexes). Protein binding may also cause alterations in the folding of the RNA. Since many of the chemical and enzymatic probes used in RNA footprinting are sensitive to RNA structure, the stabilization of weak secondary (or tertiary) structure caused

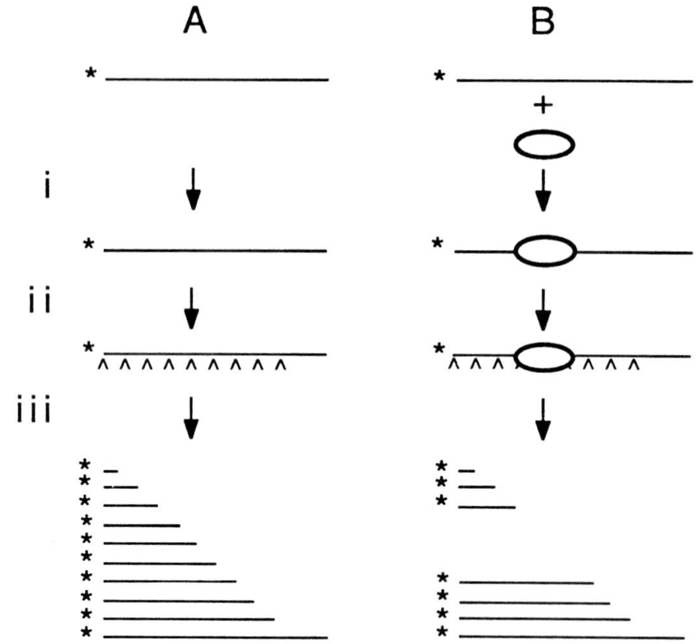

FIGURE 2. Footprinting with 5′ or 3′ end-labeled RNA. See text for details.

by binding of a protein may lead to protection of nucleotides not directly in contact with the bound protein. Enhanced reactivity of nucleotides caused by the binding of a protein may also be observed. Greater accessibility of nucleotides to chemical or enzymatic probes in bound vs. unbound RNA is also usually attributed to changes in the folding of an RNA induced by binding of a protein.

The technique of footprinting with 5′ or 3′ end-labeled RNAs has several limitations. First, and most significantly, the method can only be used for RNAs which can be obtained in pure form and have homogenous 5′ or 3′ termini which can be end labeled. Second, only chemical modifications which allow the subsequent cleavage of the RNA chain at modified bases may be detected. Finally, footprinting with end-labeled RNA is only useful for the analysis of regions of as RNA close enough to the labeled end to be well resolved by gel electrophoresis (approximately 400 bases).

C. FOOTPRINTING USING PRIMER EXTENSION

A new method useful for RNA structure determination, footprinting, and comparative sequence analysis has recently been developed. This method was first used by Qu et al.[28] to study the structure of eukaryotic 23S rRNA and uses primer extension by reverse transcriptase to detect sites of chemical modification or strand cleavage caused by either enzymatic or chemical treatment of an RNA. Reverse transcription often pauses when chemically modified nucleotides in the template RNA are encountered because the modification interferes with the formation of normal Watson-Crick base pairs.[17,29-31] Reverse transcription also terminates when the 5′ end of the template RNA is reached, and this permits the detection of enzymatic cleavage sites or sites of chemically induced strand scission, as well as 5′ ends arising from transcriptional initiation. As described below, the primer extension method is particularly useful for RNA footprinting and structural analysis because unpurified RNAs may be used. The primer extension method for footprinting has been used successfully to locate the sites contacted on rRNAs by antibiotics, tRNA, elongation factors, and ribosomal proteins,[15,32-40] and to determine the mRNA binding sites of several translational repressor proteins.[9,14,41]

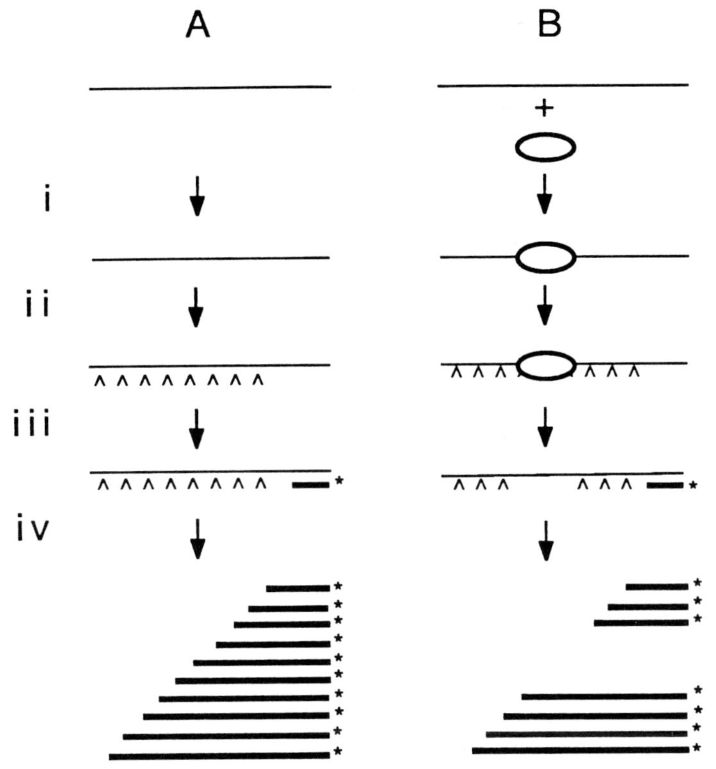

FIGURE 3. Footprinting with unlabeled RNA using the primer extension assay. RNA and DNA are represented by thin and thick lines, respectively. See text for details.

The primer extension method for RNA footprinting is schematically diagrammed in Figure 3. Unlabeled RNA is first incubated both with (B) and without (A) added protein, using conditions which allow the RNA to adopt a native conformation and permit the formation of specific protein-RNA complexes (step i). After a suitable period of time, a small amount of an enzymatic or chemical probe is added to the reactions and allowed to react with the RNA (step ii). The reactions are then stopped and the probing reagent is removed. At this point, RNA with chemical modifications which do not interfere with reverse transcription (see below) must be treated in order to hydrolyze the RNA specifically at modified bases.[17,25,29] To assay the sites of chemical modification or strand cleavage, the treated RNAs are hybridized with a short, end-labeled oligonucleotide which is complementary to a region downstream of the suspected protein binding site (step iii). The oligonucleotide-RNA hybrid is then used as a substrate for primer extension with reverse transcriptase (step iv) and the lengths of the resulting end-labeled cDNAs are analyzed by gel electrophoresis. Dideoxy sequencing reactions[42] done with untreated RNA serve as size markers.

Modified nucleotides which interfere with reverse transcription cause cDNA synthesis to stop at the base preceding the modified nucleotide, while 5′ ends of the RNA resulting from enzymatic or chemical treatment cause reverse transcription to stop at the 5′ terminal base of the RNA fragment. The amount of termination or pausing of reverse transcription at a particular base is usually proportional to the extent of that base's reaction with the probing reagent used. As in footprinting with 5′ or 3′ end-labeled RNAs, the protein binding site is located by comparing changes in the sensitivity of each base to chemical modification or enzymatic digestion in the presence and absence of the protein.

The oligonucleotides used for priming cDNA synthesis by reverse transcriptase are designed to hybridize uniquely to a sequence in the (target) RNA to be footprinted, and this allows unpurified RNAs to be used. Crude preparations of RNP complexes (e.g., ribosomes, snRNPs, etc.) may also be analyzed by this method. The primer extension method also allows the detection of chemically modified bases in RNA for which no conditions for strand scission exist, such as the 1-methyladenosine.[29] In addition, the method is not limited to nucleotides close enough to the 5′ or 3′ end of an RNA to be well resolved by gel electrophoresis; primers may be complementary to almost any portion of long RNA molecules. In analyzing large RNAs or complex protein binding sites, several primers spaced at roughly 200-nucleotide intervals are used.

There are several significant limitations to the use of the primer extension assay for footprinting. First, because the oligonucleotides used for primer extension are generally 14 to 17 bases in length, very short RNAs, and RNAs which have protein binding sites very near their 3′ ends, cannot be used. Second, some naturally occurring modified nucleotides, such as N2-methylguanosine and 2,6-dimethyladenosine found in rRNA, block reverse transcription[30,31] and may interfere with primer extension analysis. Finally, the primer extension method cannot detect secondary cleavages or modifications due to alterations in the folding of an RNA caused by a primary cleavage or modification event. Most of these problems may be resolved by using 5′ or 3′ end-labeled RNA for footprint analysis.

II. FOOTPRINTING PROCEDURES AND REAGENTS

A. REPRESENTATIVE FOOTPRINTING PROTOCOL

The conditions which permit both specific RNA-protein complex formation and reaction by the different probing reagents do not allow "cookbook" recipes for RNA footprinting to be given. As an example, the procedure used for locating the binding site(s) of the bacteriophage T4 gene *32* protein on the gene *32* mRNA[9] will be described in detail. The primer extension assay for footprinting the gene *32* protein binding site was chosen because when these experiments were initiated, no methods were available for obtaining significant quantities of pure (or homogenous) gene *32* mRNA. The gene *32* protein of bacteriophage T4 is a single-stranded DNA binding protein which binds cooperatively to single-stranded nucleic acids produced during phage DNA replication, recombination, and repair. The protein also regulates its own synthesis (reversibly) at the level of translation by specifically binding to a region near the initiation codon on the gene *32* mRNA (for details, see Reference 9). The reaction buffer used for footprinting the gene *32* protein binding site [60 m*M* NH_4Cl, 10 m*M* $Mg(OAc)_2$, 10 m*M* Tris-acetate, pH 7.4, and 6 m*M* β-mercaptoethanol] was chosen for its similarity to the buffer used to demonstrate the repression of gene *32* translation *in vitro* by exogenous gene *32* protein.[43] In this *in vitro* system, the addition of gene *32* protein to approximately 3 μ*M* caused the complete and specific repression of gene *32* translation. In the experiment described below, the protection of the gene *32* protein binding site on the gene *32* mRNA from RNAse A digestion is monitored in the presence of increasing concentrations of gene *32* protein.

The crude preparation of gene *32* mRNA used for footprinting was extracted from T4-infected cells as described in Hartz et al.[6] In this preparation, it has been estimated that the gene *32* mRNA represents approximately 1% of the total RNA by weight, corresponding to approximately 0.025 pmol of gene *32* mRNA per microgram of total RNA.[6] The most abundant RNAs in this crude preparation are the 16S and 23S rRNAs and various species of 5S RNA. Gene *32* protein was prepared as described by Bittner et al.[44] and was stored in 0.1 *M* NaCl, 20 m*M* Tris-Cl, pH 8.1, 1 m*M* EDTA, 0.5 m*M* dithiothreitol, and 10% (v/v) glycerol. The protein was judged to be substantially free of contaminating RNAse by comparing the integrity of the gene *32* mRNA preincubated with either 10 μ*M* gene *32* protein or gene *32* protein storage buffer using primer extension (see Figure 3).

Footprinting reactions (24 μl) each contained 6 μg of crude T4 RNA (approximately 0.15 pmol of gene *32* mRNA), gene *32* protein or protein storage buffer, and 1X reaction buffer. Reactions were preincubated for 10 min at 37°C to allow binding of the protein to the RNA; 6 μl of an RNAse A solution (3×10^{-6} U/μl in 1X reaction buffer) was then added and the reactions (total volume, 30 μl) incubated for 2.5 min at 37°C. Reactions were stopped by the addition of 30 μl of 1X reaction buffer containing 0.4 *M* NaOAc, pH 5.2, 20 m*M* EDTA, and 1 mg/ml *Escherichia coli* tRNA, followed immediately by two phenol extractions. The samples were then extracted twice with ether to remove residual phenol, and the digested RNAs were precipitated by adding one sixth volume of 3.0 *M* NaOAc, pH 5.2, and three volumes of ethanol. The samples were placed in a dry-ice ethanol bath for 25 min and the RNAs pelleted in an Eppendorf centrifuge (~14,000 × g) for 25 min at 4°C. The RNA pellets were washed with 0.5 ml of ice-cold 70% ethanol and centrifuged for 15 min at 4°C. The digested RNAs were then dried under vacuum, resuspended in a small volume (6 μl) of 1 m*M* EDTA, pH 8.0, and subsequently used in primer extension reactions using a gene *32* mRNA specific primer as described below.

The footprints obtained from the gene *32* mRNA in the presence of increasing concentrations of gene *32* protein were assayed by primer extension and are shown in Figures 4 and 5. A large region of the gene *32* mRNA near the initiation codon (labeled zero), extending from at least U(−53) to U(+11) is protected at the highest concentration of gene *32* protein tested (see Figure 4). At low concentrations of gene *32* protein (2 to 3 μ*M*), strong protection is limited to nucleotides U(−53) to U(−30), and this reflects the preferential binding of gene *32* protein to a site near the 5′ end of the gene *32* mRNA.[9] As the concentration of gene *32* protein is increased, the size of the protected region also increases due to the cooperative binding of the gene *32* protein to this region of the gene *32* mRNA. A control experiment showing that the protection of the gene *32* mRNA is limited to the region near the initiation codon is shown in Figure 5. Primer extension analysis using the RNA from a footprinting reaction containing 5 μ*M* gene *32* protein shows no protection of sequences in the (downstream) coding region of the gene *32* mRNA.

B. GENERAL NOTES ON FOOTPRINTING PROCEDURES

Control reactions (without added probe) containing RNA only, and RNA and protein, should be done to monitor the general background and to check for nuclease contamination of protein preparations. Pipetting larger volumes of materials (i.e., ≥2 μl) into footprinting reactions will increase the reproducibility of digestion/modification and protection. "Stop" buffers (even those containing SDS or metal chelators) may not completely stop the action of a chemical or enzymatic probe and should therefore contain a large excess of carrier RNA to reduce the action of the probe on the target RNA until it can be removed by phenol extraction or ethanol precipitation.

C. EXPERIMENTAL CONDITIONS

The experimental conditions chosen for footprinting usually represent a compromise of conditions which allow both formation of specific RNA-protein complexes and the reaction of the probing reagent. At least initially, the buffers and temperature chosen for the *in vitro* formation of protein-RNA complexes (and subsequent chemical or enzymatic probing) should be similar to those in which a meaningful biological effect can be demonstrated (e.g., in the case of gene *32* protein, translational repression). Often, protein-RNA complexes are formed at temperatures (e.g., 37°C) different from those used for reaction with chemical or enzymatic probes (e.g., 4°C). Footprinting reactions are often done at low temperatures to slow the dissociation of protein-RNA complexes and increase the amount of protection. A very short reaction time with a probe may also increase the amount of protection observed. As described below, cacodylate or HEPES buffers (near neutral pH) are compatible with

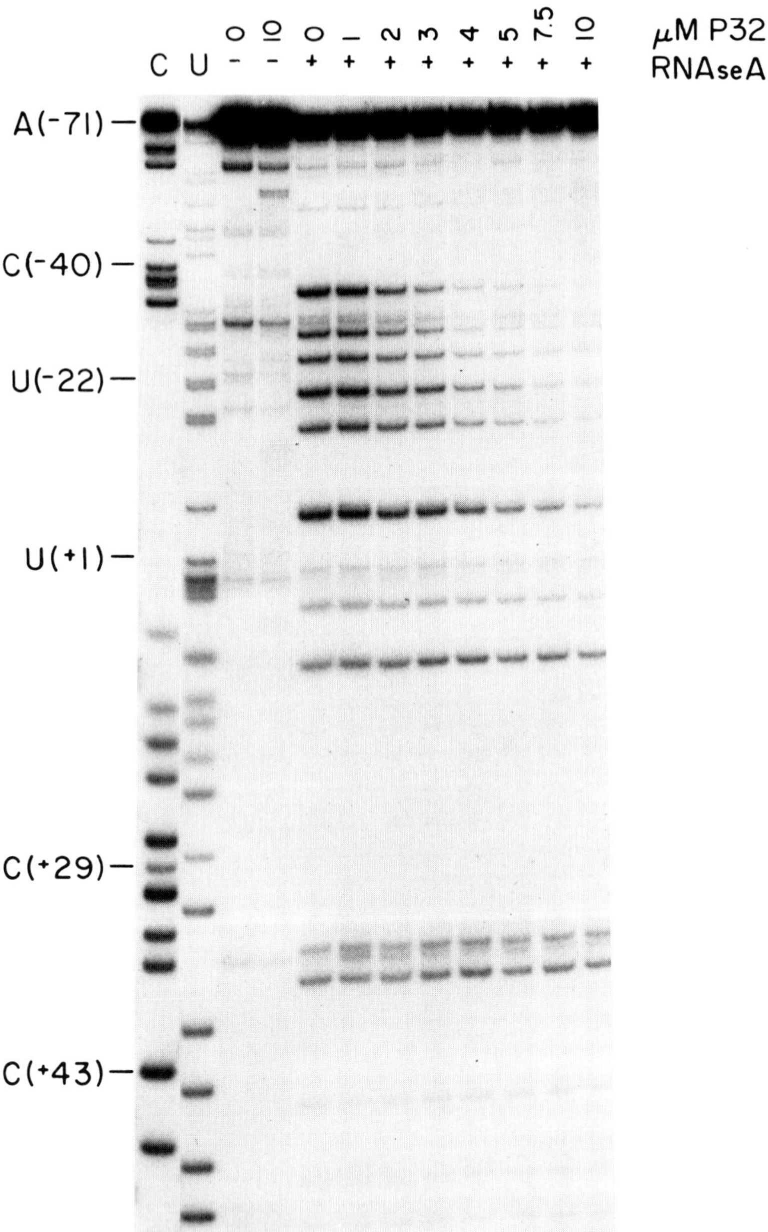

FIGURE 4. Footprinting of the gene *32* protein binding site near the initiation codon of the gene *32* mRNA using the primer extension assay. The 3′ end of the oligonucleotide primer used is located at nucleotide + 109. See text for details. (From McPheeters, D. S., Storma, G., and Gold, L., *J. Mol. Biol.*, 201, 517, 1988. With permission.)

the use of all of the chemical and enzymatic footprinting probes, and detailed footprinting studies are usually conducted in these buffers.

D. PROBES FOR FOOTPRINTING

The properties and conditions compatible with the use of several chemical and enzymatic probes for footprinting are discussed below. Overall, the use of chemical probes for footprinting is preferable to the use of enzymatic probes for several reasons. Due to the large

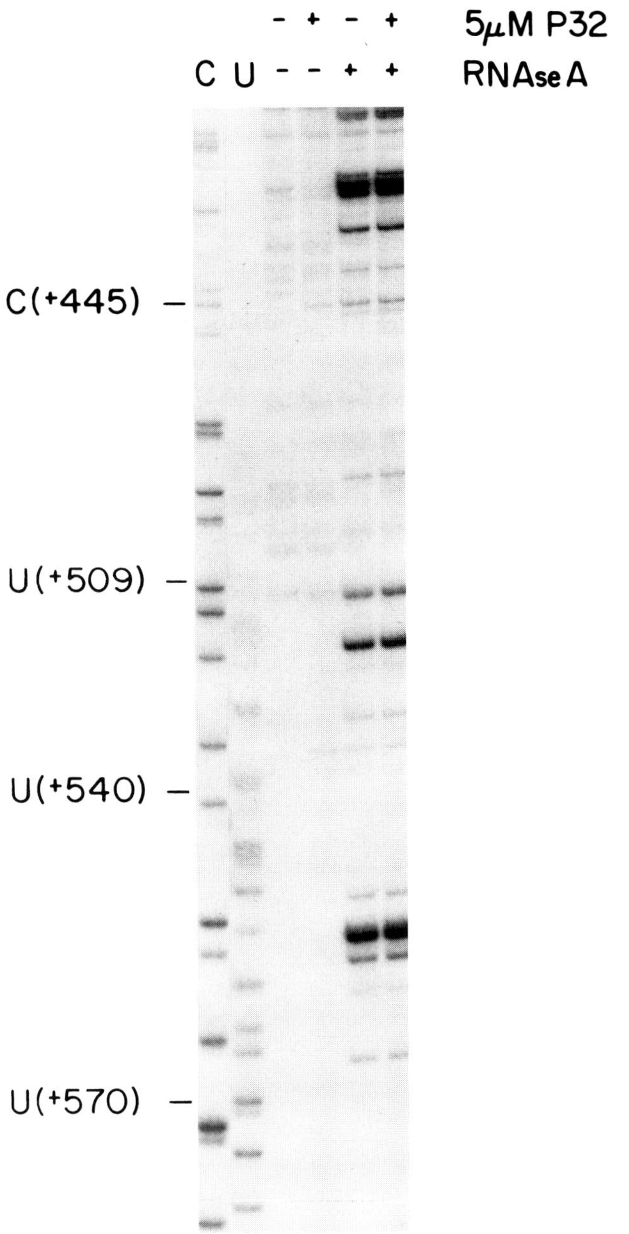

FIGURE 5. Control experiment showing that the protection by gene *32* protein seen in Figure 4 is specific for the region near the gene *32* initiation codon. The 3′ end of the nucleotide primer used is located at nucleotide +638. See text for details. (From McPheeters, D. S., Storma, G., and Gold, L., *J. Mol. Biol.*, 201, 517, 1988. With permission.)

size of nucleases, chemical probes are much less sensitive to steric hinderance and will provide more precise information on the involvement of individual atoms in base pairing and tertiary interactions, as well as protein and ionic contacts. Limited chemical modification of an RNA molecule is also less likely to alter the structure of the RNA than is similarly limited enzymatic cleavage of the RNA. However, many of the chemical probes are highly

TABLE 1
Examples of RNA Footprinting

Protein/ligand-RNA	Probes used[a]	Ref.
rProteins S5, S6, S11, S12, and S18 with 16S rRNA	DMS, kethoxal	37
S8 with 16S rRNA	DMS, CMCT, ENU T1, U2, S1, V1	34
rProteins S16, S17, and S20 with 16S rRNA	DMS, kethoxal	15
rProteins S6, S8, S15, and S18 with 16S rRNA	DMS, kethoxal, V1	39
rProteins S4 and 16S rRNA	DMS, kethoxal, V1	38
rProtein L18 with 5S rRNA	DMS, DEPC, A	12
Ribosomes with tRNA	DMS, DEPC	11
EF-G and EF-Tu with 16S rRNA	DMS, kethoxal	33
30S and 70S ribosomes	DMS, DEPC, CMCT T1, U2, S1, V1	74
30S ribosomes	DMS, CMCT, kethoxal	70
Antibiotics with 16S rRNA	DMS, kethoxal	32
AMV reverse transcriptase with $tRNA^{Trp}$	DMS, ENU, V1, N.C.	75
rProteins L5, L18, and L25 with 5S rRNA	α-Sarcin	76
rProteins S2, S3, S10, S13, and S14 with 16S rRNA	DMS, kethoxal, V1	36
rProteins S7, S9, and S19 with 16S rRNA	DMS, kethoxal, V1	35
tRNA with 16S rRNA	DMS, kethoxal, CMCT	40
R17 coat with R17 RNA	S1, A, U2	13
T4 *reg*A protein with *rIIB* mRNA	A, T1, T2	14
T4 gene *32* protein with gene *32* mRNA	A, T1	9
T4 gene *43* protein with gene *43* mRNA	A, T1	41
L10-L7/L12 with *rpl*J mRNA	DMS (DEPC)	77
TFIIIA with 5S rRNA	DMS, kethoxal, CMCT, T1, T2, V1 α-Sarcin	78 79
SRP proteins with SRP RNA	α-Sarcin	80
Aspartyl tRNA synthetase with $tRNA^{Asp}$	ENU	81
Aminoacyl tRNA synthetases with tRNA*Phe*, $tRNA^{Val}$	ENU, benzylamine	82
Valyl tRNA synthetase with $tRNA^{Val}$	ENU	64
Aminoacyl tRNA synthetases with $tRNA^{Val}$ and $tRNA^{Phe}$	V1	83
Methionyl tRNA synthetase with $tRNA_f^{Met}$	DMS, α-sarcin, A, T1	84
Arginyl tRNA synthetase with $tRNA_2^{Arg}$, $tRNA_3^{Arg}$	V1	85
RNAse P protein C5 with M1 RNA	T1	86

[a] Abbreviations used; DMS, dimethyl sulfate; CMCT, 1-cyclohexyl-3-(2-morpholinoethyl)-carbodiimide metho-*p*-toluene-sulfonate; DEPC, diethylpyrocarbonate; T1, RNAse T1; T2, RNAse T2; A, RNAse A; S1, nuclease S1; V1, V1 nuclease; U2, RNAse U2; N.C., *N. crassa* nuclease.

reactive and, in addition to reacting with nucleic acids, may react with the protein and cause dissociation of protein-RNA complexes. The small size of chemical probes and their attack only at specific atoms of a base may cause a base contacted by a bound protein to exhibit protection from some probes while remaining sensitive to others. For each chemical and enzymatic probe discussed below, individual reaction times and probe-to-substrate ratios must be varied to achieve optimal limited modification or cleavage of an RNA. Table 1 summarizes examples of RNA footprinting experiments, and the different chemical and enzymatic probes which have been used.

In general, reactions with enzymatic probes are stopped by the addition of a large excess of carrier RNA, followed immediately by at least two phenol extractions and two ether extractions. Reactions done with chemical probes are stopped by the addition of carrier RNA and immediate ethanol precipitation. Several washes of modified RNA pellets with 70% ethanol are required to insure complete removal of the chemical probe. Modified RNAs are subsequently resuspended in 0.2 *M* NaOAc, pH 5.2, extracted twice with phenol, then ether, and reprecipitated with ethanol.

1. Enzymatic Probes

RNAse A — RNAse A (mol wt, ≈13,680) is an endoribonuclease which specifically cleaves the 3′ phosphodiester bond of single-stranded pyrimidine residues and exhibits a slight preference for cleavage of the sequence 5′-UA-3′. The pH optimum of RNAse A is around 7.0 to 7.5.[45] Typically, 10^{-3} to 10^{-6} U of RNAse A will partially digest 1 μg of RNA in 2 to 30 min at 37°C. For applications of RNAse A in footprinting studies, see Table 1.

RNAse T1 — RNAse T1 (mol wt, ≈11,000) is an endoribonuclease which cleaves the 3′ phosphodiester bonds of single-stranded guanosines. The enzyme exhibits two pH optima near 7.5 and 4.5[46] and is active under a wide variety of buffer conditions. Typically, 5×10^{-3} U of RNAse T1 will partially digest 1 μg of RNA in 2 to 30 min at 37°C. For applications of RNAse T1 in footprinting studies, see Table 1.

RNAse T2 — RNAse T2 (mol wt, ≈36,000) is a single-strand specific endoribonuclease which shows a slight preference for cleaving after single-stranded adenosines. Although the pH optimum of this enzyme is around pH 4.5, it is still very active at neutral pH.[47] RNAse T2 is inhibited by some heavy metals, especially Cu.[47] At neutral pH, 10^{-3} to 10^{-5} U of RNAse T2 will partially digest 1 μg of RNA in 15 to 30 min at 37°C. For applications of RNAse T2 in footprinting studies, see Table 1.

RNAse V1 — RNAse V1 (mol wt, ≈15,900) from cobra venom is an endoribonuclease which preferentially cleaves double-stranded or structured regions of RNA in a relatively sequence-independent manner.[48,49] RNAse V1 is active between pH 6.5 and 8.5 and requires both Mg^{2+} (≥1 m*M*) and Na^+ for activity.[49,50] Typical reaction conditions for this enzyme are 25 m*M* Tris-Cl, pH 7.5, 200 m*M* NaCl, and 10 m*M* $MgCl_2$. Using these conditions, 4×10^{-2} to 5×10^{-3} U of RNAse V1 will partially digest 1 μg of RNA in a 10- to 30-min incubation at 37°C. For applications of RNAse V1 in footprinting studies, see Table 1.

α-Sarcin — α-Sarcin (mol wt, ≈17,400) is an endoribonuclease which cleaves on the 3′ side of purines in both single- and double-stranded regions of RNA.[51,52] This enzyme has been used very successfully for footprinting (see Table 1), but has the disadvantage that its activity is inhibited by both Mg^{2+} and K^+. Footprinting with α-sarcin is usually performed by forming protein-RNA complexes in the presence of low levels of Mg^{2+} and then diluting the complexes into a buffer without Mg^{2+} before addition of the enzyme.

S1 nuclease — S1 nuclease (mol wt, ≈32,000) is a nonspecific endonuclease which preferentially cuts in single-stranded regions of both RNA and DNA.[53] Although the pH optimum of this enzyme is around 4.5, several investigators have used high concentrations of S1 (0.2 of 4 U of nuclease S1 per μg of RNA) for structural[54] and footprinting (Table 1) studies at neutral pH. S1 nuclease requires Zn^{2+} for activity and is inhibited by both NaCl and phosphate.[53]

***Neurospora crassa* nuclease** — *N. crassa* nuclease (mol wt, ≈ 55,000) acts as a single-strand specific endonuclease which preferentially cuts after guanosines.[55] The enzyme contains Co^{2+}, and its activity is inhibited by both EDTA and reducing agents. The pH optimum of this enzyme is around 7.5 to 8.0, although it is 50% active at pH 5.8 and 9.3.[55] For applications of the *N. crassa* nuclease under native conditions, see Reference 56 and Table 1.

2. Chemical Probes

DMS — At neutral pH, dimethylsulfate (DMS) reacts strongly with the N7 of guanosine and the N1 of adenine and less strongly with the N3 of cytidine.[29,57] Methylation of both N1-A and N3-C is limited to single-stranded residues and can be detected directly by the primer extension assay.[23,25,29] The methylation of the N7 guanosine detects tertiary interactions, as well as protein and ionic contacts, but cannot be detected using the primer extension assay without cleavage of the modified RNA.[29] Cleavage at methylated guanosines requires reduction with sodium borohydride, followed by aniline-induced strand scission.[23,25] Reactions done with DMS work well between 0 and 90°C and at pH 4.5 to 8.5.[57] However, the pH of the buffer used for DMS reactions must be at least 7.5 unless the reaction time is less than 2 min; otherwise, the reaction will become too acidic as the DMS hydrolyzes into sulfuric acid and methanol, and this may cause random depurination.[25] DMS is very insoluble in water and is usually diluted immediately before use in either 95% ethanol or emulsified in reaction buffer. Because DMS reacts with primary amines, buffers such as Tris cannot be used. Footprinting reactions with DMS should be done in either cacodylate or HEPES buffers (see Reference 25). For footprinting reactions, typically, 0.5 μl of DMS is added to a 200-μl reaction containing 12.5 μg of RNA (target + carrier) in a 50 m*M* Na-cacodylate, pH 7.5, buffer and allowed to react for 2 min at 37°C. For applications of DMS in footprinting studies, see Table 1.

DEPC — At neutral pH, diethylpyrocarbonate (DEPC) reacts primarily with the N7 of adenosines and slightly with the N7 of guanosines. Both of these modifications can be detected directly with the primer extension assay.[29] The reaction of N7-A appears to be sensitive to stacking interactions that result from base pairing or tertiary interactions.[57] Reactions with DEPC work well between 0 and 90°C and in a variety of buffers (e.g., cacodylate or HEPES buffers, see Reference 57). Peattie et al.[12] reported that DEPC can cause the dissociation of L18-5S rRNA complexes (presumably by reacting with histidine residues), but found that a large molar excess of the L18 protein over 5S rRNA stabilized the complexes. Typical reaction conditions used for footprinting with DEPC are the same as those used for DMS (see above), except that 10 to 20 μl of fresh DEPC is used and the reactions are incubated at 37°C for 45 min. For applications of DEPC in footprinting studies, see Table 1.

Kethoxal — β-Ethoxy-α-ketobutyraldehyde (kethoxal, UpJohn) reacts with the N1 and N2 of unpaired guanosines to produce a kethoxal-guanosine adduct which is most stable at, and below, neutral pH.[58] The kethoxal-guanosine adduct is detectable directly in the primer extension assay. Typically, reactions with kethoxal (50 μl) contain 12.5 μg of RNA and are done in HEPES buffer using 1 to 5 μl of a kethoxal stock solution (37 mg/ml in 20% ethanol). Reactions are incubated at 0 to 4°C for 0.5 to 3 h. Reactions are terminated as described above, except that all buffers, including those used for primer extension, should contain 25 m*M* K-borate, pH 7.0, to stabilize the kethoxal-guanosine adduct. For applications of kethoxal in footprinting studies, see Table 1.

CMCT — At pH 8.0, 1-cyclohexyl-3-(2-morpholinoethyl)-carbodiimide metho-*p*-toluene-sulfonate (CMCT) reacts primarily with single-stranded uridines at N3 and slightly with single-stranded guanosines at N1.[59,60] CMCT modification of both uridines and guanines can be detected directly in the primer extension assay. CMCT reactions must be done at pH 8.0 because at acidic or neutral pH, CMCT attacks terminal phosphate groups and leaves bases unaffected.[59] A typical footprinting reaction with CMCT is done in 200 μl of 50 m*M* Na-borate, pH 8.0, 300 m*M* KCl, and 20 m*M* $Mg(OAc)_2$ containing 12.5 μg of RNA (target + carrier), and approximately 50 μl of a CMCT stock solution (42 mg/ml in reaction buffer) is added to each reaction. Reactions are incubated at 37°C for 15 to 30 min. For applications of CMCT in footprinting studies, see Table 1.

ENU — Ethylnitrosourea (ENU) is an alkylating reagent which reacts primarily with the phosphate groups of nucleic acids.[61,62] ENU has been used to study tRNA structure in solution,[63,64] where it has been found that all phosphates, except those involved in tertiary interactions and ionic interactions, are susceptible to ethylation. ENU also reacts with cysteine (see Reference 34). To be detected by the primer extension assay, the modified RNA must first be cleaved at ethylated phosphates. For footprinting reactions, typically, 2.5 to 5.0 μl of an ENU-saturated (700 m*M*) ethanol solution is added to a 20-μl reaction containing

12.5 μg of RNA (target + carrier) in 50 m*M* Na-cacodylate (pH 6.8 to 8.0) buffer, and the reactions are incubated at 37°C for 1 to 3 h. ENU is removed by two ethanol precipitations, and the modified RNA is resuspended in 10 μl of 0.1 *M* Tris-Cl, pH 9.0, and 1 m*M* EDTA. Incubation of this solution for 5 min at 50°C is used to split the modified RNA at ethylated phosphates. The RNA is then precipitated by the addition of 100 μl of 0.5 *M* NaOAc, pH 5.2, and 2.5 volumes of ethanol, followed by phenol extraction and ethanol precipitation as described above. For applications of ENU in footprinting studies, see Table 1.

III. PRIMER EXTENSION

A. PRIMER DESIGN

The main advantage of the primer extension assay for footprinting is that crude RNA preparations can be used because the hybridization of the oligonucleotide primer is specific for the target RNA. Statistically, primers of 12 and 17 nucleotides are long enough to hybridize uniquely within the *E. coli* and human genomes, respectively.[65] High temperatures (e.g., 48°C) are often used in primer extension reactions with AMV reverse transcriptase to avoid problems encountered with stable secondary structures in the template RNA. To assure both unique and stable hybrid formation, synthetic DNA primers 14 to 18 bases in length, with a composition of at least 40% (G + C), should be used. Primers should also be designed to anneal to a region of the target RNA at least 20 to 40 nucleotides downstream of the suspected protein binding site. Nonspecific background tends to be highest in the first five to ten nucleotides added to the primer by AMV reverse transcriptase.[87] Once a potential primer sequence has been chosen, it is useful to examine other sequences to which the primer may hybridize imperfectly using computer-assisted searches, paying particular attention to potential base pairing with the 3′ end of the primer. Imperfect hybridization to abundant RNAs, such as the ribosomal RNAs present in crude RNA preparations, is sometimes observed.

B. 5′ END-LABELING OF PRIMERS

For primer extension reactions using crude RNA preparations, it is preferable to label the cDNA by using a 5′ end-labeled primer rather than incorporating an [α^{32}P]-dNTP during the extension reaction. The use of end-labeled primers avoids potential background caused by initiation of cDNA synthesis from randomly hybridized nucleic acid fragments. Purified oligonucleotide primer (10 pmol) is 5′ end labeled in a 10-μl reaction containing 50 μCi of [γ^{32}P]-ATP (≥3000 Ci/mmol), 50 m*M* Tris-Cl, pH 7.6, 10 m*M* $MgCl_2$, 5 m*M* dithiothreitol, 0.1 m*M* spermidine, 0.1 m*M* EDTA, and 2 U of T4 polynucleotide kinase. The reaction is incubated at 37°C for 45 min. Then 40 μl of water is added to the reaction and the mixture is heated to 95°C for 2 min. The mixture is then extracted once with an equal volume of phenol to $CHCl_3$ (50:50) and extracted twice with ether. After removal of the ether, the aqueous phase can simply be dried completely under vacuum and the labeled primer resuspended in 20 μl of 1 m*M* EDTA, pH 8.0, at approximately 0.5 pmol/μl.

C. PRIMER EXTENSION

The protocol for primer extension presented below was originally described by Inoue and Cech[29] and has been modified as described in McPheeters et al.[66] The procedure for footprinting samples is simplified so that annealing and extension are done in the same tube. Using this method, at least 250 bases can be resolved on a 20-cm 8% sequencing gel.

1. Materials

AMV reverse transcriptase (15 to 25 U/μl) should be obtained from Life Sciences (St. Petersburg, FL) and stored in small aliquots at −70°C. Working aliquots of the enzyme may be stored at −20°C for up to 4 weeks without a detectable loss of activity. 10X RT(−Mg) buffer contains 0.5 *M* Tris-Cl, pH 8.3, 0.6 *M* NaCl, and 0.1 *M* dithiothreitol. 10X RT(+Mg) buffer is identical to the 10X RT(−Mg) buffer, except that it also contains 60 m*M* $MgCl_2$. Reverse transcriptase dilution buffer contains 0.2 *M* K-phosphate, pH 7.2, 2 m*M* dithiothreitol, 0.2% (v/v) Triton X-100, and 50% (v/v) glycerol. The 10X dNTP mixture contains 3.75 m*M* each dATP, dCTP, dGTP, and dTTP in 1X RT(+Mg) buffer. For sequencing reactions, four separate 10X ddNTP mixtures are prepared (one each for A, C, G, and T), each containing 2 m*M* ddNTP in 1X RT(+Mg) buffer. The ddNTP concentration of each of the 10X ddNTP mixtures may need adjustment in order to optimize the intensity of bands in the sequencing lanes. A dilution of reverse transcriptase (RT mix) is prepared immediately before use and contains 1.0 U/μl of AMV reverse transcriptase and 12% (v/v) reverse transcriptase dilution buffer in 1X RT(+Mg) buffer. Sample loading buffer for gel electrophoresis contains 94% (v/v) freshly deionized formamide, 0.04% (w/v) bromophenol blue, 0.04% (w/v) xylene cyanol, and 0.4X TBE.

2. Annealing of Primers

For each modified or cleaved RNA sample, combine 0.075 pmol of target RNA (e.g., one half of the RNAse A-treated T4 RNA sample described above) and 0.3 pmol of 5′ end-labeled primer in 5 μl of 1X RT(−Mg) buffer. The mixtures are heated to 60°C for 3 min to denature the RNA and allow annealing of the primer and then frozen in a dry-ethanol bath for 1 min. The annealing mixtures are allowed to thaw on ice, and 1 μl of 36 m*M* $MgCl_2$ in 1X RT(+Mg) buffer is added to each tube (final volume, 6 μl). The annealing procedure for sequencing reactions is identical, except a mixture containing 0.4 pmol of untreated target RNA and 1.2 pmol of 5′ end-labeled primer in 20 μl of 1X RT(−Mg) buffer is used, and 4 μl of 36 mM $MgCl_2$ in 1X RT(+Mg) buffer is added. The addition of EDTA to a final concentration of 1 m*M* in annealing mixtures helps reduce random hydrolysis of the RNA during the heating step and does not interfere with subsequent reverse transcription.

3. Extension Reactions

Primer extension reactions (final volume, 10μl) with modified or cleaved RNA samples each contain 6.0 μl of annealing mix, 1.0 μl of 10X dNTP solution, and 2.0 μl of 1X RT(+Mg) buffer. Sequencing reactions (final volume, 10 μl) are done in four separate tubes (one each for A, C, G, and T), each containing 4 μl of annealing mix, 1 μl of 10X dNTP solution, 1 μl of 10X ddNTP solution, and 3 μl of 1X RT(+Mg) buffer. Freshly prepared reverse transcriptase mix (1 μl) (RT mix) is added to each reaction and the reactions are incubated at 48°C for 15 to 30 min. Reactions are stopped by the addition of 20 μl of freshly prepared sample loading buffer and placed on ice. Immediately before electrophoresis, the samples should be denatured by heating to 95°C for 2 min and placed on ice. Approximately 5 μl from each reaction should be loaded on a gel for an overnight exposure; low amounts of template RNA in the annealing mixtures will lengthen the exposure time. The remainder of each extension reaction can be stored at −20°C and may be used for at least 1 week.

4. Notes on Primer Extension

The heat and quick-cool method for annealing primers to RNA works very well and, in general, gives better reproducibility and lower background than slow cooling methods for annealing primers. The method also works well for annealing primers within the strong hairpins of *E. coli* 16S rRNA and on low-abundance RNAs.[87] In general, a two- to tenfold molar excess of primer over template should be used; higher concentrations of primer may cause increased mishybridization.[65] Primer extension reactions done with AMV reverse transcriptase can be done at temperatures as high as 48°C; at this temperature, potential background due to the spurious annealing of primers and most problems with stable secondary structures in the template RNA are avoided.

Several peculiarities of reverse transcriptase are worth noting. First, several investigators,[67,68] have noted that reverse transcriptase may add an extra nucleotide to the 3′ end of cDNAs in a template-independent fashion. This problem occurs only rarely and is readily apparent when a base-specific chemical or enzymatic probe is used. AMV reverse transcriptase has also been observed to initiate cDNA synthesis from a primer with a mismatched 3′ nucleotide.[69] Although AMV reverse transcriptase has little problem transcribing through very stable secondary structures (see, for example, Reference 70). Turek et al.[71] have demonstrated very strong inhibition of primer extension by CUUCGG hairpins. Shelness and Williams[67] have also noted weak primary sequence effects on pausing of reverse transcription.

IV. CONCLUDING REMARKS

The methods and reagents for footprinting protein or other ligand binding sites on RNA using a primer extension assay have been described. The major advantages of the primer extension method for footprinting are that (1) it does not require purified or labeled RNA; (2) it is easily used in conjunction with studies of RNA structure and comparative sequence analysis; (3) it allows the detection of many chemically modified bases without the need for prior strand scission; and (4) it can be used to analyze RNAs of almost any length.

Footprinting is only one of the methods available for studying protein-RNA interactions. Identification of the RNA binding site of a protein by footprinting provides only limited information on what features of the RNA may be important for recognition. A combination of footprinting with other techniques, such as detailed mutational analysis and quantitative binding studies, is required to more fully understand the recognition process. Currently, the interactions between the bacteriophage R17 coat protein[3] and the *E. coli* ribosomal protein S8,[34] and their respective RNA binding sites, are the most well-understood examples of specific protein-RNA interactions. This is, in large part, due to the ability to easily purify the RNAs to which these proteins bind. New techniques for RNA synthesis *in vitro* from cloned genes with either T7 or SP6 RNA polymerases[72,73] should allow spectacular advances in the understanding of other protein-RNA interactions.

REFERENCES

1. **Shetlar, M. D.,** Crosslinking of proteins to nucleic acids by ultraviolet light, in *Photochemical and Photobiological Reviews,* Vol. 2, Smith, K. C., Ed., Plenum Press, New York, 1980, 105.
2. **Schimmel, P. R.,** Understanding the recognition of transfer RNAs by aminoacyl transfer RNA synthetases, *Adv. Enzymol.,* 49, 187, 1979.
3. **Uhlenbeck, O. C., Wu, H.-N., and Sampson, J. R.,** Recognition of RNA by proteins, in *Molecular Biology of RNA, New Perspectives,* Inouye, M. and Dudock, B. S., Eds., Academic Press, San Diego, 1987, 285.
4. **Campbell, K. and Gold, L.,** Construction of the bacteriophage T4 replication machine: regulation of synthesis of component proteins, in *Interaction of Translational and Transcriptional Controls in the Regulation of Gene Expression,* Gurnberg-Manago, M. and Safer, B., Eds., Elsevier, New York, 1982, 69.
5. **Normanly, J., Ogden, R. C., Horvath, S. J., and Abelson, J.,** Changing the identity of a transfer RNA, *Nature,* 321, 213, 1986.
6. **Hartz, D., McPheeters, D. S., Traut, R., and Gold, L.,** Extension inhibition analysis of translation initiation complexes, in *Methods in Enzymology,* Vol. 164, Noller, H. F. and Moldave, K., Eds., Academic Press, San Diego, 1988, 419.

7. **Tsugawa, A., Kurihara, T., Zuber, M., Court, D. L., and Nakamura, Y.,** *E. coli* NusA protein binds *in vitro* to an RNA sequence immediately upstream of the *boxA* signal of bacteriophage lambda, *EMBO J.*, 4, 2337, 1985.
8. **Garrett, R. A., Vester, B., Leffers, H., Sorenson, P. M., Kjems, J., Olesen, S. O., Christensen, J., and Douthwaite, S.,** Mechanisms of protein-RNA recognition and assembly in ribosomes, in *Proc. of the Alfred Benzon Symp.*, Vol. 19, Clark, B. F. C. and Petersen, H. U., Eds., Munksgaard, Copenhagen, 1983, 331.
9. **McPheeters, D. S., Stormo, G., and Gold, L.,** Autogenous regulatory site on the bacteriophage T4 gene *32* messenger RNA, *J. Mol. Biol.*, 201, 517, 1988.
10. **Thurlow, D. L., Ehresmann, C., and Ehresmann, B.,** Nucleotides in 16S rRNA that are required in unmodified form for features recognized by ribosomal protein S8, *Nucleic Acids Res.*, 11, 6787, 1983.
11. **Peattie, D. A. and Herr, W.,** Chemical probing of the tRNA-ribosome complex, *Proc. Natl. Acad. Sci. U.S.A.*, 78, 2273, 1981.
12. **Peattie, D. A., Douthwaite, S., Garrett, R. A., and Noller, H. F.,** A "bulged" double helix in a RNA-protein contact site, *Proc. Natl. Acad. Sci. U.S.A.*, 78, 7331, 1981.
13. **Carey, J., Cameron, V., de Haseth, P. L., and Uhlenbeck, O. C.,** Sequence specific interaction of R17 coat protein with its ribonucleic acid binding site, *Biochemistry,* 22, 2601, 1983.
14. **Winter, R. B., Morrissey, L., Gauss, P., Gold, L., Hsu, T., and Karam, J.,** Bacteriophage T4 regA protein binds to mRNAs and prevents translation initiation, *Proc. Natl. Acad. Sci. U.S.A.*, 84, 7822, 1987.
15. **Stern, S., Changchien, L.-M., Craven, G. R., and Noller, H. F.,** Interaction of proteins S16, S17 and S20 with 16S ribosomal RNA, *J. Mol. Biol.*, 200, 291, 1988.
16. **Turner, D. H., Sugimoto, N., Jaeger, J. A., Longfellow, C. E., Freier, S. M., and Kierzek, R.,** Improved parameters for prediction of RNA structure, *Cold Spring Harbor Symp. Quant. Biol.*, 52, 123, 1987.
17. **Ehresmann, C., Baudin, F., Mougel, M., Romby, P., Ebel, J.-P., and Ehresmann, B.,** Probing the structure of RNAs in solution, *Nucleic Acids Res.*, 15, 9109, 1987.
18. **Steitz, J. A.,** Oligonucleotide sequence of replicase initiation site in Qβ RNA, *Nature,* 236, 71, 1972.
19. **Bernardi, A. and Spahr, P.-F.,** Nucleotide sequence for coat protein on RNA of bacteriophage R17, *Proc. Natl. Acad. Sci. U.S.A.*, 69, 3033, 1972.
20. **Zimmerman, R. A., Mackie, G. A., Muto, A., Garrett, R. A., Ungewickell, E., Ehresmann, C., Stiegler, P., Ebel, J.-P., and Feller, P.,** Location and characteristics of ribosomal protein binding sites in the 16S RNA of *Escherichia coli, Nucleic Acids Res.*, 2, 279, 1975.
21. **Brownlee, G. G.,** *Determination of Sequences in RNA,* North-Holland, Amsterdam, 1972.
22. **Mount, S. M., Pettersson, I., Hinterberger, M., Karmas, A., and Steitz, J. A.,** The U1 small nuclear RNA-protein complex selectively binds a 5′ splice site *in vitro, Cell,* 33, 509, 1983.
23. **Peattie, D. A.,** Direct chemical method for sequencing RNA, *Proc. Natl. Acad. Sci. U.S.A.*, 76, 1760, 1979.
24. **Vournakis, J. N., Celantano, J., Finn, M., Lockard, R. E., Mitra, T., Pavlakis, G., Troutt, A., van den Berg, M., and Wurst, R. M.,** Sequence and structure analysis of end-labeled RNA with nucleases, in *Gene Amplification and Analysis,* Vol. 2, Chirikjian, J. G. and Papas, T., Eds., Elsevier, New York, 1981, 267.
25. **Peattie, D.,** Direct chemical method for sequencing end-labeled ribonucleic acids, in *Methods of DNA and RNA Sequencing,* Weissman, S. M., Ed., Praeger, New York, 1983, 261.
26. **RajBhandary, U. L., Lockard, R. E., and Reilly, R. M.,** Use of nucleases in RNA sequence and structural analysis, in *Nucleases,* Linn, S. M. and Roberts, R. J., Eds., Cold Spring Harbor Laboratory, Cold Spring Harbor, NY, 1982, 275.
27. **Donis-Keller, H., Maxam, A. M., and Gilbert, W.,** Mapping adenines, guanines, and pyrimidines in RNA, *Nucleic Acids Res.*, 4, 2527, 1977.
28. **Qu, L. H., Michot, B., and Bachellerie, J.-P.,** Improved methods for probing large RNAs, *Nucleic Acids Res.*, 11, 5903, 1983.
29. **Inoue, T. and Cech, T. R.,** Secondary structure of the circular form of the *Tetrahymena* rRNA intervening sequence: a technique for RNA structure analysis using chemical probes and reverse transcriptase, *Proc. Natl. Acad. Sci. U.S.A.*, 82, 648, 1985.
30. **Youvan, D. C. and Hearst, J. E.,** Reverse transcriptase pauses at N^2-methylguanine during in vitro transcription of *Escherichia coli* 16S ribosomal RNA, *Proc. Natl. Acad. Sci. U.S.A.*, 76, 3751, 1979.
31. **Hagenbuchle, O., Santer, M., Steitz, J. A., and Mans, R. J.,** Conservation of the primary structure at the 3′ end of 18S rRNA from eucaryotic cells, *Cell,* 13, 551, 1978.
32. **Moazed, D. and Noller, H. F.,** Interaction of antibiotics with functional sites in 16S ribosomal RNA, *Nature,* 327, 389, 1987.
33. **Moazed, D., Robertson, J. M., and Noller, H. F.,** Interaction of elongation factors EF-G and EF-Tu in a conserved loop of 23S RNA, *Nature,* 334, 362, 1988.

34. **Mougel, M., Eyermann, F., Westhof, E., Romby, P., Expert-Bezancon, A., Ebel, J.-P., Ehresmann, B., and Ehresmann, C.,** Binding of *Escherichia coli* ribosomal protein S8 to 16S rRNA, *J. Mol. Biol.,* 198, 91, 1987.
35. **Powers, T., Changchien, L.-M., Craven, G. R., and Noller, H. F.,** Probing the assembly of the 3′ major domain of 16S ribosomal RNA, *J. Mol. Biol.,* 200, 309, 1988.
36. **Powers, T., Stern, S., Changchien, L.-M., and Noller, H. F.,** Probing the assembly of the 3′ major domain of 16S ribosomal RNA, *J. Mol. Biol.,* 201, 697, 1988.
37. **Stern, S., Powers, T., Changchien, L.-M., and Noller, H. F.,** Interaction of ribosomal proteins S5, S6, S11, S12, S18 and S21 with 16S rRNA, *J. Mol. Biol.,* 201, 683, 1988.
38. **Stern, S., Wilson, R. C., and Noller, H. F.,** Localization of the binding site for protein S4 on 16S ribosomal RNA by chemical and enzymatic probing and primer extension, *J. Mol. Biol.,* 192, 101, 1986.
39. **Svensson, P., Changchien, L.-M., Craven, G. R., and Noller, H. F.,** Interaction of ribosomal proteins S6, S8, S15 and S18 with the central domain of 16S ribosomal RNA, *J. Mol. Biol.,* 200, 301, 1988.
40. **Moazed, D. and Noller, H. F.,** Transfer RNA shields specific nucleotides in 16S ribosomal RNA from attack by chemical probes, *Cell,* 47, 985, 1986.
41. **Andrake, M., Guild, N., Hsu, T., Gold, L., Tuerk, C., and Karam, J.,** DNA polymerase of bacteriophage T4 is an autogenous translational repressor, *Proc. Natl. Acad. Sci. U.S.A.,* 85, 7942, 1988.
42. **Sanger, F., Nicholen, S., and Coulson, A. R.,** DNA sequencing with chain-terminating inhibitors, *Proc. Natl. Acad. Sci. U.S.A.,* 74, 5463, 1977.
43. **Lemaire, G., Gold, L., and Yarus, M.,** Autogenous translational repression of bacteriophage T4 gene 32 expresison *in vitro, J. Mol. Biol.,* 126, 73, 1978.
44. **Bittner, M., Burke, R. L., and Alberts, B.,** Purification of T4 gene 32 protein free from detectable deoxyribonuclease activities, *J. Biol. Chem.,* 254, 9565, 1979.
45. **Brown, D. M. and Todd, A. R.,** Evidence on the nature of the chemical bonds in nucleic acids, in *The Nucleic Acids,* Vol. 1, Chargaff, E. and Davidson, J. N., Eds., Academic Press, New York, 1953, 409.
46. **Uchida, T. and Egami, F.,** Ribonuclease T1 from taka-diastase, *Methods Enzymol.,* 12, 228, 1967.
47. **Uchida, T. and Egami, F.,** Ribonuclease T2 from taka-diastase, *Methods Enzymol.,* 12, 239, 1967.
48. **Vassilenko, S. K., Butorin, A. S., Maev, S. P., Vityugov, F. I., Boldyreva, L. G., and Rait, V. K.,** Specificity of nuclease from oxus cobra venom for polyribonucleotide macrostructure, *Mol. Biol.,* 17, 818, 1983.
49. **Lowman, H. B. and Draper, D. E.,** On the recognition of helical RNA by cobra venom V1 nuclease, *J. Biol. Chem.,* 261, 5396, 1986.
50. **Lockard, R. E. and Kumar, A.,** Mapping tRNA structure in solution using double-stranded-specific ribonuclease V1 from cobra venom, *Nucleic Acids Res.,* 9, 5125, 1981.
51. **Endo, Y., Huber, P. W., and Wool, I. G.,** The ribonuclease activity of the cytotoxin α-sarcin, *J. Biol. Chem.,* 258, 2662, 1983.
52. **Wool, I. G.,** The mechanism of action of the cytotoxic nuclease α-sarcin and its use to analyse ribosome structure, *Trends Biochem. Sci.,* 9, 14, 1984.
53. **Rushizky, G. W.,** S1 nuclease of *Aspergillus oryzae*, in *Gene Amplification and Analysis,* Vol. 2, Chirikjian, J. G. and Papas, T., Eds., Elsevier, New York, 1981, 205.
54. **Florentz, C., Briand, J. P., Romby, P., Hirth, L., Ebel, J. P., and Giege, R.,** The tRNA-like structure of turnip yellow mosaic virus RNA: structural organization of the last 159 nucleotides from the 3′ OH terminus, *EMBO J.,* 1, 269, 1982.
55. **Shishido, K. and Ando, T.,** Single-strand specific nucleases, in *Nucleases,* Linn, S. M. and Roberts, R. J., Eds., Cold Spring Harbor Laboratory, Cold Spring Harbor, NY, 1982, 155.
56. **Tal, J.,** The cleavage of transfer RNA by a single strand specific endonuclease from *Neurospora crassa, Nucleic Acids Res.,* 2, 1073, 1975.
57. **Peattie, D. A. and Gilbert, W.,** Chemical probes for higher-order structure in RNA, *Proc. Natl. Acad. Sci. U.S.A.,* 77, 4679, 1980.
58. **Shapiro, R. and Hachmann, J.,** The reaction of guanine derivatives with 1,2-dicarbonyl compounds, *Biochemistry,* 5, 2799, 1966.
59. **Ho, N. W. Y. and Gilham, P. T.,** The reversible chemical modification of uracil, thymine, and guanine nucleotides and the action of ribonuclease on ribonucleic acids, *Biochemistry,* 6, 455, 1967.
60. **Chang, S. E., Cashmore, A. R., and Brown, D. M.,** Selective modification of uridine and guanine residues in tyrosine transfer ribonucleic acid, *J. Mol. Biol.,* 68, 455, 1972.
61. **Sun, L. and Singer, B.,** The specificity of different classes of ethylating agents towards various sites of Hela cell DNA *in vitro* and *in vivo, Biochemistry,* 14, 1795, 1975.
62. **Singer, B. and Frankel-Conrat, H.,** The specificity of different classes of ethylating agents towards various sites in RNA, *Biochemistry,* 14, 772, 1975.
63. **Vlassov, V. V., Giege, R., and Ebel, J.-P.,** Tertiary structure of tRNAs in solution monitored by phosphodiester modification with ethylnitrosourea, *Eur. J. Biochem.,* 119, 51, 1981.

64. **Vlassov, V. V., Kern, D., Giege, R., and Ebel, J. P.,** Protection of phosphodiester bonds in yeast $tRNA^{Val}$ by its cognate aminoacyl-tRNA synthetase against alkylation by ethylnitrosourea, *FEBS Lett.*, 123, 277, 1981.
65. **Smith, M.,** Synthetic oligodeoxyribonucleotides as probes for nucleic acids and as primers in sequence determination, in *Methods of DNA and RNA Sequencing,* Weissman, S. M., Ed., Praeger, New York, 1983, 23.
66. **McPheeters, D. S., Christensen, A., Young, E. T., Stormo, G., and Gold, L.,** Translational regulation of expression of the bacteriophage T4 lysozyme gene, *Nucleic Acids Res.*, 14, 5813, 1986.
67. **Shelness, G. S. and Williams, D. L.,** Secondary structure analysis of apolipoprotein II mRNA using enzymatic probes and reverse transcriptase, *J. Biol. Chem.*, 260, 8637, 1985.
68. **Inoue, T., Sullivan, F. X., and Cech, T. R.,** New reactions of the ribosomal RNA precursor of *Tetrahymena* and the mechanism of self splicing, *J. Mol. Biol.*, 189, 143, 1986.
69. **Panabieres, F., Fort, Ph., Piechaczyk, M., Blanchard, J. M., and Jeanteur, Ph.,** A warning on the use of synthetic DNA primers for initiation of reverse transcription on RNA: unexpected initiation at a mismatched nucleotide, *Gene,* 19, 321, 1982.
70. **Moazed, D., Stern, S., and Noller, H. F.,** Rapid chemical probing of conformation in 16S ribosomal RNA and 30S ribosomal subunits using primer extension, *J. Mol. Biol.*, 187, 399, 1986.
71. **Tuerk, C., Gauss, P., Thermes, C., Groebe, D. R., Gayle, M., Guild, N., Stormo, G., d'Aubenton-Carafa, Y., Uhlenbeck, O. C., Tinoco, I., Brody, E. N., and Gold, L.,** CUUCGG hairpins: extraordinarily stable RNA secondary structures associated with various biochemical processes, *Proc. Natl. Acad. Sci. U.S.A.*, 85, 1364, 1988.
72. **Lowary, P., Sampson, J., Milligan, J., Groebe, D., and Uhlenbeck, O. C.,** A better way to make RNA for physical studies, in *Structure and Dynamics of RNA,* van Knippenberg, P. H. and Hilbers, C. W., Eds., Plenum Press, New York, 1986, 69.
73. **Krieg, P. A. and Melton, D. A.,** *In vitro* RNA synthesis with SP6 RNA polymerase, *Methods Enzymol.*, 155, 397, 1987.
74. **Baudin, F., Ehresmann, C., Romby, P., Mougel, M., Colin, J., Lempereur, L., Bachellerie, J.-P., Ebel, J.-P., and Ehresmann, B.,** Higher-order structure of domain III in *Escherichia coli* 16S ribosomal RNA, 30S subunit and 70S ribosome, *Biochemie,* 69, 1081, 1987.
75. **Garret, M., Romby, P., Giege, R., and Litvak, S.,** Interactions between avian myeloblastosis reverse transcriptase and $tRNA^{Trp}$. Mapping of complexed tRNA with chemicals and nucleases, *Nucleic Acids Res.*, 12, 2259, 1984.
76. **Huber, P. W. and Wool, I. G.,** Nuclease protection analysis of ribonucleoprotein complexes: use of the cytotoxic ribonuclease α-sarcin to determine the binding sites for *Escherichia coli* ribosomal proteins L5, L18, and L25 on 5S rRNA, *Proc. Natl. Acad. Sci. U.S.A.*, 81, 322, 1984.
77. **Climie, S. C. and Friesen, J. D.,** *In vivo* and *in vitro* structural analysis of the *rplJ* mRNA leader of *Escherichia coli*: protection by bound L10-L7/L12, *J. Biol. Chem.*, 263, 15166, 1988.
78. **Christiansen, J., Brown, R. S., Sproat, B. S., and Garrett, R. A.,** *Xenopus* transcription factor IIIA binds primarily at junctions between double helical stems and internal loops in oocyte 5S RNA, *EMBO J.*, 6, 453, 1987.
79. **Huber, P. W. and Wool, I. G.,** Identification of the binding site on 5S rRNA for the transcription factor IIIA: proposed structure of a common binding site on 5S rRNA and on the gene, *Proc. Natl. Acad. Sci. U.S.A.*, 83, 1593, 1986.
80. **Siegel, V. and Walter, P.,** Binding sites of the 19-kDA and 68/72-kDA signal recognition particle (SRP) proteins on SRP RNA as determined by protein-RNA "footprinting", *Proc. Natl. Acad. Sci. U.S.A.*, 85, 1801, 1988.
81. **Romby, P., Moras, D., Bergdoll, M., Dumas, P., Vlassov, V. V., Westhoff, E., Ebel, J.-P., and Giege, R.,** Yeast $tRNA^{Asp}$ tertiary structure in solution and areas of interaction of the tRAN with aspartyl-tRNA synthetase, *J. Mol. Biol.*, 184, 455, 1985.
82. **Vlassov, V. V., Kern, D., Romby, P., Giege, R., and Ebel, J.-P.,** Interaction of $tRNA^{Phe}$ and $tRNA^{Val}$ with aminoacyl-tRNA synthetases, *Eur. J. Biochem.*, 132, 537, 1983.
83. **Favorova, O. O., Fasiblo, F., Keith, G., Vassilenko, S. K., and Ebel, J.-P.,** Partial digestion of tRNA-aminoacyl-tRNA synthetase complexes with cobra venom ribonuclease, *Biochemistry,* 20, 1006, 1981.
84. **Pelka, H. and Schulman, L. H.,** Study of the interaction of *Escherichia coli* methionyl-tRNA synthetase with $tRNA^{fMet}$ using chemical and enzymatic probes, *Biochemistry,* 25, 4450, 1986.
85. **Gangloff, J., Jaozara, R., and Dirheimer, G.,** Study of the interaction of yeast arginyl-tRNA synthetase with yeast $tRNA_2^{Arg}$ and $tRNA_3^{Arg}$ by partial digestions with cobra venom ribonuclease, *Eur. J. Biochem.*, 132, 629, 1983.
86. **Vioque, A., Arnez, J., and Altman, S.,** Protein-RNA interactions in the RNAse P holoenzyme from *Escherichia coli, J. Mol. Biol.*, 202, 835, 1988.
87. **McPheeters, D. S.,** unpublished observations.

IV. GENE EXPRESSION

Chapter 13

PLASMID CLONING VECTORS FOR THE OVERPRODUCTION OF PROTEINS IN *ESCHERICHIA COLI*

Jim D. Karam

TABLE OF CONTENTS

I. INTRODUCTION

Escherichia coli[1] is host to a wide variety of well-studied bacteriophages, plasmids, and other mobile genetic elements that have been used to expand the utility of this prokaryote for the isolation, propagation, and study of expression of DNA from both prokaryotic and eukaryotic sources. Most *E. coli* cloning vectors in use today are chimeric DNA molecules consisting of segments from different plasmid, phage, cellular, or synthetic sources that have been spliced together by laboratory design. For example, in the construction of the general-purpose plasmid cloning vector pBR322, Bolivar et al.[2-4] utilized *in vitro* manipulations (restriction enzymes and ligations) and *in vivo* transpositions to combine a replication origin from the colE1-like plasmid pMB1[5] with drug resistance markers from other plasmids, i.e., pRSF2124[6] (ampicillin resistance, Amp^R) and pSC101[7] (tetracycline resistance, Tet^R). Subsequent additional manipulations of pBR322 by several investigators have yielded a vast collection of cloning vectors that are suitable for a variety of specific purposes.[8] Similarly, phage λ and other lambdoid phages have been successfully engineered for both general and specialized cloning purposes,[9] as has the single-stranded DNA phage M13,[10] which is particularly useful in situations where site-directed mutagenesis or DNA sequencing of cloned DNA is desired (see Chapters 3 and 6, this volume). Table 1 provides a simplified listing of the types of cloning vectors that are currently available for use with the *E. coli* system. Plasmids for the overproduction of proteins in this bacterium include a group of vectors carrying strong promoters that can be experimentally controlled. Many of the available *E. coli* expression plasmids utilize promoters that have been derived from phage chromosomes (e.g., λ and T7) or *E. coli* operons (lactose [*lac*], tryptophan [*trp*], galactose [*gal*], and outer lipoprotein [*lpp*]). Some such vectors have been designed to also provide translation initiation signals for use with cloned genetic sequences that require them, e.g., to allow translation of eukaryotic sequences (cDNA cloning) in this bacterium or to replace intrinsically weak signals for some prokaryotic genes. The expression of eukaryotic genes in *E. coli* requires special considerations because this organism does not recognize eukaryotic transcription and translation control signals and does not possess the machineries for specific posttranscriptional and posttranslational modification or processing of eukaryotic gene products. Usually, cDNA cloning in *E. coli* or coliphage vectors can circumvent many posttranscriptional hurdles, but the requirement for specific protein modifications (glycosylation, phosphorylation, proteolytic processing, etc.) can be much more of a problem. In this regard, eukaryotic or eukaryotic-prokaryotic shuttle vectors are often preferred in eukaryotic gene cloning for expression purposes (see Chapter 14, this volume). Also, many proteins that are not normal inhabitants of the *E. coli* cell are degraded by this bacterium's proteolytic enzymes. This is sometimes also the case for prokaryotic proteins, and the use of protease-deficient mutants of *E. coli*[11,12] as hosts for recombinant expression plasmids can help maximize yields of plasmid-encoded proteins.[13,14] It is important to recognize that more than one class of proteolytic enzymes is probably involved in foreign protein degradation in *E. coli*[12] and that different proteins may exhibit different susceptibilities to these enzymes.

II. CHOOSING THE RIGHT VECTOR FOR EXPRESSION OF CLONED GENES IN *E. COLI*

The manual by Pouwels et al.[8] provides a comprehensive list and detailed annotated descriptions of all types of prokaryotic and eukaryotic cloning vectors, including some of the most widely used expression vectors for *E. coli*. In essence, very few tightly regulated prokaryotic promoters are being used with a large number of different expression plasmids.[15] These include the leftward (PL) and rightward (PR) promoters of the phage λ, the *trp* (PTRP), *lac* (PLAC), and *lpp* (PLPP) promoters from *E. coli* and the "ϕ10" (late) promoter of phage

TABLE 1
Types of Vectors Commonly Used for Cloning Genes in *E. coli*

General-purpose vectors that allow direct selection of recombinant DNA in low or high copy number, with or without expression of the cloned DNA

Plasmids — Examples include pBR322,[3] the pUC family,[36] pACYC184,[37] and their numerous derivatives.
Lambdoid phages — Examples include the Charon family,[38,39] λgt10/11,[40] the λEMBL family,[41] the λNM family[9] (imm21 and imm434), and other insertion and substitution vectors.
M13 phages — An example is the M13mp series (currently at mp18/19[36]).
Cosmids — Examples include pHC79,[42] Homer,[43] pcosEMBL,[44] Lorist,[45] and others.
Phasmids (phagemids) — Examples include the pBGS family[46] and the pEMBL family[47] (all carrying *ori* sequences of pMB1 and f1) and pUC118/119,[48] which carry the *ori* sequences of pMB1 and M13.

Vectors that allow expression of cloned genes *in vivo* under experimentally controlled conditions

Plasmids — Examples include the λPL- and PR-, *E. coli trp-, lac-, tac-, lpp-, lpp* plus *lac-*, and other promoter-bearing plasmids listed in Table 2, as well as some of the general-purpose cloning vectors listed above (e.g., pUC plasmids).
Other vectors — Several lambdoid cloning vectors carry an active *lac* promoter that can be used to express cloned genes (e.g., λgt11), and some allow cloning and expressing genes downstream of the phage PL or PR promoters.

Vectors suitable for *in vitro* transcription of cloned DNA (These include many of the expression vectors mentioned above, but the following vectors allow selective transcription from promoters that are not recognized by *E. coli* RNA polymerase and by RNA polymerases that do not recognize *E. coli* promoters.)

Plasmids — Examples include several commercially available vectors carrying pairwise combinations of T7 and SP6, T7 and T3, and T3 and SP6 promoters that allow transcription in either orientation of the cloned DNA, e.g., the pGEM family from Promega Biotech; pSP6/T3, pSP6/T7, and pT7/T3 vectors from Bethesda Research Laboratories; and similar dual promoter plasmids from Pharmacia, Stratagene, and U.S. Biochemicals. (Also see Table 2.)
Other vectors — λ-Phages (λSWAJ series[49] and Stratagene's "Bluescript" lambda Zap II vector), cosmids (LoristB[50]), and phagemids (in the pGEM, pSP64/65, Stratagene's pBluescript II SK +/−, and other families) are available that allow transcription from promoters not recognized by *E. coli* RNA polymerase.

Expression vectors for detection of promoters or transcription termination sites in cloned DNA

Examples — These include the promoterless *galK* (galactokinase) plasmid vector pKO-1[51] and several of its derivatives, a number of promoterless *lacZα* and *lacZ*-containing plasmids (e.g., pMC1403[52,53]), pKK231/232[54] (activation of chloramphenicol-resistance) vectors, and others. Terminator probe vectors include pKG1800[51] (*galK* inactivation), pBU10[55] (Tet^R inactivation from PLAC), pKK65-10/pKK62b[56] (inactivation of a toxic insulin gene), and other vectors that utilize promoters from the *E. coli gal, trp,* and *rrnB* promoters and λPL.[8]

T7 (see Table 2 for references). There are additional ones, e.g., the *E. coli* PPHOA[16] (phosphate-regulated) and PCEA[17] (colicin E1 gene, SOS-responsive) promoters as well as promoters from other coliphages[18] that have been less frequently utilized, but that have proven to be ideal for specific purposes. Combinations and fusions between promoter sequences as well as their mutational modification have provided additional avenues for improving their utility, e.g., PTAC[19] (*lac-trp* promoter fusion), PLAC*UV5*[20] mutant promoter, and PLPP-LAC double promoter.[21] Table 2 lists a number of expression plasmids that carry some of the most widely used promoters for hyperexpression of cloned genes in *E. coli*. The listing is not intended to be an exhaustive catalog of *E. coli* expression plasmids; rather, it merely highlights commonly used options in the development of *E. coli*-based systems for expressing cloned genes of interest. *E. coli* host genotype is usually an important component of any such system, and the utility of the plasmids described in Table 2 for any specific purpose should be assessed in relation to the bacterial strains to be used for effecting expression of the cloned gene under study. The references given in Table 2 provide information about recommended hosts and growth conditions for the designated plasmids.

TABLE 2
***E. coli* Plasmid[a] Vectors for Protein Overproduction from Cloned Genes**

Plasmid designation	Controlling element(s)[b]	Cloning site(s)[b]	Other features	Ref.
			Regulation by λ-repressor (λ*c*I gene)	
pPL family[c]	λP_L	*Eco*RI, *Bam*HI, *Hin*dIII, *Bal*I	4.5 kb; Amp^R (pPLa group is also Kan^R); fusion to amino-terminal MS2 polymerase with pPLc24 (*Bam*HI and *Hin*dIII sites); RBS fusions also possible (i.e., PLc245 and derivatives)	57,58
pHUB[c]	λP_L	*Eco*RI, *Bam*HI, *Hpa*I, *Bal*I	7.6 kb; Kan^R and Tet^R; compatible with pRK248$C_I$ts (Tet^R), which can provide thermolabile λ repressor (*c*I857); similar to pPL plasmids	59
pCP[c]	λP_L	*Eco*RI, *Bam*HI, *Hin*dIII, *Pst*I, *Xba*I	7.1 kb; R1*drd*19 replicon, Amp^R; not amplifiable	60
pKC30[c]	λP_L*N*	*Hpa*I, *Bam*HI, *Sal*I	6.4 kb; Amp^R; cloning in HpaI site inactivates λ*N* gene and can generate protein fusion with amino-terminal gpN; cloning at sites distal to *N* (*Bam*HI, *Sal*I) allows antitermination of transcription	61
pAS1[c]	λP_L*c*II$_{RBS}$	*Bam*HI	5.9-kb derivative of pKC30 in which *Hpa*I-*Bam*HI interval was replaced by RBS of λc_{II} gene	61
pJL6[c] and derivatives	λP_L*c*II$_{RBS}$	*Cla*I, *Hin*dIII, and others	5.1 kb; Amp^R; fusions to amino-terminal gpC_{II} possible; similar to pAS-1	62
pGW7	λP_R*c*I857P_L*N*	*Eco*RI, *Bam*HI, *Sal*I	8.8 kb; Amp^R; self-regulation by plasmid encoded λC_I repressor offers significant host-range convenience; cloning under control of λgpN (transcription antitermination) possible for P_R (*Eco*RI site) and P_L (*Bam*HI site); cloning in HpaI site inactivates λgpN (fusions to amino-terminal λgpN possible); derivative deleted for 3′-terminal segment of λ*N* gene also available (pJM9[26])	G. Wilson (New England Biolabs); see References 23 and 26 for examples
pHE6	λ*c*I857$P_L$$N_{RBS}$	*Bam*HI, *Sma*I, *Sal*I, *Acc*I	4 kb; derived from pUC8 (pMB1 replicon, Amp^R); fusions to amino-terminal λgpN (33 or fewer amino acids) possible	63

pCQV2	λcI857P_R	*Bam*HI	5.3 kb; Amp^R; self-regulated; utilizes RBS from λ*cro* gene (ATG exposed by *Bam*HI, i.e., ATG GATCC); derivatives with MCS available[64]	65
Regulation by *E. coli trp* Repressor (*trpR* gene); Inducible by 3-β-Indolacrylic Acid[66] (Residual expression may be high because chromosomal *trpR* gene does not produce enough repressor to tightly repress *trp* operator on these multicopy plasmids.)				
pDR720	*E. coli* P_{TRP}	*Sal*I, *Bam*HI	4.1 kb; Amp^R; galK (from pKO-1[51], Table 1); does not provide RBS and AUG; fusion possible at 14, 20, and 27 bp from transcription start	Pharmacia LKB
pSTP2	*E. coli* P_{TRP} + RBS	*Cla*I and/or *Hind*III	3.7 kb; Amp^R; allows fusions at 2 to 7 bp from *trp* leader ATG	67,68 (for similar plasmids)
pTRS3	*E. coli* P_{TRP} + RBS	*Sph*I	4.0 kb; Amp^R; allows fusion directly after *trp* leader ATG	69
Regulation by *E. coli lac* Repressor (*lacI* gene); Inducible with Isopropyl-β-D-Thiogalactoside (IPTG), Lactose, and Other β-Galactosides				
pOP203 series	P_{LACUV5} + $lacZ_{RBS}$	*Eco*RI	4.9—5.6 kb; Amp^R and/or Tet^R; cloning at codons 8 and further from *lacZ* ATG; fused and unfused protein products possible	70
pMOB48, pMOB45	P_{LACUV5}	MCS	9.5 and 10.5 kb; R_1*drd*19 replicon, CM^R or CM^R and Tet^R	71
pUC family[28] derivatives	P_{LAVUV5} + $lacZ_{RBS}$	MCS	About 2.7 kb; Amp^R; cloning at codons 4 and further downstream of *lacZ* ATG; fused and unfused protein products possible	72
pKEN602	P_{LACUV5}—P_{LACUV5}	MCS	2.9 kb; Amp^R; similar to pUC vectors, but tandem *lac uv5* promoters; manipulations allow easy excision of cloned DNA from plasmid	73
$ptac_{11}$	P_{TAC}	*Eco*RI	4.6 kb; Amp^R and Tet^R; cloning at about seven nucleotides downstream of a Shine-Dalgarno sequence; no longer in frequent use	19,74
pDR540	P_{TAC} + RBS	*Bam*HI	4.1 kb; Amp^R; regulatory segment can be transferred to other vectors as a *Hind*III-*Bam*HI fragment; expression weaker than with other P_{TAC} plasmids	75
pTTQ series	P_{TAC} + $lacZ_{RBS}$ + $lacI^Q$	MCS	About 4 kb; Amp^R; self-regulation by $lacI^Q$ gene and features similar to pUC family	76

TABLE 2 (continued)
***E. coli* Plasmid[a] Vectors for Protein Overproduction from Cloned Genes**

Plasmid designation	Controlling element(s)[b]	Cloning site(s)[b]	Other features	Ref.
Regulation by λ-repressor (λ*c*I gene)				
pKK233-2	P_{TRC} + RBS	*Nco*I, *Pst*I, *Hind*III	4.6 kb; Amp^R; P_{TRC} has 17-bp spacing between the −35 and −10 sequences of the promoter (P_{TAC} has 16-bp spacing); has the transcription termination site of the *E. coli rrnB* gene downstream of cloning sites; *Nco*I exposes ATG	77
pSE380	P_{TRC} + RBS + *lacI*^{Q}	MCS	About 4.5 kb; Amp^R; derived from pKK233-2; has extended MCS ("superpolylinker"); self-regulation by *lacI*^{Q}	78
Regulation by Other Means in *E. coli* Cells				
pKH502 and others	P_{RECA}	*Bam*HI	4.2—5.0 kb; Amp^R; inducible by nalidixic acid or DNA damage	79,80
pT7-1	P_{T7}(ϕ10)	MCS of pUC12	2.4 kb; Amp^R; T7 RNA polymerase must be supplied in trans (see text)	29
pET vectors	P_{T7}(ϕ10)	MCS	4.4 kb; Amp^R; conditions for expression similar to pT7-1	30,31
pGEM and pSP70-73 vectors	P_{T7}(ϕ10) and P_{SP6}	MCS	2.4—3.2 kb; Amp^R; P_{T7} usable *in vivo* in Tabor and Richardson[29] or Studier and Moffat[30] systems (see text)	Promega; similar vectors are also available from Stratagene, Pharmacia LKB, and others
pJFK, pTG, and pDM1, 1-T3 vectors	P_{T7}(ϕ10) and P_{T3}, and T3 gene 1		2.4 to 7.8 kb; Amp^R; allow *in vivo* expression under control of T3 RNA polymerase	32

[a] Except where indicated otherwise, all the plasmids listed carry the replicon of pMB1. Also, plasmids that cannot be used to generate unfused proteins are not listed.

[b] Cloning sites not relevant to expression by the plasmid-borne promoters are not included in the listing. The abbreviation MCS (multicloning site) is used in a generic sense to denote that cleavage at several restriction sites is possible; RBS = ribosome binding site.

One concern that may influence the choice of a plasmid expression system for a cloned gene is potential toxicity of the gene product to the *E. coli* host. Even residual levels of expression under the best conditions for repression of a promoter can be a problem. In one approach to diminish such expression, Podhajska et al.[22] constructed a PLAC plasmid in which the promoter could be maintained in an orientation opposite to that of the cloned gene as long as host cells were kept at 30°C; a brief exposure to 42°C induced an inversion (by λ-integrase-mediated recombination) to the alternate orientation and allowed activation of gene expression by IPTG. In another type of approach, one can first clone the gene under study in nonexpressible form, i.e., in a "general-purpose" cloning vector (see Table 1), and subsequently transfer defined segments encompassing the cloned gene to more than one type of expression vector in order to assess toxicity of the gene product and genetic stability of the recombinant plasmid under conditions of residual and high-level expression. For example, in the case of the T4 RegA protein (which is highly toxic to *E. coli*,[23] it has been shown that residual expression leads to selection for mutations in the cloned structural gene[23-25] and that genetic stability depends, at least in part, on cloned DNA sequences that map between the plasmid-borne promoter and the protein-encoding segment of the gene.[23] Another reason that justifies trying several different expression vectors with the same gene of interest is the contribution of factors beyond transcription on final levels, activity, and recovery of the protein product. As transcription of the same gene via different heterologous promoters generates different leader sequences in the mRNA for the protein, overall stability of the mRNA and structure around the translation initiation region may differ in different constructs. Consequently, synthesis of the protein may vary with different promoters of similar strengths or even with the same promoter in different vectors.[26] Also, more is not always better. Note, for example, that heat induction of a λPL promoter at 42°C (which gave maximal inactivation of a controlling thermolabile repressor) yielded largely insoluble protein (inclusion bodies) from a cloned T4 gene *43* (DNA polymerase), whereas heat induction at a lower temperature (41°C) resulted in a lower level of gene expression, but much higher levels of recoverable enzyme.[27] A recent review by Schein[28] describes some causes for inclusion bodies and offers suggestions for dealing with them.

III. VECTORS FOR SELECTIVE EXPRESSION OF CLONED GENES

Any plasmid-borne promoter that requires *E. coli* RNA polymerase for expression is bound to find competition from the hundreds of other promoters residing in the same *E. coli* cell. Usually, the use of a high-copy-number plasmid with a strong promoter such as λPL or PTAC works in favor of plasmid-directed gene expression, but this can also mean a higher residual level of synthesis of the cloned gene product (see Table 2). Cloning vectors have been developed which carry promoters that are not recognized at all by the host RNA polymerase. Examples are promoters from coliphages T7 (PT7) and T3 (PT3) and the *Salmonella typhimurium* phage SP6 (PSP6). Such promoters have been used in the construction of a variety of dual-promoter plasmids in which a multicloning site (MCS) is placed between two apposing transcription start sites (see Table 2). Most of these vectors are particularly useful for *in vitro* synthesis of sense and anti-sense RNA corresponding to the gene of interest, but some (namely, those harboring the T7 ϕ10 promoter, i.e., PT7 and the analogous T3 phage promoter, PT3) can be used for protein hyperproduction *in vivo* as well. Two types of host systems have been developed to allow for controlled expression of the PT7 promoter *in vivo*. One system, designed by Tabor and Richardson,[29] utilizes two-plasmid conditions in which one plasmid (derived from pACYC184, P15A replicon) encodes T7 RNA polymerase under the control of a heat-inducible λPL, while the other plasmid (which must be compatible with pACYC184, e.g., pMB1 replicon) harbors the gene of interest under the control of PT7. Thus, overproduction of the desired protein is technically initiated by a temperature upshift of the two-plasmid-carrying cells. In another type of T7 expression system, which was designed by Studier and Moffatt,[30] the plasmid carrying the PT7-controlled gene is placed in an *E. coli* host (named BL21[DE3]) containing a chromosomal insertion of T7 gene 1 (T7 RNA polymerase gene) under PLAC*UV5* control. In this system, the gene cloned under PT7 control is activated by the addition of a *lac* operon inducer, which turns on the PLAC*UV5*-controlled T7 gene 1. Thus, any of a number of commercially available PT7-bearing plasmids can be used for cloning genes in standard *E. coli* hosts where the promoter is inactive (no residual expression), and then transferred to a host in which T7 RNA polymerase can be supplied. Alternatively, the PT7 can be activated via infection of cells harboring the promoter-bearing recombinant plasmid with λ phage-harboring T7 gene 1.[30] A variety of additional T7 expression plasmids (pET vectors) have been developed by the Studier laboratory;[31] some provide transcription initiation, and some provide both transcription and translation initiation sites for the cloned gene. An *in vivo* expression system that utilizes PT3 and T3 RNA polymerase has also been described.[32]

IV. DESIGNING A PROTOCOL FOR MEASURING PLASMID-DIRECTED GENE EXPRESSION

As the level of synthesis of a desired protein from the cloned gene can depend on many factors, it is advisable to examine clones for gene expression in small amounts of culture before scaling up for protein overproduction. We generally test clones for residual (constitutive) expression as well as for expression under the fully induced condition. A culture of plasmid-bearing cells (2 to 5 ml) is grown in a shaker bath under the uninduced condition in synthetic medium containing the antibiotic to which the bacteria are resistant and all amino acids, except methionine (M9-19 medium[23]). At mid-log phase (1 to 5 $\times$ 10^8 cells per milliliter), a small aliquot (0.1 to 1 ml) of culture is transferred to the condition that initiates induction, and another aliquot is maintained in the uninduced condition (see Table 2). After 20 to 30 min, ^{35}S-methionine (NEN, catalog number NEG 009A) or ^{35}S-trans label (ICN, catalog number 51006) is added to each aliquot at a concentration of about 5 μCi/ml and aeration is continued at the two conditions for another 30 min. The cultures are then chilled in an ethanol-ice slurry to stop further incorporation of isotope and the cells are subsequently harvested by centrifugation, extracted in an SDS-mercaptoethanol buffer, and analyzed by SDS-PAGE and autoradiography. Examples are described by Miller et al.[23] and Hsu et al.[33] Autoradiograms can be scanned by densitometry to quantitate induced levels. By varying conditions for induction (e.g., host, temperature, concentration of inducer, and time of exposure to inducer), one can usually obtain baseline data for optimal protein production rather quickly. Sometimes, a change of media, e.g., from M9-19 to LB broth can have effects on residual expression and inducibility with chemical inducers. Such changes can also be monitored with small amounts of culture. For example, the ^{35}S-trans label mixture sold by ICN can label bacterial proteins in a rich medium like LB to levels that are suitable for subsequent autoradiographic or fluorographic analysis. Alternatively, simple Coomassie blue staining or, if antibody is available, Western blots can be used.[23]

V. CONCLUDING REMARKS

There is no single *E. coli* plasmid expression vector that is suitable for all applications. There is also no ideal host environment for expression of plasmid-borne genes. Best conditions are usually determined by experimentation with different vectors, hosts, and growth conditions and can depend heavily on the biological and physical properties of the protein being sought. The vectors currently available either commercially or from those investigators who have pioneered in the art of plasmid design (Table 2; also see Reference 8) make it possible for all of us to get into the act, provided we take the time to understand the biology that underlies the utility of these vectors and are not reluctant to modify them to fit our specific purposes. It appears at this stage that we are reaching a saturation point relative to the number of ways we can usefully manipulate prokaryotic transcription signals for cloning purposes. Future trends in this general area will probably exploit posttranscriptional (particularly translational) controlling elements. Examples include the use of a "translational enhancer" sequence from a T7 gene[34,83] and the use of "translational coupling" to improve the frequency of initiation at weak ribosome binding sites.[35] Undoubtedly, many ingenious new ways of expressing cloned genes in *E. coli* will have been described in the literature by the time this book is published.

ACKNOWLEDGMENTS

I thank Jürgen Brosius for comments and advice on the manuscript. Work in my laboratory is supported by grants from the National Institute of General Medical Sciences (GM18842) and the National Science Foundation (DMB-8810062).

REFERENCES

1. **Neidhardt, F. C., Ingraham, J. L., Low, K. B., Magasanik, B., Schaechter, M., and Umbarger, H. E., Eds.,** *Escherichia coli and Salmonella typhimurium: Cellular and Molecular Biology,* American Society of Microbiology, Washington, D.C., 1987.
2. **Bolivar, F., Rodriguez, R. L., Betlach, M. C., and Boyer, H. W.,** Construction and characterization of new cloning vehicles. I. Ampicillin-resistant derivatives of the plasmid pMB9, *Gene,* 2, 75, 1977.
3. **Bolivar, F., Rodriguez, R. L., Greene, P. J., Betlach, M. C., Heyneker, H. L., and Boyer, H. W.,** Construction and characterization of new cloning vehicles. II. A multipurpose cloning system, *Gene,* 2, 95, 1977.
4. **Bolivar, F.,** Plasmid pBR322: the multipurpose cloning vector, *Focus,* 10, 61, 1988.
5. **Betlach, M. C., Hershfield, V., Chow, L., Brown, W., Goodman, H. M., and Boyer, H. W.,** A restriction endonuclease analysis of the bacterial plasmid controlling the EcoR I restriction and modification of DNA, *Fed. Proc.,* 35, 2037, 1976.
6. **So, M., Gill, R., and Falkow, S.,** The generation of a ColE1-Ap^r cloning vehicle which allows detection of inserted DNA, *Mol. Gen. Genet.,* 142, 239, 1976.
7. **Cohen, S. N., Chang, A. C. Y., Boyer, H. W., and Helling, R. B.,** Construction of biologically functional bacterial plasmids *in vitro, Proc. Natl. Acad. Sci. U.S.A.,* 70, 3240, 1973.
8. **Pouwels, P. H., Enger-Valk, B. E., and Brammar, W. J.,** *Cloning Vectors: A Laboratory Manual,* Elsevier, New York, 1988.
9. **Murray, N. E.,** Phage lambda and molecular cloning, in *Lambda II,* Hendrix, R., Roberts, J., Stahl, F., and Weisberg, R., Eds., Cold Spring Harbor Laboratory, Cold Spring Harbor, NY, 1983, 395.
10. **Yanisch-Perron, C., Vieira, J., and Messing, J.,** Improved M13 phage cloning vectors and host strains: nucleotide sequences of the M13mp18 and pUC19 vectors, *Gene,* 33, 103, 1985.
11. **Strauss, D. B., Walter, W. A., and Gross, C. A.,** *Escherichia coli* heat shock mutants are defective in proteolysis, *Genes Dev.,* 2, 1851, 1988.
12. **Goldberg, A. L. and Goff, S. A.,** The selective degradation of abnormal proteins in bacteria, in *Maximizing Gene Expression,* Reznikoff, W. and Gold, L., Eds., Butterworths, Stoneham, MA, 1986, 287.
13. **Goff, S. and Goldberg, A. L.,** Production of abnormal proteins in *E. coli* stimulates transcription of *lon* and other heat shock genes, *Cell,* 41, 587, 1985.
14. **Buell, G., Shulz, M. F., Selzer, G., Challet, A., Movva, N. R., Semon, D., Escanez, S., and Kawashima, E.,** Optimizing the expression in *E. coli* of a synthetic gene encoding somatomedin-C (IGF-I), *Nucleic Acids Res.,* 13, 1923, 1985.
15. **Brosius, J.,** Expression vectors employing lambda-, *trp-*, *lac-*, and *lpp*-derived promoters, in *Vectors: A Survey of Molecular Cloning Vectors and Their Uses,* Rodriguez, R. L. and Denhardt, D. T., Eds., Butterworths, Boston, 1988, 205.

16. **Oka, T., Sakamoto, S., Miyoshi, K., Fuwa, T., Yoda, K., Yamasaki, M., Tamura, G., and Miyake, T.,** Synthesis and secretion of human epidermal growth factor by *Escherichia coli*, *Proc. Natl. Acad. Sci. U.S.A.*, 82, 7212, 1985.
17. **Lupski, J. R., Ruiz, A. A., and Godson, G. N.,** Promotion, termination, and anti-termination in the *rpsU-dnaG-rpoD* macromolecular synthesis operon of *E. coli* K-12, *Mol. Gen. Genet.*, 195, 391, 1984.
18. **Stueber, D., Ibrahimi, I., Cutler, D., Dobberstein, B., and Bujard, H.,** A novel *in vitro* transcription-translation system: accurate and efficient synthesis of single proteins from cloned DNA sequences, *EMBO J.*, 3, 3143, 1984.
19. **de Boer, H. A., Comstock, L. J., and Vasser, M.,** The *tac* promoter: a functional hybrid derived from the *trp* and *lac* promoters, *Proc. Natl. Acad. Sci. U.S.A.*, 80, 21, 1983.
20. **Reznikoff, W. S. and Abelson, J. N.,** The *lac* promoter, in *The Operon*, Miller, J. H. and Reznikoff, W. S., Eds., Cold Spring Harbor Laboratory, Cold Spring Harbor, NY, 1978, 221.
21. **Nakamura, K. and Inouye, M.,** Construction of versatile expression cloning vehicles using the lipoprotein gene of *Escherichia coli*, *EMBO J.*, 1, 771, 1982.
22. **Podhajska, A. J., Hasan, N., and Szybalski, W.,** Control of cloned gene expression by promoter inversion *in vivo*: construction of the heat-pulse-activated *att-nutL-p-att-N* module, *Gene*, 40, 163, 1985.
23. **Miller, E. S., Karam, J., Dawson, M., Trojanowska, M., Gauss, P., and Gold, L.,** Translational repression: biological activity of plasmid-encoded bacteriophage T4 RegA protein, *J. Mol. Biol.*, 194, 397, 1987.
24. **Trojanowska, M., Miller, E. S., Karam, J., Stormo, G., and Gold, L.,** The bacteriophage T4 *regA* gene: primary sequence of a translational repressor, *Nucleic Acids Res.*, 12, 5979, 1984.
25. **Adari, H. Y., Rose, K., Williams, K. R., Konigsberg, W. H., Lin, T. C., and Spicer, E.,** Cloning, nucleotide sequence, and overexpression of the bacteriophage T4 regA gene, *Proc. Natl. Acad. Sci. U.S.A.*, 82, 1901, 1985.
26. **Liang, Y., Wei, R., Hsu, T., Alford, C., Dawson, M., and Karam, J.,** Autogenous regulation of the *regA* gene of bacteriophage T4: derepression of translation, *Genetics*, 119, 743, 1988.
27. **Lin, T.-C., Rush, J., Spicer, E., and Konigsberg, W. H.,** Cloning and expression of T4 DNA polymerase, *Proc. Natl. Acad. Sci. U.S.A.*, 84, 7000, 1987.
28. **Schein, C. H.,** Production of soluble recombinant proteins in bacteria, *BioTechnology*, 7, 1141, 1989.
29. **Tabor, S. and Richardson, C. C.,** A bacteriophage T7 RNA polymerase/promoter system for controlled exclusive expression of specific genes, *Proc. Natl. Acad. Sci. U.S.A.*, 82, 1074, 1985.
30. **Studier, F. W. and Moffatt, B. A.,** Use of bacteriophage T7 RNA polymerase to direct selective high-level expression of cloned genes, *J. Mol. Biol.*, 189, 113, 1986.
31. **Rosenberg, A. H., Lade, B. N., Chui, D.-S., Lin, S.-W., Dunn, J. J., and Studier, F. W.,** Vectors for selective expression of cloned DNAs by T7 RNA polymerase, *Gene*, 56, 125, 1987.
32. **Giordano, T. J., Denschle, U., Bujard, H., and McAllister, W. T.,** Regulation of coliphage T3 and T7 RNA polymerases by the *lac* repressor-operator system, *Gene*, 84, 209, 1989.
33. **Hsu, T., Wei, R., Dawson, M., and Karam, J. D.,** Identification of two new bacteriophage T4 genes that may have roles in transcription and DNA replication, *J. Virol.*, 61, 366, 1987.
34. **Olins, P. O., Devine, C. S., Rangwala, S. H., and Kavka, K. S.,** The T7 phage gene 10 RNA, a ribosome-binding site that dramatically enhances expression of foreign genes in *Escherichia coli*, *Gene*, 73, 227, 1988.
35. **Makoff, A. J. and Smallwood, A. E.,** The use of two-cistron constructions in improving the expression of a heterologous gene in *E. coli*, *Nucleic Acids Res.*, 18, 1711, 1990.
36. **Norrander, J., Kempe, T., and Messing, J.,** Construction of improved M13 vectors using oligodeoxynucleotide-directed mutagenesis, *Gene*, 26, 101, 1983.
37. **Chang, A. C. Y. and Cohen, S. N.,** Construction and characterization of amplifiable multicopy DNA cloning vehicles derived from the P15A cryptic miniplasmid, *J. Bacteriol.*, 134, 1141, 1978.
38. **De Wet, J., Daniels, D. L., Schroeder, J. L., Williams, B. G., Denniston-Thompson, K., Moore, D. D., and Blattner, F. R.,** Restriction maps for twenty-one Charon vector phages, *J. Virol.*, 33, 401, 1980.
39. **Dunn, J. S. and Blattner, F. R.,** Charons 36 to 40: multi enzyme, high capacity, recombination deficient replacement vectors with polylinkers and polystuffers, *Nucleic Acids Res.*, 15, 2677, 1987.
40. **Huynh, T. V., Young, R. A., and Davis, R. W.,** Constructing and screening cDNA libraries in λgt10 and λgt11, in *DNA Cloning*, Vol. 1, Glover, D. M., Ed., IRL Press, Oxford, 1985, 49.
41. **Frischauf, A.-M., Lehrach, H., Poustka, A., and Murray, N.,** Lambda replacement vectors carrying polylinker sequences, *J. Mol. Biol.*, 170, 827, 1983.
42. **Hohn, B. and Collins, J.,** A small cosmid for efficient cloning of large DNA fragments, *Gene*, 11, 291, 1980.
43. **Chia, W., Scott, M. R. D., and Rigby, P. W. J.,** The construction of cosmid libraries of eukaryotic DNA using the Homer series of vectors, *Nucleic Acids Res.*, 10, 2503, 1982.

44. **Poustka, A., Rackwitz, H.-R., Frischauf, A.-M., Hohn, B., and Lehrach, H.,** Selective isolation of cosmid clones by homologous recombination in *Escherichia coli, Proc. Natl. Acad. Sci. U.S.A.,* 81, 4129, 1984.
45. **Gibson, T. J., Coulson, A. R., Sulston, J. E., and Little, P. F. R.,** Lorist 2, a cosmid with transcriptional terminators insulating vector genes from interference by promoters within the insert: effect on DNA yield and cloned insert frequency, *Gene,* 53, 275, 1987.
46. **Spratt, B. G., Hedge, P. J., Heesen, S., Edelman, A., and Broome-Smith, J. K.,** Kanamycin-resistant vectors that are analogues of plasmids pUC8, pUC9, pEMBL8 and pEMBL9, *Gene,* 41, 337, 1986.
47. **Cesareni, G. and Murray, J. A. H.,** Plasmid vectors carrying the replication origin of filamentous single-stranded phages, in *Genetic Engineering,* Vol. 9, Setlow, K., Ed., Plenum Publishing Corporation, New York, 1987, 135.
48. **Vieira, J. and Messing, J.,** Production of single-stranded plasmid DNA, in *Methods of Enzymology,* Vol. 153, Wu, R., Grossman, L., and Moldave, K., Eds., Academic Press, New York, 1987, 3.
49. **Palazzolo, M. J. and Meyerowitz, E. M.,** A family of lambda phage cDNA cloning vectors, λSWAJ, allowing the amplification of RNA sequences, *Gene,* 52, 197, 1987.
50. **Cross, S. H. and Little, P. F. R.,** A cosmid vector for systematic chromosome walking, *Gene,* 49, 9, 1986.
51. **McKenney, K., Shimatake, H., Court, D., Schmeissner, U., Brady, C., and Rosenberg, M.,** A system to study promoter and terminator signals recognized by *Escherichia coli* RNA polymerase, in *Gene Amplification and Analysis,* Vol. 2, Chirikjian, J. G. and Papas, T. S., Eds., Elsevier/North-Holland, Amsterdam, 1981, 383.
52. **Casadaban, M. J., Martinez-Arias, A., Shapira, S. K., and Chou, J.,** β-galactosidase gene fusion for analyzing gene expression in *Escherichia coli* and yeast, in *Methods of Enzymology,* Vol. 100, Wu, R., Ed., Academic Press, New York, 1983, 293.
53. **Shapira, S. K., Chou, J., Richaud, F. V., and Casadaban, M. J.,** New versatile plasmid vectors for expression of hybrid proteins coded by a cloned gene fused to *lacZ* gene sequences encoding an enzymatically active carboxy-terminal portion of β-galactosidase, *Gene,* 25, 71, 1983.
54. **Brosius, J.,** Plasmid vectors for the selection of promoters, *Gene,* 27, 151, 1984.
55. **Gentz, R., Langner, A., Chang, A. C. Y., Cohen, S. N., and Bujard, H.,** Cloning and analysis of strong promoters is made possible by the downstream placement of a RNA termination signal, *Proc. Natl. Acad. Sci. U.S.A.,* 78, 4936, 1981.
56. **Brosius, J.,** Toxicity of an overproduced foreign gene product in *Escherichia coli* and its use in plasmid vectors for the selection of transcription terminators, *Gene,* 27, 161, 1984.
57. **Remaut, E., Stanssens, P., and Fiers, W.,** Plasmid vectors for high-efficiency expression controlled by the P_l promoter of coliphage lambda, *Gene,* 15, 81, 1981.
58. **Remaut, E., Stanssens, P., and Fiers, W.,** Inducible high level synthesis of mature human fibroblast interferon in *Escherichia coli, Nucleic Acids Res.,* 11, 4677, 1983.
59. **Bernard, H., Remaut, E., Hershfield, M. V., Das, H. K., and Helinski, D. R.,** Construction of plasmid cloning vehicles that promote gene expression from the bacteriophage lambda P_l promoter, *Gene,* 5, 59, 1979.
60. **Remaut, E., Tsao, H., and Fiers, W.,** Improved plasmid vectors with a thermoinducible expression and temperature-regulated runaway replication, *Gene,* 22, 103, 1983.
61. **Rosenberg, M.,** The use of pKC30 and its derivatives for controlled expression of genes, in *Methods of Enzymology,* Vol. 101, Wu, R., Ed., Academic Press, New York, 1983, 123.
62. **Lautenberger, J. A., Court, D., and Papas, T. S.,** High-level expression in *Escherichia coli* of the carboxy-terminal sequences of the avian myelocytomatosis virus (MC29) *v-myc* protein, *Gene,* 23, 75, 1983.
63. **Milman, G.,** Expression plasmid containing the λP_L promoter and cI857 repressor, in *Methods in Enzymology,* Vol. 153, Wu, R. and Grossman, L., Eds., Academic Press, New York, 1987, 482.
64. **Botterman, J. and Zabeau, M.,** A standardized vector system for manipulation and enhanced expression of genes in *Escherichia coli, DNA,* 6, 583, 1987.
65. **Queen, C.,** A vector that uses phage signals for efficient synthesis of proteins in *Escherichia coli, J. Mol. Appl. Genet.,* 2, 1, 1983.
66. **Nichols, B. P. and Yanofsky, C.,** Plasmids containing the *trp* promoters of *Escherichia coli* and *Serratia marcescens* and their use in expressing cloned genes, in *Methods of Enzymology,* Vol. 101, Wu, R., Ed., Academic Press, New York, 1983, 155.
67. **Windass, J. D., Newton, C. R., DeMaeyer-Guignard, J., Moore, V. E., Markham, A. F., and Edge, M. D.,** The construction of a synthetic *Escherichia coli trp* promoter and its use in the expression of a synthetic interferon gene, *Nucleic Acids Res.,* 10, 6639, 1982.

68. **Ikehara, M., Ohtsuka, E., Tokunaga, T., Taniyama, Y., Iwai, S., Kitano, K., Miyamoto, S., Ohgi, T., Sakuragawa, Y., Fujiyama, K., Ikari, T., Kobayashi, M., Miyake, T., Shibahara, S., Ono, A., Ueda, T., Tanaka, T., Baba, H., Miki, T., Sakurai, A., Oishi, T., Chisaka, O., and Matsubara, K.,** Synthesis of a gene for human growth hormone and its expression in *Escherichia coli, Proc. Natl. Acad. Sci. U.S.A.*, 81, 5956, 1984.
69. **Nishi, T., Sato, M., Saito, A., Itoh, S., Takaoka, C., and Taniguchi, T.,** Construction and application of a novel plasmid "ATG vector" for direct expression of foreign genes in *Escherichia coli, DNA,* 2, 265, 1983.
70. **Fuller, F.,** A family of cloning vectors containing the *lacUV5* promoter, *Gene,* 19, 43, 1982.
71. **Bittner, M. and Vapnek, D.,** Versatile cloning vectors derived from the runaway-replication plasmid pKN402, *Gene,* 15, 319, 1981.
72. **Hanna, Z., Fregeau, C., Prefontaine, G., and Brousseau, R.,** Construction of a family of universal expression plasmid vectors, *Gene,* 30, 247, 1984.
73. **Nakamura, K., Iwasaki, Y., and Hattori, T.,** An improved *Escherichia coli* expression vector for the construction and identification of full-length cDNA clones, *Gene,* 44, 347, 1986.
74. **Amann, E., Brosius, J., and Ptashne, M.,** Vectors bearing a hybrid *trp-lac* promoter useful for regulated expression of cloned genes in *Escherichia coli, Gene,* 25, 167, 1983.
75. **Russell, D. R. and Bennett, G. N.,** Construction and analysis of *in vivo* activity of *E. coli* promoter hybrids and promoter mutants that alter the -35 to -10 spacing, *Gene,* 20, 231, 1982.
76. **Stark, M. J. R.,** Multicopy expression vectors carrying the *lac* repressor gene for regulated high-level expression of genes in *Escherichia coli, Gene,* 51, 255, 1987.
77. **Amann, E. and Brosius, J.,** "ATG vectors" for regulated high-level expression of cloned genes in *Escherichia coli, Gene,* 40, 183, 1985.
78. **Brosius, J.,** Laboratory methods: superpolylinkers in cloning and expression vectors, *DNA,* 8, 759, 1989.
79. **Shirakawa, M., Tsurimoto, T., and Matsubara, K.,** Plasmid vectors designed for high-efficiency expression controlled by the portable *recA* promoter-operator of *Escherichia coli, Gene,* 28, 127, 1984.
80. **Feinstein, S. I., Chernajovsky, Y., Chen, L., Maroteaux, L., and Mory, Y.,** Expression of human interferon genes using the *recA* promoter of *Escherichia coli, Nucleic Acids Res.*, 9, 2927, 1983.
81. **Sussman, R. and Jacob, F.,** Sur un système de repression thermosensible chez le bactériophage d' *Escherichia coli, C. R. Acad. Sci.*, 254, 1517, 1962.
82. **Lieb, M.,** Studies with heat-inducible lambda bacteriophage. I. Order of genetic sites and properties of mutant phosphages, *J. Mol. Biol.*, 16, 149, 1966.
83. **Brosius, J.,** personal communication.

Chapter 14

TRANSFECTION OF CULTURED MAMMALIAN CELLS BY MAMMALIAN EXPRESSION VECTORS

Kazuo Maruyama, Huey-Mei Wang, Atsushi Miyajima, Yutaka Takebe, and Naoko Arai

TABLE OF CONTENTS

I. INTRODUCTION

The advantages of the mammalian system for expression of cloned mammalian cDNA is the availability of native proteins which are correctly processed and glycosylated. Furthermore, the cloned genes can be transferred directly into appropriate expression vectors without further manipulations. Table 1 shows several factors which are important to consider in the mammalian expression system. This system has been successfully employed to identify rare cDNA clones in cDNA libraries and to achieve overproduction of proteins of interest. Two factors are particularly pertinent to a mammalian cDNA expression system: (1) efficient introduction of DNA into mammalian cells and (2) a high level of expression of the proteins encoded by DNA. Foreign DNA (genes or cDNAs) can be introduced into cultured mammalian cells by either transient or stable transfection. In the transient expression system, the DNAs introduced into cells are not integrated into the chromosomes due to the short period of monitoring time. By extending the culture period, the transfected DNA can be maintained either as episomal DNA or integrated into chromosomal DNA to form stably transformed (permanent) cell lines.

For the introduction of DNA into cultured mammalian cells, either viral or plasmid vectors have been employed. Among viral vectors, the retrovirus system has been widely used to introduce foreign DNA into a wide variety of cells and will be discussed later. Plasmid vectors contain a bacterial replicon, together with a genetic marker such as amp^r or tet^r, which allows replication and selection of the plasmids in the bacterial hosts (usually *Escherichia coli*). For the expression of foreign genes or cDNAs in eukaryotic cells, the genes or cDNAs must be placed under the control of eukaryotic regulatory elements such as a promoter, splicing signal(s), and a polyadenylation sequence for correct processing of the transcript.

An episomal vector can replicate extrachromosomally in mammalian cells. To initiate replication, both the *cis*-acting viral replication origin and virally encoded initiator protein which activates the origin in *trans* are required. The latter includes the large T antigen for SV40, polyoma and BK virus (BKV), and EBNA for Epstein-Barr virus (EBV). Plasmids can be replicated in suitable cell lines such as COS, Raji, or ID13, which are transformed by SV40, EBV, or BPV (bovine papilloma virus), respectively. Without such a replication unit (i.e., the replication origin and the viral replication factor supplied by either the plasmid itself or the recipient cells), transfected plasmid DNA will be integrated into the host chromosome or disappear during the course of cell culture.

The episomal vectors have potential advantages as well as disadvantages, compared to the integration vectors. First, the efficiency of transformation is generally high because plasmids need not be integrated. Second, in contrast to the integrated genes whose expression may be affected by the site of integration within the chromosome, the genes cloned into episomal vectors are free from such complications. Third, recovery of the episomal DNA from mammalian host cells is much simpler than that of integrated DNA. This is particularly useful for expression cloning of cDNAs in mammalian cells. On the other hand, because they are not integrated, the copy number may vary in each cell and with the passages. The host range of each episomal vector, in general, is very limited.

At present, five types of episomal vectors which can replicate extrachromosomally in mammalian cells are used frequently (Table 2). Two of them (SV40 in COS cells[1,2] and polyoma virus in COP, MOP, or WOP cells[3-5]) are used frequently for the transient transfection system. Because of the high copy number due to efficient replication in suitable host cells (more than 10^5 molecules per cell for SV40-based vectors in COS cells), most of the transformed cells die eventually. Therefore, these replication vectors are not suitable to isolate stable transformants of their host cells. To establish stable transformants harboring an episomal replicon, BK-,[6] BPV-,[7] and EBV-based vectors[8-10] whose copy number is controlled have been developed (Table 3).

TABLE 1
Factors To Be Considered for Mammalian Transfection System

Purpose of the experiments
- Overproduction of proteins
- Cloning of novel cDNAs of interest
- Regulation of gene expression
- Study function of cDNA or gene product
- Others

Method of transfection of DNA into mammalian cells
- DEAE Dextran
- Calcium phosphate precipitate
- Electroporation
- Lipofection
- Protoplast fusion
- Microinjection
- Viral infection (retrovirus, adenovirus, herpesvirus, papova virus, or vacciniavirus vector)

Type and species of mammalian cells to be transfected
- COS (SV40 vector), COP, MOP, or WOP cells (polyoma vector)
- Fibroblasts or endothelial cells
- Lymphocytes
- Hemopoietic cells
- Others

Culture period
- Transient expression
- Stable transformation

State of transfected DNA in mammalian cells
- Integration into chromosome
- Replication as episome (SV40, polyoma, BKV, EBV, or BPV)

Construction of appropriate expression vectors
- Transcription unit (promoter, splicing, and polyadenylation signals)
- Selection markers
- Replicon and copy number control

Type of promoter
- Constitutive promoter
- Inducible (regulatable) promoter

TABLE 2
Mammalian Episomal Virus

Vector[a]	Host cells	Copy number	Suitable method for expression
SV40[1,2]			Transient
+LT	Simian		
−LT	COS	$>10^5$	
Py[3,5]			Transient
+LT	Rodent		
−LT	COP, MOP, WOP	500—2000	
BKV[6]			Stable
+LT	Human	20—120	
−LT	Human[b]		
BPV[7]			Stable
	Rodent	20—100	
EBV[8-10]			Stable
+EBNA	Human, simian, ca-	10—50	
−EBNA	nine, human[b]		

[a] + or −LT: vectors with or without large T-antigen gene; + or −EBNA: vectors with or without EBNA (EB virus nuclear antigen gene).

[b] Virus transformed cells.

TABLE 3
Comparison of Methods for Stable Transfection

Transfection method	Cell types[a] Monolayer	Suspension	Advantages	Disadvantages
Calcium phosphate precipitation	+ + +	−	No special equipment, easy to do	Poor reproducibility
Electroporation	+ +	+ + +	Large-scale transfection, very easy to do	Special equipment
Protoplast fusion	+ +	+	No special equipment	Only for plasmid DNA; not for genomic DNA
Microinjection	+ +	−	High efficiency	Only for small-scale, special equipment; skill required
Lipofection	+ +	+	No special equipment, easy to do	Special chemicals

[a] Excellent, + + +; good, + +; fair, +; poor, −.

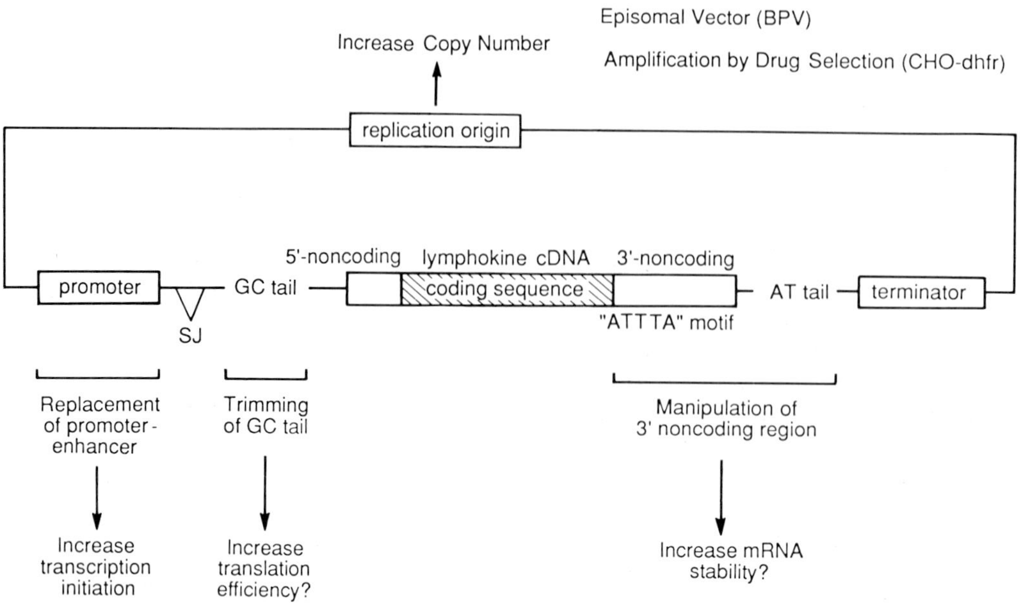

FIGURE 1. General strategy for maximizing cDNA expression in mammalian cells.

Episomal vectors transfected into host cells which do not support replication, or plasmid vectors lacking mammalian replicons, must be integrated into the host chromosome to transform cells stably. Despite a low efficiency of integration and difficulty of recovering integrated DNA, this system has been commonly employed to transfect a wide range of host cells. This procedure involves a rather simple construction of vectors, and integrated DNA in the transformed cells is generally stable. A gene of interest can be incorporated into the vector plasmid in *cis* or can be introduced into the cell simultaneously with the vector containing a selection marker (cotransfection).

Expression of cloned cDNA is influenced by several factors: (1) copy number of plasmid DNA (episomal and/or integrated); (2) efficiency of promoter, enhancer, and terminator; (3) splicing signal; (4) stability of mRNA; and (5) efficiency of translation (Figure 1). To achieve high-level expression, these factors have to be evaluated and optimized. Many of the currently used mammalian expression vectors, such as a series of pSV2[11] and pcD[12] vectors, employ transcription-controlling elements derived from SV40. The SV40 early or late promoter is well characterized and works reasonably well in many cell types. Other promoters which are efficient in mammalian cells include the major late promoter of adenovirus,[13] cytomegalovirus (CMV) promoter,[14,15] and the long terminal repeat (LTR) from Rous sarcoma virus (RSV).[16] The efficiency of the promoters is different in the various cell lines used and also depends on their differentiated state. For example, the SV40, polyoma, and Moloney murine leukemia virus (M-MuLV) promoter/enhancers are active in differentiated cells, but not in embryonal carcinoma (EC) cells.[17-19] The SRα promoter system which we have developed recently is an example of an efficient and versatile promoter system (see Section II).[20]

The other important consideration is the use of a regulatable promoter system to study the effect of an introduced gene on cell phenotype, as was the case with *lac* or *trp* promoters in *E. coli* and the *gal* promoter in yeast *Saccharomyces cerevisiae*. Since one cannot rigorously rule out the possibility that each of several independently isolated stable transformants may carry additional mutation(s), a comparison of original and transfected cells may not be sufficient to establish the function of the transfected gene. This is especially important in isolating stable transformants carrying genes or cDNA clones that regulate cell growth, since the high levels of the protein expression may be toxic to cells. Several inducible promoters currently available include mouse and human metallothionein (MT) promoters,[21] mouse mammary tumor virus (MMTV) LTR,[22] and heat-shock protein promoters.[23] However, further study is required in the development of regulatable promoters, which is a key technology in mammalian molecular biology.

II. TRANSIENT EXPRESSION: DEAE DEXTRAN METHOD

To achieve a high level of expression for efficient cDNA cloning and overproduction, we have developed a novel promoter system (SRα) which employs the SV40 early promoter and R-U5′ segment of the HTLV-I genome[20] (Figure 2). The R-U5 segment as well as U3 is required for maximum expression from the HTLV-I LTR. The stimulatory effect of the R-U5 sequence was observed even when it was linked to a heterologous promoter. The SRα promoter is one or two orders of magnitude more efficient than the SV40 early-region promoter in a wide variety of mammalian cell types and promotes high levels of expression of various lymphokine cDNA inserts in transient (Table 4) as well as stable transformants. The amount of lymphokine produced in COS-7 cells was estimated as approximately 1 mg/l and was proportional to the amount of DNA transfected within the range up to 300 ng/10-cm plate (Figure 3). The accumulation of lymphokine activity in culture supernatant continued up to 4 to 5 d after transfection.

The group at the Genetics Institute has also developed an efficient expression vector based on the adenovirus major late promoter. This vector was used for the isolation of cDNA clones, which encode several proteins, based on their biological activities.[24-26] DNA can be introduced in several ways, such as DEAE-dextran, calcium phosphate, protoplast fusion, lipofection, or a physical method such as microinjection in transient transfection systems. In this review, we will focus on the DEAE dextran method, since this method has been used most frequently for transient expression and we have evaluated this procedure extensively.

A. FIBROBLAST

Reagents

- COS-7 (CV-1 simian cells transformed by an origin-defective mutant of SV40) cells
- Growth medium: DME containing 10% fetal calf serum (FCS), L-glutamine (2 m*M*), and penicillin/streptomycin

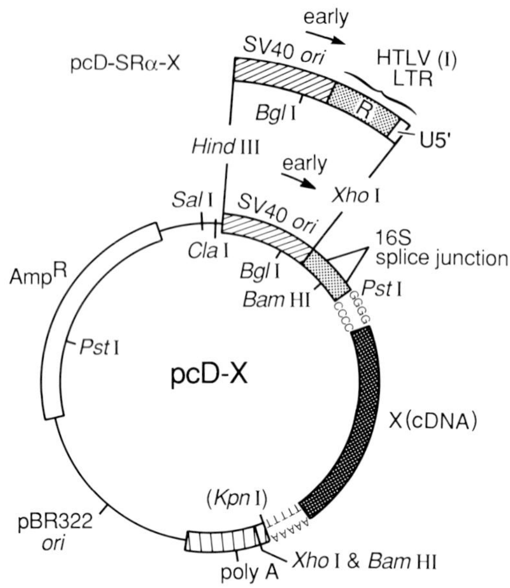

FIGURE 2. Efficient and versatile cDNA cloning and expression vector. Shown are the two alternative linker segments (*Hind*III-*Pst*I) composed of either an SV40 early or an SRα promoter (*Hind*III-*Xho*I segment) and the SV40 late gene splice junction (*Xho*I-*Pst*I segment), the dGdC bridge between the linker fragment and cloned cDNA(X), the dAdT stretch derived from the oligo dT primer and poly(A) tail of messenger RNA, and the segment carrying the SV40 late region polyadenylation signal (polyA). (From Takebe, Y., Seiki, M., Fujisawa, J., Hoy, P., Yokota, K., Arai, K., Yoshida, M., and Arai, N., *Mol. Cell. Biol.*, 8, 466, 1988. With permission.)

TABLE 4
Comparison of the Effect of SRα Promoter and SV40 Early Promoter on the Production of Various Lymphokines in COS-7 Cells

Lymphokine cDNA	Indicator cell	Lymphokine activity (U/ml)[a] SV40 early promoter	SRα promoter	SRα/SV40
Mouse IL-2	HT-2	1.1×10^3	1.3×10^5	1.2×10^2
Human IL-2	HT-2	8.3×10^3	3.1×10^5	3.7×10
Mouse IL-3	MC/9	1.4×10^5	1.2×10^6	8.6
Mouse IL-4	HT-2	3.9×10^2	4.1×10^4	1.1×10^2
Human IL-4	PBL	1.2×10^3	2.8×10^4	2.3×10
Mouse GM-CSF	NFS60 (GM)	7.2×10^3	1.2×10^5	1.7×10
Human G-CSF	NFS60	1.5×10^4	2.6×10^5	1.7×10

[a] Lymphokine activities of culture supernatants were assayed 3 d after transfection. One unit of each lymphokine activity is defined as the amount required to produce half the maximal proliferation response in the indicator cell.

- Tris-buffered serum-free medium (Tris-SFM): DME buffered with 50 m*M* Tris-HCl, pH 7.4
- Collecting medium: DME containing 4% FCS, L-glutamine (2 m*M*), and penicillin/streptomycin
- DEAE-dextran stock solution (Pharmacia, mol wt, ~500,000): 20 mg/ml in sterile water (Do not autoclave or filter.)
- Chloroquine stock solution: 10 m*M* in sterile water (Filter sterilize.)
- PBS (Dulbecco's Phosphate Buffered Saline)
- Trypsin: 0.25% as final concentration

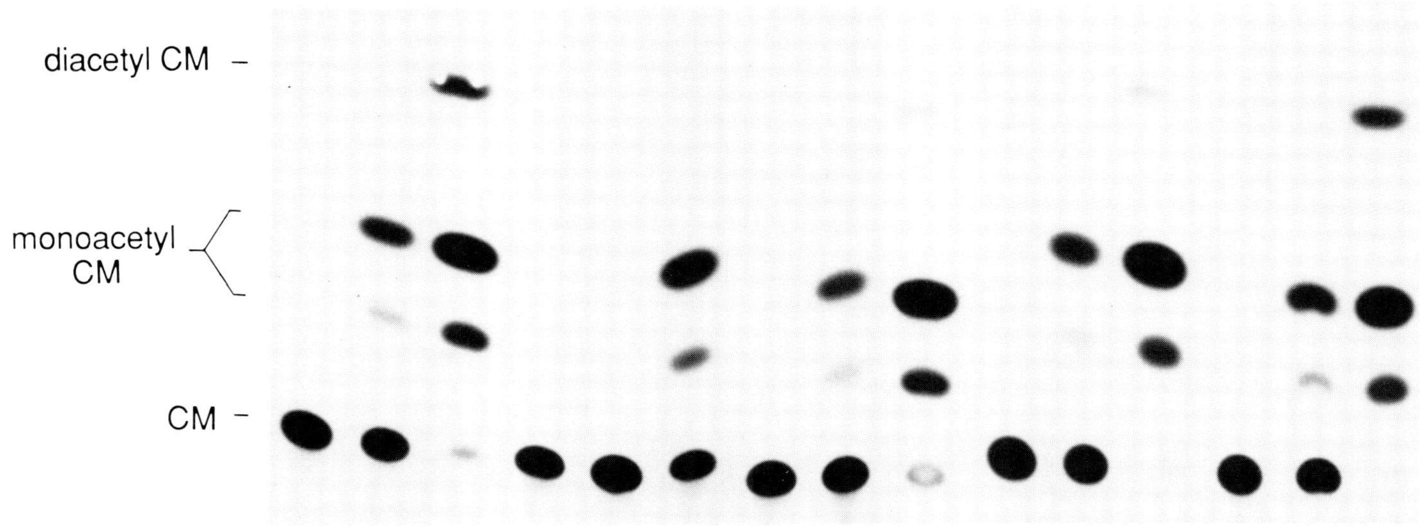

Transfected DNA	Mock	pSV2	pSRα	Mock	pSV2	pSRα	Mock	pSV2	pSRα	Mock	pSV2	pSRα	Mock	pSV2	pSRα
Recipient	Raji			Jurkat			CV1			Ltk⁻			COS 7		
% conversion to acetylated CMs	0	22.6	97.3	0	1.1	36.2	0	11.9	96.1	0	24.5	98.6	0	30.9	99.0
% conversion to diacetyl CM			10.0						1.4			2.4			12.0
	per 300 μg protein per 3 hr at 37°C									per 30 μg protein per 30 min at 37°C					

FIGURE 3. Efficiency of SRα promoter in mammalian cells derived from different species and tissues. pSV2-CAT or pSRα-CAT plasmid DNA was transfected into five different cell lines by DEAE-dextran methods, and cells were harvested 48 h after transfection. CAT activities were assayed under the indicated conditions at 37°C; the activity shown is the % conversion of [^{14}C] chloramphenicol to acetylated form.

Procedure

1. Seed 1 × 10^6 COS-7 cells in 10 ml of growth medium on 10-cm tissue culture plates and incubate at 37°C overnight. The plate should be 40 to 50% confluent by the next day.
2. On the day of transfection, prepare the transfection mixture in a sterile 15-ml Falcon test tube. Add plasmid DNA (10 to 50 μg) to 4 ml of Tris-SFM containing 80 μl of DEAE-dextran (20 mg/ml) and mix.
3. Aspirate the tissue culture medium and rinse the cells with 5 ml of PBS prewarmed to 37°C. Aspirate the PBS.
4. Add the transfection mixture and distribute evenly by tilting the plate. Incubate at 37°C for 4 h.
5. Aspirate the medium and rinse with Tris-SFM or PBS once. Add 100 μ*M* chloroquine in Tris-SFM to each plate. Incubate at 37°C for 3 h.
6. Aspirate the supernatant medium and rinse with Tris-SFM or PBS twice. Feed with 4 ml of collecting medium and incubate at 37°C.
7. Harvest the cells or supernatants after 3 to 5 d for assay.

B. LYMPHOCYTE

Reagents

- Jurkat (human T-cell leukemia cell line) cells (cell density is kept below 5 × 10^6 cells per milliliter.)
- Growth medium: RPMI 1640 medium supplemented with 10% FCS, L-glutamine (2 m*M*), and antibiotics (penicillin/streptomycin)
- Tris-buffered saline (TS): 8 g/l NaCl, 0.38 g/l KCl, 0.1 g/l $Na_2HPO_4 \cdot 7H_2O$, 3.0 g/l Tris base, 0.1 g/l $MgCl_2$, and 0.1 g/l $CaCl_2$, pH 7.4
- Chloroquine stock solution: 10 m*M* in sterile water
- DEAE-dextran stock solution (Pharmacia; mol wt, ~500,000): 20 mg/ml in sterile water (Do not autoclave or filter.)

Procedure

1. Make the DNA-DEAE-dextran mixture in TS by adding DNA (5 to 10 μg/ml) and DEAE-dextran (0.5 mg/ml).
2. Wash 5 × 10^6 cells with TS and resuspend the cells in 1 ml of TS containing DNA and DEAE-dextran.
3. After incubation for 30 min at room temperature, add prewarmed (37°C) culture medium (10 ml) containing 100 μM chloroquine and incubate at 37°C for 60 min.
4. Replace the medium with regular medium.
5. Harvest the cells or supernatant 2 to 3 d after transfection.

III. STABLE TRANSFECTION SYSTEM

Generally, transient transfection is reproducible and quicker than stable transfection. However, especially for studying the function of cloned genes which might change cell phenotype or growth property, it is necessary to isolate stable cell lines carrying the gene or cDNA of interest in either stably integrated or episomal form. Before starting the experiment, several decisions must be made according to the purpose of the experiment. The choice of cell lines is one of the most important decisions. One must choose a recipient cell

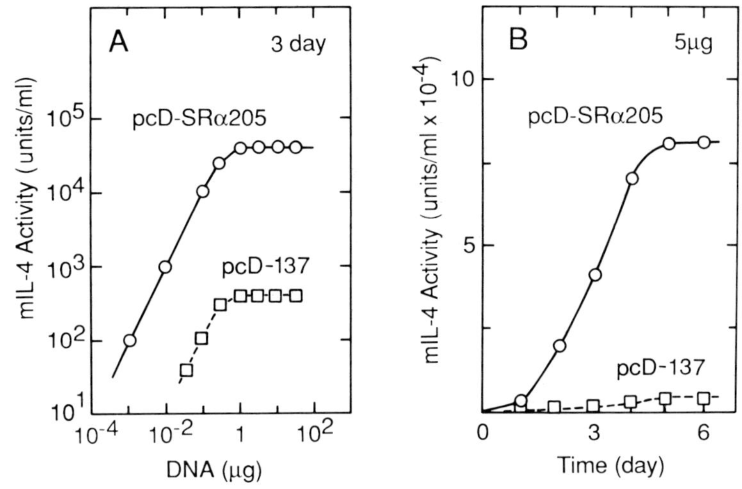

FIGURE 4. Dose-response curve and time course for transient mouse IL-4 expression in COS-7 cells. COS-7 cells were transfected with the indicated amounts of either pcD-137 or pcD-SRα205 in which mouse IL-4 cDNA is under the control of SV40 or SRα promoter, respectively. At the indicated points, the mouse IL-4 activities of culture supernatants were assayed.

line which has a suitable background phenotype for complementation by the introduced gene or selection markers. In addition, the efficiency of the promoter (either its own promoter or a heterologous promoter) will vary from cell line to cell line. Other parameters, such as selection markers (as described below) and the type of vectors which can be used for cotransfection or have a selection marker on the same plasmid, must be determined according to the cell type and method of transfection to be used. Choice of transfection method, summarized in Table 3, is also important.

A. SELECTION MARKERS

Unlike in the transient expression system, only a small fraction of transfected cells (usually less than 1%) are stably transformed. Accordingly, selection markers must be used to isolate stable transformants. Some of them can be used only with the cells which are deficient in the specific enzymes, while the others can be used as dominant selection markers. Several selection markers available are the following.

1. Hypoxanthine-Guanine Phosphoribosyl Transferase (HGPRT)[27] and Thymidine Kinase (TK)[28]

The folic acid analogue, aminopterin, blocks all reactions involving dihydrofolate reductase, including the *de novo* synthesis of IMP and the methylation of dUMP to form dTMP (Figure 4). Mammalian cells have salvage pathways in which HGPRT converts hypoxanthine and guanine to IMP and GMP, respectively, and TK converts thymidine to dTMP. Normal cells can grow in HAT medium which contains hypoxanthine, aminopterin, and thymidine. However, cells which lack HGPRT ($HGPRT^-$) or TK (TK^-) cannot grow in HAT medium, because both the *de novo* and salvage pathways are blocked. Therefore, selections with HGPRT or TK genes can be used only with $HGPRT^-$ or TK^- cells. $HGPRT^-$ cells can be isolated with the guanine analogue 6-thioguanine (by gradually increasing concentrations from 1 μg/ml to 20 μg/ml). TK^- cells can be isolated with the thymidine analogue bromodeoxyuridine (BUdR) or high-specific-activity ^3H-thymidine. Isolation of $HGPRT^-$ cells is relatively easy because this enzyme is encoded on the X chromosome in most mammals.

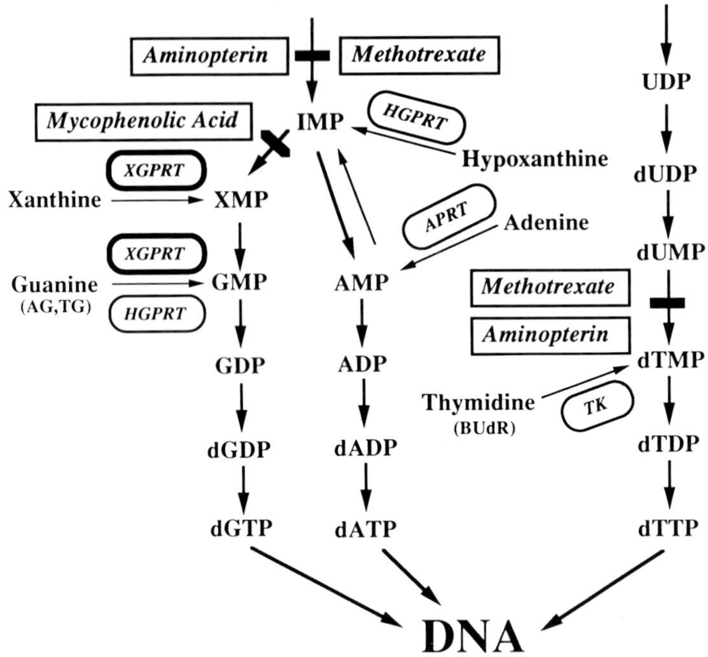

FIGURE 5. Metabolic pathways and sites of inhibition relevant to the selection of stable transfection. When the main pathways (shown with thick arrows) are blocked with inhibitors (rectangular outline), the cells must depend on the salvage pathways (thin arrows). Aminopterin and methotrexate block *de novo* synthesis of purine nucleotides from precursors at steps involving formylation. Mycophenolic acid inhibits the conversion of IMP to XMP. Hypoxanthine or guanine are converted to GMP via the enzyme HGPRT (hypoxanthine guanine phosphoribosyl transferase). APRT (adenine phosphoribosyl transferase) converts adenine to AMP. TK (thymidine kinase) converts thymidine to dTMP. *E. coli* XGPRT (xanthine guanine phosphoribosyl transferase) converts xanthine and guanine to XMP and GMP, respectively. HGPRT$^-$ cells can be selected with medium containing base analogs 6-thioguanine (TG) or 8-azaguanine (AG). TK$^-$ cells can be selected with the thymidine analog BUdR (bromodeoxyuridine); IMP, ionosine monophosphate; XMP, xanthine monophosphate.

Selection — HAT media, complete medium supplemented with 100 μ*M* hypoxanthine, 0.4 μ*M* aminopterin, and 16 μ*M* thymidine — Since aminopterin also blocks the conversion of serine to glycine, glycine (10 μ*M*) should be added to a medium which does not contain glycine, such as MEM or Fischer's. Lyophilized HAT supplement is commercially available (GIBCO or Sigma). HT medium (medium supplemented with hypoxanthine and thymidine) is used for 1 to 2 weeks after the selection as a rescue medium, which overcomes the effects of residual intracellular aminopterin. Solutions containing aminopterin should be kept in the dark.

2. Xanthine-Guanine Phosphoribosyl Transferase (XGPRT)[11]

Mycophenolic acid, an inhibitor of IMP dehydrogenase, blocks the conversion of IMP to XMP (Figure 5). The expression of the bacterial gene, *E. coli* XGPRT (Ecogpt), allows the cells to use xanthine as a substrate for the purine salvage pathway. Since mammalian cells lack this enzyme, this selection can be used as a dominant selection marker for transformation.

Selection — 250 μg/ml of xanthine, 15 μg/ml of hypoxanthine or 25 μg/ml of adenine,

150 μg/ml of L-glutamine, 10 μg/ml of thymidine, 2 μg/ml of aminopterin, and 25 μg/ml of mycophenolic acid

Stock Solution

1000X mycophenolic acid:	25 mg/ml in 0.1 *N* NaOH
1000X aminopterin:	40 mg/20 ml in 0.02 *N* NaOH
100X hypoxanthine:	150 mg/100 ml in 0.02 *N* NaOH
20X xanthine:	1 g/200 ml in 0.1 *N* NaOH

The stock solution of xanthine is basic, so after adding it to the medium, the medium should be neutralized with HCl (the same amounts as NaOH in the stock solution). Dialyzed serum may not be necessary. The amount of free guanine, which may overcome selection block, in different batches of serum varies and is usually negligible. The addition of aminopterin speeds up the selection process.

3. Aminoglycoside Phosphotransferase (APH)[29,30]

The structure of G418 resembles gentamicin, neomycin, and kanamycin, but, unlike these related compounds, G418 interferes with the function of 80S ribosomes and blocks protein synthesis in eukaryotic cells. These aminoglycoside antibiotics can be inactivated by the bacterial phosphotransferases encoded by transposons Tn5(APH [3′] II) and Tn601 (APH [3′] I).

Selection — 50 μg/ml to 2 mg/ml of G418 in complete medium — For a stock solution, dissolve 200 mg/ml of G418 (GIBCO: G418 Sulfate, Geneticin) in water and sterilize by filtration. G418 is acidic (contains two molecules of H_2SO_4 per molecule). If the pH of the selection medium affects the growth of the cells, the stock solution can be made in HEPES (100 m*M*, pH 7.3). Store the powder at room temperature and the solution at −20°C.

4. Hygromycin B Phosphotransferase (HPH)[31]

Hygromycin B (hygB) is an aminocyclitol antibiotic produced by *Streptomyces hygroscopicus* which inhibits protein synthesis in both prokaryotes and eukaryotes by interfering with ribosomal translocation and with aminoacyl-tRNA recognition. Hygromycin B phosphotransferase (HPH) is identified in *E. coli* and inactivates hygB.

Selection — 10 to 800 μg/ml of hygB in complete medium

Stock solution — 400 mg/ml in water — Store the powder at room temperature and the solution at −20°C.

5. DHFR (Dihydrofolate Reductase)[32-34]

Dihydrofolate reductase catalyzes the reduction of folic acid to tetrahydrofolate, which is the active form of the cofactor used in several biosynthetic pathways involving one-carbon transfers. DHFR is involved in the *de novo* synthesis of glycine, purine nucleotides, and thymidylate. Methotrexate (MTX) is an analogue and competitive inhibitor of DHFR. CHO (Chinese hamster ovary cell) is a DHFR-deficient cell which is used frequently as a recipient cell for cotransfection with the DHFR gene with any particular gene of interest. Transformants can be selected by their ability to grow in the absence of nucleotides such as DME medium, as well as in the presence of MTX, since mutants lacking DHFR activity behave as triple auxotrophs for glycine, hypoxanthine, and thymidine. This selection marker has been used successfully for gene amplification by increasing MTX concentrations successively. Mutant DHFR with lower affinity to MTX can also be used for transfection to wild-type cell lines,[32] since the selection condition (MTX concentration) for mutant DHFR is sufficient to inactivate the endogenous DHFR activity of wild-type cells.

Selection — 0.01 to 300 μ*M* methotrexate (MTX) — In general, APH (G418) or HPH (hygB) is the first choice for selection, because they both are dominant selection markers and, unlike for gpt selection, additional supplements are not necessary. Therefore, it is possible to select simultaneously for the expression of several markers or to select sequentially one after another with gpt, neo, and hygB. After being established, the phenotype of the cells should be stable without drug selection. However, when one wishes to use episomal vectors or when cells which receive the gene of interest might have growth disadvantages, it is better to maintain cells in the media containing the drug even after they are established.

Each cell line shows different sensitivity to the drug. Likewise, the activity of each lot of drug is different. If the concentration of the drug is too low, nontransformed cells can escape the selection. On the other hand, if the concentration is too high, even the cells carrying a selection marker gene cannot survive. The concentration of the drug affects the efficiency of transfection and the length of time required for cells to form colonies. Before starting transfection with new cell lines, one has to optimize the drug selection conditions as shown below.

Protocol

1. Plate the cells at a density of 2×10^3 cells per square centimeter with various concentrations of drug (four plates for each concentration). Change the medium every 3 to 4 d. Any size of plate can be used. Plate 10^5, 4×10^4, or 2×10^4 cells for 10-, 6-, or 3.5-cm plates, respectively. For suspension cells, a 96-well plate is the most convenient. Inoculate 10^4 and 10^3 cells per well.
2. Fix and stain the plates with Giemsa after 1 and 2 weeks (two plates each time). For suspension cells, check the cells under the microscope.

Cells may divide once or twice, even in a very high concentration of drug. Choose the concentration which kills cells within 1 week. If it takes more than 1 week to kill cells completely, even at a high concentration, choose the concentration which kills cells in 2 weeks or reduce the initial density of the cells. Usually, slower-growing cells take longer to die with drugs. In some cases, the sensitivity of cells to drugs can be increased by the use of DMSO or ionophore to change the permeability of the cell membrane.

B. STRUCTURE OF THE VECTORS FOR STABLE TRANSFECTION

Besides the essential components of the vector for mammalian expression, such as promoter/enhancer, splicing, and polyadenylation signals (see Introduction), vectors specific for stable transfection also contain selection markers. Moreover, cotransfection (with a selection marker on a separate plasmid) gives reasonable efficiency of transfection with the conventional calcium phosphate-mediated method in which transfected DNA forms large concatemers before integration into the chromosomes, when a five- to tenfold excess of plasmid DNA with the selection marker is transfected together with the plasmid having the gene of interest. A series of pSV2 vectors with selection markers such as Ecogpt, hygB, neo, and DHFR are available for this purpose. However, especially when the Chen-Okayama modified calcium phosphate precipitate method (see below) is employed, and when the gene or cDNA of interest may affect the growth property of cells, it is essential to use vectors having the selection marker and the gene of interest on the same plasmid DNA. Those are shown in Figure 6A, B, and C as examples. pcD2 vectors developed by Chen and Okayama[32] have the neo^r gene in the linker fragment to construct cDNA libraries with a selection marker (Figure 6A). We constructed the pSVneo/SRα vector into which cDNA can be transferred. The inserted cDNA will be under the control of the SRα promoter (Figure 6B). The plasmid shown in Figure 6C can be used for converting a vector without a selection marker to that with one of the three selection markers (Ecogpt, hygB, and neo).

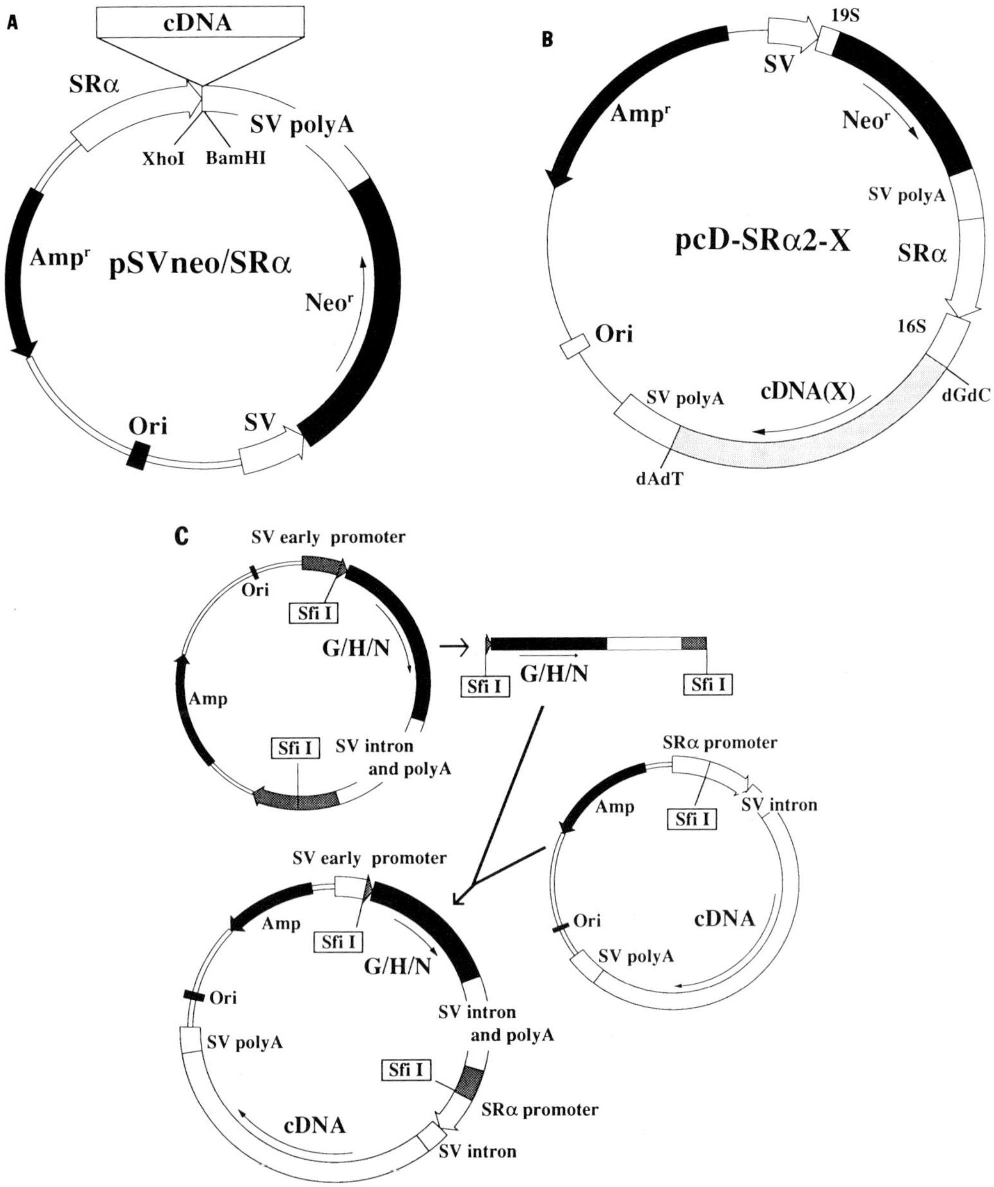

FIGURE 6. Expression vectors for stable transfection with drug selection markers. (A) pSVneo/SRα vector. pSVneo/SRα is constructed from pSV2neo and the SRα promoter. cDNA can be inserted at the unique site *Xho*I or *Bam*HI. The inserted gene is under the control of SRα promoter. The fragment between the cDNA and the neo gene contains SV40 polyadenylation signals for both orientations. (B) Generalized structure of the pcD-SRα2-X vector. This vector is derived from pcD2-X.[39] The linker segment has a neo^r gene driven by the SV40 promoter with 19S splicing and polyA signals. The cDNA library is constructed by using dT tails as primer and attached to the vector-primer segment according to the Okayama-Berg procedure.[12] (C) Construction of the vector with selection marker. The *Sfi*I fragment which contains one of the three selection marker genes (G/H/N: *Ecogpt*, hygB^r, and neo^r) and the portion of the SV40 origin sequences at both ends is excised from the cassette vector (upper left), then inserted at the *Sfi*I site of the vectors (right) with an SV40 origin. Because the recognition sequence of *Sfi*I is the octanucleotide (GGCCN$_5$GGCC) with an additional five unspecified nucleotides at the middle, the *Sfi*I restriction site is rare and the orientation of the inserted fragment is fixed to create an intact SV40 origin. The resulting vector has a selection marker under the control of the SV40 promoter and a cDNA insert under control of the SRα promoter (bottom left).

C. PROTOCOL FOR STABLE TRANSFECTION

The DEAE-dextran method which is used for transient transfection is not the method of choice to isolate stable transformants, mainly because of the toxicity of DEAE-dextran to the cells. Instead, the calcium phosphate precipitate method has been widely used to isolate stable transformants, since this method works well, especially for monolayer cells, and doesn't require any special equipment. Several other different transfection methods are available for stable transfection, as listed in Table 3. Electroporation[36] has been successfully used, especially for suspension cells which are difficult to introduce DNA into by the calcium phosphate method. Also, DNA can be introduced into the cells by lipofection,[37] microinjection,[38] and protoplast fusion.[39] If the cell line to be used is in a monolayer culture, the first choice will be calcium phosphate transfection. For suspension cells, the first choice may be electroporation. Each method has advantages and disadvantages and, furthermore, the efficiency varies from cell line to cell line (or even clone to clone). Thus, especially for the cloning of rare cDNAs, it is important to optimize transfection conditions.

1. Calcium Phosphate-Mediated Transfection

Calcium ions bind to the phosphate groups of DNA and added phosphate ions bind to the calcium ions, resulting in the formation of precipitates. Cells take up these precipitates via endocytosis. Since the initial development of the procedure by Graham and van der Eb,[40] many variations have been added to improve the efficiency of transformation. In some cell types, glycerol[41] or DMSO shock,[42] which change the permeability of cell membranes, or chloroquine treatment,[43] which inhibits lysosomal enzyme action, have been employed to increase the efficiency of transformation. Here, we describe the conventional and newly developed procedure by Chen and Okayama.[35]

a. Conventional Method

Solutions

1. 2X HBS (HEPES buffered saline): per liter

   - 10 g of HEPES (free acid)
   - 16 g of NaCl
   - 0.2 g of Na_2HPO_4

   Adjust the pH to 7.10 ± 0.05 with 2 *N* NaOH and sterilize by filtration. Dispense into small aliquots and store at −70°C.

2. 2.5 *M* $CaCl_2$

   - 18.4 g of $CaCl_2 \cdot 2H_2O$ per 50 ml of H_2O

   Sterilize by filtration, dispense into aliquots, and store at −70°C.

3. 1000X chloroquine (100 m*M*)

   - 5.16 g of chloroquine diphosphate per 100 ml of H_2O

   Sterilize by filtration. Dispense into aliquots and store at −70°C. Chloroquine is light sensitive, so store the powder at 4°C while protecting it from the light. The toxicity of chloroquine is different from lot to lot.

Procedure

1. One day before transfection, the cells are seeded at about 5×10^5 cells per 10-cm plate. *Although optimum cell density for transfection varies from cell type to cell type, in practice, cells are split 1 to 10 from a 50 to 80% confluent plate. Cells should be passaged every 3 to 4 d so that they do not reach and remain at confluence. Always include both mock transfection (DNA without a selection marker, such as pBR322) for a negative control and with a selection marker gene alone for a positive control. If the transfection efficiency is different between the selection marker alone and the selection marker with the specific genes, it suggests that transfected genes may affect the property of cell growth.*
2. Prepare the calcium phosphate-DNA precipitates using the stock solutions warmed up to room temperature as follows. The DNAs (total of 10 to 20 μg per 10-cm plate) are added to water to make 0.45 ml in a polypropylene tube (6-ml capacity, such as Falcon, catalog number 2063) and mixed by pipetting; 50 μl of 2.5 *M* $CaCl_2$ is added to the DNA solution and mixed; then this mixture is added dropwise to 0.5 ml of 2X HBS in a 6-ml polypropylene tube with continual agitation by bubbling air. Let the precipitates sit at room temperature for 20 to 40 min. *The amount of the solutions mentioned above is for one 10-cm plate; for a 6-cm plate, half the amount should be used. If many plates are to be transfected with the same DNA, the DNA mixture can be made in a large polypropylene tube, such as a 15- or 50-ml capacity; however, the efficiency may be different from those with 6-ml tubes. If the concentration of stock DNA is very low, DNA should be ethanol precipitated and then resuspended in 0.45 ml of water. Otherwise, DNA can be stored in water instead of in TE (Tris-EDTA). After adding the DNA to water, DNA solutions can be sterilized by incubation at 65°C for 15 min. This step is useful to dissolve very long DNAs, such as genomic DNA. Genomic DNA can be sheared after this step by forcing it through an 18-gauge needle, if necessary.*
3. Remove the media from the plate, add 1 ml of DNA precipitates slowly to the center of the plate, and spread the precipitates. *Mix the DNA solutions by pipetting before adding them to the plate.*
4. Let the plate sit on a level place for 20 min at room temperature. Shake the plates occasionally. Dust-like precipitates can be seen, even by the naked eye.
5. Add 10 ml of media containing 100 μ*M* chloroquine or 1 ml of 20% glycerol and incubate the plate at 37°C for 5 to 12 h or 1 min, respectively. *The toxicity of chloroquine differs with the forms of DNA precipitates, even in the same cell lines. Check the morphology of the cells under a microscope occasionally and if the cells look very sick, go to the next step.*
6. Remove the medium and wash the plates gently, twice, with medium (5 ml each) and add back 10 ml of fresh medium. Incubate for 1 to 2 d. Dust-like precipitates may still be seen after the wash; however, they will disappear the next day.
7. Forty-eight hours after transfection, split the cells 1:5 to 20.
8. Start the G418 selection after a relatively short time (16 h after transfection, compared to 48 h with mycophenolic acid selection. *This difference is due to differences in the characteristics of the drug; for example, the killing effect of G418 is very effective for dividing cells. In general, selection should start as early as possible in order to isolate independent clones. On the other hand, cells must recover from the damage caused by transfection.*
9. Change the medium every 3 to 4 d. *Within 7 to 10 d, cells in the mock transfected plate die. At this time, one can see small colonies in the transfected plates. In the selection medium, cells with newly acquired resistance grow slower than in normal medium without selection drugs. After about 2 weeks, one can see colonies with the naked eye.*

10. Pick up the colonies with cloning cylinders (Bellco, catalog number 2090-00608). *Pick colonies with different sizes and morphology, choosing colonies well isolated and preferably from different plates to avoid picking the satellite colonies. One can check whether one clone is different from the other by analyzing the integrated DNA by Southern blot hybridization, if necessary.*

b. Chen and Okayama Method

Solution

- 2X BBS
  50 m*M* BES (free acid) (BES, *N*-,*N*-bis(2-hydroxyethyl)-2-amino-ethanesulfonic acid)
  280 m*M* NaCl
  1.5 m*M* Na_2HPO_4

  Adjust the pH to 6.95 with 1 *N* NaOH.

Protocol

1. Plate the cells at 5×10^6 cells per plate and incubate at 37°C for 24 h in 5% CO_2.
2. Mix the DNA with 0.5 ml of 0.25 *M* $CaCl_2$, add 0.5 ml of 2X BBS to the DNA solution, and incubate at room temperature for 10 to 20 min. *The calcium phosphate-DNA precipitate does not form at this moment.*
3. Add 1 ml of DNA-BBS mixture to the medium (10 ml) in the plate, dropwise, and swirl the plate to mix the solution. Incubate the plate for 15 to 24 h at 35°C in 3% CO_2. *During this incubation, the medium becomes basic because the percentage of CO_2 is 3%, and precipitates form gradually.*
4. Remove the media and wash the cells twice with media or PBS and incubate for another 24 h at 37°C in 5% CO_2. *Before washing the plates, check the form of the precipitates. There is a clear correlation between the form of the precipitates and the efficiency of transfection. If the precipitates are too fine or too coarse, the efficiency is low. Only in a very narrow range can one get a very high frequency of transfection. There is also a direct correlation between the form of the precipitates and the amount of DNA in the solution. The more DNA present in the solution, the finer the precipitates' form. Thus, one can control the form of precipitates by changing the DNA concentration (10 to 50 μg) and the pH of the buffer.*
5. Split the cells (>1 to 10) and incubate for 24 h before selection is started. *With this method, more than 10% of the surviving cells can be stably transformed. However, the range of optimum conditions is very narrow.*

Optimization

We use 24-well plates (the surface area is about 1/30th of a 10-cm plate) to optimize the parameters, such as pH and the amount of DNA. The pH of the buffer may be the most critical factor which affects the efficiency. We make a buffer with the desired pH by mixing the appropriate ratio of acid and base according to the formula, pH = pKa + log [base/acid].

For BES, the pKa at 25°C is 7.07. For example, to make 1 *M* BES, pH 6.96, mix 1 volume of 1 *M* BES-acid and 0.776 volume of 1 *M* BES-sodium salt.

- 21.3 g of 1 *M* BES-acid per 100 ml of H_2O
- 1 *M* BES-sodium salt

- 21.3 g of BES-acid
- 4.0 g of NaOH per 100 ml of H_2O

1. Plate the cells at 2 × 10^4 per well. *To distribute the cells evenly in a small well, plate the cells in 5 ml of medium.*
2. Change the medium a few hours before transfection. Use 300 μl of medium per well.
3. Add 30 μl of DNA mixture per well and incubate at 35°C in 3% CO_2 for 24 h. Make the DNA mixture in 24-well plates as follows. The amount of DNA is 10 to 50 μg for a 10-cm plate and 0.3 to 1.5 μg for a 24-well plate.
4. Check the form of the precipitates under a microscope and record. Remove the media and wash the cells twice with media. *The more DNA used, the finer the precipitates' form. The more acidic the solution, the less DNA needed to get the same form of precipitates.*
5. Split the cells (one well to one 10-cm plate). Incubate for 24 h at 37°C in 5% CO_2.
6. Start the selection. Change the media every 2 to 3 d.
7. After 2 weeks, fix the cells and stain with Giemsa. *There is a good correlation between the form of the precipitates and the efficiency of transformation. If the pH of the solution is in some defined range (6.8 to 7.1), one can find the optimum amounts of DNA to give the highest efficiency (Figure 7).*

Among the various protocols for the calcium phosphate transfection method, this gives the highest efficiency of transformation (up to 60%) for mouse A9 cells. Once one makes the solutions and DNAs and sets up the CO_2 incubator, this is very reproducible because there are no difficult steps in making the precipitates as in the conventional method. Also, one can predict the efficiency by checking the form of the precipitates 1 d after transfection, because there is a very good correlation between the form of the precipitates and the efficiency of the transformation. This method does not work for linear DNA; hence, you cannot use this method for genomic transfection. Because the number of copies integrated into the host chromosome seems to be low, it is recommended to use plasmids with both a selection marker and the gene of interest. For this purpose, a pcD-type vector with a selection marker in the linker segment has been developed (see Figure 6).

IV. RETROVIRUS SYSTEM

Retrovirus vectors provide an efficient way to introduce DNA into cells which are usually poorly transfected by conventional methods, such as calcium phosphate precipitation. Because retroviral DNA is incorporated into the host genome in a precisely defined manner, the expression of genes carried by the retrovirus can be controlled.

A. PRINCIPLE

Retroviruses are a family of RNA viruses whose genetic information is transmitted by an RNA genome in a viral particle. After entry into the cell, the viral genome is converted into a linear double-stranded DNA by reverse transcriptase present in the infecting viral particle. The linear DNA is converted into circular DNA and integrated into the host chromosome through long terminal repeats (LTR) of the viral genome. The integrated viral genome has LTR at both the 5′ and 3′ ends, and between these two LTRs, the wild-type virus has three genes, gag, pol, and env, which encode the core proteins, reverse transcriptase, and envelope proteins, respectively. The promoter in the 5′ LTR drives the transcription of the integrated viral genome, and the transcription is terminated and polyadenylated in the 3′ LTR. The transcripts of the viral genome are packaged into viral particles. The viral genome has the specific sequence (ψ) which is acting in *cis* for the packaging of RNA. The

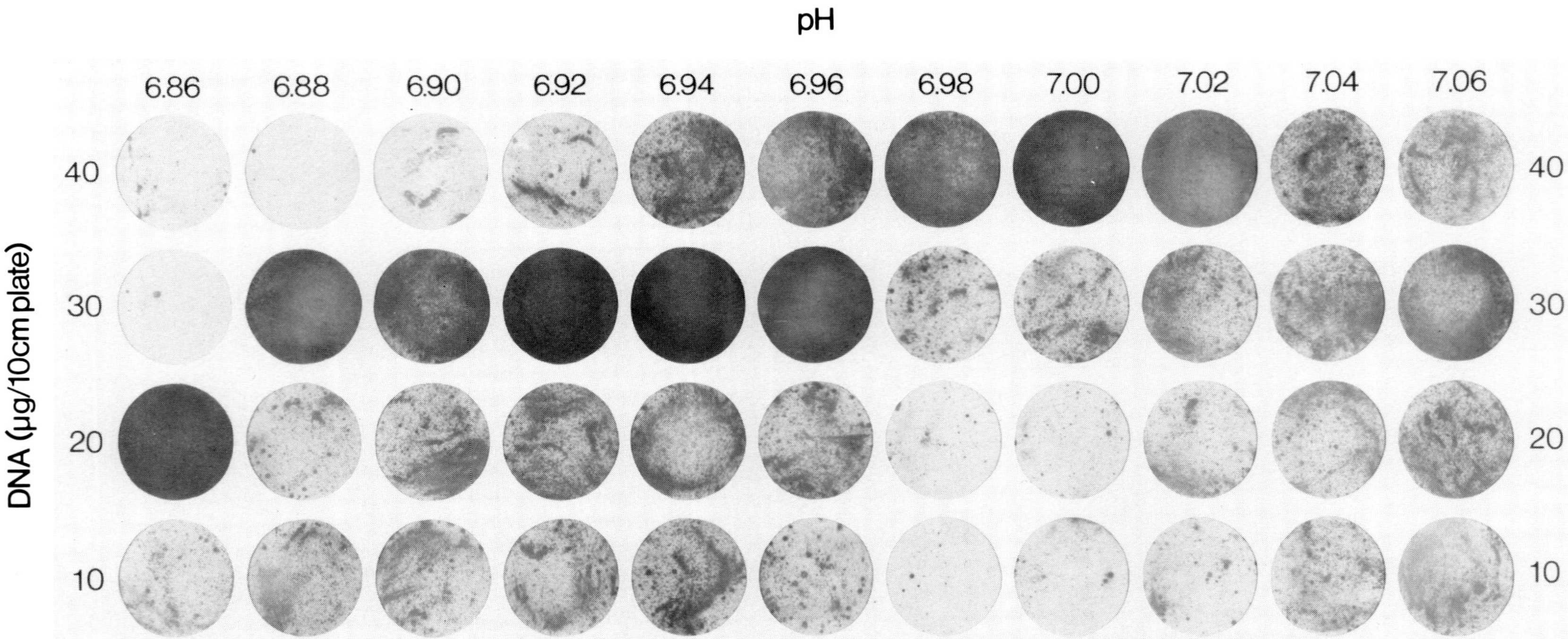

FIGURE 7. The effect of the amount of DNA and pH of BES buffer on the efficiency of transformation. A9 mouse cells are transfected with pcDneo in a 24-well plate and cells in each well are replated into one 10-cm plate (30-fold dilution). After 2 weeks of selection in G418 (0.8 mg/ml)-containing medium, cells are fixed and stained with Giemsa. The pH indicated is the calculated value of 1 *M* BES buffer at 25°C (see text for calculation). Note that the pH is not the same as that of 2X BES buffer, which includes NaCl and Na_2HPO_4. The amounts of DNA indicated are converted to the appropriate amount for using a 10-cm plate. (The actual amount using a 24-well plate is 1/30th of the amount indicated.) Slight differences of pH and the amount of DNA affect the efficiency of transformation: the lower the pH used, the smaller the amount of DNA required to get optimum transformation. If one fixes the pH of the BES buffer, the optimum condition can be found by changing the amount of DNA.

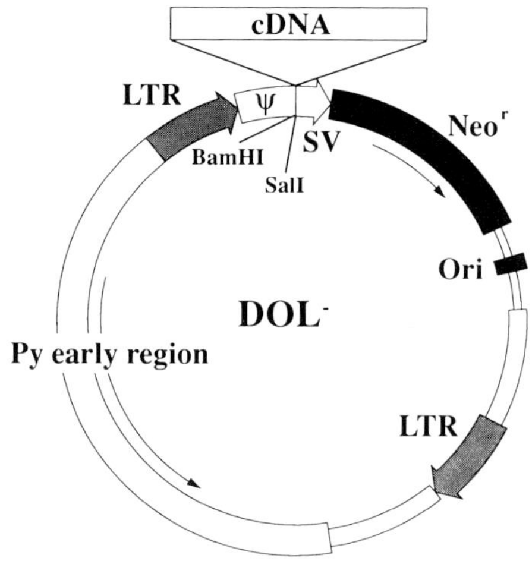

FIGURE 8. Retrovirus vector. DOL$^-$ is a retrovirus vector which contains the 5′ and 3′ LTRs and the packaging sequence ψ of MLV. The Neo^r gene is driven by the SV40 promoter and works in both mammalian cells and *E. coli*. cDNA can be inserted at the unique site *Bam*HI or *Sal*I and will be driven by the 5′ LTR. The polyoma (Py) early region increases the efficiency of rescue of infectious virus after the transfection of ψ-2 cells.

protein products encoded by the viral genome are required for viral replication and packaging, and these proteins act in *trans*. Therefore, a defective recombinant virus having foreign sequences can be produced, if the virus has elements required in *cis* and the *trans*-acting proteins are supplied from another virus (helper virus) present in the same host cell. To make infectious recombinant virus, plasmid DNA carrying the retrovirus sequence is transfected into cells which have a helper virus lacking the packaging sequence. The resultant transfectants produce mainly recombinant virus. Those helper-free recombinant virus produced in the medium can be used to infect cells of interest.[43] Alternatively, the cells can be cocultured with the virus-producing cells to obtain a higher efficiency of infection. It should be noted that because the viral genome has to be packed in the particle, there is a size limitation of the gene that can be carried by the virus.

B. STRUCTURE OF VECTORS

The essential parts of the retrovirus vectors are the 5′ and 3′ LTRs necessary for transcription initiation of the viral genome and polyadenylation of viral transcripts, as well as integration, sequences necessary for reverse transcription of the viral genome, and sequences for the encapsidation of viral RNA. To select virus-infected cells, drug-resistant genes such as neo or gpt can be incorporated into the viral genome. To express inserted cDNA, an appropriate promoter should be placed upstream of the cDNA.

Figure 8 illustrates a typical retrovirus vector, DOL$^-$,[44] which has MLV LTR and the packaging sequence (ψ). The neo gene with the SV40 promoter is placed in the same direction as the viral transcription. The viral 5′ splicing site, normally used in the production of subgenomic env-encoding mRNA, has been deleted to produce the full-length viral transcript efficiently. A DNA fragment can be inserted in the cloning site, which is located upstream of the SV40 sequences. Because the inserted gene is transcribed from the 5′ LTR as a part of the full-length viral genome, the sequences for splicing, transcription termination, and polyadenylation have to be deleted from the gene to be inserted. Inclusion of pBR322 origin sequences in the viral vector facilitates the rescue of integrated proviral sequences as bacterial plasmids. Because vectors carrying the functional polyoma tumor antigen gene and the viral origin of replication can increase the efficiency of infectious virus production, the polyoma early region is placed outside of the retroviral genome. Expression of the inserted gene can be controlled using different promoters.

C. GENERATION OF VIRUS-PRODUCING CELL LINES

The most commonly used packaging cell lines are ψ2 and ψ-AM.[45] ψ2 was established by transfecting NIH3T3 fibroblasts with pMOV-ψ^-, a plasmid DNA carrying an ecotropic Moloney murine leukemia virus (Mo-MLV) genome. pMOV-ψ^- produces all the viral gene products, but it lacks the ψ sequences necessary for encapsidation. The ecotropic viral envelope glycoprotein recognizes a receptor expressed on mouse and related rodent cells. On the other hand, ψ-AM is generated by transfection of NIH3T3 cells with a chimeric retrovirus genome containing ecotropic gag-pol sequences from Mo-MLV and envelope sequences from the amphotropic virus 4070A, but lacking ψ sequences. Therefore, virus produced from ψ-AM are capable of infecting various host cells, including human cells.

The virus-producing cells can be established by transfecting plasmid constructs in either packaging cell line using the standard calcium phosphate method. The supernatant of the individual transfectants is used to infect NIH3T3 cells to find the best virus producer. However, a more elaborate method involves the sequential use of two packaging cell lines, ψ-AM and ψ2, such that virus produced by transfection of the first cell line is used to infect the second packaging cell line. Because the packaging cell has immunity for infection with the same virus as it carries, the virus should be passed from the amphotropic packaging cell line to the ecotropic cell line or vice versa. This two-step method has several advantages over the one-step procedure. Generally, provirus DNA introduced by infection produces more virus than the one introduced by transfection. Since transfection causes multiple integration of the transfected DNA and also causes rearrangement, the virus produced by a transfected clone may be heterogeneous. On the other hand, cells derived from an infected clone produced homogeneous virus, because usually a single copy of viral DNA is integrated in the chromosome in a defined manner by virus infection. Recently, a set of packaging cell lines which are free from transfer of packaging-function and a generation of helper virus were constructed.[46]

D. PROTOCOLS

Transient Production of Virus

1. One d before transfection, plate the ψ-AM cells at 1×10^6 per 10-cm dish.
2. Make $CaPO_4$-DNA precipitates using 20 μg of DNA, as described in Section III.C.
3. Add the precipitates dropwise over the surface of the medium covering the ψ-AM cells.
4. Incubate 8 to 20 h at 37°C.
5. Remove the medium and filter through a 0.45-μm filter.

Establishment of Stable Virus-Producer

1. One d before infection, plate the ψ-2 cells at 1×10^6 per 10-cm dish.
2. Add 8 μg/ml polybrene to the viral stock (ψ-AM medium, prepared as above).
3. Remove the medium from the ψ-2 cells and add the viral stock.
4. Incubate 2.5 h at 37°C.
5. Remove the medium and add fresh medium.
6. Incubate 24 h at 37°C.
7. Replace the medium with medium containing an appropriate drug to select infected cells (if the vector has a neo gene, use 800 μg/ml G418). When the cells reach confluence, split the cells.

8. Change the medium every 3 d.
9. Clone the drug-resistant cells.
10. Titrate the virus production using NIH3T3 cells. The infection procedure of NIH3T3 is the same as above.

Infection of Suspension Cells by Cocultivation

1. One d before infection, plate the virus-producing cells at 5×10^5 cells per 10-cm plate.
2. Wash the target cells and resuspend in fresh medium.
3. Remove the medium from the virus-producing cells and add the target cells.
4. Incubate for 2 d.
5. Harvest the suspension cells and wash with fresh medium.
6. If necessary, a second round of cocultivation can be done.
7. Grow the cells in a petri dish for 1 to 2 d.
8. Culture the cells in the presence of an appropriate drug until the drug-resistant cells propagate.
9. Isolate the clones by limiting dilution, if necessary.

REFERENCES

1. **Gluzman, Y.,** SV40-transformed simian cells support the replication of early SV40 mutants, *Cell,* 23, 175, 1981.
2. **Mellon, P., Parker, V., Gluzman, Y., and Maniatis, T.,** Identification of DNA sequences required for transcription of the human α1-globin gene in a new SV40 host-vector system, *Cell,* 27, 279, 1981.
3. **Tyndall, C., Mantia, G. L., Thacker, C. M., Favaloro, J., and Kamen, R.,** A region of the polyoma virus genome between the replication origin and late protein coding sequences is required in *cis* for both early gene expression and viral DNA replication, *Nucleic Acids Res.,* 9, 6231, 1981.
4. **Muller, W. J., Naujokas, M. A., and Hassell, J. A.,** Isolation of large T antigen-producing mouse cell lines capable of supporting replication of polyomavirus-plasmid recombinants, *Mol. Cell. Biol.,* 4, 2406, 1984.
5. **Dailey, L. and Basilico, C.,** Sequences in the polyomavirus DNA regulatory region involved in viral DNA replication and early gene expression, *J. Virol.,* 54, 739, 1985.
6. **Milanesi, G. B., Brodano, G., Negrini, M., Lee, D., Corallini, A., Caputo, A., Grossi, M. P., and Ricciardi, R. P.,** BK virus-plasmid expression vector that persists episomally in human cells and shuttles into *Escherichia coli, Mol. Cell. Biol.,* 4, 1551, 1984.
7. **Law, M.-F., Byrne, J. C., and Howley, P. M.,** A stable bovine papillomavirus hybrid plasmid that expresses a dominant selective trait, *Mol. Cell. Biol.,* 3, 2110, 1983.
8. **Yates, J. L., Warren, N., and Sugden, B.,** Stable replication of plasmids derived from Epstein-Barr virus in various mammalian cells, *Nature,* 313, 812, 1985.
9. **Sugden, B., Marsh, K., and Yates, J.,** A vector that replicates as a plasmid and can be efficiently selected in B-lymphoblasts transformed by Epstein-Barr virus, *Mol. Cell. Biol.,* 5, 410, 1985.
10. **Margolskee, R. F., Kavathas, P., and Berg, P.,** Epstein-Barr virus shuttle vector for stable episomal replication of cDNA expression libraries in human cells, *Mol. Cell. Biol.,* 8, 2837, 1988.
11. **Mulligan, R. C. and Berg, P.,** Selection for animal cells that express the *Escherichia coli* gene coding for xanthine-guanine phosphoribosyltransferase, *Proc. Natl. Acad. Sci. U.S.A.,* 78, 2072, 1981.
12. **Okayama, H. and Berg, P.,** A cDNA cloning vector that permits expression of cDNA inserts in mammalian cells, *Mol. Cell. Biol.,* 3, 280, 1983.
13. **Kaufman, R. J.,** Identification of the components necessary for adenovirus translational control and their utilization in cDNA expression vectors, *Proc. Natl. Acad. Sci. U.S.A.,* 82, 689, 1985.
14. **Boshart, M., Weber, F., Jahn, G. D., Hasler, K., Fleckenstein, B., and Schaffner, W.,** A very strong enhancer is located upstream of an immediate early gene of human cytomegalovirus, *Cell,* 41, 521, 1985.
15. **Seed, B. and Aruffo, A.,** Molecular cloning of the CD12 antigen, the T-cell erythrocyte receptor, by a rapid immunoselection procedure, *Proc. Natl. Acad. Sci. U.S.A.,* 84, 3365, 1987.

16. **Gorman, C. M., Merlino, G. T., Willingham, M. C., Pastan, I., and Howard, B. H.,** The Rous sarcoma virus long terminal repeat is a strong promoter when introduced into a variety of eukaryotic cells by DNA-mediated transfection, *Proc. Natl. Acad. Sci. U.S.A.*, 79, 6777, 1982.
17. **Cremici, C. and Babinet, C.,** Negative regulation of early polyomavirus expression in mouse embryonal carcinoma cells, *J. Virol.*, 59, 761, 1986.
18. **Sleigh, M. J. and Lockett, T. J.,** SV40 enhancer activation during retinoic acid induced differentiation of F9 embryonal carcinoma cells, *EMBO J.*, 4, 3831, 1985.
19. **Linney, E., Davis, B., Overhauser, J., Chao, E., and Fan, H.,** Non-function of a Moloney murine leukemia virus regulatory sequence in F9 embryonal carcinoma cells, *Nature*, 308, 470, 1984.
20. **Takebe, Y., Seiki, M., Fujisawa, J., Hoy, P., Yokota, K., Arai, K., Yoshida, M., and Arai, N.,** SRα promoter: an efficient and versatile mammalian cDNA expression system composed of the simian virus 40 early promoter and the R-U5 segment of human T-cell leukemia virus type 1 long terminal repeat, *Mol. Cell. Biol.*, 8, 466, 1988.
21. **Mayo, K. E., Warren, R., and Palmiter, R. D.,** The mouse metallothionein-I gene is transcriptionally regulated by cadmium following transfection into human or mouse cells, *Cell*, 29, 99, 1982.
22. **Frank, L., Mulligan, R. C., Berg, P., and Ringold, G.,** Glucocorticoids regulate expression of dihydrofolate reductase cDNA in mouse mammary tumour virus chimaeric plasmids, *Nature*, 294, 228, 1981.
23. **Corces, V., Pellicer, A., Axel, R., and Meselson, M.,** Integration, transcription, and control of a *Drosophila* heat shock gene in mouse cells, *Proc. Natl. Acad. Sci. U.S.A.*, 78, 7038, 1981.
24. **Wong, G. G., Witek, J. S., Temple, P. A., Wilkens, K. M., Leary, A. C., Luxenberg, D. P., Jones, S. S., Brown, E. L., Kay, R. M., Orr, E. C., Shoemaker, C., Golde, D. W., Kaufman, R. J., Hewick, R. M., Wang, E. A., and Clark, S. C.,** Human GM-CSF: molecular cloning of the complementary DNA and purification of the natural and recombinant proteins, *Science*, 228, 810, 1985.
25. **Knopf, J. L., Lee, M.-H., Sultzman, L. A., Kriz, R. W., Loomis, C. R., Hewick, R. M., and Bell, R. M.,** Cloning and expression of multiple protein kinase C cDNA, *Cell*, 46, 491, 1986.
26. **Tsai, S.-F., Martin, D. I. K., Zon, L. I., D'Andrea, A. D., Wong, G. G., and Orkin, S. H.,** Cloning of cDNA for the major DNA-binding protein of the erythroid lineage through expression in mammalian cells, *Nature*, 339, 446, 1989.
27. **Jolly, D. J., Okayama, H., Berg, P., Esty, A. C., Filpula, D., Bohlen, P., Johnson, G. G., Shively, J. E., Hunkapillar, T., and Friedman, T.,** Isolation and characterization of a full-length expressible cDNA for human hypoxanthine phosphoribosyltransferase, *Proc. Natl. Acad. Sci. U.S.A.*, 80, 477, 1983.
28. **Wigler, M., Silverstein, S., Lee, L.-S., Pellicer, A., Cheng, Y.-C., and Axel, R.,** Transfer of purified herpes virus thymidine kinase gene to cultured mouse cells, *Cell*, 11, 223, 1977.
29. **Garapin, F., Horodniceanu, F., Kovrilsky, P., Garapin, A.-C.,** A new dominant hybrid selective marker for higher eukaryotic cells, *J. Mol. Biol.*, 150, 1, 1981.
30. **Southern, P. J. and Berg, P.,** Transformation of mammalian cells to antibiotic resistance with a bacterial gene under control of the SV40 early region promoter, *J. Mol. Appl. Genet.*, 1, 327, 1982.
31. **Blochlinger, K. and Diggelmann, H.,** Hygromycin B phosphotransferase as a selectable marker for DNA transfer experiments with higher eucaryotic cells, *Mol. Cell Biol.*, 4, 2929, 1984.
32. **Simonsen, C. C. and Levinson, A. D.,** Isolation and expression of an altered mouse dihydrofolate reductase cDNA, *Proc. Natl. Acad. Sci. U.S.A.*, 80, 2495, 1983.
33. **Urlaub, G. and Chasin, L. A.,** Isolation of Chinese hamster cell mutants deficient in dihydrofolate reductase activity, *Proc. Natl. Acad. Sci. U.S.A.*, 77, 4216, 1980.
34. **Ringold, G., Dickmann, B., and Lee, F.,** Co-expression and amplification of dihydrofolate reductase cDNA and the *Escherichia coli* XGPRT gene in Chinese hamster ovary cells, *J. Mol. Appl. Genet.*, 1, 165, 1981.
35. **Chen, D. and Okayama, H.,** High-efficiency transformation of mammalian cells by plasmid DNA, *Mol. Cell. Biol.*, 7, 2745, 1987.
36. **Chu, G., Hayakawa, H., and Berg, P.,** Electroporation for the efficient transfection of mammalian cells with DNA, *Nucleic Acids Res.*, 15, 1311, 1987.
37. **Felgner, P. L., Gadek, T. R., Holm, M., Roman, R., Chan, H. W., Wenz, M., Northrop, J. P., Ringold, G. M., and Danielsen, M.,** Lipofection: a highly efficient, lipid-mediated DNA-transfection procedure, *Proc. Natl. Acad. Sci. U.S.A.*, 84, 7413, 1987.
38. **Cappechi, M. R.,** High efficiency transformation by direct microinjection of DNA into cultured mammalian cells, *Cell*, 22, 479, 1980.
39. **Schaffner, W.,** Direct transfer of cloned genes from bacteria to mammalian cells, *Proc. Natl. Acad. Sci. U.S.A.*, 77, 2163, 1980.
40. **Graham, F. L. and van der Eb, A. J.,** A new technique for the assay of infectivity of human adenovirus 5 DNA, *Virology*, 52, 456, 1973.
41. **Parker, B. A. and Strak, G. R.,** Regulation of simian virus 40 transcription: sensitive analysis of the RNA species present early in infections by virus and viral DNA, *J. Virol.*, 31, 360, 1979.
42. **Kozak, M.,** Point mutations define a sequence flanking the AUG initiator codon that modulates translation by eukaryotic ribosomes, *Cell*, 44, 283, 1986.

43. **Luthman, H. and Magnusson, G.,** High efficiency polyoma DNA transfection of chloroquine treated cells, *Nucleic Acids Res.*, 11, 1295, 1983.

44. **Mann, R., Mulligan, R. C., and Baltimore, D.,** Construction of a retrovirus packaging mutant and its use to produce helper-free defective retrovirus, *Cell,* 33, 153, 1983.

44a. **Korman, A. J., Frantz, J. D., Strominger, J. L., and Mulligan, R. C.,** Expression of human class II major histocompatibility complex antigens using retrovirus vectors, *Proc. Natl. Acad. Sci. U.S.A.*, 84, 2150, 1987.

45. **Cone, R. D. and Mulligan, R. C.,** High-efficiency gene transfer into mammalian cells: generation of helper-free recombinant retrovirus with broad mammalian host range, *Proc. Natl. Acad. Sci. U.S.A.*, 81, 6349, 1984.

46. **Danos, O. and Mulligan, R. C.,** Safe and efficient generation of recombinant retrovirus with amphotropic and ecotropic host ranges, *Proc. Natl. Acad. Sci. U.S.A.*, 85, 6460, 1988.

Chapter 15

IMMUNOSCREENING OF cDNA EXPRESSION LIBRARIES

Steven Chao, Julie Chao, and Lee Chao

TABLE OF CONTENTS

I. INTRODUCTION

Molecular cloning of recombinant cDNA expression libraries has widespread applications in biotechnology. Expression cDNA libraries are usually constructed using bacteriophage such as λgt11 or plasmids such as pUC8/pUC9. These vectors provide a bacterial promoter and ribosome recognition sequence near a polylinker cloning site. cDNA inserts cloned into polylinker sites are subjected to the control of the bacterial promoter (e.g., lacZ), and the transcribed mRNA may produce the proper protein products, depending on the reading frame, the orientation of the inserts, and the completeness of the inserts. Identification of bacterial colonies expressing fusion proteins encoded by cloned cDNA inserts in expression libraries can be achieved with specific antibodies through immunological methods. Immunoscreening of cDNA clones is a very popular method because only specific polyclonal antibodies are required and protein sequencing and oligonucleotide construction are bypassed. The fused protein products, translated from the truncated lacZ gene and the foreign DNA, are replica plated into nitrocellulose solid support and screened with various immunological methods.[1-4] The cDNA clones are traditionally detected with an appropriate primary antibody, followed by two general types of secondary antibodies. The first group includes ^{125}I-labeled protein A or ^{125}I-labeled second antibody, and the second group consists of secondary antibodies conjugated to enzymes such as alkaline phosphatase, horseradish peroxidase,[2] or biotinated antibody, followed by avidin-conjugated horseradish peroxidase. In addition to autoradiographic and chromogenic methods, analysis of the translated products from cDNA expression libraries has been described recently by using an immunogold detection method,[5] which was originally introduced for antigen localization by electron microscopy and for Western blot analysis.[6] The immunogold detection method, employing antibodies noncovalently attached to colloidal gold particles, followed by silver enhancement treatment, is rapid, sensitive, and specific, providing an alternative to enzyme- or radionuclide-based screening procedures for cDNA expression libraries. Although the radionucleotide detection techniques are, as a general rule, more sensitive than the enzyme detection methods, we have frequently found a high number of false positives and ambiguities in the primary screening procedure. Here, we present an alternative technique using antigen overlay instead of the traditional protein A or second antibody method for immunoscreening. This procedure utilizes a radiolabeled antigen detection method described originally for the immunoblotting of plasma proteins.[7] The method is simple and has proved to be exceptionally sensitive and accurate in identifying correct clones in both primary and secondary screenings. Using this method, we have identified and isolated cDNA clones coding for rat α1-antitrypsin[8] and a rat kallikrein-binding protein from a rat liver expression cDNA library, as well as human kininogen from a human liver expression cDNA library (a gift from Dr. Savio Woo, Baylor Medical College). Screening for rat α1-antitrypsin has been used as an example to demonstrate the applicability of this technique. Comparison between Western blots of α1-antitrypsin using both labeled antigen and protein A detection methods showed that the former yielded lower background and greater sensitivity than the latter.[8] Further, this technique is at least severalfold more sensitive than the enzyme-linked chromogenic detection method and has a lower detection limit of 20 ng in Western blot analysis.

These procedures are based on the expression of the chimeric protein by cDNA clones containing the DNA inserts in the correct reading frame and orientation. Following the transfer of phage proteins to nitrocellulose membranes, bivalent antibodies bind monovalently both the nitrocellulose-bound antigen from the phage lysates and to the free, radiolabeled antigen. This improved method can be applied in general for screening expression cDNA libraries, provided the specific antiserum and labeled antigen are available.

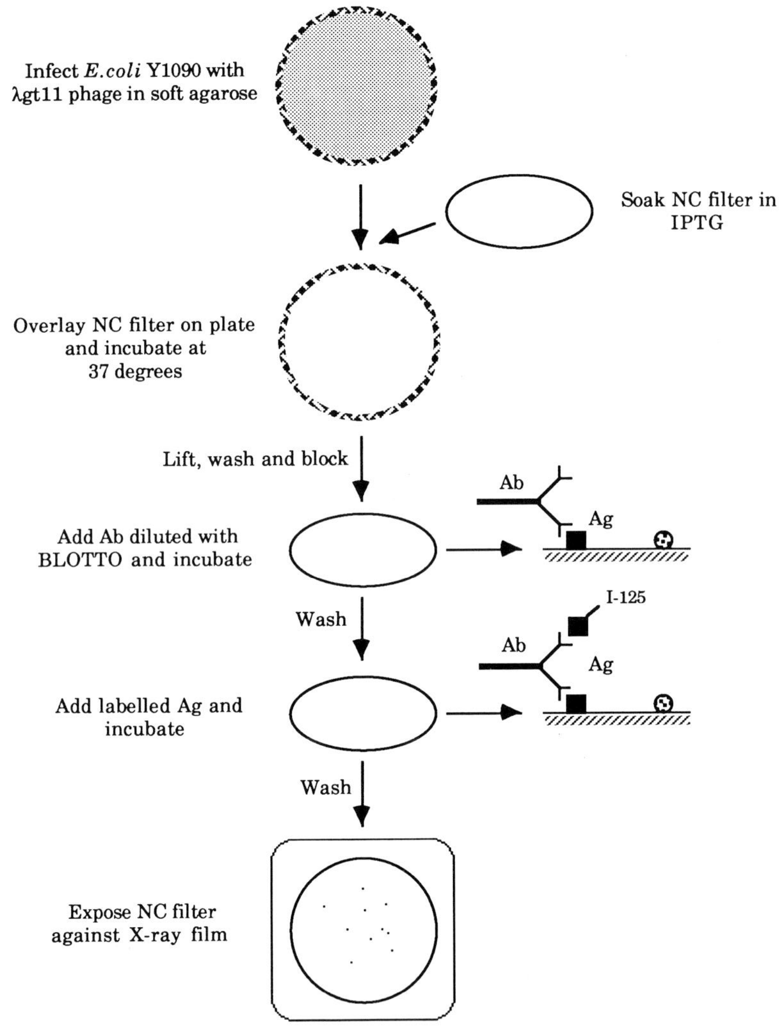

FIGURE 1. Procedure of immunoscreening cDNA clones from a λgt11 expression cDNA library. Recombinant λgt11 phage are plated on a lawn of *E. coli* Y1090. After a period of plaque growth, foreign antigen production is induced by overlaying the lawn with an IPTG-saturated NC membrane. Fused proteins released upon phage-induced lysis are immobilized on the NC and probed with primary antibody, followed by ^{125}I-labeled antigen. (From Chao, S., Chao, L., and Chao, J., *BioTechniques,* 7(1), 68, 1989. With permission.)

II. PRINCIPLES OF cDNA IMMUNOSCREENING

The present report describes a highly sensitive and specific antigen method for the immunoscreening of cDNA clones from expression cDNA libraries. The immunoscreening procedure for the identification of cDNA clones from a λgt11 expression cDNA library is outlined in Figure 1. The principle is based on the assumption that cDNA clones containing DNA inserts express specific immunoreactive antigens in the phage plaque. Following the transfer of proteins produced by phage to nitrocellulose membranes, the bivalent antibodies bind monovalently to both nitrocellulose-bound antigens and free radiolabeled antigens.

III. ANTIBODY PROBES

Polyclonal antiserum generated in rabbit or sheep[9] or a mixture of panels of monoclonal antibodies[10] have been successfully used to isolate cDNA clones from λgt11 libraries constructed from mRNA of various tissues or from certain bacterial DNA genomes. The quality of the antibody is important, since high-titer antiserum produces better signals than low-titer antiserum. Antiserum should be first titered using either a radioimmunoassay or an enzyme-linked immunoabsorbant (ELISA) assay.[9,11] Antibodies which produce good signals on a Western blot generally produce good signals in the λgt11 screening procedure. We usually use antiserum dilutions in the binding reactions of 1 to 500. However, in cases in which the antiserum is found to cross-react strongly with undesired antigens, as shown by Western blot analysis, we have used antibody dilutions of up to 1 to 10,000 to prevent saturation of these cross-reactive antigens. Shortening the duration of the binding reaction could also help to reduce cross-reactivity, but a time course study would be needed to determine the optimum binding time.

One advantage of the antigen overlay method is that the antiserum used for screening does not require further purification, while affinity-purified antibodies are used in the protein A detection method. Further, polyclonal antibodies often contain proteins which bind to natural antigens produced by *Escherichia coli* and produce a high background. Therefore, it is important to preabsorb the antiserum with *E. coli* lysate to reduce the background noise. Using the antigen overlay method, the preabsorption step becomes unnecessary, since the free radiolabeled antigen will only be bound by its specific antibody.

IV. ANTIGEN LABELING

Antigen (5 μg) is labeled with ^{125}I by the lactoperoxidase method to a specific activity of 50 to 70 μCi/μg protein.[11] Radiolabeled antigen is then separated from free ^{125}I on a Sephadex G100 column or by desalting using a spun Sephadex G50 column. We have found that, though the former procedure is more time consuming and complex, it frequently is more effective in separating free ^{125}I from radiolabeled antigen and often yields a greater quantity of radiolabeled antigen. It seems that certain proteins are difficult to separate from free ^{125}I, perhaps due to nonspecific binding to the column matrix, while others can be consistently and effectively desalted using the spin column procedure.

A. PURIFICATION OF THE ^{125}I-LABELED TRACER BY COLUMN CHROMATOGRAPHY

1. The reaction mixture contains:
   - 50 μl of 0.4 *M* sodium acetate, pH 5.6
   - 10 μl of lactoperoxidase (Sigma) (150 μg/ml, 6 U/mg) in 0.1 *M* sodium acetate, pH 5.6
   - Purified antigen: 2.5 to 5.0 μg in 5 to 10 μl
   - ^{125}I: 0.5 to 1.0 mCi (5 to 10 μl)
   - 10 μl H_2O_2 (200 μg/ml or 0.005%)
2. The reaction is initiated by the addition of H_2O_2.
3. Incubate the reaction mixture at 37°C with gentle shaking for 20 min. An additional 10 μl of H_2O_2 may be added and incubated for another 20 min to increase the amount of incorporated ^{125}I.
4. The reaction is stopped by adding 0.2 ml of 0.05 *M* sodium phosphate, pH 7.4.

5. Apply the sample directly to a Sephadex G100 column (1 × 30 cm) preequilibrated with 0.1% BSA in 0.05 *M* sodium phosphate, pH 7.4, and elute with the same buffer.
6. Collect 1-ml fractions and count 10 μl from each fraction in a γ-counter. The first radioactive peak should be the labeled antigen, and the second peak should be the free ^{125}I.

B. SEPARATION OF LABELED ANTIGEN AND FREE ^{125}I BY SPIN COLUMN DESALTING

Following steps 1 to 4 as described above; each sample is then desalted with a Sephadex G50 spin column using a modification of the procedures described in the Pharmacia manual.

5. Invert the column several times to resuspend the gel.
6. Remove the top closure first, then remove the bottom closure. Removing the closures in this order prevents the formation of air bubbles in the column, which may make draining the column buffer difficult.
7. Let the excess buffer drain from the column.
8. Place the column in a collection tube.
9. Centrifuge the column/tube assembly for 2 min at 1100 × g.
10. Discard the collected buffer in the collection tube.
11. Add 1 ml of 0.05 *M* sodium phosphate, pH 7.4, containing 0.1% BSA to the column, then repeat steps 7 to 10. This step equilibrates the column with BSA and will reduce nonspecific binding of the antigen to the column matrix. For best results, use a centrifuge with a swinging bucket or a horizontal rotor. A table-top centrifuge (Beckman Model TJ-6) is also suitable.
12. Carefully apply the sample to the center of the gel bed. Generally, a 50-μl sample volume limit is recommended. Avoid applying the sample near the sides of the column.
13. Place the column in a second collection tube.
14. Centrifuge the column/tube assembly for 4 min at 1100 × g.
15. The fluid in the collection tube contains the purified tracer.
16. Discard the used column in an appropriate fashion.

V. PROCEDURES FOR IMMUNOLOGICAL SCREENING OF A cDNA LIBRARY

1. Streak out *E. coli* Y1090 on an LB plate, pH 7.5, containing 50 μg/ml of ampicillin. Incubate at 37°C until colonies appear (usually 8 to 10 h or overnight). This plate can be stored at 4°C as a master plate.
2. Starting with a single colony, grow Y1090 to OD_{600} = 0.1 to 0.2 in LB, pH 7.5, containing 0.2% maltose at 37°C. Addition of 50 μg/ml of ampicillin to the culture is optional.
3. Mix 0.3 ml of the Y1090 culture with λ diluent (10 m*M* Tris, pH 7.5, and 10 m*M* $MgCl_2$) containing 2×10^4 to 3×10^4 PFU (plaque forming units) of the λgt11 cDNA library or 150 to 200 PFU from a primary screening plug for each plate, for primary and secondary screening, respectively. Preadsorb the phage to the cells by incubation at 37°C for 15 min.

Note: We have tried a different technique for inducing phage insert expression by isopropyl-β-D-thiogalactoside (IPTG). Instead of presoaking nitrocellulose (NC) in 10 m*M* IPTG, one can add either 25 μl (small plate) or 75 μl (large plate) of 100 m*M* IPTG to the LB soft agarose (see below) such that the final concentration is 1 m*M* IPTG. The phage insert expression begins earlier, the amount of IPTG used is less, and this method is less time consuming. The only drawback is the possibility that the translated insert might be lethal to the host cells, and thus early expression in fewer cells might be undesirable.

4. Add the cell/phage mixture to 2.5 ml (for a 90-mm plate) or 7.5 ml (for a 150-mm plate) LB soft agarose (and IPTG, see Note above), vortex, and pour onto an LB plate, pH 7.5.
5. After allowing the soft agarose to solidify for 5 to 10 min at room temperature, incubate the plates at 42°C until the phage plaques are approximately 1 mm in diameter and their edges are just beginning to touch each other, usually about 3 to 4 h.
6. Remove the plates to a 37°C incubator. Overlay each plate with a dry NC filter which has been presoaked in 10 m*M* IPTG (or just dry NC; see Note above).
7. Incubate the plates for 4 to 5 h at 37°C. If duplicate lifting is attempted, the second set of filters is then overlaid after the first has been removed (see below) and is incubated for an additional 90 min to 2 h.
8. Remove the plates from the incubator and mark the position of the filter on each plate by poking at least three asymmetric holes through the filter into the agar with a needle (usually dipped in India ink to give a permanent mark). Remove the filters carefully. From this point, the procedures for immunoscreening may follow the antigen overlay method or the conventional protein A detection method, as outlined below.

A. ANTIGEN OVERLAY METHOD

The remaining steps can be performed either at 30°C or at room temperature. Filters are washed in TBST (50 m*M* Tris, pH 8.0, 150 m*M* NaCl, and 0.05% Tween-20) three times for 5 to 10 min each and then blocked for 1 h with BLOTTO. After blocking, the NC membranes are incubated for 1 h with gentle shaking in polyclonal antiserum (1:500 to 1:10,000 dilution in BLOTTO, see Section III). The appropriate dilution of antibody for a plaque screen is similar to that necessary to obtain good signals in a Western blot. The NC sheets are washed three times in BLOTTO for 5 to 10 min each and are then incubated for a minimum of 90 min with ^{125}I-labeled antigen (250,000 cpm/ml) in BLOTTO. The NC membranes are washed three times with BLOTTO and once with phosphate-buffered saline (0.01 *M* phosphate, 0.14 *M* NaCl, pH 7.4), air dried completely, and exposed to Kodak X-Omat film. Purified antigen at concentrations of 1.5, 0.15, and 0.015 μg is spotted on NC and immunoblotted as described above for use as positive controls. Detergents, such as Triton-X100, deoxycholate, or SDS, are not required in the washing solution.

B. ^{125}I-PROTEIN A DETECTION METHOD

Following plaque lifting as described above, NC filters are washed in TS (10 m*M* Tris, pH 7.5, and 100 m*M* NaCl) buffer and then incubated in TS containing 3% bovine serum albumin for 1 h at 40°C. The NC is then incubated with the affinity-purified antibody (5 to 10 μg/ml) in TS buffer containing 0.1% NP40 for 1 h at room temperature. Following the incubation, the filters are washed again in the same buffer for 10 min, and then ^{125}I-labeled protein A (250,000 cpm/ml) is added. ^{125}I-protein A-incubated filters are once again washed in the same buffer for 10 min, air dried, and autoradiographed using Kodak X-Omat film.

The primary screening for rat α1-antitrypsin using the antigen overlay detection method yielded approximately six positive signals per plate, each of which contained 20,000 plaques. Figure 2A depicts not only the high sensitivity, but also the extremely low background of this method, compared to the previously used protein A detection method, shown in Figure 2B. Cross-reactivity between ^{125}I-protein A and nonspecific proteins in phage lysates can be virtually eliminated by using the radiolabeled antigen detection method. The autoradiographs from the primary screening were used as a guide to locate the positive clones. Sixteen 4-mm-diameter plugs containing the positive plaques as well as surrounding plaques

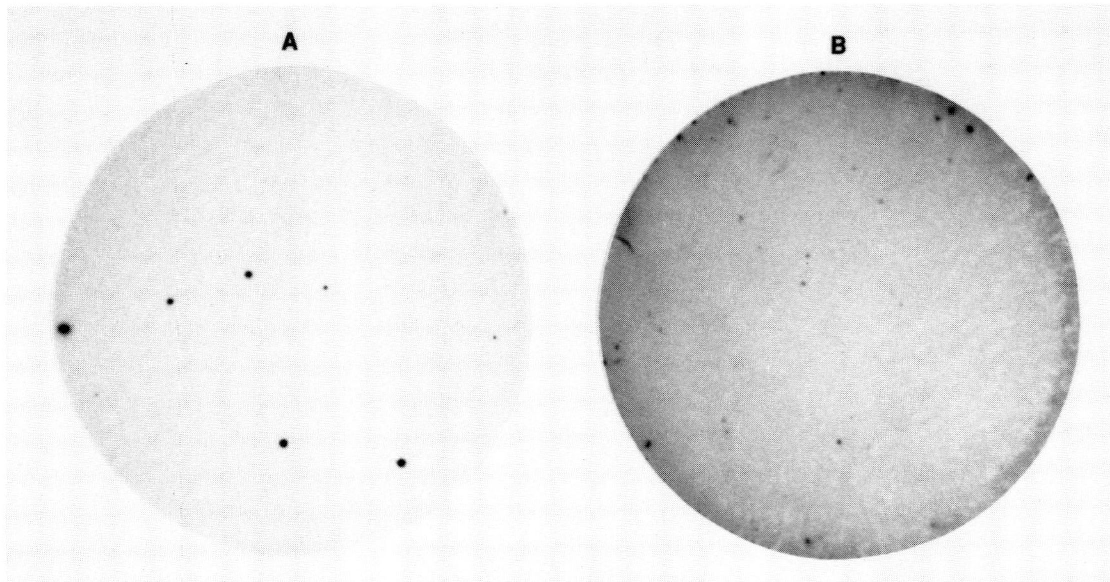

FIGURE 2. Primary immunoscreening of rat α1-antitrypsin cDNA clones. A λgt11 rat liver cDNA library was plated (20,000 PFU per plate) and plaque lifted for primary immunoscreening. Screening using (A) ^{125}I-α1-antitrypsin detection method and (B) ^{125}I-protein A detection method. (From Chao, S., Chao, L., and Chao, J., *BioTechniques,* 7(1), 68, 1989. With permission.)

were removed for secondary screening. Secondary screening revealed that 100% of the plugs selected in the primary screening contained clones which coded for α1-antitrypsin. The results of the secondary screening using the antigen overlay detection method again exhibited both high sensitivity and low background.

VI. CONFIRMATION OF THE IDENTITY OF cDNA CLONES

A. SEQUENCE ANALYSIS OF DNA INSERTS

Determination of the sequence of the cloned DNA may establish the identity of the cDNA that encodes the protein of interest if part of the antigen's primary structure is known. The positive clones containing the largest inserts (>1.2 kb) in the immunoscreening of rat α1-antitrypsin were subcloned into the M13mp19 sequencing vector. Sequence analysis was performed according to the method described by Maniatis et al.[12] Nucleic acid sequence data obtained by Sanger's dideoxy chain termination method[13] revealed that the partially sequenced rat α1-antitrypsin clone shared 75% homology with the human α1-antitrypsin sequence, confirming the identity of the cDNA clone coding for rat α1-antitrypsin.

B. WESTERN BLOT ANALYSIS OF RECOMBINANT PROTEINS

Once a single clone or group of clones are identified as positive by immunoscreening and individual clones are isolated, the fusion proteins produced by the isolated phage clones can be analyzed by Western blotting.

1. Grow Y1090 bacteria overnight in the presence of filtered sterilized ampicillin (100 μg/μl LB).
2. Dilute the bacterial culture, if necessary, so that the OD_{600} = 0.1 to 0.2. Aliquot 2.5 ml of this culture into culture tubes.
3. Add 25 μl of a 100-m*M* IPTG stock solution to each tube.

Note: Greater amounts of IPTG may be needed to induce sufficient expression for Western blot detection. However, reducing the concentration of IPTG or adding IPTG 1 or 2 h later in the growth cycle may be recommended if it is found that the induced protein is lethal to the host cells and causes early lysis.

4. Add either one fourth of the solution into which the secondary screening plug phage has been diluted or approximately 2.5 × 10^6 PFUs (plaque forming units) from an amplified phage stock into the culture. It is sometimes necessary to use a series of different phage dilutions to ensure that one of the cultures lyses between 3 to 6 h.
5. Upon seeing the first signs of the appearance of bacterial debris, part of the bacterial culture (1 to 1.5 ml) is transferred to an Eppendorf centrifuge tube, and the bacteria are recovered by centrifugation for 5 min in a microfuge.
6. The supernatant is discarded, and the bacterial pellet is resuspended in 50 μl of H_2O and can be stored at −20°C overnight; 50 μl of an extraction buffer containing 0.125 *M* Tris-HCl, pH 6.8, 30% glycerol, 40 m*M* dithiothreitol, and 5% SDS is added to each sample, and each tube is boiled for 5 min. The viscous solution is vortexed vigorously or passed through a 26-gauge needle two to three times to shear the chromosomal DNA.
7. The proteins are resolved by SDS-polyacrylamide gel electrophoresis: 5 μl of bacterial lysate is generally sufficient to visualize the proteins by Coomassie blue staining, and 10 μl can be used for Western blot analysis.
8. The fused proteins can be identified by Western blot analysis, as described below.

VII. WESTERN BLOT PROCEDURES

A. ANTIGEN OVERLAY METHOD

1. Samples (30 to 50 μl) of bacterial lysate in extraction buffer (containing 15% glycerol, 2.5% SDS, and 20 m*M* dithiothreitol in 0.0625 *M* Tris-Cl, pH 6.8) or of purified antigen at varying concentrations are boiled for 5 min, subjected to SDS-polyacrylamide gel electrophoresis on 7.5 to 15% gradient gels, and then electroblotted onto NC membranes.[6]

Note: Steps 2 through 6 can be carried out either at 30°C or at RT. Although some of the current literature recommends 30°C, we have performed the procedure many times at RT without any apparent problems.

2. The nitrocellulose membranes are blocked with BLOTTO [5% (w/v) dry milk in 0.01 *M* sodium phosphate, pH 7.4, 0.14 *M* NaCl, 1 mg/l thimerosal, 200 mg/l NaN_3, 1 μ*M* para-amidinophenylmethylsulfonyl fluoride, and 0.01% anti-foam][14] for 1 h.
3. Incubate the nitrocellulose membranes with primary antiserum (1:500 to 1:10,000 in BLOTTO; see Section III) for 1 h with gentle shaking. The diluted antibodies can be stored frozen and reused, although replenishing the antiserum periodically is necessary.
4. Wash the nitrocellulose membranes three times for 5 to 10 min each with BLOTTO.
5. Incubate the nitrocellulose membranes with ^{125}I-labeled-antigen or ^{125}I-protein A (250,000 cpm/ml) for a minimum of 90 min.
6. Wash the NC membranes three times with BLOTTO and once with phosphate-buffered saline (0.14 *M* NaCl and 0.01 *M* sodium phosphate, pH 7.4) for 5 to 10 min each. Membranes are air dried completely and exposed to Kodak X-Omat film. The relative quantities of immunoreactive antigen are measured either by densitometer scanning or by cutting the radioactive spots from the NC membranes and counting the spots in a gamma counter.

B. LACTOPEROXIDASE METHOD

Steps 1 to 4 are identical to those described in the antigen overlay method. All steps are performed at RT or 30°C.

5. Incubate the NC in a secondary antibody conjugated with horseradish peroxidase (Cappel) for 1.5 h (1:500 dilution in BLOTTO).
6. Wash the NC membranes three times in BLOTTO and then once with H_2O for 5 to 10 min each.
7. Incubate the NC membranes for 10 to 30 min until bluish signals are visualized in a developing solution containing:

    - 100 ml of 10 m*M* Tris-Cl, pH 7.4, and 0.8% NaCl
    - 60 mg 4-chloro-1-naphthol in 20 ml of methanol
    - 50 μl of 30% H_2O_2

8. Rinse the NC in distilled H_2O and store in the dark.

Western blot analysis of tissue kallikrein in the rat submandibular gland shows that the antigen overlay method (Figure 3A) is much more sensitive than the peroxidase method (Figure 3B). Comparison between immunoblots of α1-antitrypsin using the ^{125}I-α1-antitrypsin and the ^{125}I-protein A detection methods is shown in Figure 4. With use of the

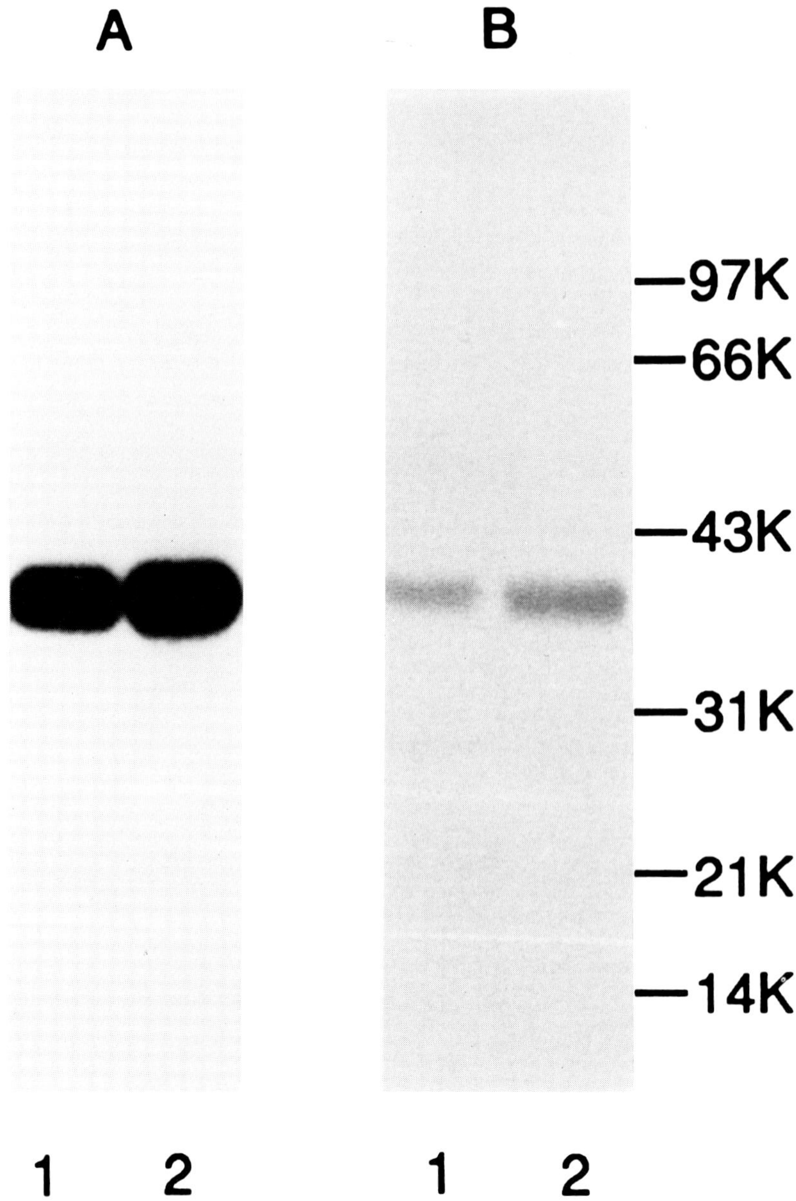

FIGURE 3. Comparison of sensitivity between immunoblots using the antigen overlay method and the enzyme-linked second antibody method. Rat submandibular extracts (10 μg each) from female (lane 1) and male (lane 2) rats were subjected to electrophoresis under reducing conditions in a 7.5 to 15% linear gradient polyacrylamide gel containing 0.1% SDS and electrotransferred to NC membranes. (A) Blots developed with sheep anti-rat urinary kallikrein (1:600 in BLOTTO), followed by ^{125}I-kallikrein. Autoradiography was for 16 h; (B) blots developed with sheep anti-rat urinary kallikrein (1:600 in BLOTTO), followed by incubation for 30 min with horseradish peroxidase-conjugated second antibody. Molecular weight markers are 97,000 (phosphorylase b); 68,000 (bovine serum albumin); 43,000 (ovalbumin); 31,000 (carbonic anhydrase); 21,000 (soybean trypsin inhibitor); and 14,000 (lysozyme). (From Chao, S., Chao, L., and Chao, J., *BioTechniques*, 7(1), 68, 1989. With permission.)

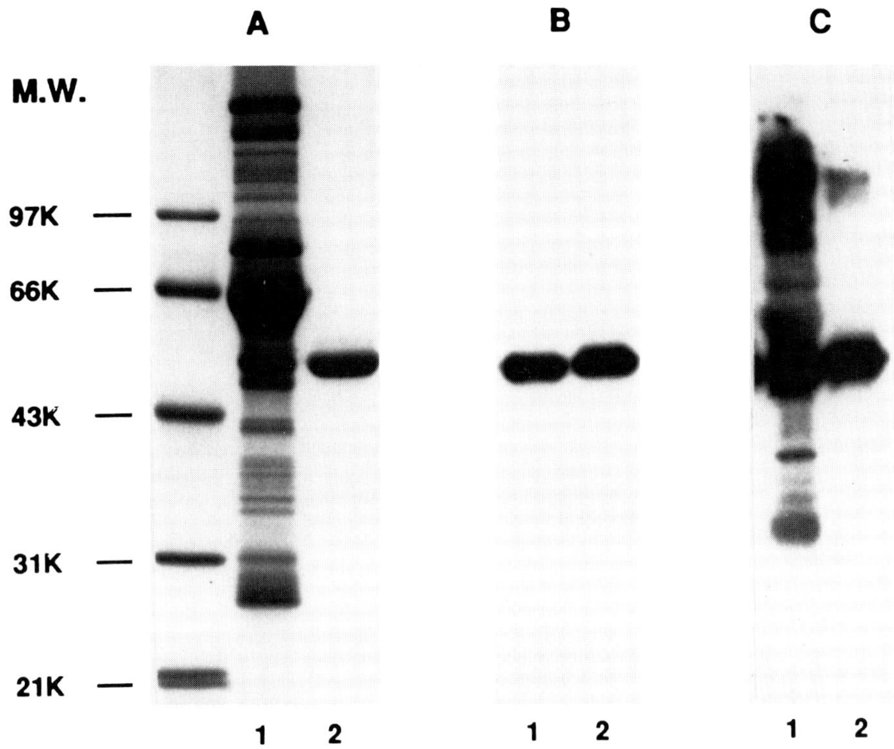

FIGURE 4. Immunoblots of rat α1-antitrypsin using ^{125}I-α1-antitrypsin and ^{125}I-protein A detection methods. Rat serum (A, 0.5 μl; B & C, 1 μl each, lane 1) or rat α1-antitrypsin (A, 7.5 μg; B & C, 1 μg each, lane 2) were subjected to electrophoresis in a 7.5 to 15% linear gradient polyacrylamide gel containing 0.1% SDS under reducing conditions and electrotransferred to NC membranes. (A) Protein staining with Coomassie blue; (B) blots developed with rabbit anti-α1-antitrypsin antiserum followed by ^{125}I-α1-antitrypsin. Autoradiography was for 16 h. (C) Blots developed with rabbit anti-α1-antitrypsin followed by ^{125}I-protein A. Autoradiography was for 16 h. (From Chao, S., Chao, L., and Chao, J., *BioTechniques,* 7(1), 68, 1989. With permission.)

antigen overlay method, only one band in total rat serum (lane 1), corresponding in molecular weight to purified rat α1-antitrypsin (lane 2), was recognized by the antiserum. The results of an experiment to analyze fusion proteins produced in bacteria containing a cDNA clone encoding human kininogen are shown in Figure 5.

VIII. CONCLUSIONS

The immunological screening of expression libraries provides a useful approach for the identification of cDNA clones. The present report presents a highly sensitive and specific immunoscreening method for the isolation of expression cDNA clones encoding proteins of interest. This immunoblotting technique with the radiolabeled antigen overlay method can detect specific proteins of interest in complex mixtures, such as whole plasma or phage lysates, with a sensitivity of less than 20 ng. Immunoscreening of cDNA clones using the antigen overlay method exhibits enhanced specificity with very low background, compared to the protein A method. BLOTTO in the blocking solution proves to be superior to bovine serum albumin, fetal calf serum, and gelatin in yielding lower signal background. With use of the antigen overlay method to search for rat α1-antitrypsin cDNA, secondary immunoscreening proved that 100% of the plugs selected in the primary screening contained an α1-antitrypsin cDNA clone, indicating the specificity of the technique. Using this method, we

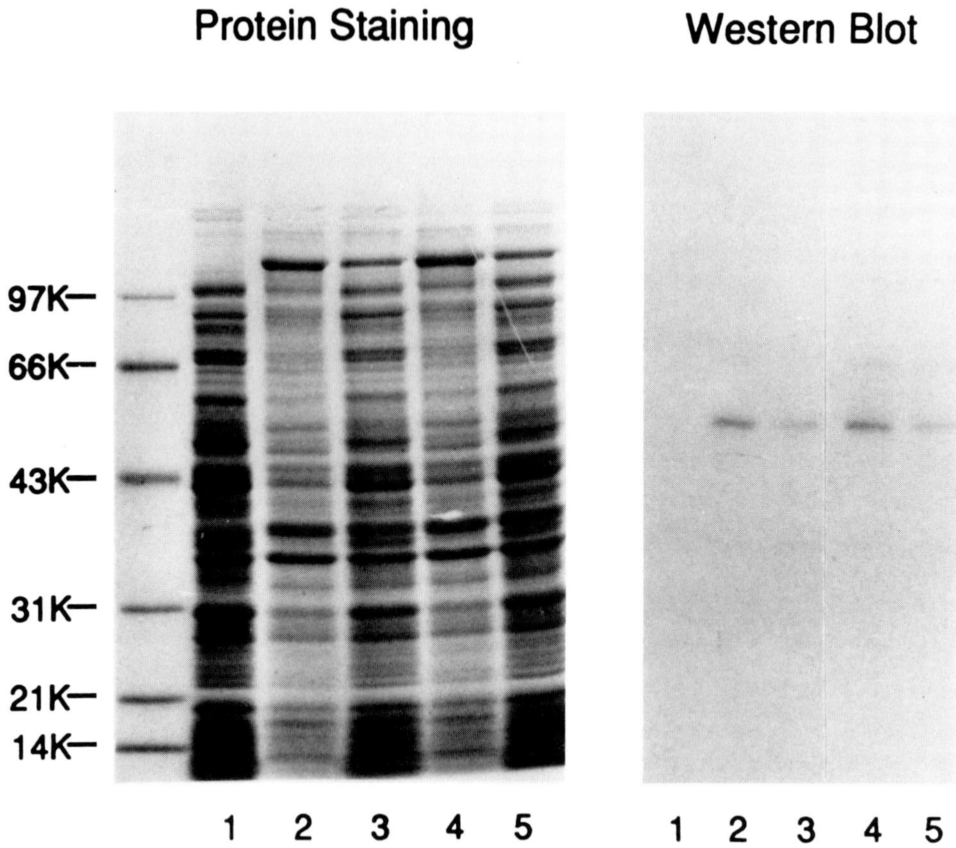

FIGURE 5. Total *E. coli* cellular extracts with foreign fusion proteins containing immunoreactive portions of human kininogen. (A) Proteins were visualized by staining with Coomassie blue; (B) proteins were electrotransferred to nitrocellulose membranes, and the blot was developed with rabbit anti-human LMW kininogen antiserum, followed by ^{125}I-human kininogen. Autoradiography was for 16 h. Molecular weight markers are the same as in Figure 3.

have isolated and characterized rat α1-antitrypsin, a rat kallikrein-binding protein, and human kininogen. This method can be applied in general for the isolation and identification of cDNA clones from expression libraries, provided the purified antigen and its specific antiserum are available.

ACKNOWLEDGMENTS

This work was supported by Grants HL29397 and HL33552 from the National Institutes of Health.

REFERENCES

1. **Young, R. A. and Davis, R. W.,** Yeast RNA polymerase II genes: isolation with antibody probes, *Science,* 222, 778, 1983.
2. **Helfman, D. M., Feramisco, J. R., Fiddes, J. C., Thomas, G. P., and Hughes, S. H.,** Identification of clones that encode chicken tropomyosin by direct immunological screening of a cDNA expression library, *Proc. Natl. Acad. Sci. U.S.A.,* 80, 31, 1983.

3. **Young, R. A. and Davis, R. W.,** Efficient isolation of genes using antibody probes, *Proc. Natl. Acad. Sci. U.S.A.*, 80, 1194, 1983.
4. **Huynh, T. V., Young, R. A., and Davis, R. W.,** *DNA Cloning: A Practical Approach,* Vol. 1, Glover, D. M., Ed., IRL Press, Oxford, 1985, 49.
5. **Pohl, H. E. W., Hock, J., and Müller-Esterl, W.,** Plaque-lift testing of expression vector λgt11 with gold-labeled immunoglobulins, *Anal. Biochem.*, 175, 414, 1988.
6. **Burnett, W. N.,** Western blotting: electrophoretic transfer of proteins from sodium dodecyl sulfate-polyacrylamide gels to unmodified nitrocellulose and radiographic detection with antibody and radioiodinated protein A, *Anal. Biochem.*, 112, 195, 1981.
7. **Lämmle, B., Berrettini, M., and Griffin, J. H.,** Enhanced specificity of immunoblotting using radiolabeled antigen overlay: studies of blood coagulation factor X and prekallikrein in plasma, *Anal. Biochem.*, 156, 118, 1986.
8. **Chao, S., Chao, L., and Chao, J.,** Enhanced specificity in immunoscreening of expression cDNA clones using radiolabeled antigen overlay, *BioTechniques,* 7(1), 68, 1989.
9. **Chao, S., Chao, L., and Chao, J.,** Sex dimorphism and inflammatory regulation of T-kininogen and T-kininogenase, *Biochem. Biophys. Acta,* 991, 477, 1989.
10. **Suzuki, K., Deyashiki, Y., Nishioka, J., Kurachi, K., Akira, M., Yamamoto, S., and Hashimoto, S.,** Characterization of a cDNA for human protein C inhibitor, a new member of the plasma serine protease inhibitor superfamily, *J. Biol. Chem.*, 262, 611, 1987.
11. **Shimamoto, K., Chao, J., and Margolius, H. S.,** The development and application of a radioimmunoassay for human urinary kallikrein, *J. Clin. Endocrinol. Metab.*, 51, 840, 1980.
12. **Maniatis, T., Fritsch, E. F., and Sambrook, J.,** *Molecular Cloning,* Cold Spring Harbor Laboratory, Cold Spring Harbor, NY, 1982, 371.
13. **Sanger, F. and Nicklen, S.,** DNA sequencing with chain-terminating inhibitors, *Proc. Natl. Acad. Sci. U.S.A.*, 74, 5463, 1977.
14. **Johnson, D. A., Gautxch, J. W., Spoasman, J. D., and Elder, J. H.,** Improved technique utilizing non-fat dry milk for analysis of proteins and nucleic acids transferred to nitrocellulose, *Gene Anal. Technol.*, 1, 3, 1984.

Chapter 16

IN VITRO TRANSCRIPTION ASSAYS FOR EUKARYOTIC GENE EXPRESSION

E. Cristy Ruteshouser, Sankar N. Maity, and Benoit de Crombrugghe

TABLE OF CONTENTS

I. INTRODUCTION

Activation and repression of specific genes play a crucial role in cellular differentiation, hormonal control, and cancer. There is ample evidence that the triggering of signaling pathways by external stimuli such as hormones and growth factors leads to changes in the transcription of specific genes. Deviation from a normal pattern of transcriptional control occurs in cancer cells and probably in other chronic diseases.

Control of the transcription of most cellular genes occurs mainly by regulating transcription initiation, i.e., allowing the enzyme RNA polymerase II to form a productive complex with the DNA at close proximity to the transcription start(s) in a specific gene. A diversity of factors, many of which directly bind to specific DNA sequences upstream or downstream from the start of transcription, participate in the formation of these preinitiation complexes and assure the spatial and temporal specificity in the activation of individual genes.

In this article, we discuss two principal methods that are commonly used in the study of the transcription of specific genes. In the first method, the so-called "run-on" assay uses intact nuclei that are allowed to pursue transcription that had already been initiated in the intact cell. The second method consists of a reconstitution experiment and uses extracts of nuclei or whole cells to perform DNA-dependent transcription initiation and elongation. The activity of a variety of wild-type and mutant DNA templates that are added to the reaction can be studied and compared to their activities in intact cells. This assay also serves to assay the function of transcriptional factors. Furthermore, the fractionation of the extracts is providing a biochemical framework to study the various components of the transcriptional machinery of eukaryotic cells.

II. NUCLEAR RUN-ON TRANSCRIPTION ASSAY

A. PRINCIPLES

The nuclear run-on transcription assay is a sensitive technique designed to measure the transcription of a specific gene or genes as a function of cell state. It is often used to demonstrate that the increase in the synthesis of a specific protein or the increase in the steady-state levels of the corresponding RNA is mediated by a transcriptional control mechanism, as opposed to changes in mRNA stability or transport into the cytoplasm. The assay directly measures the relative number of RNA polymerase molecules which have initiated transcription and are in the process of transcriptional elongation at the time of isolation of the nuclei. One assumes that the density of RNA polymerase molecules on a particular gene is proportional to the rate of transcription initiation of this gene. This assay allows the study of the transcription of a specific gene as it changes in response to states of growth or differentiation or in response to hormones, growth factors, or other treatments. In conjunction with the procedure of Northern blot analysis (which measures the relative amount of RNA from a specific gene), this technique provides information on the relative contributions of alterations in mRNA transcription and stability to changes in the level of expression of a particular gene.

In addition, this assay has been used to identify segments of the gene where transcription slows down or pauses. In certain cases, it appears that the density of RNA polymerase molecules is higher in some areas of the gene than in others.[1-3] The interpretation of this phenomenon is that the polymerases move more slowly through these segments or pause in these areas. One assumes that by using mechanisms of transcriptional antitermination, the cell can control the rate of transcription of DNA segments 3′ to the pause sites. Such control mechanisms have been extensively studied in bacteria and constitute one of the principal mechanisms for the control of gene expression in the bacteriophage λ.[4]

The nuclear run-on assay relies on the fact that RNA in the process of elongation can be labeled to high specific activity in isolated nuclei. A critical factor in success with this assay is the isolation of intact nuclei, since the use of damaged nuclei results in poor incorporation of labeled nucleoside triphosphates into RNA. Care must be taken, therefore, in the design of the procedure used to isolate nuclei. The procedures outlined below may not be optimal for nuclear isolation from some types of tissues or cultured cells, and in these cases substitution of another procedure for isolation of nuclei may be preferable. Several excellent procedures for nuclear isolation and run-on transcription have been described in the literature.[5-10] The technique includes the isolation of intact nuclei, incubation of these nuclei with radioactive and unlabeled nucleoside triphosphates, and detection of specific labeled RNAs.

B. PROCEDURES

1. Methods for Isolation of Intact Nuclei

a. Isolation of Nuclei from Tissues

1. Homogenize the freshly dissected tissue in cold buffer A (10 m*M* Tris-HCl, pH 8.0, 5 m*M* $MgCl_2$, 0.5 m*M* DTT, 0.25% Triton X-100, and 0.3 *M* sucrose) in a cold Dounce homogenizer or other tissue homogenizer.
2. Filter the homogenate through glass wool or cheesecloth.
3. Dilute the homogenized sample with a volume of cold buffer B (same as buffer A, except 2.4 *M* sucrose and 0.1% Triton X-100) equal to the volume of buffer A added in Step 1. Layer the sample over a cushion of buffer B in an ultracentrifuge tube and centrifuge at 25,000 rpm for 60 min at 4°C in a Beckman SW27 rotor (or equivalent).
4. Remove the supernatant. Gently resuspend the nuclear pellet in a small volume of buffer C (50 m*M* Tris-HCl, pH 8.0, 5 m*M* $MgCl_2$, 0.1 m*M* EDTA, 0.5 m*M* DTT, and 40% glycerol).

Note: If the nuclei do not pellet through the sucrose cushion during the centrifugation step, the concentration of sucrose in buffer B may be decreased.

5. Measure the concentration of nucleic acid by diluting an aliquot of nuclei in 10% SDS and determine the A_{260}. Adjust the concentration of nucleic acid to 1.0 to 1.5 mg/ml with buffer C.
6. Store the nuclei in 40-μl aliquots at −80°C.

b. Isolation of Nuclei from Cultured Cells

1. Remove the medium from the cells. Wash the cells three times with ice-cold phosphate-buffered saline (PBS).
2. Scrape the cells in cold PBS and centrifuge for 5 min at 500 × g at 4°C.
3. Resuspend the cells in two volumes of cold buffer D (10 m*M* Tris-HCl, pH 8.0, 5 m*M* $MgCl_2$, 2 m*M* DTT, 0.1% Triton X-100, and 0.25 *M* sucrose). Homogenize in a cold Dounce homogenizer.
4. Centrifuge the homogenate at 500 × g for 5 min at 4°C. Wash the nuclei twice in cold buffer D and recentrifuge.
5. Resuspend the nuclei in cold buffer E (same as buffer C, except 2 m*M* DTT). Measure the concentration of nucleic acid as above and adjust to a final concentration about 1.5 mg/ml.
6. Store the nuclei in 40-μl aliquots at −80°C.

2. Transcription Elongation and Purification of Labeled RNA

1. Thaw an aliquot of frozen nuclei (40 μl). Mix with 60 μl of buffer F (10 m*M* Tris-HCl, pH 8.0, 5 m*M* $MgCl_2$, 250 mM KCl, 0.42 m*M* ATP, CTP, and GTP, and 100 μCi of [α-^{32}P] UTP, 400 to 800 Ci/mmol) and incubate 30 to 45 min at 26°C.
2. To stop the reaction, add 25 μg of yeast tRNA, $CaCl_2$ to 10 m*M*, DNAse I (Worthington DPRF), and Proteinase K to 100 μg/ml each. Incubate for 20 min at 37°C.
3. Add EDTA to 15 m*M* and SDS to 0.5%. Continue the 37°C incubation for an additional 20 min.
4. Extract twice with buffered phenol/chloroform/isoamyl alcohol (25:24:1) and twice with chloroform/isoamyl alcohol (24:1).
5. Add 0.5 volume of 7.5 *M* ammonium acetate and precipitate the RNA with three volumes of ethanol.
6. Resuspend the RNA at 1 to 5 × 10^6 cpm/ml in buffer G (10 m*M* *N*-Tris [hydroxy-methyl]methyl 2-aminoethanesulfonic acid, pH 7.4, 10 m*M* EDTA, and 0.2% SDS). Using 5 × 10^6 nuclei and 100 μCi of [α^{32}P] UTP should result in 1 to 10 × 10^6 cpm in RNA.

3. Hybridization

1. Mix the labeled RNA (in buffer G) with an equal volume of buffer H (same as buffer F, except with 600 m*M* NaCl).
2. Hybridize 2 ml of RNA solution to DNA immobilized on nitrocellulose (see below) at 65°C for 36 h. Hybridizations can be performed in small (5 to 7 ml), disposable plastic scintillation vials, using coiled strips of nitrocellulose. This procedure yields more quantitative results than does hybridization in plastic bags.
3. Wash each nitrocellulose filter twice with 100 ml of 2X SSC for 1 h at 65°C.
4. Incubate each nitrocellulose filter with 8 ml of 10 μg/ml RNAse A in 2X SSC for 30 min at 37°C. This step can be performed in glass scintillation vials.
5. Wash each nitrocellulose filter with 100 ml of 2X SSC at 37°C for 1 h. Dry the nitrocellulose filters and expose to X-ray film.

4. Preparation of Nitrocellulose Filters

1. Denature the linearized DNA (about 10 μg of DNA per dot or slot) by incubating the DNA with 0.2 *M* NaOH for 30 min at room temperature.
2. Neutralize the DNA with 10 volumes of 6X SSC.
3. Spot the DNA onto nitrocellulose filters using a dot-blot or slot-blot apparatus (follow the procedure suggested by the manufacturer of the apparatus).
4. Air dry the filters, and bake for 2 h at 80°C under vacuum.
5. Store the filters under vacuum until use. Filters can be stored for at least 6 months.

C. ANALYSIS AND EVALUATION OF RESULTS

1. Quantitation

The results of this assay are easily quantitated when either a slot-blot or dot-blot apparatus is used to prepare the nitrocellulose filters. Use of a slot-blot apparatus provides the specific advantage that the results can be quantitated by densitometric scanning of the autoradiograph, a procedure that preserves the integrity of the primary data, or by counting strips of the blot in a scintillation counter. Results obtained using the dot-blots are also easily quantitated, however, by scintillation counting of individual dots. In all cases, it is extremely important that the background level of the hybridization of labeled RNA to the filter be as low as possible. For this reason, filters are washed under stringent conditions and are treated with RNAse A after hybridization.

2. Positive and Negative Controls

In order to evaluate the effects on transcription of a particular treatment, or to examine changes in transcription as a result of alterations in cell state, it is important to minimize artifactual effects of variations in the nuclear run-on assay. It is therefore essential to include the proper control DNAs in the hybridization step. Vector DNA is often included as a negative control in these assays to determine the background level of hybridization to the vector carrying the specific gene of interest. DNA from a "housekeeping" gene, defined as one whose transcription is unaffected by the treatment or cell-state alteration being examined (such as the gene for glyceraldehyde phosphate dehydrogenase), is also often included to provide an internal control for variations between individual experiments. When possible, a positive control should also be included, i.e., DNA from a gene whose expression is known to be affected by the treatment or change in cell state being studied. The proper controls will, of course, vary from one experiment to another and so must be carefully chosen for each set of conditions.

3. Evaluation of Results

In evaluating the results of this assay, certain assumptions must be made concerning the dynamics of transcriptional elongation in the cell. One assumption is that the rate of elongation of transcription is constant within a particular gene during the time of the assay. A second assumption is that the half-life, in isolated nuclei, of the specific mRNA that is synthesized is significantly greater than the assay labeling time (typically 30 to 60 min) under the conditions of the assay.

Counts per minute in individual RNA samples should be equalized prior to the hybridization step. The rationale behind this adjustment is that the overall level of RNA synthesis in the cell is not changing in response to changes in cell state. It is important that the accuracy of this assumption be tested for each set of conditions being examined. In these experiments, it is vital that the number of nuclei used in the transcription reactions be equalized between each set of conditions, since the amount of labeled nucleotide incorporated into RNA is proportional to the number of nuclei used for the labeling reaction.

Another assay that is often used to demonstrate transcriptional control is transfection of DNA into cultured mammalian cells. Strictly speaking, of course, transfection is an *in vivo*, not an *in vitro*, assay, but provides valuable information on the relative rates of transcription, within the living cell, from different transfected promoters or from promoters that have been mutagenized (by point mutations, substitution mutations, deletions, and/or additions). The advantage of this assay is that it provides a very simple way to determine relative promoter activity, by measuring the expression of a marker gene. The limitations of this assay arise from the fact that it examines transcription from promoters taken out of their normal context, i.e., truncated (at least from far upstream sequences and downstream sequences), fused to marker genes such as the gene for chloramphenicol acetyltransferase, and present in multiple copies along with vector sequences. The information derived from this assay therefore differs from, but can complement, the information provided by the nuclear run-on assay, which examines transcription from normal genes *in vitro*.

In some cases, Northern blot analysis may demonstrate an increase or decrease in the abundance of a particular RNA in response to a change in cell state, and the nuclear run-on assay will show no corresponding alteration in the transcription of the gene under the same conditions. Such a result would indicate a change in mRNA stability in response to these conditions. Posttranscriptional mechanisms have been shown to be responsible for both increases and decreases in mRNA stability. For example, destabilization of tubulin mRNAs has been shown to occur via an autoregulatory mechanism,[11] and stabilization of GM-CSF mRNA has been shown to be mediated by a tumor-specific *trans*-acting factor.[12] Attributing changes in gene expression to alterations in mRNA stability, however, requires first that changes in gene transcription be ruled out. This may best be accomplished through use of the nuclear run-on analysis. The next step in demonstrating mRNA stabilization or destabilization is to conduct a study of posttranscriptional effects, often performed with enucleated cells or with cells treated with transcriptional inhibitors such as actinomycin D or α-amanitin. mRNA half-lives can be estimated by treating cells with such inhibitors for various times, followed by mRNA extraction, poly (A) selection, and Northern blot analysis. Such studies will, in many cases, determine whether an alteration in mRNA stability is responsible for an observed change in gene expression. It is important to be aware, however, that the addition of transcriptional inhibitors may itself indirectly affect mRNA stability, e.g., by reducing the synthesis of a rapidly degraded ribonuclease. It has often been observed that the addition of protein synthesis inhibitors such as cycloheximide will cause an increase in RNA stability, presumably by reducing the synthesis of such nucleases. Such pitfalls in the interpretation of experimental data must always be considered when evaluating results from these assays.

III. *IN VITRO* TRANSCRIPTION

A. PRINCIPLE

In this assay, nuclear extracts or whole-cell extracts provide all the factors needed to initiate and elongate RNA synthesis from a DNA template that is added to the reaction. This assay is used mainly (1) to evaluate the function of *cis*-acting sequences, such as promoter elements and enhancers present in the DNA template, and (2) to test the activity of purified transcriptional factors.

B. PROCEDURES

1. Preparation of Extracts

The active components of transcription are present in the nucleus. Two different methods have been developed to extract these proteins from nuclei.

Nuclear extracts — Nuclei are first prepared; then the components of transcription are extracted from nuclei with salt.[13]

Preparation of whole-cell extracts — In the previous method, some components may leak out of the nuclei during their preparation. In this method, the transcription components are extracted in the presence of salt with simultaneous lysis of both cell and nucleus.[14] These two methods were first reported using HeLa cells. Later, these methods have been used with other cell types[15-17] and also to prepare extracts of tissues. When preparing tissue extracts, the procedure is somewhat different from the method used with cultured cells, due to the presence of larger amounts of nucleases and proteases in tissues.[18]

a. Cell Culture

HeLa cells and lymphoid cells are grown in suspension culture. HeLa cells are grown in Joklik's MEM containing 5% calf serum to a concentration of 4 to 6 $\times$ 10^5 cells per milliliter prior to harvesting. Lymphoid cells like Namalwa cells are grown to 10^6 cells per milliliter in RPMI 1640 medium supplemented with 10% fetal calf serum. Cells are harvested from media by centrifugation at 2000 rpm in a Sorval H-6000A rotor for 10 min.

Mouse NIH-3T3 fibroblast cells are grown as a monolayer culture in DMEM media supplemented with 10% calf serum. Cells are harvested at 80 to 100% confluency by scraping from dishes.

The protocols for preparation of nuclear extracts and whole-cell extracts follow previously published procedures[13,14] and have been successfully used in our laboratory. To prepare nuclear extracts from tissues, we refer the reader to Reference 18.

b. Nuclear Extract

1. Suspend the cell pellet in five packed cell volumes (PCV) of phosphate-buffered saline (PBS) at 4°C and collect by centrifugation at 2000 rpm in a HA-6000 rotor at 4°C for 10 min.
2. Resuspend the cell pellet in five PCV of buffer A (10 m*M* HEPES, pH 7.9, 1.5 m*M* $MgCl_2$, 10 m*M* KCl, and 0.5 m*M* DTT). Keep on ice for 10 min and centrifuge as before.
3. Resuspend the cells in two PCV of buffer A and lyse by ten strokes of an all-glass Dounce homogenizer (B pestle).
4. Centrifuge the homogenate as in Step 1.
5. Decant the supernatant carefully and recentrifuge the pellet for 20 min at 12,000 rpm in a Sorvall SS-34 rotor to remove residual cytoplasmic material.
6. Resuspend the nuclear pellet in 3 ml of buffer B (20 m*M* HEPES, pH 7.9, 25% glycerol, 0.42 *M* NaCl, 1.5 m*M* $MgCl_2$, 0.2 *M* EDTA, 0.5 m*M* PMSF, 0.5 m*M* DTT, 2 μg/ml leupeptin, and 2 μg/ml pepstatin) per 10^9 cells with the all-glass Dounce homogenizer (B pestle). The final NaCl concentration in the nuclei suspension is 0.28 to 0.3 *M*. Gently stir the nuclei suspension for 30 min at 4°C and then centrifuge for 30 min at 12,000 rpm in a Sorvall SS-34 rotor.
7. Dialyze the clear supernatant against 100 volumes of buffer C (20 m*M* HEPES, pH 7.9, 20% glycerol, 0.1 *M* KCl, 0.2 m*M* EDTA, 0.5 m*M* PMSF, 0.5 m*M* DTT, 2 μg/ml leupeptin, and 2 μg/ml pepstatin) for 3 h. Centrifuge the dialyzed solution at 12,000 rpm in a Sorvall SS-34 rotor for 20 min to pellet debris and precipitated protein. Freeze the supernatant as aliquots in a dry ice-ethanol bath and store at −80°C. The protein concentration is usually 6 to 8 mg/ml for HeLa and lymphoid cells and 3 to 5 mg/ml for NIH-3T3 cells.

c. Whole-Cell Extract

1. Wash the harvested cell pellet twice with five PCV of PBS.
2. Resuspend the washed cell pellet in four PCV of 10 m*M* Tris-HCl, pH 7.9, 1 m*M* EDTA, and 5 m*M* DTT. Keep on ice for 20 min.
3. Lyse the cells by homogenization in an all-glass Dounce homogenizer (B pestle), add four PCV of 50 m*M* Tris-HCl, pH 7.9, 10 m*M* $MgCl_2$, 2 m*M* DTT, 25% sucrose, and 50% glycerol. Mix gently.
4. Mix the lysed cells gently with one PCV of saturated ammonium sulfate and stir the viscous lysate gently for an additional 30 min to lyse the nuclei.
5. Centrifuge the extract for 3 h at 45,000 rpm in a SW50.2 rotor.
6. Decant the supernatant carefully, without disturbing the pellet, and add solid ammonium sulfate (0.33 g/ml of supernatant) to precipitate the proteins and nucleic acids. Add 1 *N* NaOH (0.1 ml/10 g of solid ammonium sulfate) to the suspension to neutralize the solution, and stir the suspension for an additional 30 min.
7. Collect the precipitate by centrifugation at 12,000 × g for 20 min and resuspend the pellet in 1/20 volume of the high-speed supernatant in 25 m*M* HEPES, pH 7.9, 100 m*M* KCl, 12 m*M* $MgCl_2$, 0.9 m*M* EDTA, 2 m*M* DTT, 17% glycerol, 2 μg/ml leupeptin, and 2 μg/ml pepstatin.
8. Dialyze the suspension against two changes of resuspension buffer (see step 7) (100 volumes each) for a total of 8 to 12 h.

9. Centrifuge the dialysate at 10,000 rpm for 20 min in a Sorvall SS-34 rotor to pellet the debris and precipitated protein. Freeze the whole-cell extract as small aliquots in a dry ice-ethanol bath and store at −80°C. The extract contains 10 to 15 mg/ml of protein and up to 2 mg/ml of nucleic acids.

These methods are used to prepare crude extracts. General transcription factors such as RNA polymerase II, TFII A, B, D, and E have been purified from crude nuclear extracts by conventional chromatographic procedures.[19,20] These components have been purified from HeLa cell extracts in order to examine their role in the formation of transcription preinitiation complexes.

2. Assay Conditions

Several factors affect the transcriptional activity of extracts: (1) DNA and extract concentration; (2) salt concentration; (3) divalent cation concentration and effect of polyamines; and (4) pH and temperature.

a. DNA and Extract Concentrations

At an optimal concentration of extract, there is a threshold DNA concentration below which no transcription occurs, and an inhibitory effect is observed at high concentrations of DNA. Carrier DNA like pBR322, *Escherichia coli* DNA, or alternating copolymers like poly[d(I-C):poly[d(I-C)] and poly[d(A-T)]:poly[d(A-T)] are sometimes added to the transcription reaction to titrate nucleases and nonspecific DNA binding proteins and so stimulate specific transcription. However, the addition of excess nonspecific DNA inhibits the transcription reaction.

There is a critical dependence of transcription upon extract concentration. In fact, the threshold DNA concentration depends on the protein concentration of the extract. The DNA dependence curve is less steep and the threshold DNA concentration is lower at lower protein concentrations in the extract. Thus, for each type of extract, it is necessary to titrate both DNA and extract to determine the optimal conditions.

b. Salt Concentration

Usually a high salt concentration inhibits transcription. The optimal salt concentration for transcription has been found to be approximately 60 m*M* KCl or NaCl for a series of promoters and using different extracts.

c. Divalent Cations and Polyamines

Ca^{2+}, Zn^{2+}, and Mn^{2+} inhibit transcription. Mg^{2+} is necessary for transcriptional activity. For the HeLa nuclear extract, the optimum Mg^{2+} concentration has been found to be between 18 and 12 m*M*, depending upon the template used.[13,14]

Polyamines are known to stimulate the transcriptional activity of purified RNA polymerase II[21,22] and to protect DNA from nucleases. In the presence of polyamines, a lower concentration of Mg^{2+} is required for transcription. Polyamines also allowed the detection of enhancer activity. With NIH-3T3 cell extracts, polyamines are required to detect RNA polymerase II-dependent transcription. At a particular spermidine-to-Mg^{2+} ratio, we have observed a decrease in α-amanitin-insensitive transcription and an increase in specific transcription (α-amanitin sensitive).[23]

d. pH and Temperature

There is a broad pH optimum between 7.5 and 8.5 in the case of HeLa cell extracts.[13] Most *in vitro* transcription reactions using different cell extracts have been performed at pH 7.9. There is a sharp optimum temperature around 30°C in the case of HeLa cell extracts and also with other cell and tissue extracts. In most cases, increased degradation of RNA occurs at higher temperatures, presumably due to activation of RNAses.

e. Standard *In Vitro* Transcription Conditions

1. Combine the nuclear extract (30 to 60% of the total reaction volume) with the reaction buffer and DNA. The final concentrations of buffer components should be as follows:[13] 12 m*M* HEPES (pH 7.9 at 25°C), 60 m*M* KCl, 0.3 m*M* DTT, 0.12 m*M* EDTA, 12 m*M* $MgCl_2$, 12% (v/v) glycerol, and 600 μ*M* of each of three NTPs, with the radioactive nucleoside triphosphate present at 25 μ*M*. The optimal concentration of DNA must be determined for each extract.

    When using HeLa whole-cell extract the concentration of buffer components is approximately the same, except that $MgCl_2$ should be at 7.2 m*M* and the NTPs at 50 μ*M*, and creatine phosphate should be added to 5 m*M*.[14] For NIH-3T3 nuclear extract, 0.3 m*M* PMSF, 2 m*M* spermidine, and 2.5 μg/ml leupeptin and pepstatin should be added to the reaction, and $MgCl_2$ should be present at 1 mM.[17] Similar conditions should be used for nuclear extracts from tissue.[18]
2. Incubate the reactions for 60 min at 30°C.
3. Add SDS to 0.5% and sodium acetate, pH 5.5, to 0.25 *M* to stop the reaction.
4. Extract the reactions once with phenol and once with chloroform.
5. Precipitate the nucleic acid by adding 2.5 volumes of ethanol. Centrifuge, remove the supernatant, wash the pellet with 80% ethanol, and dry the pellet under vacuum.

3. Quantitation of Transcripts

a. Direct Labeling of the RNA Using [α³²P] Nucleoside Triphosphates in the Transcription Reaction

A plasmid DNA cleaved with a restriction enzyme that cuts 3′ to the putative start of transcription is used as a template in the transcription reaction. Because the RNA polymerase II runs off the template at the site of restriction enzyme cleavage, RNA product(s) of one or more discrete sizes are obtained, depending on whether there is one or more initiation sites. RNA is labeled with 5 to 10 μCi of [α^{32}P]-UTP or -GTP. After extraction of the RNA, the product is resuspended in 80% formamide and fractionated by electrophoresis on a 7 *M* urea-polyacrylamide gel. This method is faster than the primer extension or RNAse protection methods. However, this method may produce additional nonspecific RNA species, due to the presence of DNA ends which can sequester transcription-specific proteins and to nicking of the DNA during the digestion with restriction enzymes. This problem can be overcome by linking a guanosine-free cassette to the promoter.[24] A 400-bp DNA segment which lacks cytidine residues on the transcribed strand has been used for this purpose and results in a transcript without guanosine residues. By using RNAse T1 and 3′-*O*-methyl-GTP to replace GTP in the transcription reaction, only RNA transcribed from the G-free cassette will accumulate in the reaction, eliminating all nonspecific transcripts. This RNA can be labeled using either [α^{32}P]-CTP or -UTP. When this G-free DNA fragment is placed 3′ to the start of transcription of any promoter, the specific 400-nucleotide transcript can be detected easily without linearizing the plasmid. This method is rapid and quantitative.

b. Primer Extension[25]

Primers (20 to 25 nucleotides) are made complementary to the RNA sequence 100 to 300 nucleotides 3′ to the start of transcription. Primers are labeled at their 5′ end by T4 polynucleotide kinase and [γ^{32}P]-ATP. After extracting RNA from the transcription reaction, the RNA is hybridized with the primer in the presence of 0.25 *M* KCl for 1 h at 50°C (the exact temperature depends on the length and the sequence of the primer). Primers are then extended by reverse transcriptase under standard conditions. Extended primers are analyzed by electrophoresis in 7 *M* urea-polyacrylamide gels. By this method, transcription can be performed using unmodified supercoiled plasmids (i.e., use of the G-free cassette as template is not required) and only unlabeled nucleoside triphosphates.

c. *RNAse Protection Assay*[26]

In this method, labeled antisense RNA is hybridized with the RNA synthesized with the nuclear extracts or the whole-cell extracts, and the unhybridized probe sequences are removed by digestion with RNAse A. Labeled antisense RNA is synthesized *in vitro* from a plasmid construction containing the bacteriophage SP6, T7, or T3 promoters.[27]

The protected RNA has a discrete size and is analyzed by electrophoresis on 7 *M* urea-polyacrylamide gels (see Reference 27 for a protocol). When this method is used for RNA detection, DNA can be transcribed in its supercoiled form. This method requires more time than the primer extension method, but much lower levels of transcripts can be detected.

C. ANALYSIS AND EVALUATION OF RESULTS

1. Sensitivity of the *In Vitro* Transcription Assay System

In this assay, the rate of transcription is dependent on the nature of the promoter or enhancer elements present in the template.[16,17,22,28-31] In general, the results obtained by *in vitro* transcription are consistent with those obtained by transient expression after the DNA transfection of intact cells. Mutations in specific *cis*-acting elements have comparable effects on the activity of promoters or enhancers in both systems.[17,30,32] It is important to select an appropriate promoter as an internal control in order to normalize the changes in transcription activity due to mutations or deletions in the *cis*-acting elements of the promoter under study. *Cis*-acting elements are interaction sites for DNA binding proteins which either activate or repress the transcriptional activity of promoters by binding to these sites. The role of these DNA binding factors can be studied in the *in vitro* transcription system after purification of the factors. We have, for instance, examined the effects of a heterodimeric CCAAT=binding protein on *in vitro* transcription.[17] The α2(I) collagen promoter, which contains a binding site between −84 and −80 for this CCAAT=binding protein, was used as a template. The CCAAT=binding protein was purified by chromatography of a nuclear extract over a sequence-specific DNA affinity column. Addition of this purified CCAAT=binding factor to an extract that was depleted of this factor strongly stimulated transcriptional activity of an α2(I) collagen promoter (Figure 1).

Tissue specificity has also been demonstrated *in vitro*.[16,18,30,31] This was done by using a tissue-specific promoter and nuclear extracts of a cell line in which this promoter was not expressed.[16,31] When a tissue-specific factor was added to the reaction, transcription of the tissue-specific promoter was stimulated severalfold.[16]

RNA polymerase II is associated with several different proteins which are essential for transcriptional activity. These proteins have been characterized by fractionations using different chromatographic methods. It has been shown that by combining these proteins with RNA polymerase II, the active preinitiation transcription complex can be formed.[19,20] The *in vitro* transcription assay method has therefore assisted in the understanding of the biochemical mechanism of transcription initiation.

2. Limitation of the *In Vitro* Transcription Assay

In many eukaryotic genes, *cis*-acting elements are often located at a distance of one to several kilobases from the start of transcription. One model to explain their activity is that proteins bound to these elements interact with transcription factors associated with the DNA in the immediate vicinity of the start of transcription. The DNA between these sites would bend and loop out to allow the interactions between the transcription factors. In the *in vitro*

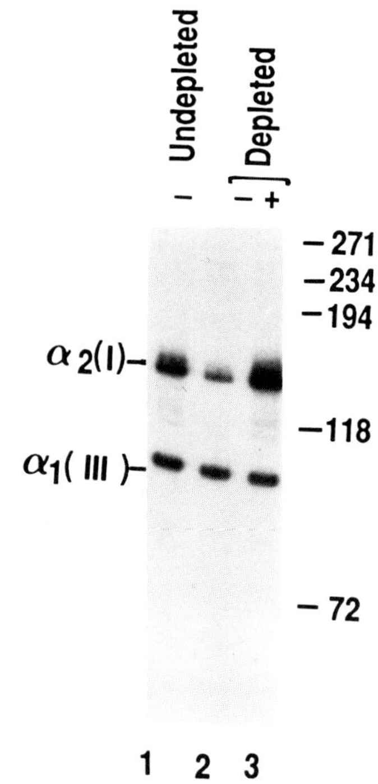

FIGURE 1. *In vitro* transcription analysis of a CCAAT-binding factor (CBF). Both the α2(I) and α1(III) collagen promoters were transcribed together *in vitro* with NIH-3T3 nuclear extracts. Transcripts were analyzed by primer extension. The α1(III) collagen promoter has no CBF binding site and was used as an internal control. CBF was purified from nuclear extracts by DNA affinity chromatography. CBF-depleted nuclear extracts were prepared by passing the extracts over the same DNA affinity resin. Both promoters were transcribed in undepleted extracts (lane 1); in CBF-depleted extracts (lane 2); and in CBF-depleted extract + 180 ng of CBF (lane 3). (From Maity, S. N., Golumbek, P. T., Karsenty, G., and de Crombrugge, B., *Science*, 241, 582, 1988. Copyright 1988 by the AAAS. With permission.)

transcription system, the activity of *cis*-acting elements can be detected when they are located close to the start of transcription.[31] If the above model is correct, it is likely that, *in vitro*, the interactions between *trans*-acting factors present on the DNA at considerable distances from each other are very inefficient. In intact cells, DNA is part of a chromatin structure with possible interactions with the nuclear matrix. These structures are not reproduced in an *in vitro* transcription reaction.

The extent of *in vitro* stimulation or repression of transcription by a specific *trans*-acting factor does not always correspond quantitatively to the effect observed *in vivo*. This difference could be due to differences in the quantities of factor which are bound to its recognition site *in vivo*, compared to the amount that is bound *in vitro*. Another possible reason for this difference is that an equilibrium exists in the distribution of *trans*-acting factors among the wide variety of binding sites inside the cells.

REFERENCES

1. **Bentley, D. L. and Groudine, M.,** A block to elongation is largely responsible for decreased transcription of c-*myc* in differentiated HL60 cells, *Nature,* 321, 702, 1986.
2. **Kao, S.-Y., Calman, A. F., Luciw, P. A., and Peterlin, B. M.,** Antitermination of transcription within the long terminal repeat of HIV-1 by *tat* gene product, *Nature,* 330, 489, 1987.
3. **Resnekov, O., Ben-Asher, E., Bengal, E., Choder, M., Hay, N., Kessler, M., Ragimov, N., Seiberg, M., Skolnik-David, H., and Aloni, Y.,** Transcription termination in animal viruses and cells, *Gene,* 72, 91, 1988.
4. **Ward, D. F. and Gottesman, M. E.,** Suppression of transcription by phage lambda, *Science,* 216, 946, 1982.
5. **Clayton, D. F. and Darnell, J. E., Jr.,** Changes in liver-specific compared to common gene transcription during primary culture of mouse hepatocytes, *Mol. Cell. Biol.,* 3, 1552, 1983.
6. **Greenberg, M. E. and Ziff, E. B.,** Stimulation of 3T3 cells induces transcription of the c-*fos* proto-oncogene, *Nature,* 311, 433, 1984.
7. **Groudine, M., Peretz, M., and Weintraub, H.,** Transcriptional regulation of hemoglobin switching in chicken embryos, *Mol. Cell Biol.,* 1, 281, 1981.
8. **McKnight, G. S. and Palmiter, R. D.,** Transcriptional regulation of the ovalbumin and conalbumin genes by steroid hormones in chick oviduct, *J. Biol. Chem.,* 254, 9050, 1979.
9. **Mulvihill, E. R. and Palmiter, R. D.,** Relationship of nuclear estrogen receptor levels to induction of ovalbumin and conalbumin mRNA in chick oviduct, *J. Biol. Chem.,* 252, 2060, 1977.
10. **Sasaki, K., Cripe, T. P., Koch, S. R., Andreone, T. L., Petersen, D. D., Beale, E. G., and Granner, D. K.,** Multihormonal regulation of phosphoenolpyruvate carboxykinase gene transcription: the dominant role of insulin, *J. Biol. Chem.,* 259, 15,242, 1984.
11. **Cleveland, D. W.,** Autoregulated instability of tubulin mRNAs: a novel eukaryotic regulatory mechanism, *Trends Biochem. Sci.,* 13, 339, 1988.
12. **Schuler, G. D. and Cole, M. D.,** GM-CSF and oncogene mRNA stabilities are independently regulated in *trans* in a mouse monocytic tumor, *Cell,* 55, 1115, 1988.
13. **Dignam, J. D., Lebovitz, R. M., and Roeder, R. G.,** Accurate transcription initiation by RNA polymerase II in a soluble extract from isolated mammalian nuclei, *Nucleic Acids Res.,* 11, 1475, 1983.
14. **Manley, J. L., Fire, A., Samuels, M., and Sharp, P.,** *In vitro* transcription: whole-cell extract, *Methods Enzymol.,* 101, 568, 1983.
15. **Mizushima-Sugano, J. and Roeder, R. G.,** Cell-type-specific transcription of an immunoglobulin κ light chain gene *in vitro, Proc. Natl. Acad. Sci. U.S.A.,* 83, 8511, 1986.
16. **Bodner, M. and Karin, M.,** A pituitary-specific trans-acting factor can stimulate transcription from the growth hormone promoter in extracts of non-expressing cells, *Cell,* 50, 267, 1987.
17. **Maity, S. N., Golumbek, P. T., Karsenty, G., and de Crombrugghe, B.,** Selective activation of transcription by a novel CCAAT binding factor, *Science,* 241, 582, 1988.
18. **Gorski, K., Carneiro, M., and Schibler, U.,** Tissue-specific *in vitro* transcription from the mouse albumin promoter, *Cell,* 47, 767, 1986.
19. **Reinberg, D. and Roeder, R. G.,** Factors involved in specific transcription by mammalian RNA polymerase II, *J. Biol. Chem.,* 262, 3310, 1987.
20. **Buratowski, S., Hahn, S., Guarente, L., and Sharp, P. A.,** Five intermediate complexes in transcription initiation by RNA polymerase II, *Cell,* 56, 549, 1989.
21. **Mandel, J. L. and Chambon, P.,** Animal DNA-dependent RNA polymerases: studies on the reaction parameters of transcription *in vitro* of simian virus 40 DNA by mammalian RNA polymerases A1 and B, *Eur. J. Biochem.,* 41, 367, 1974.
22. **Wildeman, A. G., Sassone-Corsi, P., Grundstrom, T., Zenke, M., and Chambon, P.,** Stimulation of *in vitro* transcription from the SV40 early promoter by the enhancer involves a specific trans-acting factor, *EMBO J.,* 3, 3129, 1984.
23. **Maity, S. N.,** Unpublished data, 1988.
24. **Sawadogo, M. and Roeder, R. G.,** Factors involved in specific transcription by human RNA polymerase II: analysis by a rapid and quantitative *in vitro* assay, *Proc. Natl. Acad. Sci. U.S.A.,* 82, 4394, 1985.
25. **Lee, W., Mitchell, P., and Tjian, R.,** Purified transcription factor AP-1 interacts with TPA-inducible enhancer elements, *Cell,* 49, 741, 1987.
26. **Pastorcic, M., Wang, H., Elbrecht, A., Tsai, S. Y., Tsai, M., and O'Malley, B. W.,** Control of transcription initiation *in vitro* requires binding of a transcription factor to the distal promoter of the ovalbumin gene, *Mol. Cell. Biol.,* 6, 2784, 1986.
27. **Maniatis, T., Fritsch, E. F., and Sambrook, J.,** *Molecular Cloning: A Laboratory Manual,* Cold Spring Harbor Laboratory, Cold Spring Harbor, NY, 1982.
28. **Lichtsteiner, S., Wuarin, J., and Schibler, U.,** The interplay of DNA-binding proteins on the promoter of the mouse albumin gene, *Cell,* 51, 963, 1987.

29. **Tsai, S. Y., Sagami, I., Wang, H., Tsai, M., and O'Malley, B. W.,** Interactions between a DNA-binding transcription factor (coup) and a non-DNA binding factor (s300-11), *Cell,* 50, 701, 1987.
30. **Scheidereit, C., Heguy, A., and Roeder, R. G.,** Identification and purification of a human lymphoid-specific octamer-binding protein (OTF-2) that activates transcription of an immunoglobulin promoter *in vitro, Cell,* 51, 783, 1987.
31. **Nelson, C., Albert, V. R., Elsholtz, H. P., Lu, L. I.-W., and Rosenfeld, M. G.,** Activation of cell-specific expression of rat growth hormone and prolactin genes by a common transcription factor, *Science,* 239, 1400, 1988.
32. **Karsenty, G., Golumbek, P., and de Crombrugghe, B.,** Point mutations and small substitution mutations in three different upstream elements inhibit the activity of the mouse α2(I) collagen promoter, *J. Biol. Chem.,* 263, 13, 909, 1988.

Chapter 17

MEASURING EUKARYOTIC GENE EXPRESSION *IN VITRO*: TRANSLATION ASSAYS

Albert J. Wahba and Jaydev N. Dholakia

TABLE OF CONTENTS

I. INTRODUCTION

The rapid progress in the field of molecular cloning has necessitated the use of several cell-free protein synthesizing systems for the translation of eukaryotic mRNAs. Rabbit reticulocyte lysates and wheat germ extracts have a number of advantages over other systems and have received the most attention in recent years. Extracts from a variety of cultured cells such as ascites tumor cells, Hela cells, yeast, and *Drosophila* have also been used for *in vitro* translation. The treatment of reticulocyte lysates with a calcium-dependent micrococcal nuclease[1] to remove endogenous mRNA was an important observation in the development of cell-free systems for the translation of exogenous mRNA. This chapter describes the preparation and properties of *in vitro* cell-free translation systems for eukaryotic mRNAs, considers the variety of ways in which the translational products may be characterized, and discusses the application and significance of these translational systems to answer specific questions on gene expression.

A. STEPS IN EUKARYOTIC PROTEIN SYNTHESIS

The overall mechanism of protein synthesis in eukaryotes is similar to that in prokaryotes and comprises three phases: initiation, elongation, and termination. However, some aspects of protein synthesis in eukaryotes are quite different from those in prokaryotes. This is particularly true in initiation. The eukaryotic 40S ribosomal subunit first forms a preinitiation complex with the initiator tRNA, GTP, and mRNA.[2,3] GTP is hydrolyzed only when the 60S ribosomal subunit joins the smaller ribosomal subunit to form the 80S initiation complex. In contrast to the three well-defined initiation factors of *Escherichia coli*, the eukaryotic initiation process involves a multitude of factors (designated eIFs) and their interplay is more intricate.

The first step in eukaryotic polypeptide chain initiation is the formation of a ternary complex between the Met-tRNA$_f$, GTP, and eIF-2 (Figure 1). One of the major differences between prokaryotic and eukaryotic initiation is the regulation of eIF-2 activity in eukaryotes.[2,3] Upon formation of the 80S initiation complex, GTP is hydrolyzed and eIF-2 is released as the eIF-2·GDP binary complex. In mammalian systems, this binary complex is stable in the presence of Mg^{2+}, and a second protein, the guanine nucleotide exchange factor (GEF), is required for the exchange of bound GDP for GTP.[4] It is at this point in the eIF-2 cycle that regulation appears to occur. Phosphorylation of the α subunit of eIF-2 by the rabbit reticulocyte heme-controlled repressor or the dsRNA-dependent kinase during viral

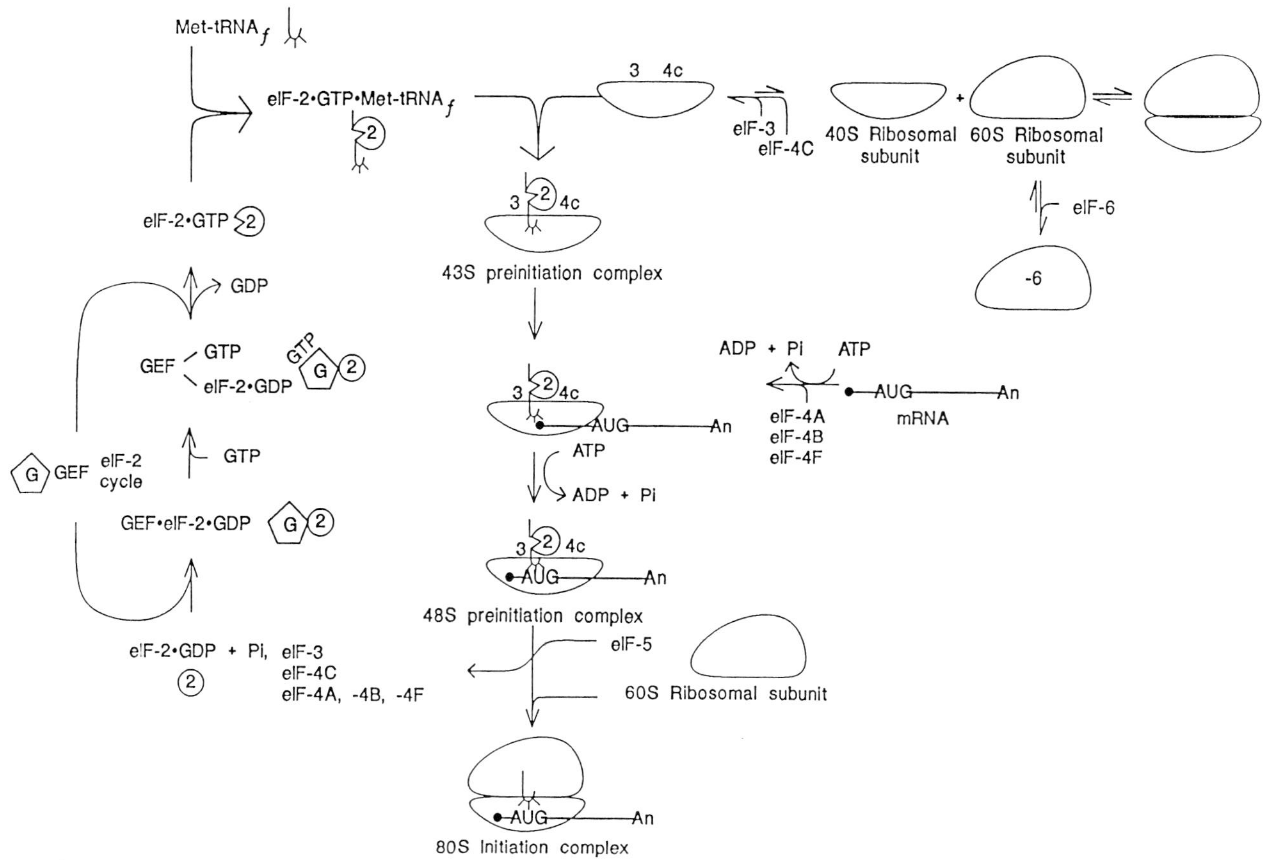

FIGURE 1. Polypeptide chain initiation in eukaryotes.

infection results in the cessation of protein synthesis and is due to the inability of GEF to catalyze the GTP/GDP nucleotide exchange reaction.[2,3] Recently, GEF was shown to catalyze the GTP/GDP nucleotide exchange reaction by a sequential mechanism.[4] The activity of GEF may be influenced directly by the phosphorylation of its M_r 82,000 subunit[5] and the redox state of the cell.[6] eIF-3, which has ribosome dissociation activity, and eIF-4C, which exhibits anti-association activity, bind to the small ribosomal subunit. As illustrated in Figure 1, the ternary complex is transferred to the 40S ribosomal subunit to form the 43S preinitiation complex. The eIF-4E component of the eIF-4F complex recognizes the "cap" structure of eukaryotic mRNA.[7] Cap binding proteins (eIF-4A, eIF-4B, and eIF-4F) bind mRNA and transfer it in an ATP-dependent reaction to the 43S complex, forming a 48S preinitiation complex in which the AUG codon is aligned with the anticodon loop of the initiator tRNA.[8] The different reactions involved in mRNA selection and binding to the ribosome are presently not well characterized. For recent reviews of the RNA binding proteins, see References 7 and 8. Eukaryotic chain initiation factor 5 (eIF-5) facilitates the binding of the 60S ribosomal subunit to the 48S initiation complex. At this point, eIF-2-bound GTP is hydrolyzed, resulting in the removal of initiation factors from the ribosomal complex and the formation of the 80S initiation complex, thereby setting the stage for polypeptide chain elongation.

Polypeptide chain elongation requires GTP and is catalyzed by the eukaryotic elongation factors, whose mode of action is similar to their prokaryotic counterparts. The different reactions involved in elongation, such as the binding of the appropriate aminoacyl-tRNA to the ribosomal A site, transpeptidation to form a peptide bond between the incoming aminoacyl moiety and peptidyl residue at the P site, and movement of the ribosomes on mRNA, are repeated until a termination codon enters the A site. The termination phase of protein synthesis also requires a GTP-dependent, codon-specific binding of a specific protein factor (RF, release factor) to the ribosomal A site, which results in the hydrolysis of peptidyl-tRNA at the P site and the release of the completed polypeptide chain.

II. PREPARATION AND OPTIMIZATION OF TRANSLATION ASSAYS

A. TRANSLATION SYSTEM FROM RABBIT RETICULOCYTES

It is often necessary to characterize the translation products of an mRNA which is extracted from a cell or tissue at a particular point during development or viral infection. Pelham and Jackson[1] developed an *in vitro* translation system from rabbit reticulocytes in which the lysate is treated with Ca^{2+}-dependent micrococcal nuclease to remove selectively messenger RNA, but not tRNA or ribosomes. This system is very active and is highly dependent on exogenous mRNA. Similar systems have been developed from other cell types, including wheat germ. The procedure described here for the preparation of nuclease-treated rabbit reticulocyte lysates is based on the one described by Pelham and Jackson.[1]

1. Preparation of Lysates

Rabbits made anemic by the injection of phenylhydrazine begin to release reticulocytes into the circulation system. These reticulocytes contain all of the components necessary for the efficient translation of endogenous or exogenous messenger RNA. Rabbit reticulocyte lysates[1,9] are prepared from male albino or New Zealand white rabbits (4 to 6 lb) by the subcutaneous injection for 4 consecutive days of either 4 ml of 1.25% acetylphenylhydrazine or 1 ml of 2% phenylhydrazine, pH 6.8 to 7.4. In the procedure described by Darnbrough et al.,[9] the acetylphenylhydrazine is prepared in a warm saline solution containing 0.13 *M* NaCl, 7.5 m*M* $MgCl_2$, and 5 m*M* KCl. The rabbits are bled by cardiac puncture on the seventh day after the initial injection. Before bleeding, the rabbits are anesthetized by injecting intramuscularly (0.5 to 0.7 ml/6- to 8-lb rabbit) a solution containing one part of

100 mg/ml xylazine and five parts of 100 mg/ml ketamine. The blood should be collected in a cold buffer containing 0.14 *M* NaCl, 0.05 *M* KCl, 5 m*M* $MgCl_2$, 1 m*M* dithiothreitol, 0.2 m*M* phenylmethylsulfonylfluoride, and 0.5 mg/ml heparin. Approximately 1 ml of this buffer should be used for each 10 ml of collected blood. Centrifuge the blood for 5 min at 5000 × g and 4°C and discard the supernatant. Care must be taken to remove as much of the supernatant as possible, since residual heparin will inhibit protein synthesis activity. Gently suspend the cells in four to five volumes of the collection buffer lacking heparin and repeat the centrifugation step. The cells should be washed at least three times. After the final washing, measure the packed cell volume and lyse the cells by the addition of an equal volume of cold double-distilled water. Remove the cell debris by centrifugation for 15 min at 20,000 × g. The lysate may be treated immediately with micrococcal nuclease or stored in 1-ml or smaller aliquots in liquid nitrogen. Once thawed for use, it is best not to refreeze lysates, although this may be done once without too much loss of activity.

2. Micrococcal Nuclease Treatment

To 1 ml of reticulocyte lysate, add 10 μl of 0.1 *M* $CaCl_2$ and 40 μl of 1 mg/ml hemin. The hemin solution is prepared by dissolving 10 mg of hemin in 0.2 ml of 0.1 *M* KOH and 0.2 ml of 1 *M* Tris base. Adjust this solution to pH 8.0 to 8.5 by the addition of 0.2 *M* HCl (approximately 0.6 ml) and, if necessary, add water to bring the volume to 1 ml. Finally, add 9 ml of ethylene glycol and store the solution in the dark at −20°C until it is needed. The optimal hemin concentration required for the translational activity of each reticulocyte lysate preparation should be determined by monitoring the synthetic activity in response to different concentrations of hemin. Add 5 μl of micrococcal nuclease (12,000 to 17,000 U/ml) and incubate 5 to 15 min at 20°C. The optimum time and level of nuclease will vary with each preparation of lysate. Terminate the reaction by placing the digest in an ice bath and adding 10 μl of 0.2 *M* EGTA, pH 7.2, and 30 μl of creatine kinase (5 mg/ml with 1 mg/ml bovine serum albumin, Type V, in 50% glycerol). The lysate may be stored in liquid N_2 in 200-μl aliquots.

3. Assay Conditions for Translation

These assays are conveniently performed in conical 0.5- to 1.5- ml polypropylene microcentrifuge tubes. Prepare an energy/salt mix containing, in 1 ml: 4 m*M* ATP, 0.8 m*M* GTP, 0.1 m*M* spermine or 3 m*M* spermidine, 32 m*M* creatine phosphate, 100 m*M* HEPES·K, 8 m*M* dithiothreitol, at least 0.32 m*M* of each amino acid except methionine, 9 m*M* $Mg(acetate)_2$, and 500 m*M* K·acetate. A 12-μl translation mixture should contain 2 μl of $[^{35}S]$methionine (10 Ci/mmol diluted with 1 m*M* methionine to 1 to 2 Ci/mmol), 3 μl of the salt/energy mix, 1 μl of messenger RNA, and 6 μl of lysate. If more mRNA and amino acids are required, they should be added to the reaction tube and dried under vacuum before the other components are pipetted in. Incubate the reactions at 30°C, removing 2-μl aliquots for assay at several time points (5 to 60 min). Place the 2-μl aliquot in a tube containing 50 μl of 0.3 *M* KOH and 2 μl of 30% H_2O_2. After incubation for 15 min at 37°C, add 0.5 ml of 10% (w/v) trichloroacetic acid containing 3% (w/v) casamino acids and transfer to an ice bath. Filter the precipitate under vacuum through a nitrocellulose membrane (0.45 μ). The membrane is washed three times with trichloroacetic acid, dried under a heat lamp, and the radioactivity is determined in a liquid scintillation counter by using the scintillation fluid suitable for nonaqueous samples.

Translation systems derived from both rabbit reticulocytes and wheat germ, either untreated or nuclease-treated, are commercially available from Amersham, Bethesda Research Laboratories, Du Pont-New England Nuclear, and Promega.

B. THE USE OF WHEAT GERM EMBRYO EXTRACTS

There are several advantages in using wheat germ extracts for an *in vitro* translation system. From a technical standpoint, wheat germ is relatively inexpensive and cell-free extracts are easy to prepare. Wheat germ systems have been reported to possess a higher level of fidelity than rabbit reticulocyte lysates in initiating translation at the correct AUG initiation codon.[10] The potential disadvantage of this system is that the optimum ionic conditions for protein synthesis depend on the nature and concentration of the mRNA and have to be determined for each mRNA species. Untoasted wheat germ may be obtained from General Mills (Minneapolis, MN) and should be stored dessicated at −20°C.

Wheat germ should be ground in small batches (1 to 15 g) in a chilled (2 to 4°C) mortar and pestle with an equal weight of either sea sand or ground glass. Crushed, sterile Pasteur pipettes may be used. Allow approximately 1 min of grinding time for each gram of wheat germ. All operations should be carried out at 0 to 4°C with prechilled glassware and buffers. After grinding, add 2 ml of extraction buffer (20 m*M* HEPES·KOH, pH 7.6, 1 m*M* Mg(acetate)$_2$, 2 m*M* $CaCl_2$, 4 m*M* dithiothreitol) for each gram of wheat germ and grind for another 0.5 min for each gram of wheat germ. Centrifuge at 2 to 4°C for 15 min at 30,000 × g. Carefully remove by aspiration the upper lipid layer and then collect the supernatant. Avoid removing any of the pelleted material. Add creatine kinase (stock solution is 5 mg/ml) to give 400 to 500 μg/ml kinase. This extract is applied to a column of Sephadex G-25 (medium) (Pharmacia/LKB) equilibrated with buffer containing 20 m*M* HEPES·KOH, pH 7.6, 120 m*M* KCl, 5 m*M* Mg(acetate)$_2$, and 4 m*M* dithiothreitol. Allow approximately 10 ml of column volume for each ml of wheat germ extract. Collect the fractions (0.5 to 1 ml) and pool those in the central one third of the peak. Place the combined fractions in an ice bath for 5 to 10 min and centrifuge for 10 min at 25,000 × g. The supernatant should be retained. These last two steps remove inhibitors of protein synthesis from the extract. Depending on the efficiency of grinding and extraction, the supernatant should have an absorbancy of 70 to 100 A_{260} U/ml and an A_{260}/A_{280} ratio of approximately 1.6. The extract is best stored in liquid nitrogen, although it should be stable for up to 3 months at −70 to −80°C. Before storage, RNAsin (Promega) should be added to a final concentration of 2 U/μl extract.

A translation assay should contain, in 12 μl: 4 μl or approximately 0.3 A_{260} wheat germ extract (note that this supplies 1.67 m*M* Mg^{2+}, 40 m*M* KCl, and creatine kinase), 3 μl of translation mix (containing 80 m*M* HEPES·KOH, pH 7.6, 2.4 m*M* spermidine, 3.3 m*M* Mg(acetate)$_2$, 36 m*M* creatine phosphate, 0.8 m*M* GTP, 4 m*M* ATP, 8 m*M* dithiothreitol, 240 m*M* potassium acetate, and 330 μ*M* of each amino acid except methionine), 2 μl of [^{35}S]methionine (10 Ci/mmol diluted with 1 m*M* methionine to 1 to 2 Ci/mmol), and up to 3 μl of mRNA (20 to 80 μg/ml final concentration) or water. Incubate at 30°C for 30 to 60 min and terminate the reaction as described in Section II.A.3. Different preparations of wheat germ extract will contain variable amounts of endogenous mRNA. Unlike the reticulocyte lysate, the endogenous protein synthesis activity of the wheat germ system is low. Translation in this system is, therefore, largely dependent on the added mRNA, and the precise requirements for optimum translation may vary with different mRNAs. The low background for amino acid incorporation of wheat germ extracts may be further reduced by treatment with micrococcal nuclease as described for rabbit reticulocyte lysates (Section II.A.2) except that hemin is omitted. The optimum concentration of potassium acetate and Mg^{2+} should be determined experimentally for each mRNA. The addition of RNAsin and spermidine increases the level of incorporation as well as the size of the polypeptide products.[11]

C. AN OVERVIEW OF THE PREPARATION OF EXTRACTS FROM OTHER TISSUES AND CELL LINES

In addition to extracts from rabbit reticulocytes and wheat germ, a variety of eukaryotic cells have been used for the preparation of cell-free *in vitro* protein synthesizing systems to study the translation of mRNAs as well as several aspects of gene expression and regulation of protein synthesis during differentiation,[12] fertilization,[13] or heat shock.[14] Most of these systems are derived from postmitochondrial supernatants which are either dialyzed or treated with Sephadex G-25 in order to have uniform ionic conditions. Alternatively, the optimal conditions for protein synthesis of these extracts may be predetermined experimentally. The cell-free extracts from Krebs II mouse ascites cells have been successfully used to translate added cellular and viral mRNAs.[15] *In vitro* translation systems with similar properties may be prepared from any one of the following: mouse L cells, Ehrlich ascites tumor cells, HeLa cells, mouse erythroleukemia cells, and Chinese hamster ovary cells. All of these may be easily grown in tissue culture in multiliter quantities. The following procedure has been successfully used for measuring *in vitro* protein synthesis activity in extracts prepared from mouse erythroleukemia (MEL) cells, and should be readily applicable to the other cell types.

MEL cells were grown to the final density of 2 to 4 $\times$ 10^6 cells per milliliter,[16] collected by centrifugation at 1000 $\times$ g for 10 min and washed twice with 20 ml of phosphate-buffered saline (10.96 g of NaCl, 1.725 g of $NaH_2PO_4{\cdot}H_2O$, and 0.46 g of EDTA, adjusted to pH 7.4 with 1 *N* NaOH and diluted to 1 l). Suspend the cells in lysis buffer (10 m*M* Tris-HCl, pH 7.6, 4 m*M* dithiothreitol, 10% v/v glycerol, 2.5 m*M* Mg^{2+}, 10 m*M* KCl, 0.33 m*M* phenylmethylsulfonyl fluoride, 20 μ*M* hemin, and 0.5% Tween 80 or Triton X-100) to approximately 10^8 cells per milliliter. In addition to reticulocyte lysates, hemin is also reported to increase cell-free protein synthesis activity in nucleated differentiating cells.[16,17] The cell suspension is held at 0°C for 5 min and gently homogenized with a stainless steel/teflon pestle (Kimble) in a 1.5-ml conical microcentrifuge tube. Nuclei and cell debris were removed by centrifugation for 10 min at 1000 rpm and 4°C. The resulting supernatant may be used immediately or treated with micrococcal nuclease[17] and frozen in small aliquots in liquid N_2. Thawed lysates should not be reused.

Translation assays were performed at 30°C and contained, in a final volume of 12 μl: 6 μl of cell extract (adjusted to 32 to 36 A_{260}/ml), 3 μl of mRNA or H_2O, and 3 μl of a master salt mix. This master salt mix contains, in 1 ml: 160 m*M* HEPES·KOH, pH 7.5, 5 m*M* $Mg(acetate)_2$, 100 μ*M* spermine, 24 m*M* creatine phosphate, 0.8 m*M* GTP, 4 m*M* ATP, 12 m*M* dithiothreitol, 150 m*M* potassium acetate, 300 μ*M* of each amino acid except methionine, 450 μg of creatine kinase (the enzyme stock is 5 mg/ml with 1 mg/ml of bovine serum albumin and 50% glycerol), and 1 to 2 mCi of [^{35}S]methionine (10 Ci/mmol).

A 3μl aliquot of the translation reaction is removed at different time intervals and placed into a tube containing 100 μl of 0.3 *M* KOH. After incubation at 30 to 37°C for 15 min, 1 ml of 10% trichloroacetic acid is added and the sample is processed as described in Section II.A.3.

III. CHARACTERIZATION OF SPECIFIC TRANSLATION PRODUCTS

Measuring gene expression *in vitro* requires initially the successful preparation of mRNA from a cloned DNA and the translation of this mRNA in an appropriate cell-free extract. In each case, it is necessary to characterize the polypeptides which are synthesized, either to confirm that a known mRNA is being translated correctly or to determine what protein is coded for by the message. The relative amounts of a certain species of mRNA may also be derived from the quantitation of translational products. The most widely used techniques are described here.

A. POLYACRYLAMIDE GEL ELECTROPHORESIS

The most versatile and widely applicable method for the analysis of translational products is polyacrylamide slab gel electrophoresis under denaturing conditions in the presence of SDS. This method resolves a large number of proteins based on their molecular weights.

The polypeptides obtained during translation may be identified by autoradiography or fluorography of the gels.

1. Sample Preparation

Dilute an appropriate aliquot (usually 5 to 10 μl) of the translation mixture five- to tenfold into the gel sample buffer to give a final concentration of 2% SDS, 10% glycerol, 10 m*M* dithiothreitol, 100 m*M* Tris-HCl, pH 6.8, and 0.001% bromophenol blue. After boiling for 5 min, the samples can be loaded on the gel or stored at −20°C until analyzed. A loss of resolution will result if the slab gels are overloaded with total protein. To calibrate the gel, run in parallel lane proteins of known molecular weights as markers. When available, radioactive markers may be added to a sample of unlabeled cell extract to ensure uniform behavior of markers and products on the gel. In each experiment, an aliquot of a control incubation with no added mRNA should also be run in a parallel lane.

2. Polyacrylamide Gels

The translational products are usually separated by employing a discontinuous buffer system gel electrophoresis. The electrophoretic techniques were recently described in a volume edited by Hames and Rickwood.[18] The choice of a gel system, the concentration of acrylamide, and the selection of either a homogeneous or a gradient resolving gel system depends on the specific application. The following recipe is adequate for the preparation of two slab (12 × 14 cm, 0.15 cm thick) gels containing a 10-cm resolving and a 2-cm stacking gel.

	Resolving gel	Stacking gel
Acrylamide (40%)/bis-acrylamide (1.08%) solution	7.5 ml	0.62 ml
1.5 *M* Tris-HCl (pH 8.8) containing 0.4% SDS	7.5 ml	—
1.0 *M* Tris-HCl (pH 6.8) containing 0.4% SDS	—	0.62 ml
60% sucrose	10.0 ml	—
H_2O	5.0 ml	3.72 ml
Degas for 2 min		
10% ammonium persulfate	90 μl	35 μl
TEMED	15 μl	3 μl

3. Electrophoresis Buffer

Dissolve 6.06 g of Tris, 28.8 g of glycine, and 2.0 g of SDS (Na dodecyl sulfate) in 2 l of water and adjust the pH to 8.5. Electrophoresis may be carried out at 200-mV constant voltage until the tracking dye (bromophenol blue) reaches the bottom of the gel.

A good separation of proteins of molecular weight 10,000 to 200,000 is obtained by this system. The polypeptides in the gels may be visualized by staining with Coomassie Brilliant Blue R-250 (1.25 g in 50% methanol and 10% acetic acid) for 15 min and destaining in an aqueous solution containing 10% methanol and 7% acetic acid. The gels are either enhanced[19] and dried or are enclosed in a plastic bag and exposed to Kodak XAR-5 film at −70°C. The analysis of translation products by two-dimensional fractionation, using either isoelectric focusing or nonequilibrium pH gradient gel electrophoresis in the first dimension and SDS-polyacrylamide electrophoresis in the second,[16,20] resolves more polypeptides and is employed for a more vigorous identification and quantitation of products.

Since the fractionation of proteins by SDS-polyacrylamide gel electrophoresis is based on molecular size alone, it is often difficult to identify conclusively the translation product by this method. Two unrelated proteins may have identical mobility in SDS-polyacrylamide gels or, conversely, two proteins may be closely related, but have different mobilities because of a precursor-product relationship or as a result of degradation during translation or storage. Therefore, in all cases, it is desirable to characterize further the translation product by either peptide mapping or by immunological techniques, as described below.

B. PEPTIDE MAPPING

Peptide mapping is usually considered as a stringent test of protein identity.[21,22] Two-dimensional tryptic fingerprinting of small amounts of protein available after analytical SDS-polyacrylamide gels is a classical method used to compare different polypeptides.[22,23] Recently, Cleveland et al.[24] described a sensitive test for protein identity based on limited proteolytic cleavage. This method is simple, does not require special equipment, and just takes a few hours to perform. Several polypeptides obtained by fractionation on SDS-polyacrylamide gel electrophoresis may be compared simultaneously to purified products. The following protocol may be used for the specific identification of a protein.

1. Initially analyze the translation assay mixture and a purified authentic protein by SDS-polyacrylamide gel electrophoresis (20 sample wells per gel and 1-mm-thick slab gels, as described in Section III.A). Briefly stain and destain the gel. The translation product may be identified by autoradiography.
2. Rinse the gel with water, place it on a glass plate over a light box, and cut out with a razor blade the part containing the polypeptides of interest. Trim the gel slices so that they fit easily into the sample wells of the second slab gel. Prepare a second slab gel according to the SDS-discontinuous system as explained in Section III.A, but with greater thickness (1.5 mm), a taller stacking gel (4 cm, 15 sample wells per gel), and the resolving gel with at least 15% acrylamide. The optimum acrylamide concentration of the resolving gel depends on the size of the polypeptide to be digested and its peptide products. Concentration gradient resolving gels may also be used.
3. Fill the sample wells with 0.1% SDS in 0.125 *M* Tris-HCl, pH 6.8, buffer and position at the bottom of the well with a microspatula each gel slice containing the protein of interest. Air bubbles surrounding the gel slices should be removed by overlayering with a buffer containing 20% glycerol. Finally, approximately 10 μl of a protease solution (20 to 200 ng dissolved in buffer containing 10% glycerol and 0.002% bromophenol blue) is added and the electrophoresis is carried out at 100 V until the tracking dye has traveled to the bottom fourth of the stacking gel. The power is then turned off for 30 min. During this period, the protease and the substrate are co-stacked and digestion occurs. The proteases which may be used include chymotrypsin, *Staphylococcus aureus* V8 protease, papain, elastase, and subtilisin. After proteolysis, electrophoresis is continued at 25 mA until the tracking dye reaches the bottom of the gel. The gels are stained, destained, and prepared for either autoradiography or fluorography.

Two proteins may not be related if the peptide profiles obtained are dissimilar. On the other hand, similar profiles, especially when using a single protease, are no proof of identity.[21] At least three different proteases should be used with this technique. The stringency of the method may be extended by digesting the polypeptide with one protease, cutting out the gel slice containing the peptide fragment, and subjecting it to digestion by a second protease. Peptide mapping by two-dimensional methods[22,23] still provides a more stringent test for the identity of polypeptides.

C. IMMUNOLOGICAL TECHNIQUES

If antibodies are available, immunological techniques will provide a powerful and definitive means for determining the identity and amounts of translation products. The most suitable method depends on the specificity of a particular antibody, and immunoprecipitation may be used to isolate and characterize a translation product.[20] For the stringent identification of the translation product, the Western blot technique is more widely employed. A mixture of proteins may be first separated by either one- or two-dimensional gel electrophoresis, transferred to a membrane, and the blot processed with antibodies.[25] This method identifies the protein in question by taking advantage of its mobility in gel electrophoresis and its antigenic properties. Protocols for both methods are described below.

1. Western Blot Analysis

1. Subject the translation mixture to either one- or two-dimensional gel electrophoresis, as previously described.
2. Polypeptides from polyacrylamide gels may be transferred to either a nitrocellulose or a polyvinylidene difluoride, PVDF, (Immobilin, Millipore) membrane. Cut the membrane to the same size of the gel. The nitrocellulose membrane should be soaked in the transfer buffer (20 m*M* Tris-HCl, 150 m*M* glycine, and 20% v/v methanol, pH 8.3) for 10 to 20 min. The PVDF membrane is extremely hydrophobic and should be treated with methanol before soaking it in the transfer buffer. Wet the PVDF membrane in methanol for 1 to 3 s, rinse with water twice (2 to 5 min each), and equilibrate in the transfer buffer. Do not allow the membrane to dry out during the entire process of Western blotting.
3. The membrane is then placed in contact with the gel and the assembly is placed in the electroblotting unit. In our experience, it is easiest if a Dacron sponge, a piece of filter paper, and the membrane are stacked in a large glass dish containing the transfer buffer. Initially the polyacrylamide gel is added on top of the membrane, followed by the filter paper and the sponge. The sandwich containing the membrane and the polyacrylamide gel is placed in a cassette, which is then closed and kept in the electroblotting chamber filled with the transfer buffer. Air bubbles should not be trapped between the gel and the membrane. The polyacrylamide gel should be placed toward the cathode and the membrane should be placed toward the anode. Electroblotting may be carried out either at 200 mV for 3 h or at 40 mV for 15 h.
4. After the transfer of proteins to the membrane, it is incubated for 2 to 4 h each in a solution containing 5% nonfat dry milk,[26] then with primary antibodies with 2.5% nonfat dry milk, and finally, with secondary antibodies with 2.5% nonfat dry milk in 10 m*M* Tris-HCl, pH 7.5, and 150 mM NaCl (Tris/saline). After each incubation, the membrane is washed for 10 min with 100 ml of Tris/saline, followed by two 15-min washes with 100 ml of Tris/saline containing 0.05% Tween-20, and, finally, for 10 min in 100 ml of Tris/saline. Either [^{125}I]-labeled *Staphylococcus aureus* protein A or enzyme (peroxidase or alkaline phosphatase)-conjugated antibodies may be used as secondary antibodies. The enzyme-conjugated secondary antibodies are available commercially. The membrane is either subjected to autoradiography when [^{125}I]-labeled protein A is used or processed for color development,[25] as described below for peroxidase- or alkaline phosphatase-conjugated secondary antibodies.
5. Color-developing reagents — For peroxidase-conjugated secondary antibodies, dissolve 30 mg of 4-chloro-1-naphthol in 10 ml of ice-cold absolute methanol. Add 50 ml of Tris/saline and 300 μl of 3% H_2O_2. Mix well and protect from light. This solution should be freshly prepared immediately prior to use. For alkaline phosphatase-conjugated secondary antibodies, dissolve 10 mg of nitroblue tetrazolium in 50 ml of buffer containing 100 m*M* Tris-HCl, pH 8.8, 100 m*M* NaCl, and 5 m*M* $MgCl_2$. Add 0.5 ml of indol phosphate (5 mg/ml in dichloroformamide). Incubate the membrane in the color-developing reagent for 10 to 30 min. Dark blue to purple bands are visualized by this process. Stop the reaction by discarding the color reagent and washing

with water and 5% acetic acid. Since the color of the bands fades quickly on the dry membrane, it is advisable to photograph the wet membrane.

2. Immunoprecipitation

1. Dilute *in vitro* translation products with 0.4 ml of buffer P containing 10 m*M* Tris-HCl, pH 7.5, 0.1% (w/v) SDS, 0.1% (v/v) Triton X-100, and 2 m*M* EDTA. Add the appropriate dilution of antibodies and the protease inhibitors (0.2 m*M* each of leupeptin, pepstatin, antipain, and chymostatin) and incubate the immunoprecipitation mixture for 2 h at 30°C.
2. Wash the *S. aureus* cells (100 μl, 10% w/v suspension, Calbiochem-Behring) three times with buffer P and add to each immunoprecipitation mixture.
3. After incubation for 30 min at 30°C, pellet the cells by centrifugation for 2 min at 9000 × g. Wash the pellets three times with buffer P listed above.
4. Elute the antigen and antibodies from the cells by suspending them in 2% (w/v) SDS. Boil the mixture for 5 min, remove the cells by centrifugation, and subject the supernatant to SDS-polyacrylamide gel electrophoresis and then to fluorography or autoradiography.

IV. SIGNIFICANCE AND APPLICATION OF *IN VITRO* TRANSLATION SYSTEMS

A. STEADY-STATE LEVELS OF mRNA IN CELLS: THE TRANSLATION OF POLY (A)$^+$ RNA EXTRACTED FROM CELLS AT DIFFERENT STAGES OF DEVELOPMENT

The population of genes being transcribed may change during the process of development or differentiation and in response to external stress. There are also other conditions in which messages are selected for translation. For example, Colbert and Young[14] demonstrated that after the heat shock of thymic lymphocytes, changes in mRNAs selected for translation are the result of both new transcription and the shift of preexisting mRNA into polysomes. It is important to know whether the appearance or disappearance of certain cellular proteins is the result of changes in the level of messenger RNA or due to selective translation of existing mRNAs. The investigator may wish to isolate total RNA, poly (A$^+$) RNA, or fractionate polysomal RNA from nontranslated mRNP messages. Whichever system is used, it will be necessary to isolate the message and translate it *in vitro*. The changes in the steady-state levels of the protein and the mRNA for eukaryotic elongation factor 1α (eEF-Tu, eEF-1α) were recently described during the development of *Artemia* embryos[20] and the differentiation of murine erythroleukemia cells.[16]

B. TRANSLATION OF SPECIFIC mRNAS FROM RNA BLOTTED ON POLY (U) PAPER

The probability of success in the preparation of a cDNA greatly depends on the selection of a RNA fraction enriched in a specific mRNA. One method which is used for nonhistone mRNAs is to prepare poly (A$^+$) RNA and electrophorese it on a preparative 1.5% agarose nondenaturing gel.[20] The RNA is then blotted onto poly (U) paper (either prepared by the method of Wreschner and Herzberg[27] or purchased commercially). A strip from the blot is hybridized with a [^{32}P]-labeled probe specific for mRNA.[20] The autoradiogram of the strip containing the hybridization is aligned with the remainder of the blot and that portion containing the desired RNA is excised from the blot. The RNA is eluted from the blot by extraction in water for 2 min at 70°C. It should be concentrated by ethanol precipitation since it will contain salts from the electroblotting procedure. Enrichment of the mRNA of interest can also be achieved by its fractionation on either sucrose or $CSCl_2$ gradient centrifugation. Before cloning, the integrity and identity of the RNA should be confirmed by the translation of the extracted mRNA in a nuclease-treated lysate derived from rabbit reticulocytes (Section II.A) or wheat germ (Section II.B).

C. TRANSLATION OF mRNAs OBTAINED FROM A cDNA CLONE PRESENT IN AN EXPRESSION VECTOR

It is often necessary to demonstrate that a cDNA clone codes for the proper protein. The clone should be in a vector where either the message or the protein may be expressed. If the protein is expressed, then the product may be identified by immunoblotting (see Section III.C). Otherwise, the mRNA may be produced in an *in vitro* transcription system by using either SP6[28] or T7[29] RNA polymerase after the insertion of the cDNA into a vector containing either the SP6 or the T7 polymerase promoter. This method will produce mRNA in good yields. However, it may be poorly translated *in vitro* due to the fact that the message may not be efficiently capped. This problem is partly alleviated by the use of a high level of mRNA in the *in vitro* translation system. The RNA is extracted with phenol, precipitated with alcohol, and suspended in buffer before translation. The product(s) may be characterized as described in Section III.

V. APPENDIX: PURIFICATION PROCEDURES

A. PURIFICATION OF POLYPEPTIDE CHAIN INITIATION FACTORS

The purification procedure for the initiation factors from rabbit reticulocyte lysates is outlined in Figure 2. All procedures are carried out at 4°C except for HPLC chromatography, which is at room temperature. Fractions from FPLC columns are collected in tubes immersed in ice. The buffer used to equilibrate gel filtration and ion-exchange columns is buffer A containing 20 m*M* Tris-HCl, pH 7.5, 2 m*M* dithiothreitol, 50 μ*M* EDTA, 10% (v/v) glycerol, and KCl, as indicated in the individual procedures. Fractions containing GEF are dialyzed against buffer A containing 50% (v/v) glycerol and are stored in liquid nitrogen. DEAE-cellulose (DE-52) and phosphocellulose (P-11) are obtained from Whatman and m^7GTP-Sepharose from Pharmacia. Mono S, Mono Q, and Mono P FPLC columns are from Pharmacia LKB Biotechnology, and they are used with either Pharmacia FPLC or ISCO HPLC systems. The ionic strength of the buffers used for column chromatography is shown in the flow diagram (see Figure 2). Protein samples for sucrose gradient fractionation are applied to 8 to 24% linear sucrose gradients (prepared in buffer A containing 500 m*M* KCl and lacking glycerol), centrifuged for 26 h at 120,000 × g. Mono S, Mono Q, and Mono P columns are equilibrated with buffer containing 100 m*M* KCl or as indicated. The columns are first washed with 2 ml of the same buffer, and the proteins are then eluted with three consecutive linear KCl gradients in buffer A: 5-ml gradient, 100 to 200 m*M* KCl; 20-ml gradient, 200 to 500 m*M* KCl; and 10-ml gradient, 500 to 1000 m*M* KCl. Fractions of 0.5 ml are collected. The purification of eIF-2,[25] GEF,[5,6] eIF-3, eIF-4A, eIF-4B, and eIF-4F[13] were previously described, and their properties are summarized in Table 1. The β subunit of eIF-2 migrates anomalously in dodecylsulfate/polyacrylamide gels. Depending upon the acrylamide/bis-acrylamide ratio of the gel, the apparent M_r of the β subunit of eIF-2 is either 48,000 or 52,000 D.[25] The M_r reported in Table 1 is derived from the corresponding cDNA sequence of the eIF-2 α- and β-polypeptides.[30,31]

B. THE USE OF PARTIAL REACTIONS OF POLYPEPTIDE CHAIN INITIATION AND ELONGATION IN THE ANALYSIS OF TRANSLATIONAL REGULATION

Initiation factors for protein synthesis may be assayed by partial reactions or in reconstituted *in vitro* translation systems. Translation assays may be simplified by using fractions containing partially purified initiation factors.

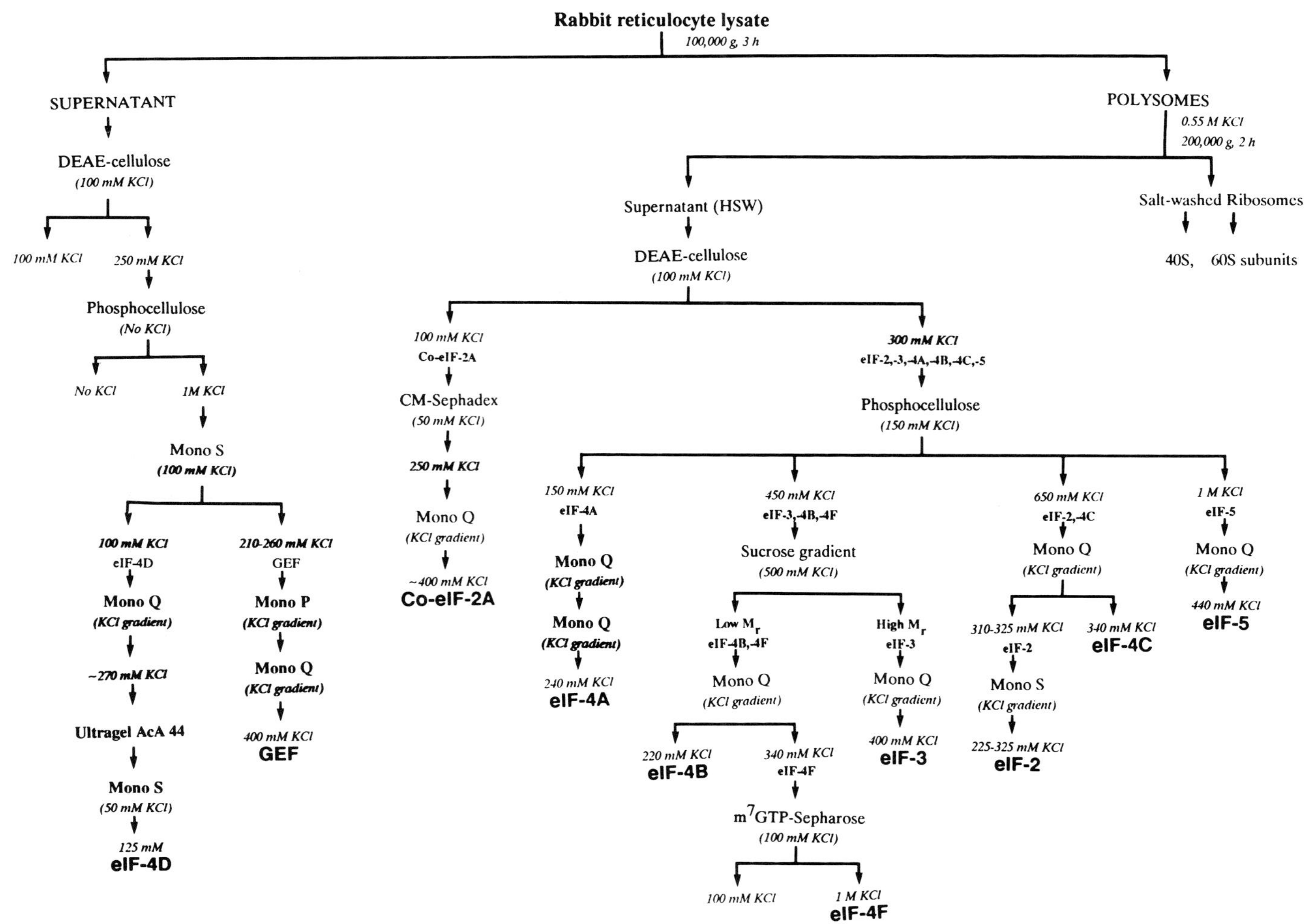

FIGURE 2. Flow chart for the purification of eukaryotic initiation factors.

TABLE 1
Properties of Eukaryotic Protein Synthesis Initiation Factors

FACTOR	ACTIVITIES	Native Protein Composition M_r	Native Protein Composition pI	Subunit Composition M_r	Subunit Composition pI	COMMENTS
eIF-2	GTP-dependent binding of Met-$tRNA_f$ to 40S ribosomal subunits	122,700	6.4	α- 36,100	5.1	Phosphorylation by HCR and DAI inhibits protein synthesis
				β- 38,400	5.4	GTP binding (Guanine moeity)
				γ- 55,000	8.9	GTP binding (γ-phosphate moeity)
GEF	Exchange of eIF-2-bound GDP for GTP	274,000		32,000		
				40,000		GTP binding
				55,000		NADPH/ATP binding
				65,000		NADPH/ATP binding
				82,000		Phosphorylation by CK II stimulates GEF activity
Co-eIF-2A	Protection of eIF-2•GTP•Met-$tRNA_f$ against inhibition by mRNA	90,000				
eIF-3	Anti-association activity against 40S and 60S ribosomal subunits	700,000	6.7	35,000		
				39,000		
				45,000		
				69,000		
				110,000		
				130,000		
eIF-4A	mRNA binding to 43S preinitiation complex. RNA-dependent ATPase	46,000	6.1	46,000	6.1	Also present in eIF-4F complex
eIF-4B	Stimulates mRNA binding to 43S preinitiation complex	80,000	6.0	80,000	6.0	Stimulates eIF-4B & eIF-4F activities

eIF-4C	Ribosome subunit joining	17,600	5.6	17,600	5.6	
eIF-4D	Stimulates Met-puromycin synthesis	16,800	5.1	16,800	5.1	^{50}Lys is modified to hypusine.
eIF-4E	Recognition of mRNA 5' cap structure	26,000	5.6	26,000	5.6	Also present in eIF-4F complex
eIF-4F	mRNA binding to 43S preinitiation complex	290,000		26,000 46,000 220,000	5.6 6.1	eIF-4E or CBP I eIF-4A Also known as CBP II
eIF-5	Hydrolysis of eIF-2-bound GTP and the release of initiation factors upon 60S ribosomal subunit joining to 43S pre-initiation complex	150,000	6.4	150,000	6.4	
eIF-6	Ribosomal subunit dissociation activity	25,000		25,000		

1. eIF-2 Assays

a. Ternary (eIF-2·GTP·Met-tRNA$_f$) Complex Formation

eIF-2 activity may be estimated by the GTP-dependent retention of [^{35}S]Met-tRNA$_f$, the initiator tRNA on a nitrocellulose membrane.[25] Each reaction mixture contains, in 75 μl: 20 m*M* Tris-HCl, pH 7.5, 1 m*M* dithiothreitol, 100 m*M* KCl, 10 μg of bovine serum albumin, 0.2 m*M* GTP, 5 pmol of [^{35}S]Met-tRNA$_f$ (approximately 15,000 cpm/pmol), and eIF-2 (1 to 3 pmols). After 10 min at 30°C, the reactions are terminated by the addition of 1 ml cold wash buffer (20 m*M* Tris-HCl, pH 7.8, and 100 m*M* KCl). Bound Met-tRNA$_f$ is measured by its retention on nitrocellulose membranes (Millipore, type HA, 0.45 μ) after filtration of the diluted reaction mixture. The tube and the membrane are rinsed with three additional 2-ml aliquots of cold wash buffer. The membranes are dried under an infrared lamp and radioactivity is determined by counting the membranes in 8 ml of a toluene-based liquid scintillation fluid. Since GDP is a potent inhibitor of ternary complex formation,[32] the use of HPLC purified GTP[25] or the incorporation of a phosphoenolpyruvate/pyruvate kinase-GTP regenerating system (1 m*M* phosphoenol pyruvate and 0.1 IU pyruvate kinase) in the reaction mixture is recommended.[33] Reactions are carried out in plastic or silicone-treated glassware since eIF-2 binds to glass surfaces. In the absence of GEF, Mg^{2+} inhibits ternary complex formation. [^{35}S]Met-tRNA$_f$ can be prepared as described in Appendix D. The specific activity of eIF-2 is defined as the number of picomoles of [^{35}S]Met-tRNA$_f$ bound per milligram of protein. Nonspecific binding of Met-tRNA$_f$ in crude extracts may be corrected by subtracting a blank value when the assay is carried out in the absence of GTP.

b. Nucleotide Binding to Initiation Factors

The principle used in ternary complex formation (eIF-2·GTP·Met-tRNA$_f$) may be applied for the binding of radioactively labeled nucleotides[4] to initiation factors. Only protein-bound nucleotides are adsorbed to nitrocellulose membranes. A typical assay mixture for [^{3}H]GDP binding to eIF-2 contains, in a volume of 75 μl: 20 m*M* Tris-HCl, pH 7.5, 100 m*M* KCl, 1 m*M* dithiothreitol, 10 μg of bovine serum albumin, 4 μ*M* [^{3}H]GDP (6000 cpm/pmol), and eIF-2. Incubation is at 30°C for 10 min, and the reaction is terminated by transferring the reaction tubes to an ice bath and adding 1 ml of wash buffer containing 1 m*M* $MgCl_2$ to each sample. The reaction mixtures may be either processed as described for ternary complex formation (with wash buffer containing 1 m*M* $MgCl_2$) or chromatographed on Mono S or phosphocellulose columns in buffer A containing 1 m*M* $MgCl_2$ when a binary complex of eIF-2·[^{3}H]GDP, free of GDP,[4,23,25] is to be isolated.

c. GEF Assay

GEF catalyzes the exchange of eIF-2-bound GDP for GTP and may be assayed by monitoring the release of [^{3}H]GDP from eIF-2·[^{3}H]GDP in the presence of added guanine nucleotides.[4,5] Each reaction mixture contains, in 75 μl: 20 m*M* Tris-Cl, pH 7.5, 1 m*M* dithiothreitol, 1 m*M* $MgCl_2$, 4 μ*M* GDP or 20 μ*M* GTP, 100 m*M* KCl, 10 μg of bovine serum albumin, 3 pmol of eIF-2·[^{3}H]GDP complex (6000 cpm/pmol), and GEF. The incubation is for 10 min at 30°C. The reaction mixtures are terminated by the addition of cold wash buffer containing 1 m*M* $MgCl_2$ and processed as described for the ternary complex assay. Nucleotide exchange is dependent on GEF only when performed in the presence of Mg^{2+}. GEF activity is expressed as the number of picomoles of [^{3}H]GDP released per milligram of protein.

2. Assays of eIF-4A, eIF-4B, and eIF-4F

The factors are most conveniently assayed by their cooperative activity in the release of [^{32}P]orthophosphate from [γ-^{32}P]ATP.[33] The hydrolysis of ATP is increased in the presence of natural or synthetic mRNA. These proteins may also be assayed by measuring the ATP-dependent binding of labeled messenger RNA to a nitrocellulose membrane.[33]

a. ATP Hydrolysis

The RNA-dependent hydrolysis of [γ-^{32}P]ATP by eIF-4A, eIF-4B, and eIF-4F is assayed at 30°C for 15 min. Prior to performing the reaction, prepare the phosphate reagent (mix 2 ml of 10 m*M* NaH_2PO_4, 2 ml of 1 *M* $HClO_4$, and 4 ml of 0.3 *M* Na_2MoO_4 in 1 *M* H_2SO_4) which is used for complexing with [^{32}P]orthophosphate. For the assay use 1.5-ml conical polypropylene tubes with snap caps.

Each assay contains, in a final volume of 38 μl: 100 m*M* KCl, 2 m*M* dithiothreitol, 30 m*M* HEPES·KOH, pH 7.4, 1 mg/ml bovine serum albumin, 2.5 m*M* Mg(acetate)$_2$, 0.1 m*M* [γ-^{32}P]ATP (2000 to 4000 cpm/pmol) and, as required, eIF-4A (1 to 2 μg), eIF-4B (1 to 2 μg), and eIF-4F (1 to 3 μg) in buffer A, and 0.125 A_{260} unit poly (A). Buffer A contains 20 m*M* Tris-Cl, pH 7.5, 2 m*M* dithiothreitol, 50 μ*M* EDTA, 10% (v/v) glycerol, and KCl, as indicated. Assay the RNA-dependent hydrolysis of [γ-^{32}P]ATP at 30°C for 15 min. The reactions are stopped by the addition of 0.4 ml of the phosphate reagent and 1 ml of *n*-butyl acetate. The contents are thoroughly mixed on a vortex shaker, then centrifuged briefly to ensure complete separation of the lower aqueous phase from the upper organic layer. The complete extraction of the ^{32}P-labeled complex into the organic phase may be monitored by the transfer of the yellow phosphate complex into the upper layer. A 0.5-ml aliquot from the top layer of each tube is placed in a vial with 5 ml of a scintillation fluid suitable for aqueous samples and counted for radioactivity. "Crude" preparations of [γ-^{32}P]ATP will produce a very high background.

b. mRNA Binding to Nitrocellulose Membranes

A natural mRNA, such as globin RNA extracted from rabbit reticulocyte lysates, may be labeled at the 3′ end by poly (A) polymerase and (α-^{35}S) ATP.[33] With polynucleotide kinase synthetic polynucleotides may be labeled at the 5′ end with ^{32}P, whereas natural messages may be labeled with ^{32}P by using RNA ligase. ^{3}H-labeled mRNA with an adequate specific activity may also be obtained by incubating cells in tissue culture with [^{3}H]uridine and isolating the poly (A^+) RNA from these cells.

Each reaction mixture contains, in a final volume of 75 μl: 100 m*M* KCl, 5m*M* Mg(acetate)$_2$, 30 m*M* HEPES·KOH, pH 7.4, 2 m*M* dithiothreitol, 0.1 mg/ml bovine serum albumin, 1 m*M* ATP, 4 m*M* phosphoenolpyruvate, 0.4 IU of pyruvate kinase, 5 to 20 μg/ml labeled mRNA, and 15 μl of buffer A containing the initiation factors (0.8 to 1.2 μg of eIF-4F, 2 to 2.5 μg of eIF-4B, and 1.5 to 2 μg of eIF-4A), as required. Incubation is at 30°C for 5 min. Stop the reaction by adding 1 ml of cold buffer A containing 100 m*M* KCl and 5 m*M* $MgCl_2$. The samples are then processed as described in the ternary complex assay.

3. eIF-5-Dependent GTP Hydrolysis

eIF-5 is assayed either by its requirement in the Met-puromycin reaction (see below) or by ribosome-dependent hydrolysis of GTP.[34] This reaction is the same as the ATPase assay described above for eIF-4A, eIF-4B, and eIF-4F, with the following modifications. The reactions also contain 0.5 to 1 μg of eIF-5, 0.8 to 1 A_{260} U of 80S ribosomes, and 20 μ*M* [γ-^{32}P]GTP at 5000 cpm/pmol. The reactions are stopped, processed, and counted as described for the ATPase reaction.

4. Methionyl-Puromycin Synthesis

The transfer of [^{35}S]Met from amino-acylated initiator tRNA to puromycin may be used to assay several factors, including eIF-2, eIF-4C, eIF-4D, and eIF-5.[35] In the presence of the antibiotic puromycin, the growing polypeptide chain positioned in the P site of the ribosome is transferred to a free NH_2 group on puromycin located in the ribosomal A site. In the presence of 80S ribosomes, the appropriate initiation factors, the initiator tRNA ([^{35}S]Met-tRNA$_f$), and an AUG template, methionine alone will be transferred to puromycin. The product may be easily extracted in an organic phase and counted for radioactivity. This offers a convenient assay for the formation of a competent 80S initiation complex.

Each reaction contains, in a final volume of 75 μl: 40 m*M* KCl, 90 m*M* K·acetate, 30 m*M* HEPES·KOH, pH 7.4, 2 m*M* dithiothreitol, 8 m*M* creatine phosphate, 15 μg of creatine kinase, 1 m*M* ATP, 0.25 m*M* GTP, 30 μ*M* spermine, 2.5 m*M* Mg(acetate)$_2$, 1.2 m*M* puromycin, 0.075 A_{260} U of AUG, 20 μl of buffer A with 100 m*M* KCl containing the initiation factors, 3 to 5 pmols of [^{35}S]Met-tRNA$_f$ (15,000 cpm/pmol), and 0.8 to 1 A_{260} U of runoff rabbit reticulocyte 80S ribosomes. As required, each reaction may also contain: 1 μg of eIF-2, 0.6 μg of eIF-5, 0.5 to 1 μg of eIF-4C, 1 to 1.5 μg of eIF-3, 0.05 μg of GEF, 0.65 μg of eIF-4B, and 1 to 1.5 μg of eIF-4D. Incubate the reactions for 20 min at 30°C and terminate by adding 0.15 ml of 0.5 *M* Tris-glycine, pH 9.0, and 1 ml of ethyl acetate. Vortex the mixture and centrifuge for 1 min at 8000 × g. Count for radioactivity a 0.5-ml aliquot of the upper, ethyl acetate layer in 5 ml of scintillation fluid suitable for aqueous samples.

Met-puromycin synthesis is dependent on the activity of several initiation factors. The eIF-4D preparation, containing 1.2 μg of the initiation factor, stimulates the reaction approximately 2.5-fold, whereas the stimulation by eIF-4C is slightly over twofold. The reaction is also dependent on eIF-2, eIF-3, and eIF-5.

5. 40S Initiation Complex Formation

One of the partial or transient reactions in polypeptide chain initiation is the binding of the initiator tRNA and mRNA to a 40S ribosomal subunit to form first a 43S, then a 48S initiation complex. This method allows the investigator to assay the components necessary for binding either Met-tRNA$_f$ or mRNA to the 40S ribosomal subunit. This assay may be used to measure the initiation factor-dependent transfer of [^{35}S]Met-tRNA$_f$ to a 40S ribosomal subunit in the presence of a template containing an AUG codon. The template may be an AUG triplet codon, poly (A, U, or G), or a natural mRNA. If the mRNA is radioactively labeled (Appendix, V.B.2.b), then its binding to the ribosomal subunit may also be monitored.

For measuring [^{35}S]Met-tRNA$_f$ binding to a 40S initiation complex, sediment the complex through a sucrose gradient and count the gradient fractions either directly in a liquid scintillation cocktail or by first filtering the fractions through a nitrocellulose membrane.[36] If the binding of labeled mRNA to the 40S subunit is to be assayed, then the sucrose gradient method must be used. The procedure described here also includes a modification of the standard sucrose gradient protocol which may be employed for measuring the binding of either Met-tRNA$_f$ or mRNA.

Each reaction contains, in a final volume of 75 μl, the same components used for the assay of Met-puromycin synthesis, with the following changes: puromycin is deleted and 0.5 A_{260} U of 40S ribosomal subunits is substituted for the 80S ribosomes. The reaction mixtures are incubated for 10 min at 30°C. At the end of the incubation, remove a 50-μl aliquot for analysis. The samples may be analyzed by sedimenting the initiation complex for 2 h at 54,000 rpm and 4°C in a Beckman SW60Ti rotor through a 5 to 30% (w/v) sucrose gradient containing 20 m*M* Tris-HCl, pH 7.5, 100 m*M* KCl, 3 m*M* $MgCl_2$, and 1 m*M* dithiothreitol. Fractionate the gradient and count the fractions for radioactivity in a scintillation cocktail suitable for aqueous samples. Alternatively, the 50-μl aliquot may be analyzed by sedimenting the initiation complex into an 80% sucrose cushion, collecting this layer with a blunt-tip microsyringe, and counting the sample in a liquid scintillation cocktail. This assay may be performed in a Beckman Airfuge (40 min at 28 psi and room temperature) or the TL-100 Ultracentrifuge (45 min at 60,000 rpm and 25°C). The volumes of the gradient layers are given for the TLA-100 rotor: 20 μl of 80% sucrose, 40 μl Of 40% sucrose, 35 μl of 30% sucrose, 30 μl of 20% sucrose, and 25 μl of 10% sucrose. These gradients may be prepared immediately before use.

6. *Sucrose Gradient Analysis of Polysomes*

An early step in the initiation of protein synthesis is the binding of a 40S ribosomal subunit to an mRNA and the subsequent binding of a 60S subunit. As elongation proceeds and the initiation site is again vacant, a second initiation step is allowed. During the course of protein synthesis, several ribosomes may be attached to the mRNA at any one time. The number of ribosomes associated with a given mRNA is a function of the length of the mRNA and the relative efficiencies of the initiation and elongation processes. Analysis of the relative distribution of ribosomes between the ribosome and the polysome pools may be determined by sucrose gradient analysis. This technique has been used to examine the polysome profile for a variety of reasons, including (1) determination of the mechanism of polypeptide chain initiation with normal and viral messages or in the presence of specific inhibitors, (2) changes that result from the regulation of protein synthesis during development, differentiation, or viral infection, and (3) the effect of stresses, such as serum deprivation and heat shock. The analysis of the polysome profile for mengovirus-infected mouse L cells reported by Pensiero and Lucas-Lenard[37] typifies the technique of analyzing the polysomes of a cell extract by sedimenting through a sucrose gradient. Harvest the cells (approximately 1×10^7 per gradient) by centrifugation at 4°C and 700 × g for 3 min. Wash the pelleted cells twice with cold phosphate-buffered saline (10 m*M* Na phosphate, pH 7.5, and 137 m*M* NaCl) containing cycloheximide (50 μg/ml) to prevent polypeptide chain elongation. Suspend the cell pellet in approximately 250 μl of buffer containing 10 m*M* HEPES·KOH, pH 7.4, 2 m*M* $Mg(acetate)_2$, 150 m*M* KCl, 0.4% (v/v) Triton X-100, and 50 μg/ml cycloheximide. Gently vortex the cells, and pellet the nonsolubilized material by centrifugation in a microcentrifuge (approximately 6000 × g) for 3 min. Layer the cell extract on a 13.2-ml, 15 to 40% (w/v) sucrose gradient in the lysis buffer minus Triton X-100, and centrifuge for 2 h at 210,000 × g (35,000 rpm in a Beckman SW41Ti rotor) and 4°C. At the completion of the centrifugation, the gradient profiles are determined at 260 nm by pumping the gradients through a UV monitor equipped with a flow cell and a chart recorder. Initially, the UV monitor should be set at either 0.5 or 1.0 U of absorbance for full-scale deflection.

C. PREPARATION OF RIBOSOMES AND RIBOSOMAL SUBUNITS

1. Rabbit Reticulocytes

For the preparation of either preincubated rabbit reticulocyte 80S ribosomes or ribosomal subunits, the following steps are employed. The crude ribosomal pellet resulting from centrifuging a rabbit reticulocyte lysate for 3 h at 100,000 × g is suspended in ribosome buffer containing 0.25 *M* sucrose, 20 m*M* Tris-Cl, pH 7.5, 100 m*M* KCl, 5 m*M* $Mg(acetate)_2$, 100 μ*M* EDTA, and 2 m*M* dithiothreitol. The ribosome concentration is adjusted to 300 to 400 A_{260} U/ml. A mixture containing 3 ml of the ribosome suspension, 50 μl of 1.5 m*M* aurin tricarboxylic acid, 13 ml of water, and 4 ml of a salt mix is incubated for 50 min at 37°C. The salt mix supplies to the final reaction: 1 m*M* ATP, 0.2 m*M* GTP, 3 m*M* phosphoenol pyruvate, 30 m*M* Tris-HCl, 3 m*M* Mg^{2+}, 100 m*M* KCl, 250 μ*M* of each amino acid, 30 μ*M* spermine, 2 m*M* dithiothreitol, 7.5 A_{260} U/ml purified reticulocyte or yeast tRNA, and 30 μg/ml pyruvate kinase. Creatine phosphate (5 to 10 m*M*) and creatine kinase (0.07 mg/ml) may be substituted for the phosphoenol pyruvate/pyruvate kinase system.

For the preparation of 80S ribosomes, the above incubation mixture is layed over a 10-ml cushion of 25% sucrose containing 75 m*M* KCl, 2 m*M* dithiothreitol, 5 m*M* Mg^{2+}, 20 m*M* Tris-Cl, and 50 μ*M* EDTA. The ribosomes are pelleted through the sucrose cushion by centrifugation overnight (18 h) at 50,000 rpm and 4°C in a Beckman 70Ti rotor. The supernatant is discarded, and the pellet is suspended in the ribosome buffer and stored in small aliquots (150 to 200 A_{260} U/ml) in liquid N_2.

Ribosomal subunits are prepared by chilling the incubation mixture in an ice bath and adding 1.75 ml of 5 *M* NH_4Cl to the 20 ml of incubation mixture. Aliquots of 3.3 ml of this ribosomal suspension are layered on a 33-ml linear sucrose gradients (15 to 35%) containing each 2 m*M* dithiothreitol, 50 m*M* Tris-HCl, 3 m*M* Mg^{2+}, 500 m*M* NH_4Cl, and 50 μ*M* spermine. After centrifugation for 16 h at 23,000 rpm and 4°C in a Beckman SW28Ti rotor, the gradients are fractionated and 1-ml fractions are collected. Fractions from the slow-sedimenting two thirds of the 40S peak are combined. Likewise, the middle two thirds of the fractions comprising the 60S peak are also combined. These pools are centrifuged for at least 16 h at 30,000 rpm and 4°C in a Beckman 45Ti rotor. To ensure proper pelleting of the ribosomal subunits, the tubes should not be more than one half filled. The ribosomal pellets are suspended in buffer containing 50 m*M* NH_4Cl, 3 m*M* Mg(acetate)$_2$, 0.25 *M* sucrose, 1 m*M* dithiothreitol, and 20 m*M* Tris-HCl, pH 7.5. The ribosomal subunits are stored in liquid N_2 in aliquots containing 75 to 125 A_{260} U/ml.

2. *Artemia* Embryos

Ribosomes isolated from dormant *Artemia* embryos are free of bound mRNA and consist almost exclusively of 80S ribosomes which are active in either partial reactions or *in vitro* protein synthesis. A 500-g (dry weight) batch of desiccated embryos (San Francisco Bay Brand) is hydrated for 3 h in 2 l of 0.5 *M* NaCl. This and all subsequent steps should be performed at 4°C. The hydrated embryos are washed by filtration through six layers of cheese cloth with cold, deionized water, then with buffer (rinse buffer) containing 50 m*M* Tris-HCl, pH 7.5, 70 m*M* KCl, 9 m*M* Mg(acetate)$_2$, 10 m*M* 2-mercaptoethanol, 0.1 m*M* EDTA, and 0.25 *M* sucrose. Batches of 100 g (wet weight) are ground for 5 min in the rinse buffer. The efficiency of grinding should be monitored by microscopic inspection. Each batch of ground embryos is dispersed in 100 ml of the rinse buffer and agitated for 15 min on a magnetic stirrer. Centrifuge the ground embryos for 20 min at 12,000 × g. Remove the resulting bright orange lipid layer by aspiration; decant and save the supernatant, and discard the pelleted material. Filter the supernatant through six layers of cheese cloth and centrifuge the filtrate for 30 min at 160,000 × g. The supernatant is filtered through six layers of cheese cloth and one layer of glass wool. Centrifuge the filtrate for 2 h at 205,000 × g. Discard the supernatant, rinse twice, and suspend the pellet in buffer containing 20 m*M* Tris-HCl, pH 7.5, 100 m*M* KCl, 9 m*M* Mg(acetate)$_2$, 50 μ*M* EDTA, and 1 m*M* dithiothreitol. Layer the ribosome suspension over 15-ml cushions containing 50% (w/v) sucrose in the suspension buffer. Centrifuge for 18 h at 50,000 rpm and remove the supernatants by aspiration. Suspend the pellets in the suspension buffer containing 0.25 *M* sucrose (storage buffer). The usual yield of ribosomes from 500 g of desiccated embryos is approximately 17,000 A_{260} U. These 80S ribosomes may be further purified by chromatography on DEAE-acrylamide (Pharmacia/LKB). Apply approximately 4000 A_{260} U of ribosomes to a 250-ml column of DEAE-acrylamide equilibrated with the ribosome buffer containing 10% (v/v) glycerol, but no sucrose. Wash the column with the application buffer to remove unadsorbed material and elute the ribosomes with the same buffer containing 250 m*M* KCl. Collect the ribosomes by centrifugation for 14 h at 115,000 × g. Suspend the ribosomes in the ribosome storage buffer. The ribosomes should be stored in liquid N_2 at a concentration of 150 to 200 A_{260} U/ml. Ribosomal subunits may be prepared as described for rabbit reticulocyte ribosomal subunits (Appendix, Section C).

D. PURIFICATION OF tRNA AND AMINOACYLATION OF EUKARYOTIC Met-tRNA$_f$

Similar results are obtained when either rabbit reticulocyte tRNA or commercial brewer's yeast tRNA, which has been further purified, is used. Brewer's yeast tRNA (Boehringer-Mannheim) is dissolved in buffer containing 20 m*M* Tris-HCl, pH 7.5, and 10 m*M* Mg(acetate)$_2$ and applied to a DEAE-acrylamide column (LKB) equilibrated with the same buffer. Approximately 1.5 to 2 mg of tRNA may be applied to each milliliter of resin. DEAE-cellulose does not work as effectively as the acrylamide form for this preparation. The column is washed extensively to elute nonadsorbed material and is then step eluted with the same buffer containing 435 m*M* NaCl. The leading three fourths of the tRNA peak eluted with NaCl is collected and precipitated by the addition of two volumes of cold ethanol. The tRNA should be allowed to precipitate overnight at −20°C, and is collected by centrifugation for 20 min at 10,000 rpm and −5°C in a Sorval GSA rotor. The pellet is washed twice with cold ethanol, dried under vacuum, and dissolved in water containing 50 μ*M* EDTA.

Crude aminoacyl tRNA synthetases are prepared from *E. coli*.[38] When stored at −20°C in a buffer containing 50% glycerol, this preparation is stable for at least 24 months. It will aminoacylate the eukaryotic initiator tRNA (tRNA$_f$), but not the species responsible for the transfer of methionine into internal positions of growing polypeptide chains. A 1.875-ml mixture containing up to 0.375 ml of either the aminoacyl tRNA synthetase preparation or its buffer, 4 mg tRNA, 50 m*M* Tris-HCl, pH 7.2, 8 m*M* Mg(acetate)$_2$, 8 m*M* KCl, 10 m*M* β-mercaptoethanol, and 0.65 m*M* ATP and 16 μ*M* methionine adjusted to 12,000 to 17,000 cpm/pmol with [^{35}S]methionine is incubated for 15 min at 37°C. The reaction is terminated by the addition of 375 μl of 1 *M* Na acetate, pH 5.0, and chilled in an ice bath. Add 2 ml of buffer (10 m*M* Na acetate, pH 5.0)-saturated phenol, vortex the mixture, and centrifuge for 5 min in a table-top centrifuge to separate the phenol and aqueous phases. Discard the phenol layer and repeat the phenol extraction. The small amounts of phenol remaining in the mixture should be removed by extraction with 2 ml of chloroform. The aqueous phase is then applied to a Sephadex G-25 column (1.5 × 16 cm) equilibrated with buffer T (20 m*M* Tricine-HCl, pH 6.6, 25 m*M* β-mercaptoethanol, and 0.05 m*M* EDTA). Fractions of 1 ml are collected and a 5-μl aliquot of each fraction is counted in 5 ml of scintillation fluid suitable for aqueous samples. The Met-tRNA$_f$ elutes in the first peak of radioactivity, and free methionine in the second. Peak fractions of Met-tRNA$_f$ are combined and mixed with 0.2 volume of 2.5 *M* Na acetate and 2.5 volumes of cold ethanol. The tRNA is allowed to precipitate overnight at −20°C, and is collected by centrifugation for 20 min at 10,000 rpm and −5°C in a Sorval SS-34 rotor. The precipitate is dried under vacuum, dissolved in buffer T, adjusted to 1 to 2 pmols of [^{35}S]Met-tRNA$_f$/μl, and stored in small aliquots at −80°C.

ACKNOWLEDGMENT

We wish to thank Charles L. Woodley for help in the organization of certain parts of this chapter. This work was supported in part by U.S. Public Health Services Grant GM 25451.

REFERENCES

1. **Pelham, R. B. and Jackson, R. J.,** An efficient mRNA-dependent translation system from reticulocyte lysate, *Eur. J. Biochem.*, 67, 247, 1976.
2. **Pain, V. M.,** Initiation of protein synthesis in mammalian cells, *Biochem. J.*, 235, 625, 1986.
3. **Moldave, K.,** Eukaryotic protein synthesis, *Annu. Rev. Biochem.*, 54, 1109, 1985.
4. **Dholakia, J. N. and Wahba, A. J.,** Mechanism of the nucleotide exchange reaction in eukaryotic polypeptide chain initiation. Characterization of the guanine nucleotide exchange factor as a GTP-binding protein, *J. Biol. Chem.*, 264, 546, 1989.

5. **Dholakia, J. N. and Wahba, A. J.,** Phosphorylation of the guanine nucleotide exchange factor from rabbit reticulocytes regulates its activity in polypeptide chain initiation, *Proc. Natl. Acad. Sci. U.S.A.,* 85, 51, 1988.
6. **Dholakia, J. N., Mueser, T. C., Woodley, C. L., Parkhurst, L. J., and Wahba, A. J.,** The association of NADPH with the guanine nucleotide exchange factor from rabbit reticulocytes: a role of pyridine dinucleotides in eukaryotic polypeptide chain initiation, *Proc. Natl. Acad. Sci. U.S.A.,* 83, 6746, 1986.
7. **Sonenberg, N.,** Cap-binding proteins of eukaryotic messenger RNA: functions in initiation and control of translation, *Prog. Nucleic Acid Res. Mol. Biol.,* 35, 173, 1988.
8. **Rhoads, R. E.,** Cap recognition and the entry of mRNA into the protein synthesis initiation cycle, *Trends Biochem. Sci.,* 13, 52, 1988.
9. **Darnbrough, C., Legon, S., Hunt, T., and Jackson, R. J.,** Initiation of protein synthesis: evidence for messenger RNA-independent binding of methionyl-transfer RNA to the 40S ribosomal subunit, *J. Mol. Biol.,* 76, 379, 1973.
10. **Kozak, M.,** Bifunctional messenger RNAs in eukaryotes, *Cell,* 47, 481, 1986.
11. **Morch, M. D., Drugeon, G., and Haenni, A. L.,** The synthesis of high-molecular weight proteins in the wheat germ translation system, *Methods Enzymol.,* 118, 154, 1986.
12. **Bader, M. and Sarre, T. F.,** A (re)initiation-dependent cell-free protein synthesis system from mouse erythroleukemia cells, *Eur. J. Biochem.,* 56, 161, 1986.
13. **Colin, A. M., Brown, B. D., Dholakia, J. N., Woodley, C. L., Wahba, A. J., and Hille, M. B.,** Evidence for simultaneous derepression of messenger: RNA and the guanine nucleotide exchange factor in fertilized sea urchin eggs, *Dev. Biol.,* 123, 354, 1979.
14. **Colbert, R. A. and Young, D. A.,** Detection of mRNAs coding for translationally regulated heat-shock protein in nonheat-shocked thymic lymphocytes, *J. Biol. Chem.,* 262, 9939, 1987.
15. **Mathews, M. B.,** Further studies on the translation of globin mRNA and encephalomyocarditis virus RNA in a cell-free system from Krebs II ascites cells, *Biochim. Biophys. Acta,* 272, 108, 1972.
16. **Roth, W. W., Bragg, P. W., Corrias, M. V., Reddy, N. S., Dholakia, J. N., and Wahba, A. J.,** Expression of a gene for mouse eucaryotic elongation factor Tu during murine erythroleukemic cell differentiation, *Mol. Cell. Biol.,* 7, 3929, 1987.
17. **Chinchar, V. G., Turner, L. A., and Kumar, G.,** Hemin and cyclic AMP stimulate messenger-dependent translation in lysates from Friend erythroleukemia cells, *Exp. Hematol.,* 17, 405, 1989.
18. **Hames, B. D. and Rickwood, D., Eds.,** *Gel Electrophoresis of Proteins: A Practical Approach,* IRL Press, Oxford, 1987.
19. **Chamberlin, J. P.,** Fluorographic detection of radioactivity in polyacrylamide gels with the water-soluble fluor, sodium salicylate, *Anal. Biochem.,* 98, 132, 1979.
20. **Daum, H. A., III, Bragg, P. W., Sittman, D. B., Dholakia, J. N., Woodley, C. L., and Wahba, A. J.,** The expression of a gene for eukaryotic elongation factor Tu in *Artemia* during development, *J. Biol. Chem.,* 260, 16347, 1985.
21. **Hames, B. D.,** Peptide mapping by limited proteolysis using SDS-polyacrylamide gel electrophoresis, in *Gel Electrophoresis of Proteins: A Practical Approach,* Hames, B. D. and Rickwood, D., Eds., IRL Press, Oxford, 1987, 219.
22. **Elder, J. H., Pickett, R., Hampton, J., and Lerner, R. A.,** Radioiodination of proteins in single polyacrylamide gel slices. Tryptic peptide analysis of all the major membranes of complex multicomponent systems using microgram quantities of total protein, *J. Biol. Chem.,* 252, 6510, 1977.
23. **Mehta, H. B., Dholakia, J. N., Roth, W. W., Parekh, B. S., Montelaro, R. C., Woodley, C. L., and Wahba, A. J.,** Structural studies on the eukaryotic chain initiation factor 2 from rabbit reticulocytes and brine shrimp *Artemia* embryos. Phosphorylation by the heme-controlled repressor and casein kinase II, *J. Biol. Chem.,* 261, 6705, 1986.
24. **Cleveland, D. W., Fischer, S. G., Kirschner, M. W., and Laemmli, U. K.,** Peptide mapping by limited proteolysis in sodium dodecyl sulfate and analysis by gel electrophoresis, *J. Biol. Chem.,* 252, 1102, 1977.
25. **Dholakia, J. N. and Wahba, A. J.,** The isolation and characterization from rabbit reticulocytes of two forms of eukaryotic initiation factor 2 having different β-polypeptides, *J. Biol. Chem.,* 262, 10164, 1987.
26. **Johnson, D. A., Gautsch, J. W., Sportsman, J. R., and Elder, J. E.,** Improved technique utilizing nonfat dry milk for analysis of proteins and nucleic acids transferred to nitrocellulose, *Genet. Anal. Technol.,* 1, 3, 1984.
27. **Wreschuer, D. H. and Herzberg, M.,** A new blotting medium for the simple isolation and identification of highly resolved messenger RNA, *Nucleic Acids Res.,* 12, 1349, 1984.
28. **Kreig, P. and Melton, D.,** *In vitro* synthesis with SP6 RNA polymerase, *Methods Enzymol.,* 155, 397, 1988.
29. **Neilsen, D. A. and Shapiro, D. J.,** Preparation of capped RNA transcripts using T7 RNA polymerase, *Nucleic Acids Res.,* 14, 5936, 1986.
30. **Ernst, H., Duncan, R. F., and Hershey, J. W. B.,** Cloning and sequencing of complementary cDNA encoding the α-subunit of translational initiation factor eIF-2, *J. Biol. Chem.,* 262, 1206, 1987.

31. **Pathak, V. K., Nielsen, P. J., Trachsel, H., and Hershey, J. W. B.,** Structure of the β-subunit of translational initiation factor eIF-2, *Cell,* 54, 633, 1988.
32. **Walton, G. M. and Gill, G. N.,** Nucleotide regulation of eukaryotic protein synthesis initiation complex, *Biochim. Biophys. Acta,* 390, 231, 1975.
33. **Abramson, R. D., Dever, T. E., Lawson, T. G., Ray, B. K., Thach, R. E., and Merrick, W. C.,** The ATP-dependent interaction of eukaryotic initiation factors with mRNA, *J. Biol. Chem.,* 262, 3826, 1987.
34. **Merrick, W. C., Kempher, W. M., and Anderson, W. F.,** Purification and characterization of homogeneous initiation factor M2A from rabbit reticulocytes, *J. Biol. Chem.,* 250, 5556, 1975.
35. **Benne, R., Brown-Leudi, M. L., and Hershey, J. W. B.,** Purification and characterization of protein synthesis initiation factors eIF-1, eIF-4C, eIF-4D and eIF-5 from rabbit reticulocytes, *J. Biol. Chem.,* 253, 3070, 1978.
36. **Das, A., Bagchi, M., Ghosh-Dastidar, P., and Gupta, N. K.,** Protein synthesis in rabbit reticulocytes. A study of peptide chain initiation using native and β-subunit-depleted eukaryotic initiation factor 2, *J. Biol. Chem.,* 257, 1282, 1982.
37. **Pensiero, M. N. and Lucas-Lenard, J. M.,** Evidence for the presence of an inhibitor on ribosomes in mouse L cells infected with mengovirus, *J. Virol.,* 56, 161, 1985.
38. **RajBhandary, U. L. and Ghosh, H. D.,** Studies on polynucleotides. Yeast methionine transfer ribonucleic acid: purification, properties, and terminal nucleotide sequences, *J. Biol. Chem.,* 224, 1104, 1969.

Chapter 18

DETECTION OF SPECIFIC RNA SEQUENCES IN TISSUE SECTIONS BY *IN SITU* HYBRIDIZATION

Alexander Awgulewitsch and Manuel F. Utset

TABLE OF CONTENTS

I. INTRODUCTION

Nucleic acid hybridization histochemistry was introduced in 1969 by Gall and Pardue,[1] who demonstrated gene amplification in amphibian oocytes by using tritium-labeled rRNA probes. Although technical progress was slow for many years, *in situ* hybridization to specific RNAs in tissue sections has recently become a well-established and widely employed method to study gene expression at the cellular level. New developments in molecular biology improved the efficiency of hybridization histochemistry by allowing the preparation of highly specific and sensitive probes. Efficient complementary DNA (cDNA) cloning methods, in combination with vector systems for the synthesis of single-stranded RNA probes,[2] achieved a sensitivity level that allows the detection of as few as 20 RNA molecules per cell.[3] In addition, the use of ^{35}S- instead of ^{3}H-labeled nucleoside triphosphates for probe synthesis contributed to a significant reduction in autoradiographic exposure times. Progress toward optimizing the preparation of different tissues has improved histological detail as well as hybridization sensitivity (for example, see Reference 4).

In situ hybridization to RNAs in tissue sections has a wide range of applications in biomedical research, where it is increasingly being used to determine patterns of tissue- and cell-specific gene expression. The rapid progress currently being made in studying genetic control mechanisms of development in the fruit fly, *Drosophila*, for instance, has taken great advantage of this method to help elucidate molecular mechanisms underlying normal and abnormal patterns of embryonic development. Our own experience stems mainly from studies of the expression of putative mammalian developmental control genes in embryonic mouse tissue sections.

We will briefly discuss the most important experimental parameters and give a comprehensive protocol describing the procedures we use for hybridization to cellular RNA in embryonic mouse sections. Our intent is not to cover completely all the theoretical and practical aspects involved, but rather, to give a concrete idea of some of the experimental options available to the investigator.

Although it may seem trivial, we would like to acknowledge that the success and efficiency of nucleic acid hybridization histochemistry methods are dependent on an optimal interplay of multiple experimental parameters. One significant variable is the tissue or cell type being tested for gene expression. Each tissue's specific biochemical composition may either favorably or adversely affect results obtained with a specific experimental protocol. The described procedures have been optimized and streamlined in order to obtain satisfactory hybridization signals with sections from early postimplantation and midgestation mouse embryos. Their application to tissues from other sources may require substantial modifications of individual experimental conditions.

II. TECHNICAL PROCEDURES

A. PRETREATMENT OF MICROSCOPE SLIDES AND SECTIONING

To ensure tissue adherence to the microscope slides during the steps of pretreatment, hybridization, and washing, the slides must be thoroughly cleaned and coated (subbed) with an adherent substance. We begin by washing the slides with detergent in water. We then treat them with chromic acid or 0.1 *M* hydrochloric acid (HCl), rinse extensively with distilled water, dip the slides into the subbing solution, and, finally, allow the slides to air dry.

We have had good results using gelatin/chrom alum (0.3% gelatin, 0.01% chromium potassium sulfate; modified from Reference 5), or poly-L-lysine (50 μg/ml) as the subbing solution. Poly-L-lysine-coated slides provide better tissue adherence, but they are associated with increased nonspecific binding of probe to the slides. This increased background can be prevented by acetic anhydride treatment,[6] as described below. Coated slides can be kept in a desiccated box at room temperature or at −20°C (poly-L-lysine-coated slides) for several weeks.

B. EMBEDDING AND SECTIONING OF TISSUES

Prior to sectioning, the specimen must be embedded in a supportive matrix. Two commonly used embedding materials that are compatible with hybridization histochemistry are paraffin (Paraplast Plus, American Scientific) and, for frozen sections, OCT compound 4583 (Miles, Elkhart, IN).

1. Paraffin Sections

The substantially more time-consuming process of paraffin embedding and sectioning is compensated by a superior morphology over frozen sections. Prior to embedding tissue in liquified paraffin at 58 to 60°C, the tissue is fixed, dehydrated through an increasing ethanol series, and cleared with xylene. There exist numerous and widely divergent protocols for tissue fixation. Good results with tissues from various sources have been obtained with paraformaldehyde as a cross-linking fixative, which we use as a 4% solution in phosphate-buffered saline (PBS) at 0 to 4°C. Optimal fixation times for different types and sizes of tissue specimens must be determined empirically, with fixation times ranging from 10 min to several days. The fixation step always involved a trade-off between the goals of optimal morphology and retention of mRNA and the need for optimal hybridization efficiency of the retained mRNA. Extended fixation may improve morphology, but decrease the hybridization efficiency to the RNA retained in the tissue, perhaps by direct modification of the RNA molecules or by decreasing tissue permeability to the probe. For mouse embryos of 8 to 9 d of gestation, we have successfully used 10 to 60 min as fixation time. Paraffin-embedded tissues can be stored at 4°C or room temperature for months without a noticeable decrease in hybridization signals.

Paraffin-embedded tissues are sectioned with a rotary microtome at a thickness of 4 to 10 μm, and the sections are floated off onto a microscope slide using degassed warm water (50°C). The sections are then adhered to the slides by air drying on a hot plate at 45 to 50°C for several hours to overnight.

2. Frozen Sections

The embedding of tissue in OCT requires no fixation or dehydration steps and is therefore considerably simpler and faster than the paraffin-embedding method. Usually, fresh dissected tissue is immersed into an aluminum or plastic mold filled with enough liquid OCT to cover the entire tissue, and then the tissue is frozen on dry ice or in liquid nitrogen. The entire procedure can be performed within a few minutes. OCT-embedded tissues can be stored at −70°C for several weeks without dramatic loss of hybridization efficiency. OCT-embedded tissues are sectioned frozen in a cryostat at −20°C and then fixed as described in Section III. According to our own experience, the hybridization signals obtained with frozen sections are generally stronger than those with paraffin sections. The lower quality in tissue morphology obtained with OCT- vs. paraffin-embedded material may be partially overcome by prefixing the tissue prior to embedding;[7] however, this may also reduce hybridization signals.

C. PRETREATMENT OF SECTIONS

The main reasons for performing the pretreatment procedures on fixed-tissue sections just prior to the hybridization reaction are (1) improving the accessibility of the target RNA to the probe; (2) maximizing retention of cellular RNA; and (3) preventing nonspecific binding of the probe to the tissue or slide. We will summarize several of these treatments, but we stress that their need varies with the tissue and its mode of preparation and should be determined empirically by the individual investigator. In general, we find that when no fixation is performed prior to sectioning, frozen sections often do not require any of the pretreatments used to increase cellular RNA accessibility. However, in our hands, sections of paraffin-embedded tissues generally require at least treatment with a proteolytic enzyme to increase hybridization efficiency.

In order to make cellular RNA more accessible for the hybridization reaction with the probe, several procedures have been used to permeabilize the tissue and to remove some of the proteins associated with the RNA. Basic proteins may be removed from RNA by treating the rehydrated sections with hydrochloric acid (typically 0.2 *M* HCl, 20 min), followed by heat treatment at 70°C in 2X SSC (0.3 *M* NaCl and 0.03 *M* sodium citrate pH 7). Alternatively, proteins may be removed by a limited digestion with Proteinase K or pronase.[4,8,9] Optimal conditions for proteolysis should be determined empirically by varying the enzyme concentration because of variability among different batches of enzyme in their activity and among different tissues in their sensitivity. The Proteinase K incubation can be combined with the preceding heat and HCl treatments. These procedures are followed by postfixation of the tissue in 4% paraformaldehyde in phosphate-buffered saline (PBS) for maximum retention of cellular RNA, which can be monitored by acridine orange staining.[4]

Nonspecific binding of the probe to basic amino acids can be prevented by acetylation of basic groups in the tissue and in the gelatin or poly-L-lysine coating of the slide. Acetylation is performed by incubating the slides in a solution of 100 m*M* triethanolamine containing 25 m*M* acetic anhydride for 10 min at room temperature.[6] The slides are then rinsed in PBS, dehydrated, and air dried. We find that this treatment is essential for poly-L-lysine-coated slides, but we are generally able to eliminate it when we use gelatin-coated slides.

D. PREPARATION OF PROBES

1. Type of Probe: DNA vs. RNA

The significant progress that has been made in recent years in the preparation of nucleic acid probes has considerably increased the sensitivity of signal detection. Highly sensitive probes should be asymmetric, representing only the anti-sense strand of the gene, and labeled to high specific activities. Such probes have been produced for *in situ* hybridization using single-stranded DNA probes prepared from an M13 derivative,[10] single-stranded RNA probes prepared from an *in vitro* transcription system,[3] or synthetic oligonucleotide probes.[11]

Most of our own experience stems from experiments that utilized asymmetric RNA probes. Several vector systems have been developed for the bidirectional synthesis of RNAs specific for DNA fragments inserted into the vector. We have successfully used the Gemini and Bluescript vector systems, both of which are commercially available through either Promega (Madison, WI) (Riboprobe® Gemini vectors) or Stratagene (La Jolla, CA) (pBLUESCRIPT®). In both vectors, a multiple cloning site containing many restriction enzyme recognition sequences in tandem is flanked by two different promoter sequences, each specific for a bacteriophage RNA-polymerase: T7 and SP6 RNA-polymerase promoters in Gemini and T7 and T3 promoters in Bluescript. By choosing the proper RNA polymerase based upon the orientation of the cloned fragment within the vector and linearizing the plasmid within the multiple cloning site at the end of the insert DNA opposite to the selected promoter, these systems allow selective synthesis of virtually pure (+)-strand (sense) or (−)-strand (anti-sense) RNA in an *in vitro* transcription system. These probes can be radioactively labeled to very high specific activities (10^8 to 10^9 cpm/μg RNA) by incorporating ^{35}S- or ^{32}P-substituted ribonucleoside triphosphates during synthesis.

Asymmetric RNA probes offer several advantages:

1. The asymmetry increases the hybridization efficiency of anti-sense RNA to the target transcripts in tissue sections, due to the absence of self-reannealing of the RNA probe in the hybridization solution.

2. The transcription systems allow the convenient synthesis of both the anti-sense RNA as a highly specific probe as well as the sense RNA as a negative control probe to determine the level of nonspecific binding to the tissue or slide.
3. Very high specific activities of ^{35}S- or ^{32}P-labeled probes may be achieved.
4. Following the hybridization reaction, nonspecifically bound probe can be efficiently removed by RNAse A treatment, greatly improving signal to noise ratios.

A detailed protocol for the preparation of RNA probes is given in Section III.

Alternatives to asymmetric RNA probes are double-stranded (ds) or single-stranded (ss) DNA probes. Double-stranded DNA probes can be prepared by nick translation[12,13] or by random primed labeling of ds DNA.[14,15] The nick translation method is based on the introduction of ss nicks into ds DNA by DNAse I and the subsequent partial replacement of the nicked strand by *Escherichia coli* DNA polymerase I with labeled deoxyribonucleoside triphosphates (for a detailed protocol, see Reference 16). The more recently introduced method of random primed labeling employs a mixture of random hexanucleotides as primers for *in vitro* DNA synthesis on denatured linear DNA templates using the large (Klenow) fragment of *E. coli* DNA polymerase I. These double-stranded probes suffer the disadvantage of decreased hybridization efficiency due to reannealing of the sense and anti-sense strands, but under some conditions might improve signals through the formation of complex networks of reannealled probe and mRNA.[17]

Single-stranded DNA probes of high specific activities can be produced *in vitro* on recombinant ss bacteriophage M13 template DNAs. DNA synthesis may be primed by standard sequencing primers using the desired labeled deoxyribonucleoside triphosphates. At the end of the reaction, the vector is cut downstream of the insert with an appropriate restriction endonuclease, and the probe is purified from the M13 template and the heterogeneous population of runoff transcripts by gel electrophoresis.[10]

Asymmetric DNA probes can also be prepared using 5′ or 3′ end-labeled synthetic oligonucleotides. Recent technical progress made in the production of synthetic oligonucleotides makes their use as probes for hybridization histochemistry increasingly economical and attractive. Oligonucleotide synthesis allows the convenient preparation of transcript-specific probes, which may be valuable for the analysis of genetic loci encoding several different mRNAs (for protocols, see References 18 and 19).

2. Labeling of Probe: Radioactive vs. Nonradioactive

The development of hybridization histochemistry depended essentially on the use of radioactively labeled probes and autoradiographic signal detection. However, nucleic acid probes conjugated with biotin molecules have also been used successfully for detecting specific transcripts in cells by immunohistochemical methods.[20,21] These methods involve the incorporation of biotin-substituted nucleotides such as biotinyl-dUTP into nucleic acid probes by enzymatic reactions identical to those described above for radioactive probes. Hybridization signals are visualized by using chromophores or enzymes that have been conjugated to biotin-specific antibodies or avidin. These probes have several advantages over radioactive probes, including increased safety, rapid signal detection, and increased resolution. However, they have not displaced the most commonly used radioactively labeled probes, even though they were introduced for hybridization histochemistry 8 years ago. Among the reasons for their lack of success are the relatively low sensitivities[16] and the difficulties in quantitating hybridization signals. Solving these problems will make nonradioactive probes significantly more attractive in the future, and some investigators are already using them with great success (for an example, see Reference 22).

The most commonly used isotopes for the preparation of radioactive probes are tritium (^{3}H), ^{35}S, and ^{32}P, which are all β-particle emitters and are used in the form of labeled ribo-

or deoxyribonucleoside triphosphates. The particle energies emitted by decaying ^3H, ^{35}S, and ^{32}P increase roughly by a factor of ten for each isotope in the following order: 0.018 MeV, ^3H; 0.167 MeV, ^{35}S; and 1.70 MeV, ^{32}P. Because of their low energies, β particles emitted by ^3H penetrate less than 1 μm into the photoemulsion, which allows precise localization of autoradiographic signals on subcellular levels. The lower cytological resolution of autoradiographic hybridization signals obtained with isotopes emitting higher-energy β particles is compensated for by significantly shorter autoradiographic exposure times due to increased autoradiographic efficiencies. ^{35}S-labeled probes offer a good compromise in this respect, combining adequate signal resolution with relatively short exposure times of a few days to weeks.

The limited resolution of ^{32}P-labeled probes is still sufficient for signal resolution on the single-cell level.[23] Because of the short time required for autoradiographic exposure, ^{32}P-labeled probes played a significant role in the more rapid optimization of experimental conditions for hybridization histochemistry some 5 to 10 years ago (for a review, see Reference 23). This included, in particular, the development of methods for *in situ* hybridization to whole-mouse sections.[24,25] Tissue- or organ-specific expression of the gene tested may be detected after overnight exposure to X-ray film. However, the more recently introduced ^{35}S-labeled probes offer the same advantages as ^{32}P in this respect, and also provide higher resolution and a longer half-life.

E. HYBRIDIZATION AND WASHING

The kinetics of hybrid formation for nucleic acids have been discussed in detail by Britten and Davidson.[26] The most important parameters determining the rate of this reaction are the temperature and composition of the reaction medium, as well as the concentration, size, and G + C content of the probe. Optimal conditions for *in situ* hybridization to RNA in tissue sections have been discussed by Angerer and co-workers,[3,27] and in tissue culture cells by Lawrence and Singer.[17]

In aqueous solutions, the thermostabilities of RNA/RNA duplexes are about 12°C greater than those of DNA/DNA duplexes.[28] Based on DNA/DNA reassociation kinetics, the optimal hybridization temperatures are about 25°C below the melting temperature (T_m) of the duplexes.[29] Since the T_m values of RNA/RNA duplexes under approximately physiological conditions (0.12 *M* sodium phosphate, pH 6.8) lie in a temperature range of ≥95°C,[3] the optimal temperature required for duplex formation would be about 70°C in the same medium. In order to avoid this relatively high temperature range, which promotes the chemical degradation of nucleic acids, hybridization reactions are commonly performed in the presence of formamide, which destabilizes hydrogen bonds and decreases T_m values. The reduction in T_m for RNA/RNA duplexes has been reported to be 0.35°C per percent of formamide in solution,[3] compared to a reduction of 0.72°C per percentage of formamide for DNA/DNA duplexes.[30] Accordingly, in standard hybridization buffer containing 50% formamide and STE (0.3 *M* NaCl, 20 m*M* Tris-HCl, pH 8.0, and 5 m*M* EDTA), the T_m of RNA/RNA duplexes is 81°C, while the T_m for duplexes formed *in situ* is further reduced to 76°C, probably due to steric hindrance of the immobilized cellular RNA. Therefore, the optimal temperature range for RNA/RNA *in situ* hybridization reactions with RNAs of typical G + C content is about 50°C in this medium. These conditions are considered to be of medium to high stringency, allowing only minimal base pair mismatches.

Optimal probe concentrations for *in situ* hybridization reactions to cellular RNAs have been determined by Cox et al.[3] and Angerer et al.[27] Accordingly, maximum sensitivities (signal-to-noise ratios) were achieved with the minimum saturating probe concentrations. Higher probe concentrations increase the level of "background" binding without a further increase of signal. For moderately abundant RNAs (about 20 to 75 copies per cell), the optimal concentration is about 0.2 to 0.3 ng/μl per kilobase of probe complexity, using asymmetric RNA probes. Under these conditions, the reaction is terminated within 16 h.

While reactions with asymmetric DNA probes would be expected to follow similar kinetics, the different reaction kinetics obtained with symmetric (ds) DNA probes result in about tenfold less hybridization efficiency.[3] This decrease is essentially due to the competition from self-reannealing of the probe.

Highest hybridization efficiencies are obtained with relatively short probes of 50 to 150 bp in length.[4,8,10] This is probably due to increased probe penetration into the tissue achieved with shorter probes. The length of asymmetric RNA probes can be adjusted by a controlled alkaline hydrolysis reaction,[3] as described in Section III.

The high thermal stabilities of RNA/RNA duplexes formed during hybridization allow for high-stringency washing conditions, which contribute to increased signal specificities. Prior to the final washes, the sections are incubated with RNAse A for the removal of single-stranded, nonspecifically bound probe. The subsequent final washes in 0.1X SSC remove probe that hybridized to heterologous RNAs due to weak homologies to the probe. These washes can be done at temperatures as high as 60 to 65°C without significant loss of signal.

F. SIGNAL DETECTION

Hybridization signals are detected either by autoradiography, when using radioactively labeled probes, or by enzymatically controlled color reactions, when using nonradioactive probes (see Section II.2). We routinely use ^{35}S-labeled RNA probes, which allow rapid monitoring of region- or organ-specific gene expression in whole-mouse sections by autoradiography to X-ray film.[31,32] This method is simple and allows the empirical determination of optimal exposure conditions for each slide by repeated autoradiographies. For higher signal resolution, the slides are then coated with photoemulsion, as outlined in Section III, and developed after appropriate exposure times. Subsequent to autoradiography, the slides may be stained with various histological dyes for optimal morphological and cytological detail. The most commonly used staining procedures include Giemsa, Hematoxylin, Eosin, or Toluidine Blue, which are all compatible the light-microscopical visualization of silver grains. We routinely stain our sections with Giemsa, as described in Section III.

G. ANALYSIS OF DATA

Of equal importance to the technical procedures needed to perform hybridization histochemistry is the critical analysis of the results obtained. Spurious positive and negative results are a common problem, and therefore all hybridization histochemistry experiments should include both positive and negative controls. Useful positive controls include genes for which the expression pattern is already known. For example, a probe derived from the α-fetoprotein gene, which is highly expressed in the developing yolk sac and liver, is a useful control for hybridization histochemistry of postimplantation mouse embryos. To test the authenticity of hybridization signals, it can also be helpful to compare results obtained with several nonoverlapping probes derived from the same gene, although different patterns can be obtained if different transcripts from one gene have different expression patterns. An additional test of the specificity of hybridization signals can be performed by using adjacent serial sections to compare the expression of two genes with highly conserved sequences but with different expression patterns, attempting to demonstrate that the technique used can distinguish their expression domains.

Important negative controls include sequences that are not expressed in the tissue of interest. Prokaryotic vector sequences are convenient negative controls. As mentioned above, when using asymmetric RNA probes, the sense strand provides another easily produced negative control, provided both strands of the gene are not transcribed. When using liquid photoemulsion, one should also control for silver grains produced by potential chemical reactions between the tissue and the emulsion by including some sections that were not treated with the probe.

Other difficulties in interpreting the results of hybridization histochemistry can arise from the laborious process of analyzing many serial sections by microscopy. A preliminary exposure of the sections to X-ray film provides a useful guide for subsequent microscopy when radioactive probes are used. In addition, the use of dark-field illumination, in which the silver grains appear bright on a dark background, is indispensable when attempting to scan rapidly a large number of sections for autoradiographic signals. Image analysis software is available (e.g., Jandel Scientific, Corte Madera, CA, or the SSRCON serial section reconstruction system described in Reference 33) to help superimpose autoradiographs onto pictures of the sections or to produce three-dimensional reconstructions of the specimen from serial sections (e.g., References 34 to 36). Quantitation of autoradiographic signals may yield important information[37] and can be done by silver-grain counting or densitometrically as, for instance, on a Zeiss IBAS 2000 image analysis system with densitometry software.

Typical results obtained by following our standard methods are illustrated in Figure 1. A frozen sagittal section of a mouse embryo of 12.5 d gestation (Figure 1A) was hybridized with a ^{35}S-labeled anti-sense RNA probe specific for the Hox-3.1 gene[31,38] and exposed to X-ray film (Figure 1B). After overnight exposure, specific hybridization signals can be detected within the cervical and thoracic spinal cord (arrow) and certain thoracic prevertebrae (arrowheads). To obtain higher resolution, the same slide was coated with photoemulsion and exposed for 4 d. By comparing brightfield (D) and darkfield (E) photomicrographs of the anterior thoracic spinal cord, one can easily detect an asymmetric distribution of silver grains along the dorsal-ventral axis of the spinal cord, with the highest concentration of silver grains located ventrally. For the exact localization of hybridization signals, the X-ray film autoradiograph was superimposed over the traced section using a Zeiss IBAS 2000 image analysis system (C).

III. STANDARD PROTOCOL

This section provides our standard protocols for *in situ* hybridization to cellular RNAs within embryonic mouse sections using asymmetric RNA probes.

A. PREPARATION OF MICROSCOPE SLIDES

Poly-L-Lysine Coating:

1. Place the slides in racks and wash in warm detergent for 30 min.
2. Rinse under running tap water for 30 min.
3. Immerse the slides in 0.2 *M* HCl for 30 min.
4. Rinse twice with deionized water for 30 min each.
5. Rinse three times with double distilled water for 30 min each; air dry.
6. Bake the slides at 150°C for 1 h and allow them to cool to room temperature.
7. Soak the slides in a solution of 50 μg/ml poly-L-lysine, 0.01 *M* Tris-HCl, pH 8, for 30 min; air dry in an upright position. Use the slides the following day or keep them for several weeks in a desiccated box at 4°C. The coating solution is prepared by diluting a 10-mg/ml poly-L-lysine (mol wt: 70 to 150 kDa; Sigma, catalog number P1274) stock solution made up in diethyl pyrocarbonate (DEPC)-treated water in sterile 10 m*M* Tris-HCl, pH 8. The stock solution is kept at −20°C.

Gelatin Coating: After step 5 above, dip the slides into a solution of 0.3% gelatin and 0.01% chromium potassium sulfate (modified from Reference 5); air dry in an upright position. Gelatin-coated slides may be stored for several weeks at room temperature.

B. TISSUE EMBEDDING AND SECTIONING

1. Paraffin Sections

1. Freshly dissected embryos are fixed in 4% paraformaldehyde (Polysciences, Warrington, PA, catalog number 0380) in PBS (0.8% NaCl, 0.02% KCl, 0.02% KH_2PO_4, and 0.115% Na_2HPO_4, pH 7.4) at 4°C for 10 min to 2 h (8 to 10 d gestation) or overnight (11 to 13 d of gestation).
2. Rinse three times in PBS at 4°C for 20 min to 1 h each.
3. Dehydrate the embryos at room temperature in an increasing ethanol series and clear in xylene at room temperature:

   - 50% ethanol, 70% ethanol, 80% ethanol, 90% ethanol, 95% ethanol (10 to 30 min each)
   - 100% ethanol (three times, 10 to 30 min each)
   - 100% ethanol: 100% xylene, 1:1 (v/v) (10 to 30 min)
   - 100% xylene (twice, 10 to 30 min each)

4. Embed the embryos in paraffin (Paraplast Plus, American Scientific Products) by continuing with the following steps:

   - 100% xylene at 58 to 60°C (10 to 30 min)
   - xylene:paraffin, 1:1 (v/v), at 58 to 60°C (3h)
   - paraffin at 58 to 60°C (3 h)
   - paraffin at 58 to 60°C (overnight)

   Transfer into plastic embedding molds with fresh paraffin at 58 to 60°C and let cool to room temperature (can be stored at 4°C or at room temperature for at least several months).
5. Mount the paraffin block onto a holder and trim the sides of the block with a razor blade, leaving a trapezoidal sectioning plane with the leading edge wider than the trailing edge.
6. Cut sections of 4- to 10-μm thickness on a rotary microtome and float individual sections or ribbons onto a layer of degassed water on poly-L-lysine-coated microscope slides at 50°C.
7. Air dry the sections overnight at 45 to 50°C on a hot plate.
8. Sections can be stored in a desiccated box at 4°C for up to several weeks before proceeding with the pretreatments.

2. Frozen Sections

1. Immerse freshly dissected embryo into an embedding mold filled with liquid OCT compound 4583 (Miles, Elkhart, IN).
2. Quick freeze on dry ice or in liquid nitrogen and store at −70°C for up to several weeks.
3. Mount the frozen OCT block by pressing the bottom onto a holder covered with a layer of liquid OCT at room temperature. This will partially melt the bottom of the block. Quickly freeze by placing the holder on dry ice.
4. Equilibrate the mounted tissue block at −20°C in a cryostat microtome for 30 min.
5. Cut sections of 4- to 20-μm thickness. If necessary, straighten the sections individually using a fine paint brush. Pick up sections individually onto poly-L-lysine or gelatin-coated slides kept at room temperature.

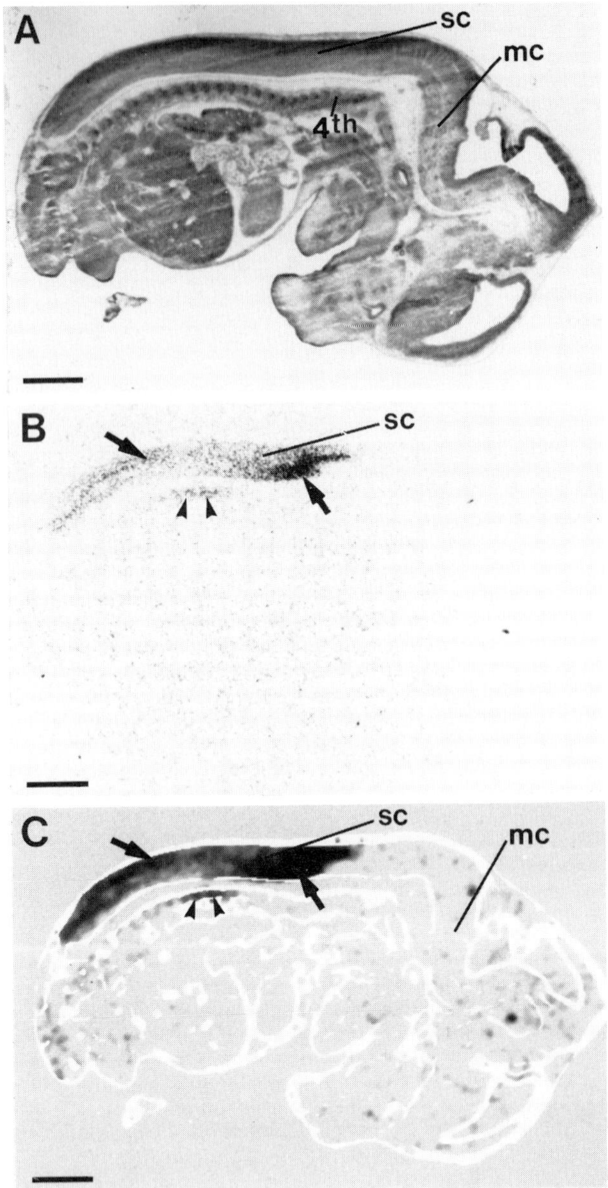

FIGURE 1. Hybridization histochemistry of the mouse Hox-3.1 gene. Sagittal sections (8 μm) of a mouse embryo of 12.5 d of gestation were hybridized with a ^{35}S-labeled Hox-3.1-specific antisense RNA probe (specific activity: 4.5 × 10^8 cpm/μg), exposed to X-ray film, exposed to photo-emulsion, and stained with Giemsa, as described in Section III for frozen sections. The template for the synthesis of the RNA probe was a Hox 3.1-specific cDNA subclone, c210-B,[41] which contains approximately 250 bp of Hox-3.1 untranslated leader sequence and about 450 bp of protein coding sequence, but lacks the highly conserved homeo box sequence.[31,42,43] (A) Micrograph of a section after hybridization, autoradiography, and staining. The section is sagittal, with anterior to the right and dorsal on top; (B) X-ray film autoradiograph obtained after overnight exposure of the section to X-ray film (Kodak X-OMAT/AR). Specific hybridization signals are observed within the spinal cord posterior to the level of the fourth prevertebra (arrows) and in certain thoracic prevertebrae (arrowheads). (C) Superimposition of the X-ray film autoradiograph over the traced section (light lines) by using a Zeiss IBAS 2000 image analysis system for the localization of the hybridization signal in the tissue. (D) higher magnification (approximately 60 ×) photomicrograph of the anterior spinal cord region using brightfield illumination; (E) darkfield photomicrograph of the region in D showing the silver grains obtained from autoradiography using Kodak NTB-2 photoemulsion; Exposure time: 4 d; sc, spinal cord; mc, myelencephalon; 4th, fourth prevertebra; scale bars, 1 mm (A to C) or 0.1 mm (D and E).

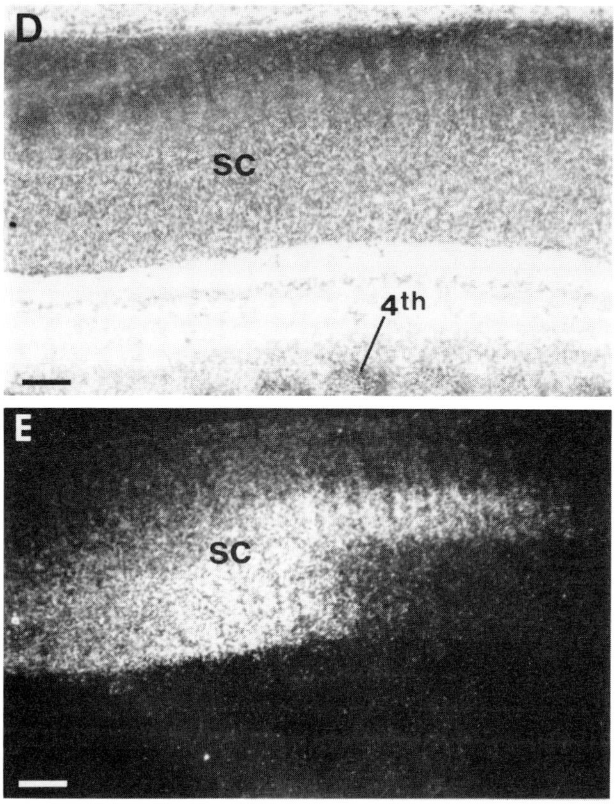

FIGURE 1 (continued).

6. Place the slides onto a hot plate at 45 to 50°C for 1 to 2 min.
7. Air dry at room temperature for 2 h.
8. Fix the sections in 4% paraformaldehyde in PBS at room temperature for 20 min.
9. Rinse in 3X PBS for 5 min at room temperature.
10. Rinse in 1X PBS (twice, 5 min each, at room temperature). If, as is preferable, the slides are to be used for *in situ* hybridization the same day, proceed directly to the pretreatment protocol by omitting steps 11 and 12.
11. Dehydrate the sections at room temperature in an increasing ethanol series: 30% ethanol, 60% ethanol, 80% ethanol, 95% ethanol, and 100% ethanol (2 min each).
12. Air dry. At this point, the slides can be stored at −20°C in a desiccated box for several days.

C. PRETREATMENT OF SECTIONS

Paraffin sections are dewaxed in xylene and rehydrated in a decreasing ethanol series at room temperature prior to the pretreatment steps: 100% xylene (20 min); 100% ethanol (10 min); 95% ethanol (10 min); 70% ethanol (2 min); 60% ethanol (2 min); 30% ethanol (2 min). Subsequently, paraffin and frozen sections are treated in the same way; all steps are carried out at room temperature.

1. Incubate the slides in 100 μg/ml Pronase E (Sigma type XIV protease, catalog number P5147), 0.05 *M* Tris-HCl, pH 7.5, and 0.005 *M* EDTA [10 min]). The working concentration must be determined for each batch of Pronase E. A stock solution of Pronase E is prepared according to Reference 4: Dissolve the pronase at a concentration

of 40 mg/ml in water and predigest at 37°C for 4 h to destroy RNAses; lyophilize 40-mg aliquots and store at −20°C.

2. Transfer the slides to a solution of 2 mg/ml glycine (0.5 min).
3. Rinse the sections in PBS (twice, 0.5 min each).
4. Fix the sections in 4% paraformaldehyde in PBS for 10 min.
5. Rinse in PBS (three times, 5 min each).
6. Acetylate the slides in 0.25% acetic anhydride in 0.1 *M* triethanolamine, pH 8, for 10 min. Make up 200 ml of solution by diluting 2.7 ml of concentrated triethanolamine (Sigma, catalog number T-1377) in 197.3 ml of DEPC-treated water and adjust the pH to 8 with approximately 0.3 ml of concentrated HCl. Add 0.5 ml of concentrated acetic anhydride (Sigma, catalog number A-6404) immediately before use and mix by rapidly moving the slide rack up and down.
7. Dehydrate the sections in an increasing ethanol series: 30% ethanol, 60% ethanol, 80% ethanol, and 95% ethanol (2 min each).

D. PREPARATION OF ASYMMETRIC RNA PROBES

1. Prepare the template DNA by cutting the recombinant plasmid (e.g., Gemini from Promega or Bluescript from Stratagene) with an appropriate restriction endonuclease such that the DNA insert lies between the restriction enzyme recognition site and the promoter region to be used for the *in vitro* transcription reaction. Check the digest for completion by running an aliquot on an agarose gel. Extract the linearized plasmid DNA with phenol/chloroform according to standard procedures[39] before ethanol precipitation. Wash the pelleted DNA twice with 70% ethanol, and dissolve the dried pellet in a small volume of water to obtain a concentration of 0.5 to 1 μg of DNA per microliter.
2. At room temperature, set up a 40-μl *in vitro* transcription reaction by using components of the Riboprobe® system (Promega) as follows:

   - 8.0 μl of 5X transcription buffer (0.2 *M* Tris-HCl, pH 7.5, 0.03 *M* $MgCl_2$, 0.01 *M* spermidine, and 0.05 *M* NaCl)
   - 4.0 μl of 0.1 *M* dithiothreitol (DTT)
   - 1.6 μl of RNasin® ribonuclease inhibitor (25 U/μl stock solution)
   - 8.0 μl of ribonucleoside triphosphate mix: 2.5 m*M* each of ATP, CTP, and GTP
   - 0.5 μl of 1 m*M* UTP (1:10 dilution of 10 m*M* stock solution)
   - 2.0 μl of linearized template DNA (0.5 to 1.0 μg/μl)
   - 15.0 μl of ^{35}S-UTP (1 to 1.25 Ci/mmol, 12.5 μCi/μl)
   - 25 units of appropriate RNA polymerase (either T3, T7, or SP6 RNA-polymerase)
   - Add DEPC-treated water to a final volume of 40 μl

3. Incubate at 37°C for 1 to 2 h.
4. Add 1 U of RQ1 DNAse (Promega) per microgram of DNA to destroy the template DNA; incubate at 37°C for 15 min.
5. Bring the volume to 100 μl with TE buffer (10 m*M* Tris-HCl and 1 m*M* EDTA, pH 7.4) and remove 1 μl to determine the total radioactivity.
6. Prepare a spin column in a 1.0-ml syringe with Sephadex® G50 fine (e.g., Boehringer-Mannheim, catalog number 100411) and equilibrate with TE buffer as described in Reference 39. Remove the unincorporated ribonucleoside triphosphates by passing the remaining reaction mix over the spin column in a swinging bucket rotor for 4 min at 1060 to 8800 × g.
7. Remove 1 μl from the RNA probe solution to determine the incorporated radioactivity. Determine the ratio of incorporated to unincorporated ^{35}S-UTP and calculate the ap-

proximate amount of RNA synthesized by assuming the same ratio for the cold UTP used. We routinely obtain 70 to 90% incorporation of labeled UTP.

8. Adjust the probe length by limited alkaline hydrolysis, according to Reference 3, using a solution of 40 m*M* $NaHCO_3$ and 60 m*M* Na_2CO_3, pH 10.2, as follows: 100 μl of RNA probe solution and 100 μl of 0.2 *M* carbonate buffer (80 m*M* $NaHCO_3$ and 120 m*M* Na_2CO_3, pH 10.2; make up fresh or store frozen in small aliquots). Incubate at 60°C for the required time according to the following formula: $t = (L_o - L_f)/k\, L_oL_f$, where t = incubation time in min, L_o = initial transcript length in kilobases, L_f = desired probe length in kilobases (0.1 kb), and k = reaction rate for hydrolysis, which is approximately 0.11 kilobases per min. Neutralize the sample by adding 6 μl of 3 *M* sodium acetate, pH 6, and 10 μl of 10% (v/v) glacial acetic acid.
9. Add 50 μg of *E. coli* tRNA (Sigma, catalog number R-1753) and precipitate with 2.5 volumes of absolute ethanol at −20°C for a minimum of 1 h. After pelleting by centrifugation, wash the precipitate twice with 70% ethanol and dry the pellet.
10. Dissolve the RNA probe in TE buffer, 10 m*M* DTT at a concentration of 2 to 3 ng/μl per kilobase of probe complexity, and store at −20°C. This is a 10X probe stock solution.

E. HYBRIDIZATION

The composition of hybridization solution is 0.3 *M* NaCl, 0.01 *M* Tris-HCl, pH 7.5, 0.001 *M* EDTA, 0.05 *M* DTT, 1X Denhardt's solution,[40] 0.5 mg/ml tRNA, 50% formamide, 10% dextran sulfate (optional), and 0.2 to 0.3 ng of probe per μl per kilobase of complexity.

1. Prepare 10 ml of 10X salt solution by combining the following components: 5 *M* NaCl (6 ml), 1 *M* Tris-HCl, pH 7.5 (1 ml), 0.5 *M* EDTA (0.2 ml), bovine serum albumin (Sigma, BSA fraction V) (20 mg), polyvinylpyrrolidone (20 mg), Ficoll (mol wt = 400,000) (20 mg). Store at −20°C.
2. Preparation of 1 ml of hybridization mix:

   - 100 μl of 10X salt solution
   - 500 μl of formamide (deionized)
   - 50 μl of *E. coli* tRNA (10 mg/ml)
   - 50 μl of 1 *M* Dithiothreitol (DTT)
   - 200 μl of 50% dextran sulfate (Pharmacia) (optional)
   - 100 μl 10X probe stock solution (2 to 3 ng/μl); optional: boil aliquoted probe stock solution for 3 min and cool on ice before adding
   - For hybridizations without dextran sulfate, add water to 1 ml

3. Set up the hybridization by pipetting 50 μl of probe onto the microscope slides. Spread the probe over the sections with a 22 × 40 mm cover slip. (Cover slips are cleaned with 0.1 *M* HCl, rinsed twice in DEPC-treated water, and air dried before use). After sealing the edges of the cover slips with rubber cement, place the slides in a closed chamber on filter paper saturated with 1X salt and 50% formamide, and incubate at 50°C for 5 to 16 h (overnight for convenience).

F. WASHING

1. Carefully peel off the rubber cement and float off the cover slips by immersing the slides in 2X SSC, 50% formamide, and 1% β-mercaptoethanol at 50°C for up to 15 min.
2. Wash the slides in fresh 2X SSC, 50% formamide, and 1% β-mercaptoethanol at 50°C for 30 min.

3. Equilibrate in RNAse A buffer containing 0.5 *M* NaCl and 10 m*M* Tris-HCl, pH 8, at 37°C for 10 min.
4. Remove the unhybridized probe by incubating the slides in 20 μg/ml RNAse A in 0.5 *M* NaCl and 10 m*M* Tris-HCl, pH 8, at 37°C for 30 min to up to several hours. RNAse A stock solution: dissolve RNAse A (Sigma type XII-A, catalog number R-5500) at a concentration of 10 mg/ml in 10 m*M* Tris-HCl, pH 7.5, and 15 m*M* NaCl, and boil for 15 min. Allow to cool slowly to room temperature and store aliquoted at −20°C.
5. Proceed with the following washing schedule:

   - 0.5 M NaCl and 10 m*M* Tris-HCl, pH 8, at 37°C for 10 min
   - 2X SSC, 50% formamide, and 0.1% β-mercaptoethanol at 55°C for 30 min
   - 2X SSC and 0.1% β-mercaptoethanol in 2 l of volume at room temperature for 15 min
   - 0.1X SSC and 0.1% β-mercaptoethanol in 2 l of volume at room temperature for 15 min
   - 0.1X SSC and 0.1% β-mercaptoethanol in 2 l of volume at 60 to 65°C for 15 min
   - 0.1X SSC and 0.1% β-mercaptoethanol in 2 l of volume at room temperature for 30 min

6. Dehydrate the sections by immersing the slides in an increasing ethanol series containing 300 m*M* ammonium acetate: 30% ethanol, 50% ethanol, 70% ethanol, and 95% ethanol (10 s each); air dry.

G. AUTORADIOGRAPHY

1. Exposure to X-Ray Film

For a quick determination of region- and organ-specific hybridization signals in whole-mouse sections, slides are exposed to fast X-ray film (e.g., Kodak X-OMAT/AR) for 1 to 5 d. Slides may be conveniently re-exposed several times to obtain autoradiographic signals of optimal strength. In order to obtain a higher resolution, slides may subsequently be coated with photoemulsion.

2. Exposure to Photoemulsion

All manipulations involving photoemulsion are carried out in an absolutely light-tight darkroom equipped with a proper safelight, 15-W bulb with a Wratten red filter 1.

Preparation of Photoemulsion:

1. Melt Kodak NTB-2 photoemulsion in a 45°C water bath and dilute with an equal volume of 600 m*M* ammonium acetate and 0.2% glycerol, prewarmed to 45°C for about 45 min. Mix by gentle stirring with a Pasteur pipette or repeated inversion of the closed container. Do not shake.
2. Dispense 10-ml aliquots of the diluted emulsion into plastic scintillation vials.
3. Wrap individual vials with aluminum foil and store in a light-tight box at 4°C.

Coating of Slides with Photoemulsion:

1. Melt 10 ml of the diluted emulsion in a 45°C water bath for 30 to 45 min.
2. Pour the emulsion into a dipping chamber (e.g., Cyto® slide mailer, American Scientific) situated in the water bath and allow the emulsion to sit for about 10 min to let air bubbles rise.

3. Remove air bubbles from the surface by dipping one to two blank slides.
4. Dip the hybridized slides by immersing and withdrawing each slide slowly and smoothly; drain the excess emulsion by blotting the bottom edge of each slide on a filter paper for 1 to 2 s and allow the emulsion to air dry by placing the slides in a vertical position for 1 to 2 h. It is important to use a consistent dipping technique to obtain consistent autoradiographic results from different slides.
5. Transfer the slides to an exposure box containing silica gel and leave the box sealed with a light-tight adhesive tape overnight at room temperature.
6. Continue the exposure at 4°C the following day.

Developing Exposed Slides:

Develop the slides from one experiment at serial time points ranging from 2 d to several weeks, as follows:

1. To avoid condensation on the emulsion, allow the exposure box to equilibrate to room temperature for about 15 min before opening. Equilibrate all solutions at 15°C. Slight temperature differences between different solutions can cause the emulsion to crack.
2. Remove a pair of slides from the box and develop in Kodak D19 developer at 15°C for 2.5 min.
3. Stop in 1% acetic acid at 15°C for 15 to 30 s.
4. Fix in Kodak fixer at 15°C for 4 to 5 min.
5. Rinse in distilled water at 15°C for 15 min.
6. Rinse in cold, running tap water for 30 min.

At this stage, slides may be air dried for preliminary monitoring of hybridization signals under the microscope prior to histological staining. Alternatively, the slides may be stained immediately.

H. GIEMSA-STAINING AND MOUNTING

1. Prepare fresh staining solution by diluting concentrated Giemsa stock solution (Sigma Accustain™, catalog number GS-500) 1:20 in 5 m*M* Tris-HCl, pH 7.5.
2. Stain the slides for 15 to 30 min.
3. Rinse in water for 15 to 30 s.
4. Dehydrate by immersing the slides in an increasing ethanol series: 30% ethanol, 50% ethanol, 70% ethanol, 95% ethanol, and 100% ethanol (10 s each). Air dry vertically.
5. Place one to several drops of Permount on the slide and lower a clean cover slip at an angle onto the mounting medium by using a pair of fine forceps.
6. Harden the Permount by placing the slides on a slide heater at 40°C for several hours.

ACKNOWLEDGMENTS

We gratefully acknowledge the generous support of Dr. Frank Ruddle, in whose laboratory most of this work was carried out. We thank Drs. William McGinnis, Michael Levine, and Katherine Gordon for technical discussions. We thank Dr. Jim Karam for critically reading the manuscript and Donna Jacobs for excellent technical assistance. This work was supported in part by Grant GM 43334-01 awarded by the National Institute of General Medical Services, Department of Health and Human Services, to Alexander Awgulewitsch, as well as MUSC Institutional Research Funds and a corporate contribution by Westvaco to the Health Sciences Foundation. Manuel Utset is a recipient of a Life and Health Insurance Medical Research Fund Scholarship.

REFERENCES

1. **Gall, J. G. and Pardue, M. L.,** Formation and detection of RNA-DNA hybrid molecules in cytological preparations, *Proc. Natl. Acad. Sci. U.S.A.*, 63, 378, 1969.
2. **Melton, P., Krieg, P., Rebagliati, M. R., Maniatis, T., Zinn, K., and Green, M. R.,** Efficient *in vitro* synthesis of biologically active RNA and RNA hybridization probes from plasmids containing a bacteriophage SP6 promoter, *Nucleic Acids Res.*, 12, 7035, 1984.
3. **Cox, K. H., DeLeon, D. V., Angerer, L. M., and Angerer, R. C.,** Detection of mRNAs in sea urchin embryos by *in situ* hybridization using asymmetric RNA probes, *Dev. Biol.*, 101, 485, 1984.
4. **Hafen, E., Levine, M., Garber, R. L., and Gehring, W. J.,** An improved *in situ* hybridization method for the detection of cellular RNAs in *Drosophila* tissue sections and its application for localizing transcripts of the homeotic *Antennapedia* gene complex, *EMBO J.*, 2, 617, 1983.
5. **Gall, J. G. and Pardue, M. L.,** Nucleic acid hybridization in cytological preparations, *Methods Enzymol.*, 21, 470, 1971.
6. **Hayashi, W., Gillam, I. C., Delaney, A. D., and Tener, G. M. J.,** Acetylation of chromosome squashes of *Drosophila melanogaster* decreases the background in autoradiographs from hybridization with [^{25}I]-labeled RNA, *Histochem. Cytochem.*, 26, 677, 1978.
7. **Hogan, B. M., Constantini, F., and Lacy, E.,** *Manipulating the Mouse Embryo: A Laboratory Manual,* Cold Spring Harbor Laboratory, Cold Spring Harbor, NY, 1986, Sect. F.
8. **Brahic, M. and Haase, A. T.,** Detection of viral sequences of low reiteration frequency by *in situ* hybridization, *Proc. Natl. Acad. Sci. U.S.A.*, 75, 6125, 1978.
9. **Brigati, D. J., Myerson, D., Leary, J. J., Spalholtz, B., Travis, S. Z., Fong, C. K. Y., Hsiung, G. D., and Ward, D. C.,** Detection of viral genomes in cultured cells and paraffin-embedded tissue sections using biotin-labeled hybridization probes, *Virology,* 126, 32, 1983.
10. **Akam, M. E.,** The location of *Ultrabithorax* transcripts in *Drosophila* tissue sections, *EMBO J.*, 2, 2075, 1983.
11. **Bandtlow, C. E., Heumann, R., Schwab, M. E., and Thoenen, H.,** Cellular localization of nerve growth factor synthesis by *in situ* hybridization, *EMBO J.*, 6, 891, 1987.
12. **Maniatis, T., Jeffrey, A., and Kleid, D.,** Nucleotide sequence of the rightward operator of phage λ, *Proc. Natl. Acad. Sci. U.S.A.*, 72, 1184, 1975.
13. **Macgregor, H. C. and Mizuno, S.,** *In situ* hybridization of "nick-translated" ^{3}H-ribosomal DNA to chromosomes from salamanders, *Chromosoma,* 54, 15, 1976.
14. **Feinberg, A. P. and Vogelstein, B.,** A technique for radiolabeling DNA restriction endonuclease fragments to high specific activity, *Anal. Biochem.*, 132, 6, 1983.
15. **Feinberg, A. P. and Vogelstein, B.,** A technique for radiolabeling DNA restriction endonuclease fragments to high specific activity, *Anal. Biochem.*, 137, 266, 1984.
16. **Pardue, M. L.,** *In situ* hybridization, in *Nucleic Acid Hybridization: A Practical Approach,* Hames, B. D. and Higgins, S. J., Eds., IRL Press, Oxford, 1985, 179.
17. **Lawrence, J. B. and Singer, R. H.,** Quantitative analysis of *in situ* hybridization methods for the detection of actin gene expression, *Nucleic Acids Res.*, 13, 1777, 1985.
18. **Young, W. S., III,** Single-stranded end-labeled probes, in *In Situ Hybridization in Brain,* Uhl, G. R., Ed., Plenum Press, New York, 1986, 228.
19. **Davies, L. G., Lewis, M. E., and Baldino, F., Jr.,** Single-stranded end-labeled probes, in *In Situ Hybridization in Brain,* Uhl, G. R., Ed., Plenum Press, New York, 1986, 230.
20. **Singer, R. H. and Ward, D. C.,** Actin gene expression visualized in chicken muscle tissue culture by using *in situ* hybridization with a biotinated nucleotide analog, *Proc. Natl. Acad. Sci. U.S.A.*, 79, 7331, 1982.
21. **Varndell, I. M., Polak, J. M., Sikri, K. L., Minth, C. D., Ploom, S. R., and Dixon, J. E.,** Visualization of messenger RNA directing peptide synthesis by *in situ* hybridisation using a novel single-stranded cDNA probe, *Histochemistry,* 81, 597, 1984.
22. **Lawrence, J. B. and Singer, R. H.,** Intracellular localization of messenger RNAs for cytoskeletal proteins, *Cell,* 45, 407, 1986.
23. **Coghlan, J. P., Aldred, P., Haralambidis, J., Niall, H. D., Penschow, J. D., and Tregar, G. W.,** Hybridization histochemistry, *Anal. Biochem.*, 149, 1, 1985.
24. **Southern, P. J., Blount, P., and Oldstone, B. A.,** Analysis of persistent virus infections by *in situ* hybridization to whole-mouse sections, *Nature,* 312, 555, 1984.
25. **Rall, L. B., Scott, J., Bell, G. I., Crawford, R. J., Penschow, J. D., Niall, H. D., and Coghlan, J. P.,** Mouse prepro-epidermal growth factor synthesis by the kidney and other tissues, *Nature,* 313, 228, 1985.
26. **Britten, R. J. and Davidson, E. H.,** Hybridization strategy, in *Nucleic Acid Hybridization: A Practical Approach,* Hames, B. C. and Higgins, S. J., Eds., IRL Press, Oxford, 1985, 3.

27. **Angerer, L. M., Cox, K. H., and Angerer, R. C.,** Demonstration of tissue-specific gene expression by *in situ* hybridization, *Methods Enzymol.,* 152, 649, 1987.
28. **Burnett, J. P., Frank, B. H., and Douthart, R. J.,** Ionic strength effects on the stability and conformation of *Penicillium chrysogenum* mycophage double stranded RNA, *Nucleic Acids Res.,* 2, 759, 1975.
29. **Wetmur, J. G. and Davidson, N. J.,** Kinetics of renaturation of DNA, *Mol. Biol.,* 31, 349, 1968.
30. **McConaughy, B. L., Laird, C. D., and McCarthy, B. J.,** Nucleic acid reassociation in formamide, *Biochemistry,* 8, 3289, 1969.
31. **Awgulewitsch, A., Utset, M. F., Hart, C. P., McGinnis, W., and Ruddle, F. H.,** Spatial restriction in expression of a mouse homeo box locus within the central nervous system, *Nature,* 320, 328, 1986.
32. **Utset, M. F., Awgulewitsch, A., Ruddle, F. H., and McGinnis, W.,** Region-specific expression of two mouse homeo box genes, *Science,* 235, 1379, 1987.
33. **Shepherd, A. M., Perkins, W. J., Green, R. J., and Clark, S. A.,** A study of oesophageal structure in a nematode, *Aphetenchoides blastophthorus* (Aphelenchida), using computer graphics reconstruction from serial thin sections, *J. Zool.,* 204, 271, 1984.
34. **Wilkinson, D. G., Bailes, J. A., and McMahon, A. P.,** Expression of the proto-oncogene *int*-1 is restricted to specific neural cells in the developing mouse embryo, *Cell,* 50, 79, 1987.
35. **Rogers, W. T. and Schwaber, J. S.,** Computer-aided mapping of brain tissue, *Appl. Opt.,* 26, 3384, 1987.
36. **Bogarad, L. D., Utset, M. F., Awgulewitsch, A., Miki, T., Hart, C. P., and Ruddle, F. H.,** The developmental expression pattern of a new murine homeo box gene: Hox-2.5, *Dev. Biol.,* 133, 537, 1989.
37. **Akam, M. E. and Martinez-Arias, A.,** The distribution of *Ultrabithorax* transcripts in *Drosophila* embryos, *EMBO J.,* 4, 1689, 1985.
38. **Breier, G., Bucan, M., Francke, U., Colberg-Poley, A. M., and Gruss, P.,** Sequential expression of murine homeo box genes during F9 EC cell differentiation, *EMBO J.,* 5, 2209, 1986.
39. **Maniatis, T., Fritsch, E. F., and Sambrook, J.,** *Molecular Cloning: A Laboratory Manual,* Cold Spring Harbor Laboratory, Cold Spring Harbor, NY, 1982, 458.
40. **Denhardt, D. T.,** A membrane-filter technique for the detection of complementary DNA, *Biochem. Biophys. Res. Commun.,* 23, 641, 1966.
41. **Awgulewitsch, A. and Jacobs, D.,** Differential expression of Hox 3.1 protein in subregions of the embryonic and adult spinal cord, *Development,* 108, 411, 1990.
42. **LeMouellic, H., Condamine, H., and Brûlet, P.,** Pattern of transcription of the homeo gene *Hox-3.1* in the mouse embryo, *Genes Dev.,* 2, 125, 1988.
43. **Breier, G., Dressler, G. R., and Gruss, P.,** Primary structure and developmental expression pattern of *Hox* 3.1, a member of the murine *Hox* 3 homeobox gene cluster, *EMBO J.,* 7, 1329, 1988.

V. APPENDIX

APPENDIX

SOURCES OF REAGENTS AND SUPPLIES FOR DIDEOXY DNA SEQUENCING AND OTHER APPLICATIONS

Barton E. Slatko

Note: It is advisable, before ordering, to check with the specific supplier for product number changes or updated information. General biochemical reagents can be obtained from American Bioanalytical, Inc., American Laboratory Supply, Fisher Scientific, Sigma Chemical Co., or VWR Scientific, among others.

TABLE 1
Reagents

Reagent	Supplier	Catalog no.
Acrylamide		
Solid	Accurate Chemical Supply	A10675
	American Bioanalytical	AB-100
	Amresco	0341
	Beckman	265600
	Bio-Rad	161-0100
	Boehringer-Mannheim	100-137
	Clontech Labs	8002-1
	GIBCO BRL	5512UA
	Hoefer Scientific	B44313-4A
	IBI	IB70020
	ICN/Schwartz-Mann	814329
	Integrated Separation Systems	SE141000
	Polysciences	00019
	Pharmacia LKB	1820-101
	Promega	V3111
	Research Organics	1155A
	Stratagene	300011
	U.S. Biochemical	32800
Liquid	Amresco	0132
	Boehringer-Mannheim	100-670
	Hoefer Scientific	B44354-5P
	IBI	IB70010
Bis-acrylamide		
Solid	Accurate Chemical and Scientific	29196
	American Bioanalytical	AB-200
	Amresco	0172
	Bio-Rad	161-0200
	Boehringer-Mannheim	100-140
	Clontech Labs	8000-1
	GIBCO BRL	5516UA
	Hoefer Scientific	B44300-2M
	IBI	IB70100
	ICN/Schwartz-Mann	800175
	Integrated Separation Systems	SE141020
	Pharmacia LKB	1820-102
	Polysciences	00719
	Promega	V3141
	Research Organics	0120M
	Stratagene	300033
	U.S. Biochemical	19199

TABLE 1 (continued)
Reagents

Reagent	Supplier	Catalog no.
Liquid	American Bioanalytical	AB-201
	Amresco	0832
	Hoefer Scientific	B44355-5R
	IBI	IB70012
Acrylamide/bis-acrylamide premixed (19:1)		
Solid	American Bioanalytical	AB-102
	Amresco	0673
	Boehringer-Mannheim	100-150
	Bio-Rad	161-0120
	Clontech Labs	8003-1
	ICN/Schwartz-Mann	800656
	Pharmacia LKB	1820-201
	Polysciences	1745
	Research Organics	1153A
	IBI	IB70023
Liquid	American Bioanalytical	AB-104
	Amresco	0729
	GIBCO BRL	5543UA (6%)[a]
		5545UA (8%)[a]
	Boehringer-Mannheim	100-693
	Integrated Separation Systems	SE141061
	IBI	IB70014
Hydrolink (liquid)	AT Biochem	HL-SEQ
Ammonium persulfate		
Solid	Accurate Chemical and Scientific	GC2010
	American Bioanalytical	AB-112
	Amresco	0486
	AT Biochem	HL-ACT
	Beckman	265602
	Bio-Rad	161-0700
	Boehringer-Mannheim	100-190
	Clontech Labs	8801-1
	GIBCO BRL	5523UA
	Hoefer Scientific	B44307-3E
	IBI	IB70080
	ICN/Schwartz-Mann	190556
	Integrated Separation Systems	SE141034
	Pharmacia	1820-103
	Polysciences	00086
	Promega	V3131
	Research Organics	1404A
	Stratagene	300022
	U.S. Biochemical	32810
TEMED		
Liquid	Accurate Chemical and Scientific	GC1010
	American Bioanalytical	AB-2020
	Amresco	0761
	AT Biochem	HL-INI
	Bio-Rad	161-0800
	Boehringer-Mannheim	100-139
	Clontech Labs	8840-1
	GIBCO BRL	5524UB
	Hoefer Scientific	B44308-2F

[a] Indicates inclusion of urea, TEMED, water, and buffer.

TABLE 1 (continued)
Reagents

Reagent	Supplier	Catalog no.
	IBI	IB70120
	ICN/Schwartz-Mann	195516
	Integrated Separation Systems	SE141035
	Pharmacia LKB	1820-104
	Polysciences	08036
	Promega	V3161
	Research Organics	3009T
	Stratagene	300161
	U.S. Biochemical	32825
Urea (ultrapure)	Accurate Chemical and Scientific	24524
	American Bioanalytical	AB-2100
	Amresco	0568
	Beckman	265634
	Boehringer-Mannheim	100-164
	Bio-Rad	161-0730
	Clontech Labs	8050-1
	GIBCO BRL	5505-UX
	Hoefer Scientific	B45204-3W
	IBI	IB72060
	ICN/Schwartz-Mann	821519
	Integrated Separation Systems	SE141030
	Pharmacia LKB	1830-601
	Polysciences	02232
	Promega	V3171
	Research Organics	4110U
	Stratagene	300191
	U.S. Biochemical	23040
Formamide (liquid)	Accurate Chemical and Scientific	21710
	Ambion	9342
	American Bioanalytical	AB-600
	Amresco	0606
	Beckman	265634
	Boehringer-Mannheim	100-144
	Clontech Labs	8011-1, 8010-1
	GIBCO BRL	5515UA
	Hoefer Scientific	B44254-2Y
	IBI	IB72020
	ICN/Schwartz-Mann	800686
	Pharmacia LKB	1830-602
	U.S. Biochemical	15899
Dithiothreitol	Accurate Chemical and Scientific	20697
	American Bioanalytical	AB-400
	Amresco	0281
	Beckman	265607
	Bio-Rad	161-0610
	Boehringer-Mannheim	100-032
	GIBCO BRL	5508UA
	Hoefer Scientific	B44149-2L
	ICN/Schwartz-Mann	856126
	IBI	IB02140
	Integrated Separation Systems	SE141043
	Pharmacia-LKB	1830-402
	Promega	V3151
	Research Organics	219OD
	U.S. Biochemical	15397

TABLE 1 (continued)
Reagents

Reagent	Supplier	Catalog no.
X-gal	Accurate Chemical and Scientific	YWA 02307851
	Ambion	9942
	American Bioanalytical	AB-2400
	Amresco	0428
	Bio-Rad	170-3455
	Boehringer-Mannheim	100-081
	Clontech Labs	8060-081
	5′ → 3′	5306-115455
	GIBCO BRL	5520UB
	Gold Biotechnology	X4281
	IBI	IB02260
	ICN	150001
	Polysciences	17136
	Promega	V3941
	Research Organics	2190D
	Stratagene	300200
	U.S. Biochemical	12385
IPTG	Ambion	9942
	American Bioanalytical	AB-7091
	Amresco	0487
	Bio-Rad	170-3201
	Boehringer-Mannheim	724-815
	Clontech Labs	8070-1
	5′ → 3′	5306-333555
	GIBCO BRL	5529UA
	Gold Biotechnology	I2481C
	IBI	IB02100
	ICN/Schwartz-Mann	102101
	Pharmacia LKB	27-3054-01
	Promega	V3951
	Research Organics	7054I
	Stratagene	300127
	U.S. Biochemical	17884

TABLE 2
DNA Sequencing Kits

Kit	Supplier	Product no.
M13 sequencing kit	Amersham	N.4502
M13 sequencing kit (Klenow)	Bio-Rad	170-3401
M13 sequencing kit (AMV-RT)	Bio-Rad	170-3403
Bst sequencing kit	Bio-Rad	170-3404
pUC sequencing kit	Boehringer-Mannheim	1013-106
M13 sequencing kit	Boehringer-Mannheim	1013-092
Kilobase sequencing system (Klenow), $^{32}P/^{35}S$	GIBCO BRL	8185SA
Klenow, ^{32}P	Du Pont	NEK023A/Z[a]
Klenow, ^{35}S	Du Pont	NEK023B/Z[a]
Maxam-Gilbert sequencing kit	Du Pont	NEK010
M13 DNA sequencing system (Klenow), $^{32}P/^{35}S$	GIBCO BRL	8236SA
Klenow, 7-dG	IBI	IB77180
Taq polymerase, $^{32}P/^{35}S$ (7-dG)	IBI	IB77190
Klenow, ^{32}P	New England Biolabs	405
Klenow, ^{35}S	New England Biolabs	406
Klenow, ^{32}P	Pharmacia LKB	27-1680-01
^{32}P (7-d-G)	Pharmacia LKB	27-1685-01
T7 sequencing kit	Pharmacia LKB	27-1682-01
Klenow, ^{35}S	Pharmacia LKB	27-1681-01
^{35}S (7-d-G)	Pharmacia LKB	27-1684-01
Klenow/AMV-RT	Promega	Q5280
Taq polymerase, $^{32}P/^{35}S$	Promega	Q5500
Taq polymerase, $^{32}P/^{35}S$ (7-dG)	Promega	Q5530
T7 sequencing kit	Promega	Q5780
T7 sequencing kit (7-dG)	Promega	Q5790
AMV-RT^{32}P	Stratagene	200315
AMV-RT,^{35}S	Stratagene	200316
Klenow, ^{32}P	Stratagene	200311
Klenow, ^{35}S	Stratagene	200312
Taq polymerase, $^{32}P/^{35}S$	Stratagene	200320
ExoMeth sequencing kit	Stratagene	200317
Sequenase (v1) sequencing kit, $^{32}P/^{35}S$ (dITP/dGTP)	U.S. Biochemical	70700
Sequenase (v2) sequencing kit $^{32}P/^{35}S$(dITP/dGTP)	U.S. Biochemical	70770
M13 sequencing kit	U.S. Biochemical	70050
Taq polymerase $^{32}P/^{35}S$	U.S. Biochemical	771060

Includes radioactive label.

TABLE 3
Cloning, Deletion, Subcloning, or Other Kits

Kit	Supplier	Product no.
5′ end-labeling kits	Boehringer-Mannheim	702-757
	Du Pont	NEP101
		NEK006[a]
	IBI	IB19500
	Promega	Q5500
RNA sequencing kits	Boehringer-Mannheim	1134-477
	Pharmacia LKB	27-0997-01
	Promega	Q5240
M13 (and other single-stranded DNA) cloning kits	Amersham	N.4503
	Boehringer-Mannheim	1013076
	Bio-Rad	170-3402
	GIBCO BRL	8235SA
	IBI	IB77240
	New England Biolabs	408
	Promega	P2230
	U.S. Biochemical	70055
pUC cloning kit	Boehringer-Mannheim	1013084
Sequex-35	Gold Biotechnology	Z3510
Cyclone deletion system (T4 polymerase)	IBI	IB77200
ExoIII deletion system	Pharmacia LKB	27-1691-01
Erase-a-base (ExoIII/S1 nuclease)	Promega	E5750
ExoIII/Mung Bean nuclease kit	Stratagene	200330

[a] Includes radioactive label.

TABLE 4
Enzymes

Enzyme	Supplier	Product no.
AMV-RT	Amersham	T.2610Y
	Bio-Rad	170-3110
	Boehringer-Mannheim	109-118
	Du Pont	NEE127
	GIBCO BRL	8020SA
	IBI	IB20180
	ICN	152020
	Pharmacia LKB	27-0922-01
	Promega	M5101
	Stratagene	600081
	U.S. Biochemical	70041
Mu-MLV-RT	Boehringer-Mannheim	1062-603
	GIBCO BRL	8025SA
	New England Biolabs	253
	Pharmacia LKB	27-0925-01
	Stratagene	600083
	U.S. Biochemical	70456
Klenow fragment	Accurate Chemical and Scientific	JZ 2606771
	Ambion	2012
	Amersham	T.2141Y
	Bio-Rad	170-3420
	Boehringer-Mannheim	104-523
	Du Pont	NEE102
	5′ → 3′	5306-332565

TABLE 4 (continued)
Enzymes

Enzyme	Supplier	Product no.
	GIBCO BRL	8021SA
	IBI	IB20140
	ICN	151009
	New England Biolabs	210
	Pharmacia LKB	27-0928-01
	Promega	M2201
	Stratagene	600071
	U.S. Biochemical	70015
T4 polynucleotide kinase	Accurate Chemical and Scientific	JZ 2606872
	Ambion	2310
	Amersham	T.2020Y
	Boehringer-Mannheim	117-645
	Du Pont	NEE101
	GIBCO BRL	8004SA
	IBI	IB20080
	ICN	800708
	Invitrogen	E108-01
	New England Biolabs	201
	Pharmacia LKB	27-0734-01
	Promega	M4101
	Stratagene	600103
	U.S. Biochemical	70031
Taq polymerase	Amersham	T.0303Y
	Beckman	267300
	Epicentre Technology	T84250
	GIBCO BRL	8038SA
	IBI	IB19010
	Perkin Elmer Cetus	N801-0043
	Promega	M1861
	Stratagene	600131
	U.S. Biochemical	71053
Bst polymerase	Bio-Rad	170-3406
T7 polymerase	Accurate Chemical and Scientific	JZ 2607276
	New England Biolabs	255, 256
	Pharmacia LKB	27-0985-01
	Promega	M2091
	Stratagene	600161
	U.S. Biochemical	70775, 70722
VENT™ polymerase	New England Biolabs	

TABLE 5
Sequencing Vectors and Associated Sequencing Primers

Supplier	Product description
Vectors	
Amersham	pAM18/19*, pTTQ*, M13, pUC*
ATTC	λgt11*, pUC118/119, pEMBL18/19, pICEM18/19, pSP64/65, pfdA/B, pBR322*, M13
Boehringer-Mannheim	M13mp18/19, pUC18/19*, pBR322*, pBR328*, T7/SP6, pEMBL12/18/19, λgt10/11*
Bio-Rad	M13/pUC*, pBR322*, pTZ18U/19U
Clontech	pT3/T7-3, pUC18/19*, GUS plant vectors*
GIBCO BRL	M13mp18/19, pUC18/19*, pT7/T3α-18/19, pBR322*
IBI	M13/pUC*, pBR322*, pIB124 (−25)
ICN	λgt10/11*, M13/pUC*, pAT153*, pBR322*
Invitrogen	pcDNAII, pTZ18R, pBR322*, λgt10/11*
New England Biolabs	pUG19*, pBR322*
Pharmacia LKB	M13/pUC*, pT7 T3 18U/19U
Promega	pBR322*, pGEM3Z/4Z/5z/7z/9z/11z f(+/−), λGEM2/4/11/12*, λgt10/11*, SP72/73*
Stratagene	BluescriptII series, pBS +/−, pBC series
U.S. Biochemical	pUC*/M13, pT71/2, pTZ18/19
Sequencing primers	
Ambion	M13/pUC, T7, T3, SP6
Amersham	M13/pUC, tg130/131
Bio-Rad	M13/pUC
Boehringer-Mannheim	M13, pUC, λgt11, pBR322, T7, T3, SP6
Clontech	λgt10/gt11, SP6, T7, T3
Du Pont	SP6, T7, M13
GIBCO BRL	T7, SP6, M13/pUC
Gold Biotechnology	SEQUEX-20
IBI	M13/pUC, RD24, RD-29, T3
ICN	M13/pUC, SP6, T3, T7, λgt11, T3
Midland Certified Reagent	M13/pUC, pBR322, T7, T3
New England Biolabs	λgt10/gt11, pBR322, M13/pUC, T7, T3, SP6, IS1, RD24/29
Pharmacia LKB	M13/pUC, T7, T3, SP6, pBR322, pKK223-3
Promega	pBR322, SP6/T7, pUC/M13
Stratagene	T3, T7, M13/pUC, pSK/KS (bluescript)
U.S. Biochemical	M13/pUC, SP6, T7
Custom primers	
Clontech	
DNA Factory	
GENOSYS	
Invitrogen	
Lofstrand Labs	
Midland Certified Reagent	
New England Biolabs	
OCS Research Products	
Pharmacia	
Promega	
Research Genetics	
Stratagene	
Synthecell	
Synthetic Genetics	

Note: Listed are vectors for which primers are commercially available and are the latest in each of its series; all have extensive polylinkers and can produce single-stranded DNA, except those marked with an asterisk.

TABLE 6
Nucleotides

Product	Supplier	Product no.
Nucleotides (unlabeled solutions)		
dATP, dGTP, dCTP, dTTP	Bio-Rad	170-3425—170-3428
	Boehringer-Mannheim	1051-440, 1051-458 1051-466, 1051-482
	GIBCO BRL	8252SA, 8353SA, 8254SA, 8255SA
	New England Biolabs	421
	Pharmacia LKB	27-2035-01, 27-2050-01, 27-2070-01, 27-2060-01, 27-2080-01
	Promega	U1240
	U.S. Biochemical	14244, 14279, 14314, 22324
7-deaza-dGTP	Boehringer-Mannheim	988-537
	ICN	151464
	New England Biolabs	425
	Pharmacia LKB	27-2090-01
dITP	Boehringer-Mannheim	1051-474
	ICN	101490
	New England Biolabs	426
	Pharmacia LKB	27-2095-01
ddATP, ddGTP, ddCTP, ddTTP	American Bioanalytical	Inquire
	Amresco	0553, 0564, 0570, 0592
	Boehringer-Mannheim	1000-101, 1000-098 1000-128, 1000-080
	Bio-Rad	170-3435, 170-3439
	GIBCO BRL	8243SC, 8244SC, 8245SC, 8246SC
	New England Biolabs	422
	Pharmacia LKB	27-2045-01
ddITP	Boehringer-Mannheim	1000-136
	New England Biolabs	427
Nucleotides (powders)		
dATP, dGTP, dCTP, dTTP	American Bioanalytical	AB-340, AB-341, AB-342, AB-343
	Amresco	0503, 0518, 0526, 0532
	Boehringer-Mannheim	103-977, 104-035 104-094, 104-264
	Chemical Dynamics	26-9260-00, 26-8461-00, 26-7590-00, 26-4295-00
	ICN	100549, 100558, 100556, 152140
	Pharmacia LKB	27-5500-01, 27-5600-01 27-5701-01, 27-5800-01
7-deaza-GTP	U.S. Biochemical	70065
dITP	Boehringer-Mannheim	104-116
	U.S. Biochemical	14335
ddATP, ddCTP, ddGTP, ddTTP	Boehringer-Mannheim	606-146, 606-162, 606-154, 348-732
	Pharmacia LKB	27-4819-01, 27-4831-01, 27-4817-01, 27-4823-01
	U.S. Biochemical	14707, 14703, 14704, 14706
ddITP	Boehringer-Mannheim	633-500
	Chemical Dynamics	26-9395-00
	Pharmacia LKB	27-4843
Nucleotides (radioactively labeled, solutions)		
α-^{32}P dATP	Amersham	PB.10384
	Du Pont	NEG012H
	ICN	33002H
α-^{32}P dCTP	Amersham	PB.10205
	Du Pont	NEG013H

TABLE 6 (continued)
Nucleotides

Product	Supplier	Product no.
	ICN	33004H
α-^{35}S dATP	Amersham	SJ.1304
	Du Pont	NEG034S (Klenow)
		NEG0344 (T7 polymerase)
α-^{35}S dCTP	Amersham	SJ.1305
	Du Pont	NEG036S (Klenow)
		NEG0364 (T7 polymerase)
γ-^{32}P rATP	Amersham	PB.10168
	Du Pont	NEG002A
	ICN	35001X
γ-^{35}S rATP	Amersham	SJ.1305
	Du Pont	NEG027H
Sequetide (α-^{35}s dATP, dCTP, dGTP, dTTP) for T7 polymerase	Du Pont	NEG034N

TABLE 7
Equipment, Accessories, and Other Supplies

Equipment	Supplier	Product no.
Sequencing gel apparatus (all include necessary clips, spacers, etc.)		
Standard	American Bioanalytical	SG series
	Ann Arbor Plastics	Q1, QA1, Q2
	Bio-Rad	165-3600
	GIBCO BRL	1105SC, 1196KC
	Hoefer Scientific	SE1500, SE1550
	IBI	IB80000, IB78600
	Jordan Scientific	JVP-150, JSQ-250
	Owl Scientific Plastics	S1S, S2S, ADJ3
	Pharmacia LKB	2010-S1, 2010-S2
	Schleicher and Schuell	Profile 3545, 3565
	West Coast Scientific	EVSQ 36, 40
Gel dryers	Bio-Rad	165-1745
	Hoefer Scientific	SE1160
	Savant	SS5
	Gene Vac	SGD3545
Power Supplies	Accurate Chemical and Scientific	HNB200
	American Bioanalytical	EC series
	Bio-Rad	165-0554
	E. C. Apparatus	EC series
	Fotodyne	74200
	GIBCO BRL	1067PS
	Hoefer Scientific	PS2500
	IBI	IB92800
	Integrated Separation Systems	SS110604
	Owl Scientific	OSP 4000, 8000, 500
	Pharmacia LKB	18-0501-01
	Schleicher and Schuell	PS5500, PS4000
	Stratagene	400600, 400601
	West Coast Scientific	EC series
Gel tape	American Bioanalytical	GT7210 (—7215)
	GIBCO BRL	1032ST
	Hoefer Scientific	SE1510
	IBI	IB81630
	Owl Scientific Plastics	TAPE

TABLE 7 (continued)
Equipment, Accessories, and Other Supplies

Equipment	Supplier	Product no.
Combs and/or spacers[a]	American Bioanalytical	SGC, SGS series
	Ann Arbor Plastics	SQ-1
	Bio-Rad	165-3680 (—165-3720)
	GIBCO BRL	SE, SH, UE, and SC series
	Hoefer Scientific	SE series
	IBI	IB80424, IB80380 series
	Jordan Scientific	PC, SP series
	Owl Scientific Plastics	ADJ2, S1S, S2S series
	Pharmacia LKB	2101-401 series
	Schleicher and Schuell	Profile series
	West Coast Scientific	Watson series
Radiation shields	Accurate Chemical and Scientific	APO939BW
	Ann Arbor Plastics	Custom
	American Bioanalytical	RD series, custom
	American Scientific Products	S-1386-1
	Diamond Research Products	O1RS-25 series
	Dynalabs	7229, 7230, 7231
	Hoefer Scientific	RS series
	IBI	IB81300
	Jordan Scientific	SHS180; SHB series, PF-150
	Owl Scientific Plastics	BT1 series
	Schleicher and Schuell	β-safe series
	Stratagene	400025
	West Coast Scientific	SHD series
Sequencing trays/microtiter trays	American Bioanalytical	77163320
	Beckman	373660
	Costar	3799, 3894, 3790
	Flow Laboratories	76-013-05
	Pharmacia LKB	2010-700
	Sarstedt	82.1582.101
	Stratagene	420001
Centrifugal vacuum evaporators	Gene Vac	CE100
	Peninsula Labs	CC180
	Savant	SS1
DNA sequencing gel pipettes and tips	American Bioanalytical	701753, 900050
	Bio-Rad	223-35165, 223-3013
	Costar	4854, 4895
	Drummond Scientific	3-00-203
	Dynalab	6020
	Hoefer Scientific	SYPD3
	IBI	IB81010, IB81000
	Integrated Separation Systems	SA104002, 104003
	Marsh Biomedical	T4000
	Pharmacia LKB	2010-150, 2010-151
	Schleicher and Schuell	Profile series
	Stratagene	410004
	West Coast Scientific	Apex series
Gel transfer paper	American Bioanalytical	3M, Gel Blot
	Schleicher and Schuell	GB-002, GB-003, GB-004, custom size
	Whatman	1114-917
Autoradiography enhancers		
Spray	Amersham	NPN.16
	Du Pont	NEF 970
	EM Corp.	7040

[a] All suppliers provide a wide variety of choices of spacer and comb lengths and widths, including tapered spacers and sharks-tooth combs.

TABLE 7 (continued)
Equipment, Accessories, and Other Supplies

Equipment	Supplier	Product no.
Liquid	American Bioanalytical	Fluorohance
	Amersham	NAMP.100
	Du Pont	NEF981, NEF974
	EM Corp.	7030
	Integrated Separation Systems	SE 122004
Membranes for genomic sequencing	American Bioanalytical	Zetabind, MagnaGraph, MagnaNylon, Nytran, Westran
	Amersham	RPN.1500
	Bio-Rad	162-0153—162-0166
	Cuno	NM802-01-045-SP
	Du Pont	NEF972, NEF983
	MSI	NJ4H4320F5
	Integrated Separation Systems	Inquire
	Millipore	MBBUINS01
	Pall	Inquire
	Pharmacia LKB	2005-109
	Schleicher and Schuell	BA, BA-S, Nytran series
	West Coast Scientific	Sartorius series
Lighted viewing boxes	Accurate Chemical and Scientific	SPEC series
	American Bioanalytical	50-275-100 series
	E. M. Parker	S & S series
	Fotodyne	3-3700
	Hoefer Scientific	PH46
	West Coast Scientific	Inquire
Automated DNA sequencers and associated reagent kits	Applied Biosystems	470A
	E. I. DuPont	Genesis 2000
	EG & G Biomolecular	Acugen 402
	Milligen/Biosearch	Inquire
	Pharmacia LKB	ALF system
Automated gel readers or digitizer systems	BioImage	Visage60
	BioMed Instruments	SLR-DNA
	Bio-Rad	170-3545
	Hoefer Scientific	GP100, GP200
	IBI	IB81450, IB81406
	Intelligenetics	SpeedReader
	Pharmacia LKB	2017-S1
	Gold Biotechnology	WB-3520
Darkroom and autoradiography equipment		
Film holders, film, developer, fixer, cassettes, phosphorescent marking systems, etc.	Eastman Kodak distributor	
	Fugi film distributor	
	Local photography supply distributor (American Bioanalytical, Fotodyne, Owl Scientific, Integrated Separation Systems, Schleicher and Schuell, IBI, etc.)	
Custom DNA sequencing	Advanced Biotechnologies	
	Ambion	
	Bionetics Research Institute	
	GENOSYS	
	GenoType	
	Lark Biotechnology	
	Lofstrand Labs	
DNA sequencing workshops	GIBCO BRL	
	IBI	
	New England Biolabs	

TABLE 8
Names, Addresses, and Telephone Numbers for Suppliers

Accurate Chemical and Scientific Co., 300 Shames Dr., Westbury, NY 11590; (800) 645-6264
Advanced Biotechnologies, Inc., 9108 Guilford Rd., Columbia, MD 21046; (800) 426-0764
Ambion, 2130 Woodward St., Suite 200, Austin, TX 78744; (800) 888-8804
American Bioanalytical, Inc., 10 Huron Dr., Natick, MA 01760; (800) 443-0600
American GeneVac, Inc., 35 Demarst Mill Rd., West Nyack, NY 10994; (914) 623-2892
American Scientific Products, 1430 Waukegan Rd., McGraw Park, IL 60085; (800) 282-3550
American Type Culture Collection (ATCC), 12301 Parklawn Dr., Rockville, MD 20852; (301) 231-5524
Amresco, Inc., 30175 Solon Industrial Pkwy., Solon, OH 44139; (800) 448-4442
Ann Arbor Plastics, 2285 S. State Rd., Ann Arbor, MI 48104; (313) 994-3674
Amersham Corporation, 2636 S. Clearbrook Dr., Arlington Heights, IL 60005; (800) 323-9750
Applied Biosystems, 800 Lincoln Centre Dr., Foster City, CA 94404; (415) 570-6667
AT Biochem, 74 Great Valley Pkwy., Malvern, PA 19355; (215) 889-9300
Beckman Instruments, Bioanalytical Systems Group, 2500 Harbor Blvd., P. O. Box 3100, Fullerton, CA 92634; (800) 742-2345
BioImage, 1460 Eisenhower Pl., Ann Arbor, MI 48108; (313) 971-7500
BioMed Instruments, 1020 S. Raymond Ave., Suite B, Fullerton, CA 92631; (800) 662-2566
Bionetics Research Institute, 1330A Piccard Dr., Rockville, MD 20850; (301) 258-5200
Bio-Rad, Inc., 3300 Regatta Blvd., Richmond, CA 94804; (415) 232-7000
Boehringer-Mannheim, Biochemical Division, P. O. Box 50816, Indianapolis, IN 46250; (800) 428-5433
Clontech Labs, 4030 Fabian Way, Palo Alto, CA 94303; (415) 474-8222
Costar Scientific, 1 Alewife Center, Cambridge, MA 02140; (617) 868-6200
Chemical Dynamics, Inc., P. O. Box 395, 3001 Hadley Rd., South Plainfield, NJ 07080; (201) 753-5000
CUNO, Inc., 400 Research Pkwy., Meriden, CT 06450; (203) 238-8970
Drummond Scientific Co., 500 Parkway, Broomall, PA 19008; (800) 523-7480
Dynalab Corp., P. O. Box 112, Rochester, NY 14692; (800) 828-6595
Diamond Research Products, P. O. Box 6416, Raleigh, NC 27628; (919) 821-1688
E. C. Apparatus Co., 3831 Tyrone Blvd. North, St. Petersburg, FL 33709; (800) 327-2643
E. I. Du Pont Co., NEN Research Products, 549 Albany St., Boston, MA 02118; (800) 551-2121
EG & G Biomolecular, 4 Tech Cir., Natick, MA 01760; (508) 651-3250
EM Corp., P. O. Box 285, Chestnut Hill, MA 02167; (800) 528-5250
E. M. Parker, 646 Brookline Ave., Brookline, MA 02147; (800) 225-9177
Epicentre Technologies, 1202 Ann St., Madison, WI 53713; (608) 251-8474
Fisher Scientific, 711 Forbes Ave., Pittsburgh, PA 15226; (412) 562-2692
5′ → 3′, Inc., 117 Brandywine Pkwy., West Chester, PA 19380; (800) 533-5703
Flow Laboratories, Inc., 655 Old Spring House Rd., McLean, VA 22102; (800) 386-3569
Fotodyne, 16700 W. Victor Rd., New Berlin, WI 53151; (800) 362-3686
GIBCO BRL, Research Products Division, Life Technologies, Inc., P. O. Box 6009, Gaithersburg, MD 20877; (800) 828-6686
GENOSYS Biotechnologies, Inc., 8701 New Trails Dr., The Woodlands, TX 77381; (800) 234-5362
GenoType, Inc., P. O. Box 82042, Lincoln, NE 68501; (402) 472-6083
Gold Biotechnology, 10143 Paget St., St. Louis, MO 63132; (800) 248-7609; (314) 432-4058
Hoefer Scientific Instruments, 654 Minnesota St., San Francisco, CA 94107; (800) 227-4750
ICN Biomedicals, 3300 Hyland Ave., Costa Mesa, CA 92626; (800) 854-0530
International Biotechnologies, Inc. (IBI), 25 Science Park, New Haven, CT 06511; (800) 243-2555
Invitrogen, 11588 Sorrento Valley Rd., San Diego, CA 92121; (800) 544-4684
Integrated Separation Systems, 1 Westinghouse Plaza, Hyde Park, MA 02136; (800) 433-6433
Jordan Scientific, 4315 S. State Rd. 446, Bloomington, IN 47401; (800) 222-2092
Lark Sequencing Technologies, Ltd., Texas Medical Center, 1709 Dryden, Suite 604, Houston, TX 77030; (713) 799-9193
Lofstrand Labs, 7961 Cessna Ave., Gaithersburg, MD 20879; (800) 541-0362
Marsh Biomedical Products, 274 N. Goodman St., Rochester, NY 14607; (800) 445-2812
Midland Certified Reagent Co., 3112-A West Cuthbert Ave., Midland, TX 79701; (800) 247-8766
Millipore/Milligen-Biosearch, Inc., 186 Middlesex Turnpike, Burlington, MA 01803; (617) 229-2870
MSI, 135 Flanders Rd., P. O. Box 1046, Westborough, MA 01581; (508) 366-5840
New England Biolabs, Inc., 32 Tozer Rd., Beverly, MA 01915; (800) 632-5227
OCS Research Products, P.O. Box 2067, Denton, TX 76202; (800) 227-0627
Owl Scientific Plastics, P. O. Box 566, Cambridge, MA 02139; (800) 242-5560
Pall, Inc., 77 Crescent Beach Rd., Glen Cove, NY 11542; (800) 645-6246

TABLE 8 (continued)
Names, Addresses, and Telephone Numbers for Suppliers

Peninsula Labs, 611 Taylor Way, Belmont, CA 94022; (415) 592-5392
Perkin Elmer Cetus, Inc., 761 Main Ave., Norwalk, CT 06859; (203) 834-6722
Pharmacia-LKB, 800 Centennial Pkwy., Piscataway, NJ 08855; (800) 922-0318
Polysciences, 400 Valley Rd., Warrington, PA 18976; (800) 523-2575
Promega, 2800 S. Fish Hatchery Rd., Madison, WI 53711; (800) 832-6362
Research Genetics, 2130 S. Memorial Pkwy., Huntsville, AL 35801; (800) 533-4363
Research Organics, 4353 E. 49th St., Cleveland, OH 44125; (800) 321-0570
Savant Instruments, Inc., 110-103 Bi-County Blvd., Farmingdale, NY 11735; (800) 634-8886
Sarstedt, Inc., P. O. Box 4090, Princeton, NJ 08543; (800) 257-5101
Schleicher and Schuell, Inc., 10 Optical Ave., Keene, NH 03431; (800) 245-4024
Stratagene, Inc., 11099 N. Torrey Pines Rd., La Jolla, CA 92037; (800) 424-5444
Synthecell, 2096 Gaither Rd., Rockville, MD 20850; (301) 896-7455
Synthetic Genetics, 10455 Roselle St., San Diego, CA 92121; (800) 562-5544
The DNA Factory, P. O. Box 210226, San Diego, CA 92121; (619) 558-9570
U.S. Biochemical Corp., 26111 Miles Rd., Cleveland, OH 44128; (800) 321-9322
VWR Scientific, 1400 N. Providence Rd., Suite 100, Media, PA 19063; (215) 891-2770
West Coast Scientific, 2542 Barrington Ct., Hayward, CA 94545; (800) 367-8462

Note: Many sources have non-U.S. distributors; only U.S. addresses and telephone numbers are listed.

ACKNOWLEDGMENTS

I wish to thank my friends and colleagues for their help in (and review of) the preparation of the source list of supplies and suppliers; in particular, thanks to M. Kashdan (NEN, Du Pont); W. Gagnon (American Bioanalytical); G. Magnant and Sarah Crocker (Owl Scientific); P. Dimond (American Biotechnology); G. Davies and B. Janosen (Boehringer-Mannheim); G. Sams (EG & G Biomolecular); P. Gold (Gold Biotechnology); L. Smith and P. Fong (Clontech Labs); S. Burns (Amersham); G. Kleinstuber (Drummond Scientific); N. Sasavage (Gibco BRL); W. Schick and A. Lackey (Hoefer Scientific); L. Morris ($5' \rightarrow 3'$); S. Jones (Stratagene); E. Hernandez and M. Lorence (Bio-Rad); M. Weisblatt (U.S. Biochemical); P. McOsker (Pharmacia LKB); M. Schaeffer (ICN); D. Titus (Promega); N. Berry (IBI); E. Leach and B. Appel (Integrated Separation Systems); R. Robertson (Savant, Inc.); L. Perry and M. Ward (Amresco); G. Merry and A. Creasey (Milligen); C. Smith (AT Biochem); J. Fernandez (Invitrogen); T. Walker (Sarstedt); R. Robertson (Savant Instruments); A. Wood (MSI); S. Cohen (Research Organics); M. Lambert (Cuno); D. Asa (West Coast Scientific); C. Sotwick (Nunc); J. Smoot (Fotodyne); A. Mahoney (Costar); A. Hoyt (E. M. Parker); T. Stio (Jordan Scientific Co.); M. Tarker (Polysciences); J. Besant (Accurate Chemical and Scientific Co.); D. J. Witklewtz and J. Lewis (Dynalab); L. Linkkila (Schleicher and Schuell); G. Gummere (Ambion); and R. Case (Midland Certified Reagent Co.). All errors, of course, remain mine.

INDEX

INDEX

D

E

F

K

L

M

N

R

U

V

W

X

Y